T0338244

Near-Capacity Variable-Length Coding

Near-Capacity Variable-Length Coding

Regular and EXIT-Chart-Aided Irregular Designs

Lajos Hanzo, *University of Southampton, UK*

Robert G. Maunder, *University of Southampton, UK*

Jin Wang, *Aeroflex, UK*

Lie-Liang Yang, *University of Southampton, UK*

A John Wiley and Sons, Ltd, Publication

This edition first published 2011

© 2011 John Wiley & Sons Ltd

Registered office

John Wiley & Sons Ltd, The Atrium, Southern Gate, Chichester, West Sussex, PO19 8SQ, United Kingdom

For details of our global editorial offices, for customer services and for information about how to apply for permission to reuse the copyright material in this book please see our website at www.wiley.com.

Library of Congress Cataloging-in-Publication Data

Near-capacity variable length coding : regular and exit-chart aided
irregular designs / L. Hanzo ... [et al.].
 p. cm.
Includes bibliographical references and index.
ISBN 978-0-470-66520-6 (cloth)
1. Coding theory. I. Hanzo, Lajos, 1952-
 TK5102.92.N43 2011
 003'.54–dc22

                                                                                         2010004481

A catalogue record for this book is available from the British Library.

ISBN - 978-0-470-66520-6

Set in 10/12pt Times by Sunrise Setting Ltd, Torquay, UK.
Printed and bound in Singapore by Markono Print Media Pte Ltd

We dedicate this monograph to the numerous contributors of this field, many of whom are listed in the Author Index.

Contents

About the Authors   xv

Other Wiley and IEEE Press Books on Related Topics   xvii

Preface   xix

Acknowledgements   xxi

Chapter 1   Introduction   1
   1.1   Historical Overview 1
        1.1.1   Source Coding and Decoding 1
              1.1.1.1   Variable-Length Coding 1
              1.1.1.2   Variable-Length Decoding 3
              1.1.1.3   Classification of Source-Decoding Algorithms 5
        1.1.2   Joint Source–Channel Decoding 6
        1.1.3   Iterative Decoding and Convergence Analysis 7
   1.2   Applications of Irregular Variable-Length Coding 11
        1.2.1   Near-Capacity Operation 11
        1.2.2   Joint Source and Channel Coding 15
        1.2.3   Unequal Error Protection 17
   1.3   Motivation and Methodology 17
   1.4   Outline of the Book 18
   1.5   Novel Contributions of the Book 22

Chapter 2   Information Theory Basics   25
   2.1   Issues in Information Theory 25
   2.2   Additive White Gaussian Noise Channel 28
        2.2.1   Background 28
        2.2.2   Practical Gaussian Channels 30
        2.2.3   Gaussian Noise 30

2.3   Information of a Source 32

2.4   Average Information of Discrete Memoryless Sources 34

     2.4.1   Maximum Entropy of a Binary Source 34

     2.4.2   Maximum Entropy of a q-ary Source 36

2.5   Source Coding for a Discrete Memoryless Source 37

     2.5.1   Shannon–Fano Coding 37

     2.5.2   Huffman Coding 39

2.6   Entropy of Discrete Sources Exhibiting Memory 43

     2.6.1   Two-State Markov Model for Discrete Sources Exhibiting Memory 43

     2.6.2 N-State Markov Model for Discrete Sources Exhibiting Memory . 45

2.7   Examples 46

     2.7.1   Two-State Markov Model Example 46

     2.7.2   Four-State Markov Model for a Two-Bit Quantizer 48

2.8   Generating Model Sources 49

     2.8.1   Autoregressive Model 49

     2.8.2   AR Model Properties 49

     2.8.3   First-Order Markov Model 51

2.9   Run-Length Coding for Discrete Sources Exhibiting Memory 51

     2.9.1   Run-Length Coding Principle 51

     2.9.2   Run-Length Coding Compression Ratio 52

2.10   Information Transmission via Discrete Channels 54

     2.10.1   Binary Symmetric Channel Example 55

     2.10.2   Bayes' Rule 56

     2.10.3   Mutual Information 59

     2.10.4   Mutual Information Example 61

     2.10.5   Information Loss via Imperfect Channels 62

     2.10.6   Error Entropy via Imperfect Channels 63

2.11   Capacity of Discrete Channels 70

2.12   Shannon's Channel Coding Theorem 73

2.13   Capacity of Continuous Channels 75

     2.13.1   Practical Evaluation of the Shannon–Hartley Law 79

     2.13.2   Shannon's Ideal Communications System for Gaussian Channels . . 82

2.14   Shannon's Message for Wireless Channels 82

2.15   Summary and Conclusions 85

I   Regular Concatenated Codes and Their Design   87

List of Symbols in Part I   89

Chapter 3   Sources and Source Codes   91

3.1   Introduction 91

3.2   Source Models 91

     3.2.1   Quantization 92

     3.2.2   Memoryless Sources 93

     3.2.3   Sources with Memory 94

3.3 Source Codes 95
   3.3.1 Definitions and Notions 96
   3.3.2 Some Properties of Variable-Length Codes 99
   3.3.3 Huffman Code 99
   3.3.4 Reversible Variable-Length Codes 100
      3.3.4.1 Construction of RVLCs 101
      3.3.4.2 An Example Construction of a RVLC 102
      3.3.4.3 Experimental Results 103
   3.3.5 Variable-Length Error-Correcting Code 106
      3.3.5.1 Construction of VLEC Codes 107
      3.3.5.2 An Example Construction of a VLEC Code 109
      3.3.5.3 Experimental Results 111
3.4 Soft Decoding of Variable-Length Codes 112
   3.4.1 Trellis Representation 112
      3.4.1.1 Symbol-Level Trellis 112
      3.4.1.2 Bit-Level Trellis 114
   3.4.2 Trellis-Based VLC Decoding 116
      3.4.2.1 Sequence Estimation Using the Viterbi Algorithm 116
         3.4.2.1.1 Symbol-Level Trellis-Based Sequence Estimation. 116
         3.4.2.1.2 Bit-Level Trellis-Based Sequence Estimation. . . . 120
      3.4.2.2 Symbol-by-Symbol MAP Decoding 122
         3.4.2.2.1 Symbol-Level Trellis-Based MAP Decoding. . . . 122
         3.4.2.2.2 Bit-Level Trellis-Based MAP Decoding. 128
      3.4.2.3 Concatenation of MAP Decoding and Sequence Estimation 130
   3.4.3 Simulation Results 130
      3.4.3.1 Performance of Symbol-Level Trellis-Based Decoding . . . 131
      3.4.3.2 Performance of Bit-Level Trellis-Based Decoding 134
      3.4.3.3 Comparison of Symbol-Level Trellis and Bit-Level
           Trellis-Based Decoding 135
3.5 Summary and Conclusions 139

Chapter 4 Iterative Source–Channel Decoding 143
4.1 Concatenated Coding and the Turbo Principle 144
   4.1.1 Classification of Concatenated Schemes 144
      4.1.1.1 Parallel Concatenated Schemes 144
      4.1.1.2 Serially Concatenated Schemes 144
      4.1.1.3 Hybrid Concatenated Schemes 146
   4.1.2 Iterative Decoder Activation Order 146
4.2 SISO APP Decoders and their EXIT Characteristics 146
   4.2.1 A Soft-Input Soft-Output APP Module 146
      4.2.1.1 The Encoder 146
      4.2.1.2 The Decoder 147
   4.2.2 EXIT Chart 150
      4.2.2.1 Mutual Information 150
      4.2.2.2 Properties of the $J(\cdot)$ Function 152

4.2.2.3   Evaluation of the EXIT Characteristics of a SISO APP
Module 152
4.2.2.4   Simplified Computation of Mutual Information 154
4.2.2.5   Examples 156
4.2.2.5.1   The Intermediate Code. 156
4.2.2.5.2   The Inner Code. 156
4.2.2.5.3   The Outer Code. 157
4.3   Iterative Source–Channel Decoding Over AWGN Channels 158
4.3.1   System Model 158
4.3.2   EXIT Characteristics of VLCs 159
4.3.3   Simulation Results for AWGN Channels 161
4.4   Iterative Channel Equalization, Channel Decoding and Source Decoding . . 165
4.4.1   Channel Model 167
4.4.2   Iterative Channel Equalization and Source Decoding 170
4.4.2.1   System Model 170
4.4.2.2   EXIT Characteristics of Channel Equalizer 171
4.4.2.3   Simulation Results 172
4.4.2.4   Performance Analysis Using EXIT Charts 173
4.4.3   Precoding for Dispersive Channels 175
4.4.3.1   EXIT Characteristics of Precoded Channel Equalizers . . . 176
4.4.3.2   Performance Analysis 177
4.4.4   Joint Turbo Equalization and Source Decoding 181
4.4.4.1   System Model 181
4.4.4.2   Simulation Results 182
4.5   Summary and Conclusions 182

Chapter 5   Three-Stage Serially Concatenated Turbo Equalization   187
5.1   Introduction 187
5.2   Soft-In Soft-Out MMSE Equalization 188
5.3   Turbo Equalization Using MAP/MMSE Equalizers 190
5.3.1   System Model 190
5.3.2   EXIT Chart Analysis 190
5.3.3   Simulation Results 191
5.4   Three-Stage Serially Concatenated Coding and MMSE Equalization 191
5.4.1   System Model 191
5.4.2   EXIT Chart Analysis 192
5.4.2.1   Determination of the Convergence Threshold 192
5.4.2.2   Optimization of the Outer Code 196
5.4.2.3   Optimization of the Activation Order 196
5.4.3   BER Performance 199
5.4.4   Decoding Trajectories 200
5.4.5   Effects of Interleaver Block Length 203
5.5   Approaching the Channel Capacity Using EXIT-Chart Matching and IRCCs 205
5.5.1   Area Properties of EXIT Charts 205
5.5.2   Analysis of the Three-Stage System 208

5.5.3   Design of Irregular Convolutional Codes 210

5.5.4   Simulation Results 212

5.6   Rate Optimization of Serially Concatenated Codes 214

5.7   Joint Source-Channel Turbo Equalization Revisited 218

5.8   Summary and Conclusions 220

II   Irregular Concatenated VLCs and Their Design   223

List of Symbols in Part II   224

Chapter 6   Irregular Variable-Length Codes for Joint Source and Channel Coding   231

6.1   Introduction 231

6.2   Overview of Proposed Scheme 234

6.2.1   Compression 234

6.2.2   VDVQ/RVLC Decomposition 237

6.2.3   Serial Concatenation and Iterative Decoding 237

6.3   Transmission Frame Structure 239

6.3.1   Frame Difference Decomposition 239

6.3.2   VDVQ/RVLC Codebook 241

6.3.3   VDVQ/RVLC-Induced Code Constraints 242

6.3.4   VDVQ/RVLC Trellis Structure 244

6.4   VDVQ/RVLC Encoding 246

6.5   APP SISO VDVQ/RVLC Decoding 248

6.6   Simulation Results 250

6.7   Summary and Conclusions 256

Chapter 7   Irregular Variable-Length Codes for EXIT-Chart Matching   259

7.1   Introduction 259

7.2   Overview of Proposed Schemes 262

7.2.1   Joint Source and Channel Coding 262

7.2.2   Iterative Decoding 266

7.3   Parameter Design for the Proposed Schemes 268

7.3.1   Scheme Hypothesis and Parameters 268

7.3.2   EXIT-Chart Analysis and Optimization 271

7.4   Simulation Results 276

7.4.1   IrCC-Based Benchmark 276

7.4.2   Iterative Decoding Convergence Performance 278

7.4.3   Interleaver Length and Latency 280

7.4.4   Performance During Iterative Decoding 281

7.4.5   Complexity Analysis 282

7.4.6   Unequal Error-Protection Performance 284

7.5   Summary and Conclusions 287

Chapter 8   Genetic Algorithm-Aided Design of Irregular Variable-Length Coding
               Components   291
    8.1   Introduction 291
    8.2   The Free Distance Metric 295
    8.3   Overview of the Proposed Genetic Algorithm 300
    8.4   Overview of Proposed Scheme 303
          8.4.1   Joint Source and Channel Coding 306
          8.4.2   Iterative Decoding 307
    8.5   Parameter Design for the Proposed Scheme 308
          8.5.1   Design of IrVLC Component VLEC Codebook Suites 309
          8.5.2   Characterization of Component VLEC Codebooks 313
          8.5.3   Suitability of IrVLC Component Codebook Suites 318
          8.5.4   Parameterizations of the Proposed Scheme 320
          8.5.5   Interleaver Length 327
    8.6   Simulation Results 330
    8.7   Summary and Conclusions 335

Chapter 9   Joint EXIT-Chart Matching of Irregular Variable-Length Coding and
               Irregular Unity-Rate Coding   339
    9.1   Introduction 339
    9.2   Modifications of the EXIT-Chart Matching Algorithm 342
    9.3   Joint EXIT-Chart Matching 344
    9.4   Overview of the Transmission Scheme Considered 345
          9.4.1   Joint Source and Channel Coding 345
          9.4.2   Iterative Decoding 347
    9.5   System Parameter Design 349
          9.5.1   Component VLEC Codebooks 349
          9.5.2   Component URC Codes 350
          9.5.3   EXIT-Chart Matching 356
          9.5.4   Parameterizations of the Proposed Scheme 358
    9.6   Simulation Results 365
    9.7   Summary and Conclusions 370

III   Applications of VLCs   373

Chapter 10   Iteratively Decoded VLC Space–Time Coded Modulation:
               Code Construction and Convergence Analysis   375
               S. X. Ng, M. Tao, J. Wang, L.-L. Yang, L. Hanzo

    10.1   Introduction 375
    10.2   Space–Time Coding Overview 377
    10.3   Two-Dimensional VLC Design 378
    10.4   VL-STCM Scheme 381
    10.5   VL-STCM-ID Scheme 384
    10.6   Convergence Analysis 386

10.7   Simulation Results 391
10.8   Non-Binary VL-STCM 393
       10.8.1   Code Design 394
       10.8.2   Benchmarker 396
       10.8.3   Decoding 396
       10.8.4   Simulation results 397
10.9   Conclusions 398

Chapter 11   Iterative Detection of Three-Stage Concatenated IrVLC FFH-MFSK   401
Dr S. Ahmed, R. G. Maunder, L.-L. Yang, L. Hanzo
11.1   Introduction 401
11.2   System Overview 403
       11.2.1   Joint Source and Channel Coding 403
       11.2.2   FFH-MFSK Modulation 404
       11.2.3   The Channel 405
       11.2.4   FFH-MFSK Demodulation 405
11.3   Iterative Decoding 406
       11.3.1   Derivation of Soft Information 406
       11.3.2   Two-Stage and Three-Stage Concatenated Schemes 408
       11.3.3   EXIT-Chart Analysis 409
       11.3.4   VLC Decoding 412
11.4   System Parameter Design and Results 413
11.5   Conclusion 419

Chapter 12   Conclusions and Future Research   421
12.1   Chapter 1: Introduction 421
12.2   Chapter 2: Information Theory Basics 421
12.3   Chapter 3: Sources and Source Codes 422
12.4   Chapter 4: Iterative Source–Channel Decoding 423
12.5   Chapter 5: Three-Stage Serially Concatenated Turbo Equalization 426
12.6   Chapter 6: Joint Source and Channel Coding 428
12.7   Chapters 7–9: EXIT-Chart Matching 430
12.8   Chapter 8: GA-Aided Design of Irregular VLC Components 434
12.9   Chapter 9: Joint EXIT-Chart Matching of IRVLCs and IRURCs 436
12.10 Chapter 10: Iteratively Decoded VLC Space–Time Coded Modulation 443
12.11 Chapter 11: Iterative Detection of Three-Stage Concatenated IrVLC
       FFH-MFSK 443
12.12 Future Work 444
12.13 Closing Remarks 447

Appendix A   VLC Construction Algorithms   449
A.1   First RVLC Construction Algorithm 449
A.2   Second RVLC Construction Algorithm 450
A.3   Greedy Algorithm and Majority Voting Algorithm (MVA) 451

Appendix B   SISO VLC Decoder   453

Appendix C   APP Channel Equalization   457

Bibliography   461

Glossary   483

Subject Index   487

Author Index   491

About the Authors

Lajos Hanzo FREng, FIEEE, FIET, DSc received his degree in electronics in 1976 and his doctorate in 1983. During his career he has held various research and academic posts in Hungary, Germany and the UK. Since 1986 he has been with the School of Electronics and Computer Science, University of Southampton, UK, where he holds the chair in telecommunications. He has co-authored 19 books on mobile radio communications totaling in excess of 10 000 pages, published in excess of 900 research papers, acted as TPC Chair of IEEE conferences, presented keynote lectures and been awarded a number of distinctions. Currently he is directing an academic research team, working on a range of research projects in the field of wireless multimedia communications sponsored by industry, the Engineering and Physical Sciences Research Council (EPSRC) UK, the European IST Programme and the Mobile Virtual Centre of Excellence (VCE), UK. He is an enthusiastic supporter of industrial and academic liaison and he offers a range of industrial courses. He is also an IEEE Distinguished Lecturer as well as a Governor of both the IEEE ComSoc and the VTS. He is the acting Editor-in-Chief of the IEEE Press. For further information on research in progress and associated publications please refer to http://www-mobile.ecs.soton.ac.uk

Robert G. Maunder has studied with the School of Electronics and Computer Science, University of Southampton, UK, since October 2000. He was awarded a first class honours BEng in Electronic Engineering in July 2003, shortly before beginning his current PhD studies in the Communications Research Group at the same university. His research interests include video coding, joint source–channel coding and iterative decoding. He has published a number of IEEE papers in these areas. In 2007 he joined the academic staff.

Jin Wang received the BE degree from the University of Science and Technology of China (USTC), Hefei, China, in 1999 and the ME degree in video signal processing from the Graduate School of the Chinese Academy of Sciences (GSCAS), Beijing, China, in 2002. He completed his PhD degree with the Communications Research Group at the School of Electronics and Computer Science, University of Southampton, UK, where he conducted research on source coding, channel coding and joint source–channel coding, as well as on iterative detection and decoding designed for digital communication systems. His research results were published in a dozen or so IEEE journal and conference papers. Upon completing his PhD he joined Imagination Technologies in the UK, and in 2008 he transferred to Aeroflex, Cambridge, UK, working on 3GPP LTE solutions.

Lie-Liang Yang received his BEng degree in communication engineering from Shanghai TieDao University, Shanghai, China in 1988, and his MEng and PhD degrees in communications and electronics from Northern Jiaotong University, Beijing, China, in 1991 and 1997 respectively. From June 1997 to December 1997 he was a visiting scientist at the Institute of Radio Engineering and Electronics, Academy of Sciences of the Czech Republic. Since December 1997 he has been with the Communications Research Group, School of Electronics and Computer Science, University of Southampton, UK, where he was first a Postdoctoral Research Fellow (December 1997–August 2002) and then a Lecturer (September 2002–February 2006), and currently holds the academic post of Reader. Dr Yang's research has covered a wide range of areas in telecommunications, which include error control coding, modulation and demodulation, spread-spectrum communications and multiuser detection, synchronization, space-time processing and adaptive wireless systems, as well as wideband, broadband and ultra-wideband code-division multiple-access (CDMA). He has published around 150 papers in journals and conference proceedings, co-authored two Wiley–IEEE Press books and published his research monograph on Multicarrier Communications in 2009. He was awarded the Royal Society Sino-British Fellowship in 1997 and the EPSRC Research Fellowship in 1998. Dr Yang is currently an associate editor for both the *Journal of Communications and Networks* (JCN) and the *Journal of Communications* (JCM).

Other Wiley and IEEE Press Books on Related Topics[1]

- R. Steele, L. Hanzo (Ed): *Mobile Radio Communications: Second and Third Generation Cellular and WATM Systems*, John Wiley and IEEE Press, 2nd edition, 1999, 1064 pages

- L. Hanzo, T. H. Liew, B. L. Yeap: *Turbo Coding, Turbo Equalisation and Space-Time Coding*, John Wiley and IEEE Press, 2002, 751 pages

- L. Hanzo, C. H. Wong, M. S. Yee: *Adaptive Wireless Transceivers: Turbo-Coded, Turbo-Equalised and Space-Time Coded TDMA, CDMA and OFDM Systems*, John Wiley and IEEE Press, 2002, 737 pages

- L. Hanzo, L.-L. Yang, E.-L. Kuan, K. Yen: *Single- and Multi-Carrier CDMA: Multi-User Detection, Space-Time Spreading, Synchronisation, Networking and Standards*, John Wiley and IEEE Press, June 2003, 1060 pages

- L. Hanzo, M. Münster, T. Keller, B.-J. Choi, *OFDM and MC-CDMA for Broadband Multi-User Communications, WLANs and Broadcasting*, John-Wiley and IEEE Press, 2003, 978 pages

- L. Hanzo, S.-X. Ng, T. Keller and W. T. Webb, *Quadrature Amplitude Modulation: From Basics to Adaptive Trellis-Coded, Turbo-Equalised and Space-Time Coded OFDM, CDMA and MC-CDMA Systems*, John Wiley and IEEE Press, 2004, 1105 pages

- L. Hanzo, T. Keller: *An OFDM and MC-CDMA Primer*, John Wiley and IEEE Press, 2006, 430 pages

- L. Hanzo, F. C. A. Somerville, J. P. Woodard: *Voice and Audio Compression for Wireless Communications*, John Wiley and IEEE Press, 2007, 858 pages

- L. Hanzo, P. J. Cherriman, J. Streit: *Video Compression and Communications: H.261, H.263, H.264, MPEG4 and HSDPA-Style Adaptive Turbo-Transceivers* John Wiley and IEEE Press, 2007, 680 pages

---

[1] For detailed contents and sample chapters please refer to http://www-mobile.ecs.soton.ac.uk

- L. Hanzo, J. S. Blogh, S. Ni: *3G, HSDPA, HSUPA and FDD Versus TDD Networking: Smart Antennas and Adaptive Modulation* John Wiley and IEEE Press, 2008, 564 pages
- L. Hanzo, O. Alamri, M. El-Hajjar, N. Wu: *Near-Capacity Multi-Functional MIMO Systems: Sphere-Packing, Iterative Detection and Cooperation,* IEEE Press – John Wiley, 2009
- L. Hanzo, J. Akhtman, M. Jiang, L. Wang: *MIMO-OFDM for LTE, WIFI and WIMAX: Coherent versus Non-Coherent and Cooperative Turbo-Transceivers,* John Wiley and IEEE Press, 2010, 608 pages

Preface

Variable-Length Codes (VLCs) have as long a history as that of information theory. Shannon's original lossless entropy codes adopted the conceptually appealing philosophy that a stream of binary source data may be partitioned into fixed-length data segments, where the segments having a low probability of occurrence may be assigned long codewords, while the high-probability ones must have short codewords assigned to them. This facilitates the reduction of the average number of bits that has to be transmitted. A further design criterion is that no shorter codeword may constitute a prefix, i.e. the initial segment of a longer codeword.

Naturally, the compressed data stream becomes more susceptible to the loss of synchronization in the presence of transmission errors, in the sense that the decoder cannot recognize a corrupted codeword and hence it is bound to read past the end of the original error-free codeword. This phenomenon potentially prevents the decoder from recognizing the next legitimate codeword, and hence can result in an avalanche-like error propagation. In order to circumvent this potential problem, numerous design alternatives have emerged. Examples include Variable-Length Error Correction (VLEC) codes and reversible VLCs (RVLCs) (which contain redundancy in the sense that they are constituted by a bit-pattern that is symmetric with respect to the middle, and hence may be decoded from both ends).

The family of VLCs has also found its way into virtually all video and audio compression standards, which are used by widespread consumer products, and yet there has been no dedicated book on the subject of VLCs. This volume aims to fill that gap, with the objective of reporting on the most recent research advances, including iterative turbo decoding and irregular designs based on EXtrinsic Information Transfer (EXIT) charts. The proposed code design principles are generic, and hence they may be applied to arbitrary code design problems.

Our intention with the book is:

- First, to pay tribute to all the researchers, colleagues and valued friends who have contributed to the field. Hence this monograph is dedicated to them, since without their quest for improved VLCs it could not have been conceived. They are too numerous to name here, but they appear in the author index. Our hope is that the conception of this monograph on the topic will provide an adequate portrayal of the community's research and will further fuel this innovation process.

- Second, to stimulate further research by exposing open research problems, and by collating a range of practical problems as well as design issues for the practitioners. The coherent further efforts of the wireless research community are expected to lead to the solution of a range of outstanding problems, ultimately providing us with flexible joint source- and channel-coded wireless transceivers exhibiting a performance close to the information theoretical limits.

Acknowledgements

We are indebted to our many colleagues who have enhanced our understanding of the subject. These colleagues and valued friends, too numerous to be mentioned individually, have influenced our views concerning the subject of the book. We thank them for the enlightenment gained from our collaborations on various projects, papers and books. We are particularly grateful to our academic colleagues Prof. Sheng Chen and Dr Soon-Xin Ng. We would also like to express our appreciation to Osamah Alamri, Dr Sohail Ahmed, Andreas Ahrens, Jos Akhtman, Jan Brecht, Jon Blogh, Nicholas Bonello, Marco Breiling, Marco del Buono, Peter Cherriman, Stanley Chia, Byoung Jo Choi, Joseph Cheung, Jin-Yi Chung, Peter Fortune, Thanh Nguyen Dang, Sheyam Lal Dhomeja, Lim Dongmin, Dirk Didascalou, Mohammed El-Hajjar, Stephan Ernst, Eddie Green, David Greenwood, Chen Hong, Hee Thong How, Bin Hu, Ming Jiang, Thomas Keller, Lingkun Kong, Choo Leng Koh, Ee Lin Kuan, W. H. Lam, Wei Liu, Kyungchun Lee, Xiang Liu, Fasih Muhammad Butt, Matthias Münster, Song Ni, C. C. Lee, M. A. Nofal, Xiao Lin, Chee Siong Lee, Tong-Hooi Liew, Noor Shamsiah Othman, Raja Ali Raja Riaz, Vincent Roger-Marchart, Redwan Salami, Prof. Raymond Steele, Shinya Sugiura, David Stewart, Clare Sommerville, Tim Stevens, Shuang Tan, Ronal Tee, Jeff Torrance, Spyros Vlahoyiannatos, Jin Wang, Li Wang, William Webb, Chun-Yi Wei, Hua Wei, Stefan Weiss, John Williams, Seung-Hwang Won, Jason Woodard, Choong Hin Wong, Henry Wong, James Wong, Andy Wolfgang, Nan Wu, Lei Xu, Chong Xu, Du Yang, Wang Yao, Bee-Leong Yeap, Mong-Suan Yee, Kai Yen, Andy Yuen, Jiayi Zhang, Rong Zhang, and many others with whom we have enjoyed an association.

We also acknowledge our valuable associations with the Virtual Centre of Excellence in Mobile Communications, in particular with its chief executive, Dr Walter Tuttlebee, and other members of its Executive Committee, namely Dr Keith Baughan, Prof. Hamid Aghvami, Prof. Mark Beach, Prof. John Dunlop, Prof. Barry Evans, Prof. Peter Grant, Dr Dean Kitchener, Prof. Steve MacLaughlin, Prof. Joseph McGeehan, Dr Tim Moulsley, Prof. Rahim Tafazolli, Prof. Mike Walker and many other valued colleagues. Our sincere thanks are also due to John Hand and Andrew Lawrence of EPSRC, UK for supporting our research. We would also like to thank Dr Joao Da Silva, Dr Jorge Pereira, Bartholome Arroyo, Bernard Barani, Demosthenes Ikonomou, and other valued colleagues from the Commission of the European Communities, Brussels, Belgium.

The authors of this book are particularly indebted to Peter Cherriman and Jurgen Streit for generously allowing them to use Chapter 1 of Hanzo, Cherriman and Streit: Wireless Video Communications, IEEE Press 2001 as Chapter 1 of this monograph. This chapter might be found helpful by readers, who are new to information theory. The inclusion of this chapter into this book was facilitated by transferring the copyright of the above-mentioned original source from the IEEE Press to the authors, namely to Hanzo, Cherriman and Streit.

Similarly, our sincere thanks are due to Mark Hammond, Sarah Tilley and their colleagues at Wiley in Chichester, UK. Finally, our sincere gratitude is due to the numerous authors listed in the Author Index – as well as to those whose work was not cited owing to space limitations – for their contributions to the state of the art, without whom this book would not have materialized.

Lajos Hanzo, Robert G. Maunder, Jin Wang and Lie-Liang Yang
School of Electronics and Computer Science
University of Southampton, UK

Chapter 1

Introduction

1.1 Historical Overview

Multimedia transmission over time-varying wireless channels presents a number of challenges beyond the existing capabilities of Third-Generation (3G) networks. The prevalent design principles stemming from Shannon's source and channel separation theorem [1] have to be reconsidered. The separation theorem, stating that optimal performance bounds of source and channel decoding can be approached as closely as desired by independently designing source and channel coding, holds only under asymptotic conditions, where both codes may have a length and a complexity tending to infinity, and under the conditions of stationary sources. In practice, the design of the system is heavily constrained in terms of both complexity and delay, and hence source and channel codecs are typically suboptimal [2]. Fortunately, the 'lossy' nature of the audio and video codecs may be concealed with the aid of perceptual masking by exploiting the psycho-acoustic and psycho-visual properties of the human ear and eye. The output of practical source encoders, such as speech codecs [3] and image/video codecs [4], may contain a significant amount of redundancy. Moreover, having a nonzero residual channel error rate may lead to dramatically increased source symbol error rates [2]. In these circumstances, the assumptions of Shannon's separation theorem no longer hold; neither can they be used as a good approximation. One may potentially improve the attainable performance by considering the source and channel code designs jointly, allowing us to exploit both the residual redundancy of the source-coded stream and the redundancy introduced by the channel code, or optimally allocate redundancy (in the best form) within the source and channel coding chain [2, 5–7].

1.1.1 Source Coding and Decoding

1.1.1.1 *Variable-Length Coding*

Source coding constitutes a mapping from a sequence of symbols of an information source to a sequence of codewords (usually binary bits), so that the source symbols can be exactly recovered from the binary bits (lossless source coding) or recovered within some distortion (lossy source coding). Claude Shannon initiated this subject (along with other subjects in the foundations of information theory) in his ground-breaking paper [1]. The key concept of entropy was formulated here. Shannon then filled the concept of entropy with practical significance by proving his noiseless source coding theorem. Shannon's theorem [1, Theorem 9] states that the entropy rate of a stationary ergodic finite-alphabet Markov source is the optimum rate at which the source can be encoded using 'almost noiseless' fixed-rate block codes or using noiseless variable-rate block codes. To prove the 'fixed-rate half' of the theorem, Shannon constructed fixed-rate block codes using an auxiliary theorem which later became known as the Shannon–McMillan Theorem [8]. In the proof of the 'variable-rate half' of the theorem, Shannon devised a noiseless variable-rate block code (nowadays referred to as a Shannon–Fano code), which encodes a source symbol having a probability of occurrence p into a codeword having a length within one bit of $-\log_2 p$.

The Shannon–Fano coding algorithm is capable of producing fairly efficient variable-length prefix codes based on a set of symbols and their probabilities (estimated or measured), but it is nonetheless suboptimal. As a remedy, Huffman coding [9], named after its inventor, D. A. Huffman, is always capable of producing optimal prefix codes, which approach the lower bound of the expected codeword length under the constraint that each symbol is represented by a codeword formed of an integer number of bits. This constraint, however, is often unnecessary, since the codewords will be packed end-to-end in long sequences. If we consider groups of codes at a time, symbol-by-symbol Huffman coding is only optimal if the probabilities of the symbols are independent and assume a value of $1/2^n$, where n is an integer. In most situations, arithmetic coding [10] is capable of producing a higher overall compression than either Huffman or Shannon–Fano coding, since it is capable of encoding in terms of fractional non-integer numbers of bits, which more closely approximate the actual information content of the symbol. However, arithmetic coding has not superseded Huffman coding, as Huffman coding superseded Shannon–Fano coding, partly because arithmetic coding is more computationally expensive and partly because it is covered by multiple patents. An excellent tutorial on the history of source coding can be found in [11].

In addition to coding efficiency, the robustness of the variable-length codes (VLCs) to transmission errors is another design concern. Takishima *et al.* [12] proposed Reversible Variable-Length Codes (RVLCs) for providing both forward and backward decoding capability so that the correct data can be recovered with as high a probability as possible from the received data when transmission errors have occurred. For example, a decoder can commence by processing the received bit stream in the forward direction, and upon detecting an error proceed immediately to the end of the bit stream and decode in the reverse direction. This technique and a range of other related strategies can be used for significantly reducing the effects of bit errors on the fidelity of the decoded data. Wen and Villasenor [13, 14] proposed a new class of asymmetric RVLCs having the same codeword length distribution as the Golomb–Rice codes [15, 16] as well as exp-Golomb codes [17], and applied them to the video coding framework of the H.263+ [18] and MPEG-4 [19] standards. Later on, in the quest to construct optimal RVLCs, further improved algorithms were proposed in [20–24]. More particularly, Lakovic and Villasenor [22] designed a family of new RVLCs having a free distance of $d_f = 2$, which exhibit a better error-correcting performance than the RVLCs and Huffman codes having a lower free distance. The conditions for the existence of RVLCs and their properties were also studied in [25–27].

For the sake of attaining an even stronger error-correction capability, Buttigieg and Farrell [28, 29] proposed a new class of Variable-Length Error-Correction (VLEC) codes, which generally have higher free distances ($d_f > 2$). It was shown in [29] that the free distance is the most important parameter that determines the decoding performance at high E_b/N_0 values. Detailed treatment of various VLEC code properties, their construction, decoding and performance bounds can be found in Buttigieg's PhD thesis [30]. Lamy and Paccaut [31] also investigated a range of low-complexity construction techniques designed for VLEC codes. In fact, both Huffman codes and RVLCs may be viewed as special VLEC codes having a free distance of $d_f < 2$ [32]. Hence a generic construction algorithm may be applied for designing both RVLCs and VLEC codes [32]. For the reader's convenience, we have summarized the major contributions on Variable-Length Coding (VL-coding) in Table 1.1, where they can be seen at a glance.

1.1.1.2 Variable-Length Decoding

Owing to their prefix property, source messages encoded by Huffman codes as well as RVLCs and VLEC codes may be decoded on a bit-by-bit basis with the aid of their code trees or lookup tables [2, 30, 33]. Given the wide employment of the ITU-T H.26x and ISO/IEC Motion Picture Experts Group (MPEG) series video coding standards, numerous efficient hardware architectures have been proposed [34, 35] for achieving low-power, memory-efficient and high-throughput decoding. These methods aim for low-latency, near-instantaneous decoding and generally assume having an error-free input bit stream. However, owing to their variable-length nature, their decoding is prone to the loss of synchronization when transmission errors occur, and the errors are likely to propagate during the decoding of the subsequent bit stream. In other words, should a decoding error be encountered in the current codeword, a relatively large number of subsequent erroneously decoded codewords are likely to follow, even if in the meantime there were no further errors imposed by the channel. Hence, although these codes may be decoded instantaneously, it is clear that if we were to wait for several codewords before making a decision, a better decoding performance (fewer decoding errors) could be expected. Indeed, the decoding of VLCs in error-prone environments is jointly a segmentation (i.e. recovering source symbol boundaries) and an estimation problem.

Numerous source-decoding algorithms capable of generating a source estimate based on a sequence of received data have been developed [7, 28, 29, 36–46]. In parallel, various trellis representations [29, 37, 38, 40, 41, 44] have been devised for describing the source (coder). In [28, 29] a Maximum Likelihood (ML) sequence estimation algorithm was proposed, which selected the most likely symbol sequence instead of symbol-by-symbol decoding. More specifically, the decoder invoked a modified form of the Viterbi Algorithm (VA) [47] based on the trellis structure of VLEC codes, where each state of the trellis corresponded to the number of bits produced so far and each transition between the states represented the transmission of a legitimate codeword (or source symbol). Maximum *A Posteriori* Probability (MAP) sequence estimation based on the same trellis structure was also studied in [30]. Both algorithms are based on hard decoding, where the inputs of the source decoder are binary zeros as well as ones and the path metric is calculated based on the Hamming distance between the transmitted bits and the received bits. Soft information may be used for improving the attainable decoding performance, resulting in Soft-Input Soft-Output (SISO) decoding algorithms. The term 'soft'

Table 1.1: Major contributions on variable-length coding

Year	Author(s)	Contribution
1948	Shannon [1]	Invented the first variable-length coding algorithm, referred to as Shannon–Fano coding, along with the creation of information theory.
1952	Huffman [9]	Proposed the optimal prefix codes, Huffman codes.
1979	Rissanen [10]	Proposed arithmetic coding.
1994	Buttigieg and Farrell [28–30]	Proposed VLEC codes and investigated various VLEC code properties, construction, decoding and performance bounds.
1995	Takishima *et al.* [12]	Proposed RVLCs for providing both a forward and a backward decoding capability.
1997	Wen and Villasenor [13, 14]	Proposed a new class of asymmetric RVLCs and applied them to the video coding framework of the H.263+ [18] and MPEG-4 [19] standards.
2001	Tsai and Wu [20, 21]	Proposed the Maximum Symmetric-Suffix Length (MSSL) metric [20] for designing symmetric RVLCs and the Minimum Repetition Gap (MRG) [21] metric for asymmetric RVLCs.
2002	Lakovic and Villasenor [22]	Proposed a new family of RVLCs having a free distance of $d_f = 2$.
2003	Lakovic and Villasenor [23]	Proposed the so-called 'affix index' metric for constructing asymmetric RVLCs.
	Lamy and Paccaut [31]	Proposed low-complexity construction techniques for VLEC codes.

implies that the decoder is capable of accepting as well as of generating not only binary ('hard') decisions, but also reliability information (i.e. probabilities) concerning the bits.

The explicit knowledge of the total number of transmitted source symbols or transmitted bits can be incorporated into the trellis representations as a termination constraint in order to assist the decoder in resynchronizing at the end of the sequence. Park and Miller [37] designed a symbol-based trellis structure for MAP sequence estimation, where the number of transmitted bits is assumed to be known at the receiver, resulting in the so-called bit-constrained directed graph representation. Each state of the trellis corresponds to a specific source symbol and the number of symbols produced so far. By contrast, when the total number of transmitted symbols is assumed to be known at the receiver, Demir and Sayood [38] proposed a symbol-constrained directed graph representation, where each state of the trellis corresponds to a specific symbol and the corresponding number of bits that have been produced. Moreover, when both constraints are known at the receiver, Bauer and Hagenauer [40] designed another symbol-trellis representation for facilitating symbol-by-symbol MAP decoding of VLCs, which is a modified version of the Bahl–Cocke–Jelinek–Raviv (BCJR) algorithm [48]. Later they [42] applied their BCJR-type MAP decoding technique to a bit-level trellis representation of VLCs that was originally devised by Balakirsky [41], where the trellis was derived from the corresponding code tree by mapping each tree node to a specific trellis state. The symbol-based trellis has the drawback that for long source sequences the number of trellis states increases quite drastically, while the bit-based trellis has a constant number of states, which only depends on the depth of the VLC tree.

For sources having memory, the inter-symbol correlation is usually modeled by a Markov chain [2], which can be directly represented by a symbol-based trellis structure [36–38, 43]. In order to fully exploit all the available *a priori* source information, such as the inter-symbol correlation or the knowledge of the total number of transmitted bits and symbols, a three-dimensional symbol-based trellis representation was also proposed by Thobaben and Kliewer in [44]. In [45], the same bit trellis as that in [42] was used, but the *A Posteriori* Probability (APP) decoding algorithm was modified for the sake of exploiting the inter-symbol correlation. As a low-complexity solution, sequential decoding [49, 50] was also adopted for VLC decoding. In addition to classic trellis-based representations, Bayesian networks were also employed for assisting the design of VLC decoders [46], where it was shown that the decoding of the Markov source and that of the VLC can be performed separately.

1.1.1.3 Classification of Source-Decoding Algorithms

Generally, the source-decoding problem may be viewed as a hidden Markov inference problem, i.e. that of estimating the sequence of hidden states of a Markovian source/source-encoder through observations of the output of a memoryless channel. The estimation algorithms may employ different estimation criteria and may carry out the required computations differently. The decision rules can be optimum either with respect to a sequence of symbols or with respect to individual symbols. More explicitly, the source-decoding algorithms can be broadly classified into the following categories [7].

- **MAP/ML Sequence Estimation:** The estimate of the entire source sequence is calculated using the MAP or ML criterion based on all available measurements. When the *a priori* information concerning the source sequence is uniformly distributed, implying that all sequences have the same probability, or simply ignored in order to reduce the computational complexity, the MAP estimate is equivalent to the ML estimate. The algorithms outlined in [28, 29, 36–38, 41, 46] belong to this category.

- **Symbol-by-Symbol MAP Decoding:** This algorithm searches for the estimate that maximizes the *a posteriori* probability for each source symbol rather than for the entire source sequence based on all available measurements. The algorithms characterized in [40, 42–46] fall into this category.

- **Minimum Mean Square Error (MMSE) Decoding:** This algorithm aims at minimizing the expected distortion between the original signal and the reconstructed signal, where the expected reconstructed signal may be obtained using APP values from the MAP estimation. The algorithm presented in [39] falls into this category.

1.1.2   Joint Source–Channel Decoding

Let us consider a classic communication chain, using a source encoder that aims at removing the redundancy from the source signal, followed by a channel encoder that aims to reintroduce the redundancy in the transmitted stream in an efficiently controlled manner in order to cope with transmission errors [51] (Chapter 1, page 2). Traditionally, the source encoder and the channel encoder are designed and implemented separately according to Shannon's source and channel coding separation theorem [1], which holds only under idealized conditions. Therefore Joint Source–Channel Decoding (JSCD) has gained considerable attention as a potentially attractive alternative to the separate decoding of source and channel codes. The key idea of JSCD is to exploit jointly the residual redundancy of the source-coded stream (i.e. exploiting the suboptimality of the source encoder) and the redundancy deliberately introduced by the channel code in order to correct the bit errors and to find the most likely source symbol estimates. This requires modeling of all the dependencies present in the complete source–channel coding chain.

In order to create an exact model of dependencies amenable to optimal estimation, one can construct the amalgam of the source, source encoder and channel encoder models. This amalgamated model of the three elements of the chain has been proposed in [52]. The set of states in the trellis-like joint decoder graph contains state information about the source, the source code and the channel code. The amalgamated decoder may then be used to perform a MAP, ML or MMSE decoding. This amalgamated approach allows for optimal joint decoding; however, its complexity may remain excessive for realistic applications, since the state-space dimension of the amalgamated model explodes in most practical cases. Instead of building the Markov chain of the amalgamated model, one may consider the serial or parallel connection of two Hidden Markov Models (HMMs), one for the source as well as the source encoder and one for the channel encoder, in the spirit of serial and parallel turbo codes. The dimension of the state space required for each of the separate models is then reduced. Furthermore, the adoption of the turbo principle [53] led to the design of iterative estimators operating alternately on each factor of the concatenated model. The estimation performance attained is close to the optimal performance only achievable by the amalgamated model [7].

In [37, 38, 40–46, 54, 55], the serial concatenation of a source and a channel encoder was considered for the decoding of both fixed-length and variable-length codes. *Extrinsic* information was iteratively exchanged between the channel decoder and the source decoder at the receiver. The convergence behavior of iterative source–channel decoding using fixed-length source codes and a serially concatenated structure was studied in [56], and in [45] for VLCs. The gain provided by the iterations is very much dependent on the amount of correlation present at both sides of the interleaver. The variants of the algorithms proposed for JSCD of VLC-encoded sources relate to various trellis representation forms derived for the source encoder, as described in Section 1.1.1.2, as well as to the different underlying assumptions with respect to the knowledge of the length of the symbol or bit sequences. Moreover, a parallel-concatenated source–channel coding and decoding structure designed for VLC-encoded sources is described in [57]. In contrast to a turbo code, the explicit redundancy inherent in one of the component channel codes is replaced by the residual redundancy left in the source-compressed stream after VLC encoding. It was shown in [57] that the parallel concatenated structure is superior to the serial concatenated structure for the RVLCs considered, and for the Additive White Gaussian Noise (AWGN) channel experiencing low Signal-to-Noise Ratios (SNRs).

Another possible approach is to modify the channel decoder in order to take into account the source statistics and the model describing both the source and the source encoder, resulting in the so-called Source-Controlled Channel Decoding (SCCD) technique [58]. A key idea presented in [58] is to introduce a slight modification of a standard channel-decoding technique in order to take advantage of the source statistics. This idea has been explored [58] first in the case of Fixed-Length Coding (FLC), and has been validated using convolutional codes in the context of transmitting coded speech frames over the Global System of Mobile telecommunication (GSM) [59]. Source-controlled channel decoding has also been combined with both block and convolutional turbo codes, considering FLC for hidden Markov sources [60] or images [61–63]. Additionally, this approach has been extended to JSCD of VLC sources, for example in the source-controlled convolutional decoding of VLCs in [64, 65] and the combined decoding of turbo codes and VLCs in [64–67]. The main contributions on JSCD of VLCs are summarized in Table 1.2.

1.1.3 Iterative Decoding and Convergence Analysis

The invention of turbo codes over a decade ago [69] constitutes a milestone in error-correction coding designed both for digital communications and for storage. The parallel concatenated structure, employing interleavers for decorrelating the constituent encoders, and the associated iterative decoding procedure, exchanging only extrinsic information between the constituent decoders, are the key elements that are capable of offering a performance approaching the previously elusive Shannon limit, although they also impose an increasing complexity upon inching closer to the limit, in particular beyond the so-called Shannonian cut-off [70]. Reliable communication for all channel capacity rates slightly in excess of the source entropy rate becomes approachable. The practical success of the iterative turbo decoding algorithm has inspired its adaptation to other code classes, notably serially concatenated codes [71], Turbo Product Codes (TPCs) which may be constituted from Block Turbo Codes (BTCs) [72–74], and Repeat–Accumulate (RA) codes [75], and has rekindled interest [76–78] in Low-Density Parity-Check (LDPC) codes [79], which constitute a historical milestone in iterative decoding.

The serially concatenated configuration of components [71] commands particular interest in communication systems, since the 'inner encoder' of such a configuration can be given more general interpretations [53, 80, 81], as seen in Figure 1.1, such as an 'encoder' constituted by a dispersive channel or by the spreading codes used in Code-Division Multiple Access (CDMA). The corresponding iterative decoding algorithm can then be extended into new areas, giving rise to Turbo Equalization (TE) [82], Bit-Interleaved Coded Modulation (BICM) [83, 84], serial concatenated trellis-coded modulation [85, 86], turbo multiuser detection [87, 88], turbo synchronization [89] and coded CDMA [90]. Connection of turbo decoding algorithms to the belief propagation algorithm [91] has also been discussed in [92]. Moreover, factor graphs [93, 94] were employed [95, 96] to provide a unified framework that allows us to understand the connections and commonalities among seemingly different iterative receivers.

Table 1.2: Major contributions on JSCD of VLCs

Year	Author(s)	Contribution
1991	Sayood and Borkenhagen [4]	Considered the exploitation of the residual redundancy found in the fixed-length coded quantized-source indices in the design of joint source–channel coders.
1994	Phamdo and Farvardin [68]	Studied the optimal detection of discrete Markov sources over discrete memoryless channels.
1995	Hagenauer [58]	Proposed the first SCCD scheme with application in speech transmissions.
1997	Balakirsky [41]	Proposed a bit-level trellis representation of VLCs and applied it in a joint source–channel VLC sequence estimation scheme.
1998	Murad and Fuja [52]	Proposed a joint source–channel VLC decoding scheme based on a combined model of the source, the source encoder and the channel encoder.
	Miller and Park [39]	Proposed a sequence-based approximate MMSE decoder for JSCD.
	Demir and Sayood [38]	Investigated a JSCD scheme using MAP sequence estimation based on trellis structure constrained by the number of transmitted bits.
1999	Park and Miller [37]	Studied a JSCD scheme using both exact and approximate MAP sequence estimation based on a trellis structure constrained by the number of transmitted symbols.
2000	Bauer and Hagenauer [40]	Proposed a symbol-level trellis representation of VLCs and derived a symbol-by-symbol MAP VLC decoding algorithm.
	Guivarch et al. [64]	Considered SCCD of Huffman codes combined with turbo codes.
	Peng et al. [63]	Investigated a SCCD scheme using turbo codes for image transmission.
2001	Bauer and Hagenauer [42]	Investigated MAP VLC decoding based on the bit-level trellis proposed in [41].
	Guyader et al. [46]	Studied JSCD schemes using Bayesian networks.
	Garcia-Frias and Villasenor [60]	Applied SCCD to the decoding of hidden Markov sources.

Table 1.2: Continued

Year	Author(s)	Contribution
2002	Kliewer and Thobaben [43]	Adopted the JSCD scheme of [40] to the decoding of correlated sources.
2003	Thobaben and Kliewer [44]	Proposed a 3D symbol-based trellis for the sake of fully exploiting the *a priori* source information.
2004	Jaspar and Vandendorpe [67]	Proposed a SCCD scheme using a hybrid concatenation of three SISO modules.
2005	Thobaben and Kliewer [45]	Proposed a modified BCJR algorithm [48] incorporating inter-symbol correlation based on the bit-level trellis of [41].
	Jeanne *et al.* [65]	Studied SCCD schemes using both convolutional codes and turbo codes.

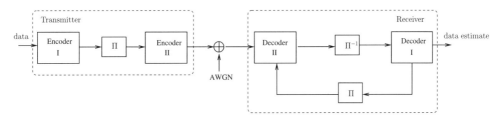

Configuration	En-/Decoder I (Outer Code)	En-/Decoder II (Inner Code)
Serial Concatenated Code	FEC En-/Decoder	FEC En-/Decoder
LDPC Codes	Check Nodes	Variable Nodes
Turbo Equalization	FEC En-/Decoder	Multipath Channel/Equalizer
Turbo BICM	FEC En-/Decoder	Mapper/Demapper
Turbo Source–Channel Decoding	Source En-/Decoder	FEC En-/Decoder
Turbo CDMA	FEC En-/Decoder	Spreading Code/MUD
Turbo MIMO	FEC En-/Decoder	Mapper/MIMO Detector

Figure 1.1: A serial concatenated system with iterative detection/decoding [97], where Π denotes the interleaver and Π^{-1} represents the de-interleaver. (MIMO: Multiple-Input Multiple-Output; MUD: Multi-User Detection; FEC: Forward Error Correction.)

Typically, Bit-Error Ratio (BER) charts of iterative decoding schemes can be divided into three regions [98]:

- the region of low E_b/N_0 values with negligible iterative BER reduction;
- the 'turbo-cliff' region, which is also referred to as the 'waterfall' region, with consistent iterative BER reduction over many iterations;

- the BER floor region often recorded at moderate to high E_b/N_0 values, where a rather low BER can be reached after just a few iterations.

Efficient analytical bounding techniques have been found for moderate to high E_b/N_0 values in [71, 99–102], where design rules were also derived for the sake of lowering the BER floors. On the other hand, a long-standing open problem was understanding the convergence behavior of iterative decoding algorithms in the turbo-cliff region. A successful approach to this problem was pioneered in [77, 78], which investigated the Probability Density Functions (PDFs) of the information communicated within the iterative decoding algorithms. Based on the evolution of these PDFs during iterative decoding, convergence thresholds have been derived for investigating the iterative decoding of LDPC codes [77] with the aid of the related technique, which is referred to as *density evolution*. Simplified approaches were also proposed by assuming that the PDFs considered above are Gaussian [103], or by observing a single parameter of them such as the SNR measure in [104], a reliability measure in [105] and the mutual information measure in [98, 106]. The convergence prediction capabilities of six different measures were compared in [107], where the mutual information measure used in the EXtrinsic Information Transfer (EXIT) chart [98, 106] analysis was found to have a high accuracy, regardless of the specific iterative decoding schemes considered. Additionally, the cross-entropy [108] was employed by Luo and Sweeney in [109] as a measure used for determining the convergence threshold of turbo decoding algorithms, where neither the knowledge of the source information nor the Gaussian approximation was required, while the relationship between the Minimum Cross-Entropy decoding algorithm and the iterative decoding algorithm was discussed in [110].

EXIT charts have been successfully applied to predict the convergence behavior of various concatenated systems, such as parallel [98, 111] and serially concatenated codes [112], turbo equalization [113], Bit-Interleaved Coded Modulation using Iterative Decoding (BICM-ID) [106, 114] and JSCD [45, 56]. A unified framework derived to apply EXIT chart analysis for various serial concatenation schemes was presented in [115]. EXIT functions were defined with the aid of a universal decoding model in [116] and several properties of the related functions were derived for erasure channels. A specific property allows us to quantify the area under an EXIT function in terms of the conditional entropy. A useful consequence of this result is that the design of capacity-approaching codes reduces to a curve-fitting problem in the context of EXIT charts [116]. For example, EXIT charts have been applied for designing Irregular Convolutional Codes (IrCCs) [80, 81] in order to approach the AWGN channel's capacity, and for designing irregular LDPC codes [117] and Irregular Repeat–Accumulate (IRA) codes [118] for approaching the Multiple-Input Multiple-Output (MIMO) fading channel's capacity.

EXIT charts have then also been extended from 2D to 3D for the sake of analyzing the convergence behavior of multi-stage iterative decoding schemes consisting of more than two components, such as, for example, Parallel Concatenated ('turbo') Codes (PCCs) consisting of three constituent codes [119], three-stage turbo equalization [120] and three-stage iterative decoding and demodulation [121, 122]. A 3D to 2D projection technique was presented in [120–122], which simplifies the convergence analysis of multi-stage systems. Optimizing the activation order of the component decoders was also investigated in [121, 122]. Furthermore, the philosophy of EXIT chart analysis has been extended from binary decoders to non-binary decoders [123, 124]. The major contributions on iterative decoding and its convergence analysis are summarized in Tables 1.3 and 1.4.

1.2 Applications of Irregular Variable-Length Coding

In this monograph we introduce the novel concept of Irregular Variable-Length Codes (IrVLCs) and investigate their applications, characteristics and performance in the context of digital multimedia broadcast telecommunications. More specifically, we consider the employment of IrVLCs to represent or encode multimedia source signals, facilitating their transmission to, and reconstruction by, remote receivers. Explicitly, IrVLCs are defined as encoding schemes that represent particular components or segments of the source signals with different sets of codewords, which comprise various numbers of bits. This may be contrasted with regular Variable-Length Codes (VLCs), which encode all components or segments of the source signals using the same codebook of codewords.

Three particular aspects of IrVLCs designed for digital multimedia telecommunications are investigated in this monograph, namely near-capacity operation, joint source and channel coding, and Unequal Error Protection (UEP).

1.2.1 Near-Capacity Operation

In his seminal contribution [133], Shannon considered source signals that contain some redundancy. By definition, the redundant content of a particular source signal may theoretically be accurately predicted, given a perfect knowledge of its unpredictable information content. Provided that unimpaired knowledge of the source signal's information content can be reliably conveyed to the receiver by the channel, the transmitter is allowed to employ source coding in order to remove all the source's redundancy, without jeopardizing the receiver's ability to reconstruct it. When communicating over a perfectly noiseless channel, Shannon [133] showed that the minimum number of bits that are required to reliably convey perfect knowledge of the source signal's information content to the receiver is given by the source's entropy. Note however that the computational complexity and latency imposed by a source codec typically escalates in optimal entropy coding, where all redundancy is removed from the source signal.

Unfortunately, when communicating over noisy channels, which impose uncertainty on the received signal, the reliable transmission of a source signal's information content can never be guaranteed. However, Shannon [133] also showed that if a source signal's information content is conveyed over a noisy channel at a rate (expressed in bits per second) that does not exceed the channel's capacity, then it is theoretically possible to reconstruct it with an infinitesimally low probability of error. This motivates the employment of channel coding that introduces redundancy into the transmitted signal in a specifically designed manner. This redundancy may be exploited in the receiver to mitigate any channel-induced errors within the original non-redundant information content.

As with optimal entropy coding, the computational complexity and latency imposed by a channel codec escalates when approaching the channel's capacity [134]. Indeed, Jacobs and Berlekamp [135] showed that the decoding trees employed by the then state-of-the-art sequential decoding algorithm [136–139] for Convolutional Codes (CCs) [140] have an escalating computational complexity, when operating at an information rate above the so-called cut-off rate of the channel. This cut-off rate has an upper bound, which is equal to

Table 1.3: Major contributions on iterative decoding and convergence analysis (Part 1)

Year	Author(s)	Contribution
1962	Galager [79]	Invented LDPC codes.
1966	Forney [125]	Proposed the first concatenated coding scheme using a convolutional code as the inner code and a Reed–Solomon code as the outer code.
1993	Berrou *et al.* [69]	Invented turbo codes, which achieved near-Shannon limit performance by employing iterative decoding.
1995	Douillard *et al.* [82]	Proposed turbo equalization, which was capable of eliminating channel-induced inter-symbol interference by performing channel equalization and channel decoding iteratively.
	Divsalar and Pollara [126]	Extended the turbo principle to multiple parallel concatenated codes.
1996	Benedetto and Montorsi [99, 127, 128]	Proposed an analytical bounding technique for evaluating the bit error probability of a parallel concatenated coding scheme [99], and also extended the turbo principle to serially concatenated block and convolutional codes [127, 128].
	Hagenauer *et al.* [72]	Studied the log-likelihood algebra of iterative decoding algorithms, which were generalized for employment in any linear binary systematic block code.
	MacKay and Neal [76]	'Re-discovered' LDPC codes [79] and proposed a new class of LDPC codes constructed on the basis of sparse random parity check matrices, which was capable of achieving near-Shannon limit performance.
1997	Benedetto *et al.* [85]	Investigated iterative decoding of an outer convolutional decoder and an inner Trellis-Coded Modulation (TCM) decoder, originally devised by Ungerboeck in 1982 [129].
1998	Benedetto *et al.* [71, 130]	Derived design guidelines for serially concatenated codes with the aid of the union-bound technique [71], and also extended the turbo principle to multiple serially concatenated codes [130].

Table 1.3: Continued

Year	Author(s)	Contribution
1998	Moher [87]	Proposed an iterative multiuser decoder for near-capacity communications.
	Reed *et al.* [88]	Investigated iterative multiuser detection designed for Direct-Sequence Code Division Multiple Access (DS-CDMA).
	Kschischang and Frey [95]	Studied iterative decoding based on factor graph models.
	Divsalar *et al.* [75]	Proposed Repeat–Accumulate (RA) codes, where turbo-like performance was achieved at a low complexity.
	Pyndiah [73]	Proposed turbo product codes.
	ten Brink *et al.* [84]	Studied iterative decoding between an outer channel decoder and an inner soft-demapper in the context of BICM.
	McEliece *et al.* [92]	Discovered the connection between turbo decoding algorithms and Pearl's belief propagation algorithm [91].
1999	Wang and Poor [90]	Proposed iterative interference cancelation and channel decoding for coded CDMA schemes.

the channel's capacity. Throughout the 1960s and 1970s the cut-off rate was regarded as a practical design objective, while approaching the channel capacity was deemed to be an unrealistic design goal. The escalating complexity of CC decoding trees may be avoided by employing the more convenient trellises [141] to represent CCs and by applying the Bahl–Cocke–Jelinek–Raviv (BCJR) [48] algorithm. This forms the basis of turbo codes designed by Berrou *et al.* [142] in 1993, which employ a pair of iteratively decoded CCs and facilitate operation near the channel's capacity without an escalating computational complexity, as described in Section 4.1. The advent of turbo coding led to the 'rediscovery' of Low Density Parity Check (LDPC) codes [143–145], which apply iterative decoding techniques to bipartite graphs [146] in order to facilitate operation near the channel's capacity. Recently, irregular coding techniques [147–149] have enabled 'very-near-capacity' operation, as described in Section 5.5. Indeed, in Chapters 7–9 of this volume we demonstrate a novel application of IrVLCs to near-capacity operation in this manner.

Near-capacity operation is of particular interest in digital multimedia broadcast applications. This is because the channel's capacity reduces with the channel's Signal-to-Noise Ratio (SNR), which depends on the transmit power, on the distance to the remote receiver and on the amount of noise that is imposed upon the signal. Hence, multimedia broadcast schemes having a particular information transmission rate are associated with a minimum threshold

Table 1.4: Major contributions on iterative decoding and convergence analysis (Part 2)

Year	Author(s)	Contribution
1999	ten Brink [106]	Proposed EXIT charts for analyzing the convergence behavior of iterative decoding and demapping [106].
2000	ten Brink [111, 112, 115]	Applied EXIT charts to the convergence analysis of various iterative decoding algorithms, such as iterative decoding of PCCs [111] and Serially Concatenated Codes (SCCs) [112], and to turbo equalization [115]; also proposed a common framework of EXIT chart analysis in [115].
2001	Richardson and Urbanke [77]	Proposed the density evolution algorithm for analyzing the convergence behavior of LDPC decoding.
	Richardson et al. [78]	Employed the density evolution algorithm for the sake of designing capacity-achieving irregular LDPCs.
	Kschischang et al. [93]	Studied the sum–product algorithm with the aid of factor graphs.
	ten Brink [119]	Extended EXIT charts from 2D to 3D for analyzing the convergence behavior of three-stage parallel concatenated codes.
	Grant [123]	Extended EXIT charts to non-binary decoders, where mutual information was calculated based on histogram measurements.
2002	Tüchler and Hagenauer [80]	Proposed irregular convolutional codes as well as a simplified method for computing mutual information in EXIT chart analysis.
	Tüchler [120]	Applied 3D EXIT charts to the convergence analysis of three-stage serially concatenated turbo equalization schemes.
	Tüchler et al. [113, 131]	Proposed turbo MMSE equalization [131] and applied EXIT chart analysis to it [113].
2003	Brännström et al. [121, 122]	Proposed a 3D-to-2D projection technique for facilitating the EXIT chart analysis of multiple serially concatenated codes. Additionally, a search algorithm was proposed for finding the optimal decoder activation order.

Table 1.4: Continued

Year	Author(s)	Contribution
2004	Ashikhmin *et al.* [116]	Presented a universal model for conducting EXIT chart analysis as well as various properties of EXIT functions, which form the theoretical foundation of EXIT-curve matching techniques.
	Land *et al.* [132]	Provided a low-complexity mutual information estimation method, which significantly simplifies the EXIT chart analysis.
2006	Kliewer *et al.* [124]	Proposed an efficient and low-complexity method of computing non-binary EXIT charts from index-based *a posteriori* probabilities, which may be considered as a generalization of the approach presented in [132].

SNR at which the channel's capacity drops below the particular information transmission rate to be maintained. If a multimedia broadcast scheme can be designed to support near-capacity operation, then the threshold SNR required to achieve high-quality reception is reduced.

1.2.2   Joint Source and Channel Coding

Shannon's source and channel coding separation theorem [133] states that the removal of undesirable redundancy during source coding and the reintroduction of specifically designed intentional redundancy during channel coding can be performed separately, without jeopardizing our ability to achieve an infinitesimally low probability of transmission errors, while maintaining a near-capacity information transmission rate. However, Shannon's findings are only valid under a number of idealistic assumptions [150], namely that the information is transmitted over an uncorrelated non-dispersive narrowband Additive White Gaussian Noise (AWGN) channel, while potentially imposing an infinite decoding complexity and buffering latency. These assumptions clearly have limited validity for practical finite-delay transmissions over realistic fading wireless channels [151].

Additionally, Shannon assumed that the source is stationary and is losslessly encoded with the aid of entropy-encoded symbols having equal significance and identical error sensitivity. These assumptions have a limited validity in the case of multimedia transmission, since video, image, audio and speech information is typically non-stationary, having characteristics that vary in time and/or space [151, 152]. Furthermore, typically lossy multimedia coding [151, 152] is employed in order to achieve a high level of compression and a concomitant low bandwidth requirement, while exploiting the psycho-visual properties of the human vision or hearing. Finally, the components of the encoded multimedia information typically have varying perceptual significance and a corresponding unequal error sensitivity, since they are often generated using several diverse encoding techniques. For example, video coding typically employs the Discrete Cosine Transform (DCT) [153], Motion Compensation

(MC) [154] and entropy coding [155], as exemplified in the MPEG-1 [156], MPEG-2 [157], MPEG-4 [158], H.261 [159], H.263 [160] and H.264 [161] video codecs.

Hence, the employment of joint source and channel coding [162, 163] is motivated. This may be achieved using diverse methods, which we now briefly discuss.

'Channel-optimized' source coding [163] may be employed to reduce the reconstruction error that results when the channel decoder is unable to correct all transmission errors. Here, the source encoder is designed with special consideration of the transmission errors that are most likely to occur, namely those causing a particular binary codeword to be confused with another similar codeword. In this way, channel-optimized source encoding allocates pairs of similar codewords to represent similar reconstructed source parameter values. For example, this may be applied to scalar quantization [164, 165], where real-valued source samples are represented by one of a number of discrete quantization levels, which are indexed by binary codewords. In this way, the allocation of binary codeword pairs having a low Hamming distance to represent quantization level pairs having a low Euclidean distance was demonstrated in [166, 167]. Similarly, the authors of [168, 169] proposed Channel-Optimized Vector Quantization (COVQ), which employs the channel-optimized indexing of Vector Quantization (VQ) tiles [170]. Note that both scalar and vector quantization are discussed in greater detail in Section 3.3.

Joint source and channel coding can also be beneficially employed if some of the source correlation is not removed during source encoding, resulting in the manifestation of residual redundancy [171] within the resultant source-encoded signal. In the receiver, a model of the source correlation may be employed to exploit the residual redundancy in order to provide an error-correction capability, which may be employed to mitigate any transmission errors that could not be eliminated during channel decoding. The error-correction capability constituted by the source-encoded signal's residual redundancy may be invoked during source decoding in order to provide a Minimum Mean Square Error (MMSE) estimate of the source signal [172–175]. Alternatively, a Maximum *A Posteriori* Probability (MAP) estimate [172, 173, 176] of the source-coded signal may be obtained prior to source decoding by exploiting the aforementioned residual redundancy. Note that MMSE and MAP decoders have the ability to consider the entire source-coded signal sequence at once [175–177], or they may consider the individual symbols of the source-coded signal sequence separately [172–174].

In the same way that residual source redundancy can be exploited in the receiver to provide an error-correction capability, so can redundancy that is intentionally introduced during source encoding. For example, this may be achieved by imposing correlation on the source signal with the aid of Trellis-Coded Quantization (TCQ) [178], as exemplified in [177]. Alternatively, Variable-Length Error Correction (VLEC) coding [179] may be employed to incorporate redundancy within the source-encoded signal. This may be exploited to provide an error correction capability in the receiver [180–183], as detailed in Section 3.3. In Section 4.1 we detail the turbo principle [142], which may be employed for supporting iterative joint source and channel decoding [184–186], where the redundancy introduced by both source and channel coding is alternately exploited so that each provides *a priori* information for the other concerning the source-encoded signal. A novel scheme employing this approach is described in Chapter 6, and hence the detailed literature review of this topic is postponed to Section 6.1.

A further approach to joint source and channel coding that we highlight here employs sophisticated rate allocation in order to jointly optimize the amount of redundancy that is retained or introduced during source encoding and the amount that is introduced during channel encoding. In the scenario where a lossy source codec is employed, an increased degree of lossy compression may be achieved by employing coarser quantization, for example. While this results in additional quantization-induced reconstruction distortion, the associated reduction in the amount of resultant source-coded information facilitates the employment of a lower-rate channel codec, without increasing the overall transmission rate. Since lower-rate channel codecs are associated with higher error-correction capabilities, they are capable of mitigating more channel-induced reconstruction distortion. Hence, rate allocation may be employed to optimize the trade-off between quantization- and channel-induced reconstruction distortion [167, 187, 188].

1.2.3 Unequal Error Protection

As mentioned above, this book considers the application of IrVLCs for UEP. In a manner similar to that of [189–191], for example, UEP may be employed to appropriately protect audio-, speech-, image- and video-encoded bit sequences, which are typically generated using diverse encoding techniques and exhibit various error sensitivities. For example, video coding typically employs the DCT and MC, as described above. As noted in [151], typically a higher degree of video reconstruction distortion is imposed by transmission errors that affect the motion vectors of MC than from those inflicted on the DCT-encoded information. Hence, UEP may be employed to protect all MC-related information with a relatively strong error-correction capability, whilst employing a relatively weak error-correction code to protect the DCT-encoded information. This approach may hence be expected to yield a lower degree of video reconstruction distortion than equal protection, as noted in [189–191], for example. In Chapter 7 we demonstrate the novel application of IrVLCs for UEP by employing different sets of VLEC codewords having various error-correction capabilities to appropriately protect particular components of the source signal that have different error sensitivities.

1.3 Motivation and Methodology

In the spirit of the remarkable performance of turbo codes, as described in Section 1.1, the turbo principle has been successfully applied to various generalized decoding/detection problems, resulting in diverse turbo transceivers [70]. In the meantime new challenges have arisen: for example, how to choose the parameters of a turbo transceiver consisting of multiple components so that near-capacity performance can be achieved. Owing to iterative decoding, different components of the turbo transceivers are combined and affect each other during each iteration. Hence, in contrast to the design of conventional non-iterative transceivers, the optimization should be considered jointly rather than individually. Moreover, as argued in Section 1.1, EXIT charts have shown their power in analyzing the convergence behavior of iterative receivers. Therefore, in this treatise we mainly rely on this semi-analytical tool as well as on the conventional Monte Carlo simulation method in our attempt to tackle the problem of optimizing multi-stage turbo transceivers and JSCD schemes.

1.4   Outline of the Book

Having briefly reviewed the literature of source coding, joint source–channel decoding, concatenated coding and iterative decoding, as well as convergence analysis, let us now outline the organization of the book.

- **Chapter 2: Information Theory Basics**

  In Chapter 2 we commence by laying the information-theoretic foundations for the rest of the book.

- **Chapter 3: Sources and Source Codes**

  This chapter introduces various entropy coding schemes designed for memoryless sources having known alphabets, along with several trellis-based soft-decision decoding methods in contrast to conventional hard-decision decoding. Section 3.2 describes the source models used in this treatise; various source coding schemes designed for memoryless sources are introduced in Section 3.3, such as Huffman codes, Reversible Variable-Length Codes (RVLCs) and Variable-Length Error Correction (VLEC) codes. A generic algorithm contrived for efficiently constructing both RVLCs and VLEC codes having different free distances is also presented. In Section 3.4.1 two different trellis representations of source codes are introduced, together with their conventional binary tree representation. Based on these trellis representations, the Maximum Likelihood Sequence Estimation (MLSE) and MAP decoding algorithm are derived for soft-decision-aided source decoding in Sections 3.4.2.1 and 3.4.2.2 respectively. Finally, the source Symbol-Error Ratio (SER) performance of these decoding algorithms is presented in Section 3.4.3.

- **Chapter 4: Iterative Source–Channel Decoding**

  Based on the soft VLC decoding algorithms described in Chapter 3, various Iterative Source–Channel Decoding (ISCD) schemes may be formed. The SER performance of these schemes contrived for communicating over both AWGN channels and dispersive channels is quantified and their convergence behavior is analyzed with the aid of EXIT charts. More explicitly, Section 4.1 provides an overview of various code concatenation schemes and the so-called 'turbo principle.' A generic SISO *A Posteriori* Probability (APP) decoding module, which serves as the 'kernel' of iterative decoding schemes, is described in Section 4.2. Additionally, the semi-analytical EXIT chart tool for convergence analysis is introduced in Section 4.2.2. Section 4.3 evaluates the performance of ISCD schemes designed for transmission over AWGN channels, along with the EXIT chart analysis of their convergence behavior. More particularly, the effects of different source codes are considered. Furthermore, the performance of ISCD schemes is investigated for transmission over dispersive channels in Section 4.4, which necessitates the employment of channel equalization. Both scenarios, with and without channel precoding, are considered in Sections 4.4.2 and 4.4.3 respectively. A three-stage iterative channel equalization, channel decoding and source decoding scheme is presented in Section 4.4.4.

- **Chapter 5: Three-Stage Serially Concatenated Turbo Equalization**

  In this chapter a three-stage serially concatenated turbo MMSE equalization scheme is investigated. The different components of the iterative receiver are jointly optimized with the aid of EXIT charts for the sake of achieving a near-capacity performance.

More specifically, Section 5.2 provides a short introduction to SISO MMSE equalization. The conventional two-stage turbo equalization scheme is described in Section 5.3, while Section 5.4 presents a three-stage turbo equalization scheme. Both 3D and 2D EXIT chart analysis are carried out in Section 5.4.2, where the channel code is optimized for obtaining the lowest possible E_b/N_0 convergence threshold and the activation order of the component decoders is optimized for achieving the fastest convergence, while maintaining a low decoding complexity. Various methods of visualizing the three-stage iterative decoding process are described in Section 5.4.4. The effects of different interleaver block lengths on the attainable system performance are discussed in Section 5.4.5. Section 5.5 demonstrates the technique of EXIT chart matching and the design of IRregular Convolutional Codes (IRCCs), which results in a capacity-approaching three-stage turbo equalization scheme. In addition to unity-rate intermediate codes, the employment of non-unity-rate intermediate codes in three-stage schemes is investigated in Section 5.6.

– **Chapter 6: Irregular Variable-Length Codes for Joint Source and Channel Coding** In Chapter 6 we demonstrate the application of IrVLCs for the joint source and channel coding of video information. The proposed scheme employs the serial concatenation and iterative decoding of a video codec with a channel codec. Our novel video codec represents the video information using Variable-Dimension Vector Quantization (VDVQ) tiles. The VDVQ tiles employed are represented using the corresponding RVLC codewords selected from the VDVQ/RVLC codebook. However, the legitimate use of the VDVQ tiles and their corresponding RVLC codewords is limited by a number of code constraints, which ensure that the VDVQ tiles employed perfectly tessellate, among other desirable design objectives. As a result, different subsets of the RVLC codewords are available at different points during the encoding of the video information, and the proposed approach adopts an IrVLC philosophy.

In the video codec of Chapter 6, the VDVQ/RVLC-induced code constraints are uniquely and unambiguously described by a novel VDVQ/RVLC trellis structure, which resembles the symbol-based VLEC trellis of [181, 192]. Hence, the employment of the VDVQ/RVLC trellis structure allows the consideration of all legitimate transmission frame permutations. This fact is exploited in the video encoder in order to perform novel MMSE VDVQ/RVLC encoding, using a novel variant of the Viterbi algorithm [193].

Additionally, the employment of the VDVQ/RVLC trellis structure during video decoding guarantees the recovery of legitimate – although not necessarily error-free – video information. As a result, the video decoder never has to discard video information. This is unlike conventional video decoders, where a single transmission error may render an entire transmission frame invalid. Furthermore, a novel version of the BCJR algorithm [48] is employed during APP SISO VDVQ/RVLC decoding in order to facilitate the iterative exchange of soft information with the serially concatenated channel decoder, and in order to perform the soft MMSE reconstruction of the video sequence. Finally, since the VDVQ/RVLC trellis structure describes the complete set of VDVQ/RVLC-induced code constraints, all of the associated redundancy is beneficially exploited with the aid of the modified BCJR algorithm.

Owing to its aforementioned benefits and its employment of a joint source and channel coding philosophy, the video transmission scheme of Chapter 6 is shown to outperform the corresponding benchmarks that employ a separate source and channel coding philosophy. Our findings were originally published in [194, 195].

– **Chapter 7: Irregular Variable-Length Codes for EXIT-Chart Matching**
In Chapter 7, we investigate the application of IrVLCs for Unequal Error Protection (UEP). Here, a number of component VLC codebooks having different error-correction capabilities are employed to encode various fractions of the source symbol frame. When these fractions have different error sensitivities, this approach may be expected to yield a higher reconstruction quality than does equal protection, as noted in [189–191], for example.

Chapter 7 also investigates the application of IrVLCs to near-capacity operation. Here, a number of component VLC codebooks having different inverted EXIT functions are employed to encode various fractions of the source symbol frame. We show that the inverted IrVLC EXIT function may be obtained as a weighted average of the inverted component VLC EXIT functions. Additionally, the EXIT-chart matching algorithm of [149] is employed to shape the inverted IrVLC EXIT function to match the EXIT function of a serially concatenated inner channel code and to create a narrow but still open EXIT chart tunnel. In this way, iterative decoding convergence to an infinitesimally low probability of error is facilitated at near-capacity SNRs.

Furthermore, in Chapter 7 the UEP and near-capacity operation of the described scheme is assessed using novel plots that characterize the computational complexity of iterative decoding. More specifically, the average number of Add, Compare and Select (ACS) operations required to reconstruct each source symbol with a high quality is plotted against the channel SNR. These plots are employed to compare the novel IrVLC-based scheme with a suitably designed IrCC and regular VLC-based benchmarks, quantifying the advantages of the IrVLCs. Furthermore, these plots demonstrate that the complexity associated with the bit-based VLEC trellis is significantly lower than that of the corresponding symbol-based trellis. Our findings were originally published in [196, 197] and we proposed attractive near-capacity IrVLC schemes in [198–203].

– **Chapter 8: Genetic Algorithm-Aided Design of Irregular Variable-Length Coding Components**
In Chapter 8 we introduce a novel Real-Valued Free Distance Metric (RV-FDM) as an alternative to the Integer-Valued Free Distance (IV-FD) for the characterization of the error-correction capability that is associated with VLEC codebooks. Unlike the IV-FD lower bound, the RV-FDM assumes values from the real-valued domain, hence allowing the comparison of the error-correction capability of two VLEC codebooks having equal IV-FD lower bounds. Furthermore, we show that a VLEC codebook's RV-FDM affects the number of inflection points that appear in the corresponding inverted EXIT function. This complements the property [204] that the area below an inverted VLEC EXIT function equals the corresponding coding rate, as well as the property that a free distance of at least two yields an inverted VLEC EXIT function that reaches the top right-hand corner of the EXIT chart.

These properties are exploited by a novel Genetic Algorithm (GA) in order to design beneficial VLEC codebooks having arbitrary inverted EXIT function shapes. We will demonstrate that this is in contrast to the methods of [205–207], which are incapable of designing codebooks having specific EXIT function shapes without imposing a significant level of 'trial-and-error'-based human interaction. This novel GA is shown to be attractive for the design of IrVLC component codebooks for EXIT-chart matching, since Chapter 8 also demonstrates that our ability to create open EXIT-chart tunnels at near-capacity channel SNRs depends on the availability of a sufficiently diverse suite of component codes having a wide variety of EXIT function shapes.

Finally, a suite of component VLEC codebooks designed by the proposed GA is found to facilitate higher-accuracy EXIT-chart matching than a benchmark suite designed using the state-of-the-art method of [207]. Our novel RV-FDM and GA were originally published in [200, 201].

- **Chapter 9: Joint EXIT-Chart Matching of Irregular Variable-Length Coding and Irregular Unity-Rate Coding**
  In Chapter 9, we propose a novel modification of the EXIT-chart matching algorithm introduced in [149] that additionally seeks a reduced APP SISO decoding complexity by considering the complexities associated with each of the component codes. Another novel modification introduced in this chapter facilitates the EXIT-chart matching of irregular codes that employ a suite of component codes having the same coding rate.

  Additionally, Chapter 9 demonstrates the joint EXIT-chart matching of two serially concatenated irregular codecs, namely an outer IrVLC and an inner Irregular Unity-Rate Code (IrURC). This is achieved by iteratively matching the inverted outer EXIT function to the inner EXIT function and vice versa. By employing an irregular inner code in addition to an irregular outer code, we can afford a higher degree of design freedom than the proposals of [149], which employ a regular inner code. Hence, the proposed approach is shown to facilitate even nearer-capacity operation, which is comparable to that of Irregular Low-Density Parity Checks (IrLDPCs) and irregular turbo codes. Our findings were originally published in [202, 203] and we additionally demonstrated the joint EXIT-chart matching of serially concatenated irregular codecs in [208].

- **Chapter 10: Iteratively Decoded VLC Space–Time Coded Modulation**
  In this chapter an Iteratively Decoded Variable-Length Space–Time Coded Modulation (VL-STCM-ID) scheme capable of simultaneously providing both coding and iteration gain as well as multiplexing and diversity gain is investigated. Non-binary unity-rate precoders are employed for assisting the iterative decoding of the VL-STCM-ID scheme. The discrete-valued source symbols are first encoded into variable-length codewords that are spread to both the spatial and the temporal domains. Then the variable-length codewords are interleaved and fed to the precoded modulator. More explicitly, the proposed VL-STCM-ID arrangement is a jointly designed iteratively decoded scheme combining source coding, channel coding and modulation as well as spatial diversity/multiplexing. We demonstrate that, as expected, the higher the source correlation, the higher the achievable performance gain of the scheme becomes. Furthermore, the performance of the VL-STCM-ID scheme is about 14 dB better than

that of the Fixed-Length STCM (FL-STCM) benchmark at a source symbol error ratio of 10^{-4}.

– **Chapter 11: Iterative Detection of Three-Stage Concatenated IrVLC FFH-MFSK**
Serially concatenated and iteratively decoded Irregular Variable-Length Coding (IrVLC) combined with precoded Fast Frequency Hopping (FFH) M-ary Frequency Shift Keying (MFSK) is considered. The proposed joint source and channel coding scheme is capable of operating at low Signal-to-Noise Ratio (SNR) in Rayleigh fading channels contaminated by Partial Band Noise Jamming (PBNJ). By virtue of its inherent time and frequency diversity, the FFH-MFSK scheme is capable of mitigating the effects of Partial Band Noise Jamming (PBNJ). The IrVLC scheme comprises a number of component variable-length coding codebooks employing different coding rates for encoding particular fractions of the input source symbol stream. These fractions may be chosen with the aid of EXtrinsic Information Transfer (EXIT) charts in order to shape the inverted EXIT curve of the IrVLC codec so that it can be matched with the EXIT curve of the inner decoder. We demonstrate that, using the proposed scheme, an infinitesimally low bit error ratio may be achieved at low SNR values. Furthermore, the system employing a unity-rate precoder yields a substantial SNR gain over that dispensing with the precoder, and the IrVLC-based scheme yields a further gain over the identical-rate single-class VLC-based benchmark.

– **Chapter 12: Conclusions and Future Research**
The major findings of our work are summarized in this chapter along with our suggestions for future research.

1.5   Novel Contributions of the Book

Whilst it may appear somewhat unusual to outline the novel contribution of a book in a style similar to that found in research papers, this research monograph does indeed collate a large body of original contributions to the current state of the art. This justifies the inclusion of the following list.

- A novel and high-efficiency algorithm is proposed for constructing both RVLCs and VLEC codes [32]. For RVLCs the algorithm results in codes of higher efficiency and/or shorter maximum codeword length than the algorithms previously disseminated in the literature. For VLEC codes it significantly reduces the construction complexity imposed, in comparison with the original algorithm of [29, 30].
- Soft VLC decoding based on both symbol-level and bit-level trellis representation is investigated. The performance of different decoding criteria such as MLSE and MAP is evaluated. In particular, the effect of different free distances of the source codes is quantified.
- A novel iterative channel equalization, channel decoding and source decoding scheme is proposed [209, 210]. It is shown that without using channel coding the redundancy in the source code alone is capable of eliminating the channel-induced Inter-Symbol Interference (ISI), provided that the channel equalization and source decoding are performed iteratively at the receiver. When employing channel coding, a three-stage joint source–channel detection scheme is formed, which is capable of outperforming the separate source–channel detection scheme by 2 dB in terms of the E_b/N_0 value

required for achieving the same BER. Furthermore, with the aid of EXIT charts, the systems' convergence thresholds and the design of channel precoders are optimized.

- A novel three-stage serially concatenated MMSE turbo equalization scheme is proposed [211]. The performance of conventional two-stage turbo equalization schemes is generally limited by a relatively high BER 'shoulder' [212]. However, when employing a unity-rate recursive convolutional code as the intermediate component, the three-stage turbo equalization scheme is capable of achieving a steep BER turbo 'cliff,' beyond which an infinitesimally low BER is obtained.

- The concept of EXIT modules is proposed, which significantly simplifies the EXIT chart analysis of multiple concatenated systems, casting them from the 3D domain to the 2D domain. As a result, the outer constituent code of the three-stage turbo equalization scheme is optimized for achieving the lowest possible convergence threshold. The activation order of the component decoders is optimized for obtaining the convergence at the lowest possible decoding complexity using the lowest total number of iterations or decoder activations.

- A near-channel-capacity three-stage turbo equalization scheme is proposed [213]. The proposed scheme is capable of achieving a performance within 0.2 dB of the dispersive channel's capacity with the aid of EXIT-chart matching and IRCCs. This scheme is further generalized to MIMO turbo detection [214].

- The employment of non-unity-rate intermediate codes in a three-stage turbo equalization scheme is also investigated. The maximum achievable information rates of such schemes are determined and the serial concatenation of the outer constituent code and the intermediate constituent code is optimized for achieving the lowest possible decoding convergence threshold.

- As an application example, a novel vector quantized RVLC-TCM scheme is proposed for the iterative joint source and channel decoding of video information, and its RVLC symbol-based trellis structure is investigated. Furthermore, a novel version of the BCJR algorithm is designed for the APP SISO VDVQ/RVLC decoding of the video information.

- IrVLC schemes are designed for near-capacity operation and their complexity versus channel SNR characteristics are recorded. The corresponding relationships are also parameterized by the achievable reconstruction quality.

- The error-correction capability of VLECs having the same Integer-Valued Free Distance (IV-FD) is characterized and the relationship between a VLEC's IV-FD and the shape of its inverted EXIT function is established.

- A novel Genetic Algorithm (GA) is devised for designing VLECs having specific EXIT functions.

- A suite of VLECs that are suitable for a wide range of IrVLC applications are found.

- A novel EXIT-chart matching algorithm is devised for facilitating the employment of component codes having the same coding rate.

- This EXIT-chart matching algorithm is further developed to additionally seek a reduced APP SISO-decoding computational complexity.

- A joint EXIT-chart matching algorithm is developed for designing schemes employing a serial concatenation of an irregular outer and inner codec. This results in IrVLC–IrURC schemes designed for near-capacity joint source and channel coding.

- A number of VLC-based system design examples are provided, including a three-stage concatenated VLC-aided space–time scheme.
- As a second VLC-based design example, an IrVLC-coded and unity-rate precoder-assisted noncoherently detected Fast Frequency Hopping M-ary Frequency Shift Keying (FFH-MFSK) system communicating in an uncorrelated Rayleigh fading channel contaminated by PBNJ is conceived and investigated. This system may find application in low-complexity noncoherent detection-aided cooperative systems.
- We investigate the serial concatenation of the FFH-MFSK demodulator, unity-rate decoder and IrVLC outer decoder with the aid of EXIT charts, in order to attain a good performance even at low SNR values.
- We contrast the two-stage iterative detection scheme, exchanging extrinsic information between the inner rate-1 decoder and the IrVLC outer decoder, with the three-stage extrinsic information exchange among the demodulator, inner decoder and outer decoder.

Chapter 2

Information Theory Basics

2.1 Issues in Information Theory

The ultimate aim of telecommunications is to communicate information between two geographically separated locations via a communications channel, with adequate quality. The theoretical foundations of information theory accrue from Shannon's pioneering work [133, 215–217], and hence most tutorial interpretations of his work over the past fifty years rely fundamentally on [133, 215–217]. This chapter is no exception in this respect. Throughout the chapter we make frequent references to Shannon's seminal papers and to the work of various authors offering further insights into Shannonian information theory. Since this monograph aims to provide an all-encompassing coverage of video compression and communications, we begin by addressing the underlying theoretical principles using a light-hearted approach, often relying on worked examples.

Early forms of human telecommunications were based on smoke, drum or light signals, bonfires, semaphores and the like. Practical *information sources* can be classified as analog and digital. The output of an analog source is a continuous function of time, such as, for example, the air pressure variation at the membrane of a microphone due to someone talking. The roots of Nyquist's sampling theorem are based on his observation of the maximum achievable telegraph transmission rate over *bandlimited* channels [218]. In order to be able to satisfy Nyquist's sampling theorem the analog source signal has to be bandlimited before sampling. The analog source signal has to be transformed into a digital representation with the aid of time- and amplitude-discretization using *sampling* and *quantization*.

The output of a digital source is one of a finite set of ordered, discrete symbols often referred to as an alphabet. Digital sources are usually described by a range of characteristics, such as the *source alphabet*, the *symbol rate*, the *symbol probabilities*, and the *probabilistic interdependence of symbols*. For example, the probability of 'u' following 'q' in the English language is $p = 1$, as in the word 'equation.' Similarly, the joint probability of all pairs of consecutive symbols can be evaluated.

In recent years, electronic telecommunications have become prevalent, although most information sources provide information in other forms. For electronic telecommunications, the source information must be converted to electronic signals by a *transducer*. For example, a microphone converts the air pressure waveform $p(t)$ into voltage variation $v(t)$, where

$$v(t) = c \cdot p(t - \tau), \tag{2.1}$$

and the constant c represents a scaling factor, while τ is a delay parameter. Similarly, a video camera scans the natural three-dimensional scene using optics, and converts it into electronic waveforms for transmission.

The electronic signal is then transmitted over the *communications channel* and converted back to the required form; this may be carried out, for example, by a loudspeaker. It is important to ensure that the channel conveys the transmitted signal with adequate quality to the receiver in order to enable information recovery. Communications channels can be classified according to their ability to support analog or digital transmission of the source signals in a *simplex*, *duplex* or *half-duplex* fashion over *fixed* or *mobile* physical channels constituted by pairs of wires, Time Division Multiple Access (TDMA) time-slots, or a Frequency Division Multiple Access (FDMA) frequency slot.

The *channel impairments* may include superimposed, unwanted random signals, such as thermal noise, crosstalk via multiplex systems from other users, and interference produced by human activities (for example, car ignition or fluorescent lighting) and natural sources such as lightning. Just as the natural sound pressure wave between two conversing persons will be impaired by the acoustic background noise at a busy railway station, the reception quality of electronic signals will be affected by the unwanted electronic signals mentioned above. In contrast, distortion manifests itself differently from additive noise sources, since no impairment is explicitly added. Distortion is more akin to the phenomenon of reverberating loudspeaker announcements in a large, vacant hall, where no noise sources are present.

Some of the channel impairments can be mitigated or counteracted; others cannot. For example, the effects of unpredictable additive random noise cannot be removed or 'subtracted' at the receiver. They can be mitigated by increasing the transmitted signal's power, but the transmitted power cannot be increased without penalties, since the system's nonlinear distortion rapidly becomes dominant at higher signal levels. This process is similar to the phenomenon of increasing the music volume in a car parked near a busy road to a level where the amplifier's distortion becomes annoyingly dominant.

In practical systems, the *Signal-to-Noise Ratio* (SNR), quantifying the wanted and unwanted signal powers at the channel's output, is a prime *channel parameter*. Other important channel parameters are its *amplitude* and *phase response*, determining its usable bandwidth (B), over which the signal can be transmitted without excessive distortion. Among the most frequently used statistical noise properties are the *Probability Density Function* (PDF), the *Cumulative Density Function* (CDF), and the *Power Spectral Density* (PSD).

The fundamental communications system design considerations are whether a high-fidelity (hi-fi) or just acceptable video or speech quality is required from a system (this predetermines, among other factors, its cost, bandwidth requirements and the number of channels available, and has implementational complexity ramifications), and, equally important, the issues of robustness against channel impairments, system delay and so on. The required transmission range and worldwide roaming capabilities, the maximum available transmission speed in terms of symbols per second, information confidentiality, reception

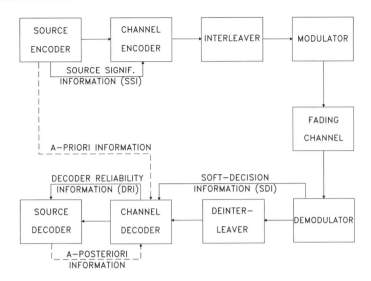

Figure 2.1: Basic transmission model of information theory.

reliability, and the requirement for a convenient lightweight, solar-charged design, are similarly salient characteristics of a communications system.

Information theory deals with a variety of problems associated with the performance limits of an information transmission system, such as the one depicted in Figure 2.1. The components of this system constitute the subject of this monograph; hence they will be treated in greater depth later in this volume. Suffice it to say at this stage that the transmitter seen in Figure 2.1 incorporates a source encoder, a channel encoder, an interleaver and a modulator, and their inverse functions at the receiver. The *ideal source encoder* endeavors to remove as much redundancy as possible from the information source signal without affecting its source representation fidelity (i.e., distortionlessly), and it remains oblivious of such practical constraints as a finite delay and limited signal processing complexity. In contrast, a practical source encoder will have to retain a limited signal processing complexity and delay while attempting to reduce the source representation bit rate to as low a value as possible. This operation seeks to achieve better transmission efficiency, which can be expressed in terms of bit-rate economy or bandwidth conservation.

The channel encoder re-inserts redundancy or parity information but in a controlled manner, in order to allow error correction at the receiver. Since this component is designed to ensure the best exploitation of the re-inserted redundancy, it is expected to minimize the error probability over the most common channel, namely the so-called *Additive White Gaussian Noise* (AWGN) channel, which is characterized by a memoryless, random distribution of channel errors. However, over wireless channels, which have recently become prevalent, the errors tend to occur in bursts due to the presence of deep received signal fades induced by the destructively superimposed multipath phenomena. This is why our schematic of Figure 2.1 contains an interleaver block, which is included in order to randomize the bursty channel errors. Finally, the modulator is designed to ensure the most bandwidth-efficient transmission of the source- and channel-encoded interleaved information stream, while maintaining the lowest possible bit error probability.

The receiver simply carries out the corresponding inverse functions of the transmitter.

Observe in the figure that, besides the direct interconnection of the adjacent system components, there are a number of additional links in the schematic. These will require further study before their role can be highlighted. At the end of this chapter we will return to this figure and guide the reader through its further intricate details.

Some fundamental problems transpiring from the schematic of Figure 2.1, which have been addressed in depth in a range of studies by Shannon [133, 215–217], Nyquist [218], Hartley [219], Abramson [220], Carlson [221], Raemer [222], Ferenczy [223] and others, are as follows.

- What is the true information generation rate of our information sources? If we know the answer, the efficiency of coding and transmission schemes can be evaluated by comparing the actual transmission rate used with the source's information emission rate. The actual transmission rate used in practice is typically much higher than the average information delivered by the source, and the closer these rates are, the better is the coding efficiency.
- Given a noisy communications channel, what is the maximum reliable information transmission rate? The thermal noise induced by the random motion of electrons is present in all electronic devices, and if its power is high it can seriously affect the quality of signal transmission, allowing information transmission only at low rates.
- Is the information emission rate the only important characteristic of a source, or are other message features, such as the probability of occurrence of a message and the joint probability of occurrence for various messages, also important?

In a wider context, the topic of this whole monograph is related to the blocks of Figure 2.1 and to their interactions, but in this chapter we lay the theoretical foundations of source and channel coding as well as transmission issues, and define the characteristics of an ideal Shannonian communications scheme.

Although there are numerous excellent treatises on these topics that treat the same subjects with a different flavor [221, 223, 224], our approach is similar to that of the above-mentioned classic sources; since the roots are in Shannon's work, references [133, 215–217, 225, 226] are the most pertinent and authoritative sources.

In this chapter we consider mainly discrete sources, in which each source message is associated with a certain probability of occurrence, which might or might not be dependent on previous source messages. Let us now give a rudimentary introduction to the characteristics of the AWGN channel, which is the predominant channel model in information theory due to its simplicity. The analytically less tractable wireless channels will be modeled mainly by simulations in this monograph.

2.2 Additive White Gaussian Noise Channel

2.2.1 Background

In this section we consider the communications channel, which exists between the transmitter and the receiver, as shown in Figure 2.1. Accurate characterization of this channel is essential if we are to remove the impairments imposed by the channel using signal processing at the

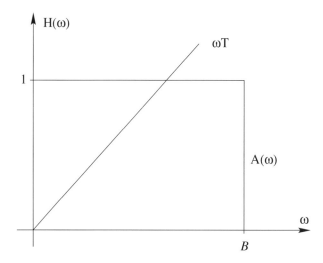

Figure 2.2: Ideal, distortion-free channel model having a linear phase and a flat magnitude response.

receiver. Here we initially consider only fixed communications links, where both terminals are stationary, although mobile radio communications channels, which change significantly with time, are becoming more prevalent.

We define fixed communications channels to be those between a fixed transmitter and a fixed receiver. These channels are exemplified by twisted pairs, cables, wave guides, optical fiber and point-to-point microwave radio channels. Whatever the nature of the channel, its output signal differs from the input signal. The difference might be deterministic or random, but it is typically unknown to the receiver. Examples of channel impairments are dispersion, nonlinear distortions, delay and random noise.

Fixed communications channels can often be modeled by a linear transfer function, which describes the channel dispersion. The ubiquitous additive white Gaussian noise is a fundamental limiting factor in communications via Linear Time-Invariant (LTI) channels. Although the channel characteristics might change due to factors such as aging, temperature changes and channel switching, these variations will not be apparent over the course of a typical communication session. It is this inherent time invariance that characterizes fixed channels.

An ideal, distortion-free communications channel would have a flat frequency response and linear phase response over the frequency range of $-\infty \cdots +\infty$, although in practice it is sufficient to satisfy this condition over the bandwidth (B) of the signals to be transmitted, as seen in Figure 2.2. In this figure, $A(\omega)$ represents the magnitude of the channel response at frequency ω, and $\phi(\omega) = wT$ represents the phase shift at frequency ω due to the circuit delay T.

Practical channels always have some linear distortions due to their bandlimited, nonflat frequency response and nonlinear phase response. In addition, the group-delay response of the channel, which is the derivative of the phase response, is often given.

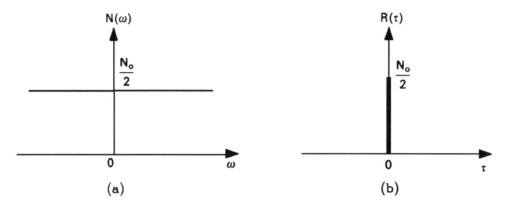

Figure 2.3: Power spectral density and autocorrelation of WN.

2.2.2 Practical Gaussian Channels

Conventional telephony uses twisted copper wire pairs to connect subscribers to the local exchange. The bandwidth is approximately 3.4 kHz, and the waveform distortions are relatively benign.

For applications requiring a higher bandwidth, coaxial cables can be used. Their attenuation increases approximately with the square root of the frequency. Hence, for wideband, long-distance operation, they require channel equalization. Typically, coaxial cables can provide a bandwidth of about 50 MHz, and the transmission rate they can support is limited by the so-called skin effect.

Point-to-point microwave radio channels typically utilize high-gain directional transmit and receive antennas in a line-of-sight scenario, where free-space propagation conditions may be applicable.

2.2.3 Gaussian Noise

Regardless of the communications channel used, random noise is always present. Noise can be broadly classified as natural or synthetic. Examples of synthetic (artificial) types of noise are those due to electrical appliances and fluorescent lighting, and the effects of these can usually be mitigated at the source. Natural noise sources affecting radio transmissions include galactic star radiations and atmospheric noise. There exists a low-noise frequency window in the range of 1–10 GHz, where the effects of these sources are minimized.

Natural thermal noise is ubiquitous. This is due to the random motion of electrons, and it can be reduced by reducing the temperature. Since thermal noise contains practically all frequency components up to some 10^{13} Hz with equal power, it is often referred to as White Noise (WN) in an analogy to white light, which contains all colors with equal intensity. This WN process can be characterized by its uniform Power Spectral Density (PSD) $N(\omega) = N_0/2$, shown together with its Autocorrelation Function (ACF) in Figure 2.3.

The power spectral density of any signal can be conveniently measured with the help of a selective narrowband power meter tuned across the bandwidth of the signal. The power measured at any frequency is then plotted against the measurement frequency. The autocorrelation function $R(\tau)$ of the signal $x(t)$ gives an average indication of how predictable the signal $x(t)$ is after a period of τ seconds from its present value. Accordingly, it is defined as:

$$R(\tau) = \lim_{T \to \infty} \frac{1}{T} \int_{-\infty}^{\infty} x(t)x(t + \tau)\, dt. \tag{2.2}$$

For a periodic signal $x(t)$, it is sufficient to evaluate the above equation for a single period T_0, yielding:

$$R(\tau) = \frac{1}{T_0} \int_{-T_0/2}^{T_0/2} x(t)x(t + \tau)\, dt. \tag{2.3}$$

The basic properties of the ACF are as follows.

- The ACF is symmetric: $R(\tau) = R(-\tau)$.
- The ACF is monotonically decreasing: $R(\tau) \leq R(0)$.
- For $\tau = 0$ we have $R(0) = \overline{x^2}(t)$, which is the signal's power.

The ACF and the PSD form a Fourier transform pair. This is formally stated as the Wiener–Khintchine theorem, as follows:

$$\begin{aligned}
R(\tau) &= \frac{1}{2\pi} \int_{-\infty}^{\infty} N(\omega)e^{j\omega\tau}\, d\omega \\
&= \frac{1}{2\pi} \int_{-\infty}^{\infty} \frac{N_0 e^{j\omega\tau}}{2}\, d\omega \\
&= \frac{1}{2\pi} \frac{N_0}{2} \int_{-\infty}^{\infty} e^{j\omega\tau}\, d\omega = \frac{N_0}{2}\delta(\tau),
\end{aligned} \tag{2.4}$$

where $\delta(\tau)$ is the Dirac delta function. Clearly, for any timed-domain shift $\tau > 0$, the noise is uncorrelated.

Bandlimited communications systems bandlimit not only the signal but the noise as well, and this filtering limits the rate of change of the time-domain noise signal, introducing some correlation over the interval of $\pm 1/2B$. The stylized PSD and ACF of bandlimited white noise are displayed in Figure 2.4.

After bandlimiting, the autocorrelation function becomes:

$$\begin{aligned}
R(\tau) &= \frac{1}{2\pi} \int_{-B}^{B} \frac{N_0}{2} e^{j\omega\tau}\, d\omega = \frac{N_0}{2} \int_{-B}^{B} e^{j2\pi f\tau}\, df \\
&= \frac{N_0}{2} \left[\frac{e^{j2\pi f\tau}}{j2\pi\tau} \right]_{-B}^{B} \\
&= \frac{1}{j2\pi\tau}[\cos 2\pi B\tau + j\sin 2\pi B\tau - \cos 2\pi B\tau + j\sin 2\pi B\tau] \\
&= N_0 B \frac{\sin(2\pi B\tau)}{2\pi B\tau},
\end{aligned} \tag{2.5}$$

which is the well-known sinc function seen in Figure 2.4.

In the time domain, the amplitude distribution of the white thermal noise has a normal or Gaussian distribution, and since it is inevitably added to the received signal, it is usually

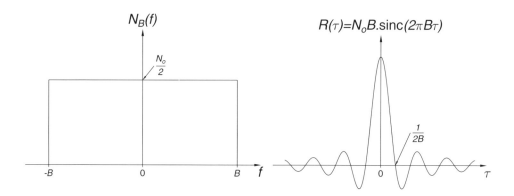

Figure 2.4: Power spectral density and autocorrelation of bandlimited WN.

referred to as Additive White Gaussian Noise (AWGN). Note that AWGN is therefore the noise generated in the receiver. The Probability Density Function (PDF) is the well-known bell-shaped curve of the Gaussian distribution, given by

$$p(x) = \frac{1}{\sigma\sqrt{2\pi}} e^{-(x-m)/2\sigma^2}, \tag{2.6}$$

where m is the mean and σ^2 is the variance. The effects of AWGN can be mitigated by increasing the transmitted signal power and thereby reducing the relative effects of noise. The Signal-to-Noise Ratio (SNR) at the receiver's input provides a good measure of the received signal quality. This SNR is often referred to as the channel SNR.

2.3   Information of a Source

Based on Shannon's work [133, 215–217, 225, 226], let us introduce the basic terms and definitions of information theory by considering a few simple examples. Assume that a simple eight-bit Analog-to-Digital Converter (ADC) emits a sequence of mutually independent source symbols that can take the values $i = 1, 2, \ldots, 256$ with equal probability. One may wonder how much information can be inferred upon receiving one of these samples. It is intuitively clear that this inferred information is definitely proportional to the 'uncertainty' resolved by the reception of one such symbol, which in turn implies that the information conveyed is related to the number of levels in the ADC. More explicitly, the higher the number of legitimate quantization levels, the lower is the relative frequency or probability of receiving any one of them, and hence it is the more 'surprising' when any one of them is received. Therefore, less probable quantized samples carry more information than their more frequent, more likely counterparts.

Not surprisingly, one could resolve this uncertainty by simply asking a maximum of 256 questions, such as 'Is the level 1?,' 'Is it 2?' ... 'Is it 256?' Following Hartley's approach [219], a more efficient strategy would be to ask eight questions, for example: 'Is the level larger than 128?' No. 'Is it larger than 64?' No ... 'Is it larger than 2?' No. 'Is it larger than 1?' No. Clearly in this case the source symbol emitted was of magnitude one, provided that the zero level was not a possibility. We can therefore infer that $\log_2 256 = 8$ 'Yes/No' binary answers would be needed to resolve any uncertainty as regards the source symbol's level.

In more general terms, the information carried by any one symbol of a q-level source, where all the levels are equiprobable with probabilities of $p_i = 1/q$, $i = 1 \cdots q$, is defined as

$$I = \log_2 q. \tag{2.7}$$

Rewriting Equation (2.7) using the message probabilities $p_i = \frac{1}{q}$ yields a more convenient form:

$$I = \log_2 \frac{1}{p_i} = -\log_2 p_i, \tag{2.8}$$

which now is also applicable in the case of arbitrary, unequal message probabilities p_i, again implying the plausible fact that the lower the probability of a certain source symbol, the higher the information conveyed by its occurrence. Observe, however, that for unquantized analog sources, where for the number of possible source symbols we have $q \to \infty$, and hence the probability of any analog sample becomes infinitesimally low, these definitions become meaningless.

Let us now consider a sequence of N consecutive q-ary symbols. The number of different values that his sequence can take is q^N, delivering q^N different messages. Therefore, the information carried by one such sequence is:

$$I_N = \log_2(q^N) = N \log_2 q, \tag{2.9}$$

which is in perfect harmony with our expectation, delivering N times the information of a single symbol, which was quantified by Equation (2.7). Doubling the sequence length to $2N$ carries twice the information, as suggested by:

$$I_{2N} = \log_2(q^{2N}) = 2N \cdot \log_2 q. \tag{2.10}$$

Before we proceed, let us briefly summarize the basic *properties of information* following Shannon's work [133, 215–217, 225, 226].

- If for the probability of occurrences of the symbols j and k we have $p_j < p_k$, then as regards the information carried by them we have $I(k) < I(j)$.
- If in the limit we have $p_k \to 1$, then for the information carried by the symbol k we have $I(k) \to 0$, implying that symbols whose probability of occurrence tends to unity carry no information.
- If a symbol's probability is in the range $0 \le p_k \le 1$, then as regards the information carried by it we have $I(k) \ge 0$.
- For independent messages k and j, their joint information is given by the sum of their information: $I(k, j) = I(k) + I(j)$. For example, the information carried by the statement 'My son is 14 years old and my daughter is 12' is equivalent to that of the sum of these statements: 'My son is 14 years old' and 'My daughter is 12 years old'.
- In harmony with our expectation, if we have two equiprobable messages 0 and 1 with probabilities $p_1 = p_2 = \frac{1}{2}$, then from Equation (2.8) we have $I(0) = I(1) = 1$ bit.

2.4   Average Information of Discrete Memoryless Sources

Following Shannon's approach [133,215–217,225,226], let us now consider a source emitting one of q possible symbols from the alphabet $s = s_1, s_2, \ldots, s_i, \ldots, s_q$ having symbol probabilities of p_i, $i = 1, 2, \ldots, q$. Suppose that a long message of N symbols constituted by symbols from the alphabet $s = s_1, s_2, \ldots, s_q$ having symbol probabilities of p_i is to be transmitted. Then the symbol s_i appears in every N-symbol message on average $p_i \cdot N$ number of times, provided the message length is sufficiently long. The information carried by symbol s_i is $\log_2 1/p_i$ and its $p_i \cdot N$ occurrences yield an information contribution of

$$I(i) = p_i \cdot N \cdot \log_2 \frac{1}{p_i}. \tag{2.11}$$

Upon summing the contributions of all the q symbols, we acquire the total information carried by the N-symbol sequence:

$$I = \sum_{i=1}^{q} p_i N \cdot \log_2 \frac{1}{p_i} \text{ [bits]}. \tag{2.12}$$

Averaging this over the N symbols of the sequence yields the average information per symbol, which is referred to as the source's *entropy* [215]:

$$H = \frac{I}{N} = \sum_{i=1}^{q} p_i \cdot \log_2 \frac{1}{p_i} = -\sum_{i=1}^{q} p_i \log_2 p_i \text{ [bit/symbol]}. \tag{2.13}$$

Then the *average source information rate* can be defined as the product of the information carried by a source symbol, given by the entropy H and the source emission rate R_s:

$$R = R_s \cdot H \text{ [bits/second]}. \tag{2.14}$$

Observe that Equation (2.13) is analogous to the discrete form of the first moment, or in other words the mean of a random process with a PDF of $p(x)$, as in

$$\overline{x} = \int_{-\infty}^{\infty} x \cdot p(x) \, dx, \tag{2.15}$$

where the averaging corresponds to the integration, and the instantaneous value of the random variable x represents the information $\log_2 p_i$ carried by message i, which is weighted by its probability of occurrence p_i quantified for a continuous variable x by $p(x)$.

2.4.1   Maximum Entropy of a Binary Source

Let us assume that a binary source, for which $q = 2$, emits two symbols with probabilities $p_1 = p$ and $p_2 = (1 - p)$, where the sum of the symbol probabilities must be unity. In order to quantify the maximum average information of a symbol from this source as a function of the symbol probabilities, we note from Equation (2.13) that the entropy is given by:

$$H(p) = -p \cdot \log_2 p - (1 - p) \cdot \log_2(1 - p). \tag{2.16}$$

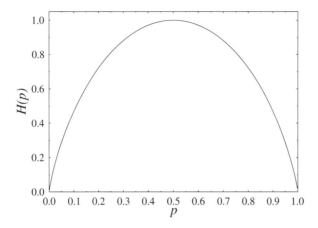

Figure 2.5: Entropy versus message probability p for a binary source. © Shannon [215], BSTJ, 1948.

As in any maximization problem, we set $\partial H(p)/\partial p = 0$, and upon using the differentiation chain rule of $(u \cdot v)' = u' \cdot v + u \cdot v'$ as well as exploiting that $(\log_a x)' = \frac{1}{x} \log_a e$ we arrive at

$$\frac{\partial H(p)}{\partial p} = -\log_2 p - \frac{p}{p} \cdot \log_2 e + \log_2(1-p) + \frac{(1-p)}{(1-p)} \log_2 e = 0$$
$$\log_2 p = \log_2(1-p)$$
$$p = (1-p)$$
$$p = 0.5.$$

This result suggests that the entropy is maximal for equiprobable binary messages. Plotting Equation (2.16) for arbitrary p values yields Figure 2.5, in which Shannon suggested that the average information carried by a symbol of a binary source is low if one of the symbols has a high probability while the other has a low probability.

Example: Let us compute the entropy of the binary source having message probabilities of $p_1 = \frac{1}{8}, p_2 = \frac{7}{8}$.

The entropy is expressed as:

$$H = -\frac{1}{8} \log_2 \frac{1}{8} - \frac{7}{8} \log_2 \frac{7}{8}.$$

Exploiting the following equivalence:

$$\log_2(x) = \log_{10}(x) \cdot \log_2(10) \approx 3.322 \cdot \log_{10}(x) \qquad (2.17)$$

we have:

$$H \approx \frac{3}{8} - \frac{7}{8} \cdot 3.322 \cdot \log_{10} \frac{7}{8} \approx 0.54 \text{ [bit/symbol]},$$

again implying that if the symbol probabilities are rather different, the entropy becomes significantly lower than the achievable 1 bit/symbol. This is because the probability of encountering the more likely symbol is so close to unity that it carries hardly any information, which cannot be compensated by the more 'informative' symbol's reception. For the even more unbalanced situation of $p_1 = 0.1$ and $p_2 = 0.9$ we have

$$
\begin{aligned}
H &= -0.1 \log_2 0.1 - 0.9 \cdot \log_2 0.9 \\
&\approx -(0.3322 \cdot \log_{10} 0.1 + 0.9 \cdot 3.322 \cdot \log_{10} 0.9) \\
&\approx 0.3322 + 0.1368 \\
&\approx 0.47 \text{ [bit/symbol]}.
\end{aligned}
$$

In the extreme case of $p_1 = 0$ or $p_2 = 1$ we have $H = 0$. As stated before, the *average source information rate* is defined as the product of the information carried by a source symbol, given by the entropy H and the source emission rate R_s, yielding $R = R_s \cdot H$ [bits/second]. Transmitting the source symbols via a perfect noiseless channel yields the same received sequence without loss of information.

2.4.2   Maximum Entropy of a q-ary Source

For a q-ary source the entropy is given by

$$
H = -\sum_{i=1}^{q} p_i \log_2 p_i, \tag{2.18}
$$

where, again, the constraint $\sum p_i = 1$ must be satisfied. When determining the extreme value of the above expression for the entropy H under the constraint of $\sum p_i = 1$, the following term has to be maximized:

$$
D = \sum_{i=1}^{q} -p_i \log_2 p_i + \lambda \cdot \left[1 - \sum_{i=1}^{q} p_i \right], \tag{2.19}
$$

where λ is the so-called Lagrange multiplier. Following the standard procedure of maximizing an expression, we set

$$
\frac{\partial D}{\partial p_i} = -\log_2 p_i - \frac{p_i}{p_i} \cdot \log_2 e - \lambda = 0
$$

leading to

$$
\log_2 p_i = -(\log_2 e + \lambda) = \text{Constant for } i = 1, \ldots, q,
$$

which implies that the maximum entropy of a q-ary source is maintained if all message probabilities are identical, although at this stage the value of this constant probability is not explicit. Note, however, that the message probabilities must sum to unity, and hence

$$
\sum_{i=1}^{q} p_i = 1 = q \cdot a, \tag{2.20}
$$

Table 2.1: Shannon–Fano Coding Example based on Algorithm 2.1 and Figure 2.6

		Coding Steps				
Symbol	Probability	1	2	3	4	Codeword
S_0	0.27	0	0			00
S_1	0.20	0	1			01
S_2	0.17	1	0	0		100
S_3	0.16	1	0	1		101
S_4	0.06	1	1	0	0	1100
S_5	0.06	1	1	0	1	1101
S_6	0.04	1	1	1	0	1110
S_7	0.04	1	1	1	1	1111

where a is a constant, leading to $a = 1/q = p_i$, implying that the entropy of any q-ary source is maximum for equiprobable messages. Furthermore, H is always bounded according to

$$0 \leq H \leq \log_2 q. \tag{2.21}$$

2.5 Source Coding for a Discrete Memoryless Source

Interpreting Shannon's work [133, 215–217, 225, 226] further, we see that source coding is the process by which the output of a q-ary information source is converted to a binary sequence for transmission via binary channels, as seen in Figure 2.1. When a discrete memoryless source generates q-ary equiprobable symbols with an average information rate of $R = R_s \log_2 q$, all symbols convey the same amount of information, and efficient signaling takes the form of binary transmissions at a rate of R bits per second. When the symbol probabilities are unequal, the minimum required source rate for distortionless transmission is reduced to

$$R = R_s \cdot H < R_s \log_2 q. \tag{2.22}$$

Then the transmission of a highly probable symbol carries little information, and hence assigning $\log_2 q$ number of bits to it does not use the channel efficiently. What can be done to improve transmission efficiency? *Shannon's source coding theorem* suggests that by using a *source encoder* before transmission the efficiency of the system with equiprobable source symbols can be arbitrarily approached.

Coding efficiency can be defined as the ratio of the source information rate and the average output bit rate of the source encoder. If this ratio approaches unity, implying that the source encoder's output rate is close to the source information rate, the source encoder is highly efficient. There are many source encoding algorithms, but the most powerful approach suggested has been Shannon's method [133], which is best illustrated by means of the example described below and portrayed in Table 2.1, Algorithm 2.1 and Figure 2.6.

2.5.1 Shannon–Fano Coding

The Shannon–Fano coding algorithm is based on the simple concept of encoding frequent messages using short codewords and infrequent ones by long codewords, while reducing

Algorithm 2.1 (Shannon–Fano Coding) This algorithm summarizes the Shannon–Fano coding steps. (See also Figure 2.6 and Table 2.1.)

1. The source symbols $S_0 \ldots S_7$ are first sorted in descending order of probability of occurrence.
2. Then the symbols are divided into two subgroups so that the subgroup probabilities are as close to each other as possible. This division is indicated by the horizontal partitioning lines in Table 2.1.
3. When allocating codewords to represent the source symbols, we assign a logical zero to the top subgroup and logical one to the bottom subgroup in the appropriate column under 'coding steps.'
4. If there is more than one symbol in the subgroup, this method is continued until no further divisions are possible.
5. Finally, the variable-length codewords are output to the channel.

the average message length. This algorithm is part of virtually all treatises dealing with information theory, such as, for example, Carlson's work [221]. The formal coding steps listed in Algorithm 2.1 and in the flowchart of Figure 2.6 can be readily followed in the context of a simple example in Table 2.1. The average codeword length is given by weighting the length of any codeword by its probability, yielding:

$$(0.27 + 0.2) \cdot 2 + (0.17 + 0.16) \cdot 3 + 2 \cdot 0.06 \cdot 4 + 2 \cdot 0.04 \cdot 4 \approx 2.73 \text{ [bits]}.$$

The entropy of the source is:

$$H = -\sum_i p_i \log_2 p_i$$

$$= -(\log_2 10) \sum_i p_i \log_{10} p_i$$

$$\approx -3.322 \cdot [0.27 \cdot \log_{10} 0.27 + 0.2 \cdot \log_{10} 0.2$$
$$+ 0.17 \cdot \log_{10} 0.17 + 0.16 \cdot \log_{10} 0.16$$
$$+ 2 \cdot 0.06 \cdot \log_{10} 0.06 + 2 \cdot 0.04 \cdot \log_{10} 0.04]$$

$$\approx 2.691 \text{ [bit/symbol]}. \tag{2.23}$$

Since the average codeword length of 2.73 bits/symbol is very close to the entropy of 2.691 bit/symbol, a high coding efficiency is predicted, which can be computed as:

$$E \approx \frac{2.691}{2.73} \approx 98.6\%.$$

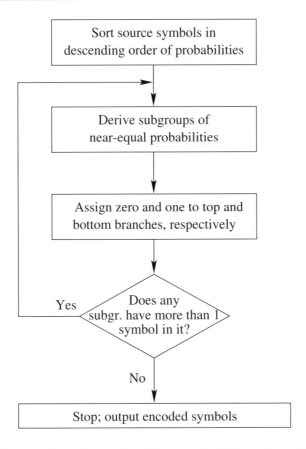

Figure 2.6: Shannon–Fano Coding Algorithm (see also Table 2.1 and Algorithm 2.1).

The straightforward 3 bits/symbol Binary-Coded Decimal (BCD) assignment gives an efficiency of

$$E \approx \frac{2.691}{3} \approx 89.69\%.$$

In summary, Shannon–Fano coding allowed us to create a set of uniquely invertible mappings to a set of codewords, which facilitate a more efficient transmission of the source symbols than straightforward BCD representations would. This was possible with no coding impairment (i.e., losslessly). Having highlighted the philosophy of the Shannon–Fano noiseless or distortionless coding technique, let us now concentrate on the closely related Huffman coding principle.

2.5.2 Huffman Coding

The Huffman Coding algorithm is best understood by referring to the flowchart in Figure 2.7 and to the formal coding description in Algorithm 2.2. A simple practical example is portrayed in Table 2.2, which leads to the Huffman codes summarized in Table 2.3. Note that

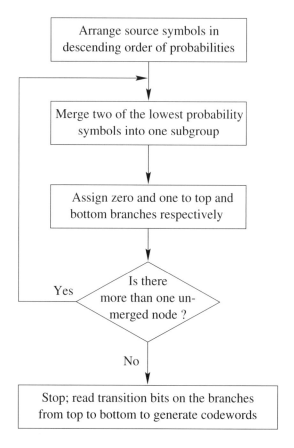

Figure 2.7: Huffman coding algorithm (see also Algorithm 2.2 and Table 2.2).

we have used the same symbol probabilities as in our Shannon–Fano coding example, but the Huffman algorithm leads to a different codeword assignment. Nonetheless, the code's efficiency is identical to that of the Shannon–Fano algorithm.

The symbol-merging procedure can also be conveniently viewed using the example of Figure 2.8, where the Huffman codewords are derived by reading the associated 1 and 0 symbols from the end of the tree backwards; that is, towards the source symbols $S_0 \cdots S_7$. Again, these codewords are summarized in Table 2.3.

In order to arrive at a fixed average channel bit rate, which is convenient in many communications systems, a long buffer might be needed, causing storage and delay problems. Observe from Table 2.3 that the Huffman coding algorithm gives codewords that can be uniquely decoded, which is a crucial prerequisite for its practical employment. This is because no codeword can be a prefix of any longer one. For example, for the sequence of codewords . . . , $00, 10, 010, 110, 1111, \ldots$ the source sequence of . . . $S_0, S_1, S_2, S_3, S_8 \ldots$ can be uniquely inferred from Table 2.3.

Algorithm 2.2 (Huffman Coding) This algorithm summarizes the Huffman coding steps.

```
1. Arrange the symbol probabilities pᵢ in decreasing order
   and consider them as 'leaf-nodes,' as suggested by
   Table 2.2.
2. While there is more than one node, merge the two nodes
   having the lowest probabilities, and assign 0 and 1 to
   the upper and lower branches respectively.
3. Read the assigned 'transition bits' on the branches from
   top to bottom in order to derive the codewords.
```

Table 2.2: Huffman Coding Example Based on Algorithm 2.2 and Figure 2.7 (for final code assignment see Table 2.3)

Symbol	Probability	Steps 1 and 2		Steps 3 and 4			
		Code	Probability	Code	Probability	Group	Code
S_0	0.27					S_0	–
S_1	0.20					S_1	–
S_2	0.17			0	0.33	S_{23}	0
S_3	0.16			1			1
S_4	0.06	0	0.12	0			00
S_5	0.06	1		0	0.20	S_{4567}	01
S_6	0.04	0	0.08	1			10
S_7	0.04	1		1			11

Symbol	Probability	Steps 5 and 6		Step 7		Codeword
		Code	Probability	Code	Probability	
S_{23}	0.33	0	0.6	0		00
S_0	0.27	1			1.0	01
S_1	0.20	0	0.4	1		10
S_{4567}	0.20	1				11

In our discussions so far, we have assumed that the source symbols were completely independent of each other. Such a source is usually referred to as a memoryless source. By contrast, sources where the probability of a certain symbol also depends on what the previous symbol was are often termed *sources exhibiting memory*. These sources are typically bandlimited sample sequences, such as, for example, a set of correlated or 'similar-magnitude' speech samples or adjacent video pixels. Let us now consider sources that exhibit memory.

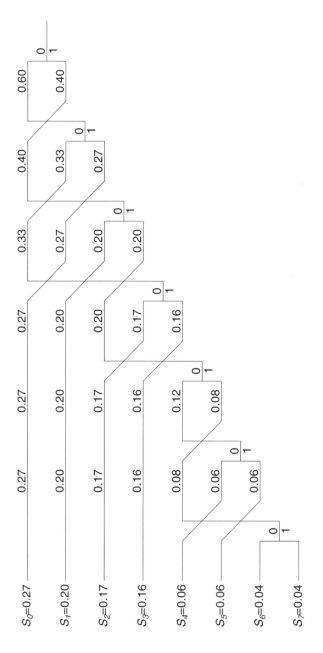

Figure 2.8: Tree-based Huffman coding example.

Table 2.3: Huffman Coding Example Summary of Table 2.2

Symbol	Probability	BCD	Huffman Code
S_0	0.27	000	01
S_1	0.20	001	10
S_2	0.17	010	000
S_3	0.16	011	001
S_4	0.06	100	1100
S_5	0.06	101	1101
S_6	0.04	110	1110
S_7	0.04	111	1111

2.6 Entropy of Discrete Sources Exhibiting Memory

Let us invoke Shannon's approach [133, 215–217, 225, 226] in order to illustrate sources with and without memory. Let us therefore consider an uncorrelated random White Gaussian Noise (WGN) process, which has been passed through a low-pass filter. The corresponding ACFs and Power Spectral Density (PSD) functions were portrayed in Figures 2.3 and 2.4. Observe in the figures that through low-pass filtering a WGN process introduces correlation by limiting the rate at which amplitude changes are possible, smoothing the amplitude of abrupt noise peaks. This example suggests that all bandlimited signals are correlated over a finite interval. Most analog source signals, such as speech and video, are inherently correlated, owing to physical restrictions imposed on the analog source. Hence all practical analog sources possess some grade of memory, a property that is also retained after sampling and quantization. An important feature of sources with memory is that they are predictable to a certain extent; hence, they can usually be more efficiently encoded than unpredictable sources having no memory.

2.6.1 Two-State Markov Model for Discrete Sources Exhibiting Memory

Let us now introduce a simple analytically tractable model for treating sources that exhibit memory. Predictable sources that have memory can be conveniently modeled by *Markov processes*. A source having a memory of one symbol interval directly 'remembers' only the previously emitted source symbol, and it emits one of its legitimate symbols with a certain probability, which depends explicitly on the state associated with this previous symbol. A one-symbol-memory model is often referred to as a *first-order* model. As an example, if in a first-order model the previous symbol can take only two different values, we have two different states, and this simple two-state first-order Markov model is characterized by the state transition diagram of Figure 2.9. Previously, in the context of Shannon–Fano and Huffman coding of memoryless information sources, we used the notation of S_i, $i = 0, 1, \ldots$, for the various symbols to be encoded. In this section we are dealing with sources exhibiting memory, and hence for the sake of distinction we use the symbol notation of X_i, $i = 1, 2, \ldots$. If, for the sake of illustration, the previous emitted symbol was X_1, the state machine of Figure 2.9 is in state X_1, and in the current signaling interval it can generate one

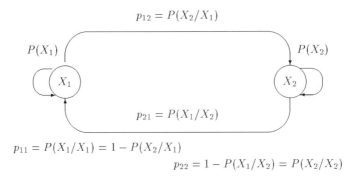

$$p_{11} = P(X_1/X_1) = 1 - P(X_2/X_1)$$
$$p_{22} = 1 - P(X_1/X_2) = P(X_2/X_2)$$

Figure 2.9: Two-state first-order Markov model.

of two symbols, namely, X_1 and X_2, whose probability depends explicitly on the previous state X_1. However, not all two-state Markov models are as simple as that of Figure 2.9, since the transitions from state X_1 to X_2 are not necessarily associated with emitting the same symbol as the transitions from state X_2 to X_1. A more elaborate example will therefore be considered later in this chapter.

Observe in Figure 2.9 that the corresponding transition probabilities from state X_1 are given by the conditional probabilities $p_{12} = P(X_2/X_1)$ and $p_{11} = P(X_1/X_1) = 1 - P(X_2/X_1)$. Similar findings can be observed as regards state X_2. These dependencies can also be stated from a different point of view as follows. The probability of occurrence of a particular symbol depends not only on the symbol itself, but also on the previous symbol emitted. Thus, the symbol entropy for states X_1 and X_2 will now be characterized by means of the conditional probabilities associated with the transitions merging in these states. Explicitly, the symbol entropy for state $X_i, i = 1, 2$ is given by

$$H_i = \sum_{j=1}^{2} p_{ij} \cdot \log_2 \frac{1}{p_{ij}} \quad i = 1, 2$$
$$= p_{i1} \cdot \log_2 \frac{1}{p_{i1}} + p_{i2} \cdot \log_2 \frac{1}{p_{i2}}, \tag{2.24}$$

yielding the symbol entropies (that is, the average information carried by the symbols emitted) in states X_1 and X_2 respectively as

$$H_1 = p_{11} \cdot \log_2 \frac{1}{p_{11}} + p_{12} \cdot \log_2 \frac{1}{p_{12}}$$
$$H_2 = p_{21} \cdot \log_2 \frac{1}{p_{21}} + p_{22} \cdot \log_2 \frac{1}{p_{22}}. \tag{2.25}$$

The symbol entropies, H_1 and H_2, are characteristic of the average information conveyed by a symbol emitted in states X_1 and X_2 respectively. In order to compute the overall entropy H of this source, they must be weighted by the probabilities of occurrence, P_1 and P_2, of these states:

$$H = \sum_{i=1}^{2} P_i H_i$$

$$= \sum_{i=1}^{2} P_i \sum_{j=1}^{2} p_{ij} \log_2 \frac{1}{p_{ij}}. \tag{2.26}$$

Assuming a highly predictable source having high adjacent sample correlation, it is plausible that once the source is in a given state it is more likely to remain in that state than to traverse into the other state. For example, assuming that the state machine of Figure 2.9 is in state X_1 and the source is a highly correlated, predictable one, we are likely to observe long runs of X_1. Conversely, once it is in state X_2, long strings of X_2 symbols will typically follow.

2.6.2 N-State Markov Model for Discrete Sources Exhibiting Memory

In general, assuming N legitimate states, (i.e., N possible source symbols) and following similar arguments, Markov models are characterized by their state probabilities $P(X_i)$, $i = 1, \ldots, N$, where N is the number of states, as well as by the transition probabilities $p_{ij} = P(X_i/X_j)$, where p_{ij} explicitly indicates the probability of traversing from state X_j to state X_i. Their further basic feature is that they emit a source symbol at every state transition, as shown in the context of an example presented in Section 2.7. Similarly to the two-state model, we define the entropy of a source having memory as the weighted average of the entropy of the individual symbols emitted from each state, where the weighting takes into account the probabilities of occurrence of the individual states, namely P_i. In analytical terms, the symbol entropy for state X_i, $i = 1, \ldots, N$ is given by:

$$H_i = \sum_{j=1}^{N} p_{ij} \cdot \log_2 \frac{1}{p_{ij}} \quad i = 1, \ldots, N. \tag{2.27}$$

The averaged, weighted symbol entropies give the source entropy:

$$H = \sum_{i=1}^{N} P_i H_i$$

$$= \sum_{i=1}^{N} P_i \sum_{j=1}^{N} p_{ij} \log_2 \frac{1}{p_{ij}}. \tag{2.28}$$

Finally, assuming a source symbol rate of v_s, the average information emission rate R of the source is given by:

$$R = v_s \cdot H \text{ [bits/second]}. \tag{2.29}$$

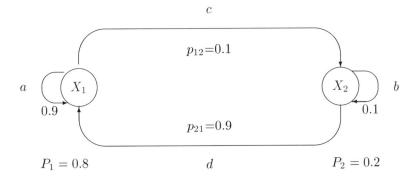

Figure 2.10: Two-state Markov model example.

2.7 Examples

2.7.1 Two-State Markov Model Example

As mentioned in the previous section, we now consider a slightly more sophisticated Markov model, where the symbols emitted upon traversing from state X_1 to X_2 are different from those when traversing from state X_2 to X_1.

More explicitly, consider a discrete source that was described by the two-state Markov model of Figure 2.9, where the transition probabilities are

$$p_{11} = P(X_1/X_1) = 0.9 \quad p_{22} = P(X_2/X_2) = 0.1$$
$$p_{12} = P(X_1/X_2) = 0.1 \quad p_{21} = P(X_2/X_1) = 0.9,$$

while the state probabilities are

$$P(X_1) = 0.8 \quad \text{and} \quad P(X_2) = 0.2. \tag{2.30}$$

The source emits one of four symbols, namely, a, b, c, and d, upon every state transition, as seen in Figure 2.10. Let us find:

- the source entropy;
- the average information content per symbol in messages of one, two, and three symbols.

Message Probabilities

Let us consider two sample sequences acb and aab. As shown in Figure 2.10, the transitions leading to acb are $(1 \rightsquigarrow 1)$, $(1 \rightsquigarrow 2)$ and $(2 \rightsquigarrow 2)$. The probability of encountering this sequence is $0.8 \cdot 0.9 \cdot 0.1 \cdot 0.1 = 0.0072$. The sequence aab has a probability of zero because the transition from a to b is illegal. Further path (i.e., message) probabilities are given in Table 2.4 along with the information $I = -\log_2 P$ conveyed by all of the legitimate messages.

Table 2.4: Message probabilities of example

Message probabilities	Information conveyed (bits/message)
$P_a = 0.9 \times 0.8 = 0.72$	$I_a = 0.474$
$P_b = 0.1 \times 0.2 = 0.02$	$I_b = 5.644$
$P_c = 0.1 \times 0.8 = 0.08$	$I_c = 3.644$
$P_d = 0.9 \times 0.2 = 0.18$	$I_d = 2.474$
$P_{aa} = 0.72 \times 0.9 = 0.648$	$I_{aa} = 0.626$
$P_{ac} = 0.72 \times 0.1 = 0.072$	$I_{ac} = 3.796$
$P_{cb} = 0.08 \times 0.1 = 0.008$	$I_{cb} = 6.966$
$P_{cd} = 0.08 \times 0.9 = 0.072$	$I_{cd} = 3.796$
$P_{bb} = 0.02 \times 0.1 = 0.002$	$I_{bb} = 8.966$
$P_{bd} = 0.02 \times 0.9 = 0.018$	$I_{bd} = 5.796$
$P_{da} = 0.18 \times 0.9 = 0.162$	$I_{da} = 2.626$
$P_{dc} = 0.18 \times 0.1 = 0.018$	$I_{dc} = 5.796$

Source Entropy

According to Equation (2.27), the entropies of symbols X_1 and X_2 are computed as

$$H_1 = -p_{12} \cdot \log_2 p_{12} - p_{11} \cdot \log_2 p_{11}$$

$$= 0.1 \cdot \log_2 10 + 0.9 \cdot \log_2 \frac{1}{0.9}$$

$$\approx 0.469 \text{ bits/symbol} \tag{2.31}$$

$$H_2 = -p_{21} \cdot \log_2 p_{21} - p_{22} \cdot \log_2 p_{22}$$

$$\approx 0.469 \text{ bits/symbol.} \tag{2.32}$$

Then their weighted average is calculated using the probability of occurrence of each state in order to derive the average information per message for this source:

$$H \approx 0.8 \cdot 0.469 + 0.2 \cdot 0.469 \approx 0.469 \text{ bit/symbol.}$$

The average information per symbol I_2 in two-symbol messages is computed from the entropy h_2 of the two-symbol messages, which is

$$h_2 = \sum_1^8 P_{symbol} \cdot I_{symbol}$$

$$= P_{aa} \cdot I_{aa} + P_{ac} \cdot I_{ac} + \cdots + P_{dc} \cdot I_{dc}$$

$$\approx 1.66 \text{ bits/2 symbols,} \tag{2.33}$$

giving $I_2 = h_2/2 \approx 0.83$ bits/symbol information on average when receiving a two-symbol message.

There are eight two-symbol messages; hence, the maximum possible information conveyed is $\log_2 8 = 3$ bits/2 symbols, or 1.5 bits/symbol. However, since the symbol probabilities of $P_1 = 0.8$ and $P_2 = 0.2$ are fairly different, this scheme has a significantly lower conveyed information per symbol, namely $I_2 \approx 0.83$ bits/symbol.

Similarly, one can find the average information content per symbol for arbitrarily long messages of concatenated source symbols. For one-symbol messages we have:

$$
\begin{aligned}
I_1 = h_1 &= \sum_1^4 P_{symbol} \cdot I_{symbol} \\
&= P_a \cdot I_a + \cdots + P_d \cdot I_d \\
&\approx 0.72 \times 0.474 + \cdots + 0.18 \times 2.474 \\
&\approx 0.341 + 0.113 + 0.292 + 0.445 \\
&\approx 1.191 \text{ bits/symbol.}
\end{aligned}
\tag{2.34}
$$

We note that the maximum possible information carried by one-symbol messages is $h_{1\,max} = \log_2 4 = 2$ bits/symbol, since there are four one-symbol messages in Table 2.4.

Observe the important tendency in which, when sending longer messages from dependent sources, the average information content per symbol is reduced. This is due to the source's memory, since consecutive symbol emissions are dependent on previous ones and hence do not carry as much information as independent source symbols would. This becomes explicit by comparing $I_1 \approx 1.191$ and $I_2 \approx 0.83$ bits/symbol.

Therefore, expanding the message length to be encoded yields more efficient coding schemes, requiring a lower number of bits, if the source has a memory. This is the essence of Shannon's source coding theorem.

2.7.2 Four-State Markov Model for a Two-Bit Quantizer

Let us now augment the previously introduced two-state Markov model concepts with the aid of a four-state example. Let us assume that we have a discrete source constituted by a two-bit quantizer, which is characterized by Figure 2.11. Assume further that due to bandlimitation only transitions to adjacent quantization intervals are possible, since the bandlimitation restricts the input signal's rate of change. The probabilities of the signal samples residing in intervals 1–4 are given by

$$
P(1) = P(4) = 0.1, \quad P(2) = P(3) = 0.4.
$$

The associated state transition probabilities are shown in Figure 2.11, along with the quantized samples a, b, c and d, which are transmitted when a state transition takes place – that is, when taking a new sample from the analog source signal at the sampling rate f_s.

Although we have stipulated a number of simplifying assumptions, this example attempts to illustrate the construction of Markov models in the context of a simple practical problem. Next we construct a simpler example for use in augmenting the underlying concepts, and set aside the above four-state Markov model example as a potential exercise for the reader.

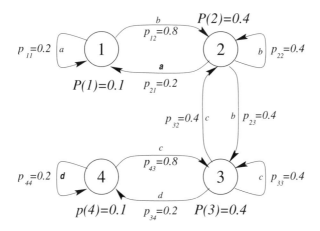

Figure 2.11: Four-state Markov model for a two-bit quantizer.

2.8   Generating Model Sources

2.8.1   Autoregressive Model

In evaluating the performance of information processing systems such as encoders and predictors, it is necessary to have 'standardized' or easily described model sources. Although a set of semi-standardized test sequences of speech and images is widely used in preference to analytical model sources by researchers in codec performance testing, real speech or image sources cannot be used in analytical studies. A widely used analytical model source is the *Autoregressive (AR) model*. A zero-mean random sequence $y(n)$ is called an AR process of order p if it is generated as follows:

$$y(n) = \sum_{k=1}^{p} a_k y(n-k) + \varepsilon(n), \quad \forall n, \tag{2.35}$$

where $\varepsilon(n)$ is an uncorrelated zero-mean, random input sequence with variance σ^2; that is,

$$E\{\varepsilon(n)\} = 0$$
$$E\{\varepsilon^2(n)\} = \sigma^2$$
$$E\{\varepsilon(n) \cdot y(m)\} = 0. \tag{2.36}$$

Equation (2.35) states that an AR system recursively generates the present output from p previous output samples given by $y(n-k)$ and the present random input sample $\varepsilon(n)$.

2.8.2   AR Model Properties

AR models are very useful in studying information processing systems, such as speech and image codecs, predictors and quantizers. They have the following basic properties:

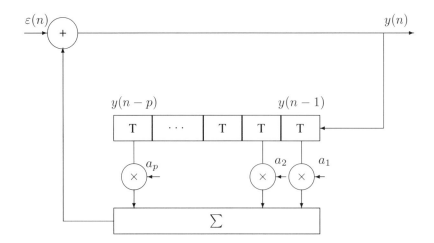

Figure 2.12: Markov model of order p.

- The first term of Equation (2.35), which is repeated here for convenience,

$$\hat{y}(n) = \sum_{k=1}^{p} a_k y(n - k)$$

  defines a predictor, giving an estimate $\hat{y}(n)$ of $y(n)$, which is associated with the minimum mean squared error between the two quantities.

- Although $\hat{y}(n)$ and $y(n)$ depend explicitly only on the past p samples of $y(n)$, through the recursive relationship of Equation (2.35) this entails the entire past of $y(n)$. This is because each of the previous p samples depends on its predecessors.

- Equation (2.35) can therefore be written in the form of

$$y(n) = \hat{y}(n) + \varepsilon(n),\tag{2.37}$$

  where $\varepsilon(n)$ is the *prediction error* and $\hat{y}(n)$ is the minimum variance prediction estimate of $y(n)$.

- Without proof, we state that for a random Gaussian distributed prediction error sequence $\varepsilon(n)$ these properties are characteristic of a pth order Markov process, as portrayed in Figure 2.12. When this model is simplified for the case of $p = 1$, we arrive at the schematic diagram shown in Figure 2.13.

- The PSD of the prediction error sequence $\varepsilon(n)$ is that of a random 'white noise' sequence, containing all possible frequency components with the same energy. Hence, its ACF is the Kronecker delta function, given by the Wiener–Khintchine theorem:

$$E\{\varepsilon(n) \cdot \varepsilon(m)\} = \sigma^2 \delta(n - m).\tag{2.38}$$

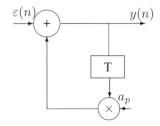

Figure 2.13: First-order Markov model.

2.8.3   First-Order Markov Model

A variety of practical information sources are adequately modeled by the analytically tractable first-order Markov model depicted in Figure 2.13, where the prediction order is $p = 1$. With the aid of Equation (2.35) we have

$$y(n) = \varepsilon(n) + ay(n - 1),$$

where a is the adjacent sample correlation of the process $y(n)$. Using the following recursion:

$$y(n - 1) = \varepsilon(n - 1) + a_1 y(n - 2)$$

$$\vdots \quad \vdots \quad \vdots$$

$$y(n - k) = \varepsilon(n - k) + a_1 y(n - k - 1) \tag{2.39}$$

we arrive at:

$$y(n) = \varepsilon(n) + a_1[\varepsilon(n - 1) + ay(n - 2)]$$
$$= \varepsilon(n) + a_1 \varepsilon(n - 1) + a^2 y(n - 2),$$

which can be generalized to:

$$y(n) = \sum_{j=0}^{\infty} a^j \varepsilon(n - j). \tag{2.40}$$

Clearly, Equation (2.40) describes the first-order Markov process with the help of the adjacent sample correlation a_1 and the uncorrelated zero-mean random Gaussian process $\varepsilon(n)$.

2.9   Run-Length Coding for Discrete Sources Exhibiting Memory

2.9.1   Run-Length Coding Principle [221]

For discrete sources having memory (i.e., possessing intersample correlation), the coding efficiency can be significantly improved by predictive coding, allowing the required transmission rate and hence the channel bandwidth to be reduced. Particularly amenable to run-length coding are binary sources with inherent memory, such as black-and-white documents, where the predominance of white pixels suggests that a Run-Length Coding (RLC) scheme,

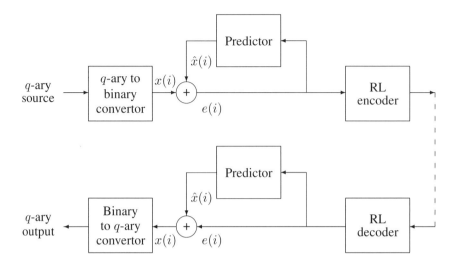

Figure 2.14: Predictive run-length codec scheme © Carlson [221].

which encodes the length of zero runs rather than repeating long strings of zeros, provides high coding efficiency.

Following Carlson's interpretation [221], a predictive RLC scheme can be constructed as shown in Figure 2.14. The q-ary source messages are first converted to binary bit format. For example, if an 8-bit ADC is used, the 8-bit digital samples are converted to binary format. This bit stream, $x(i)$, is then compared with the output signal of the predictor, $\hat{x}(i)$, which is fed with the prediction error signal $e(i)$. The comparator is a simple mod-2 gate, outputting a logical one whenever the prediction fails; that is, the predictor's output is different from the incoming bit $x(i)$. If, however, $x(i) = \hat{x}(i)$, the comparator indicates this by outputting a logical zero. For highly correlated signals from sources with significant memory the predictions are usually correct, and hence long strings of zero runs are emitted, interspersed with an occasional one. Thus, the prediction error signal $e(i)$ can be efficiently run-length encoded by noting and transmitting the length of zero runs.

The corresponding binary run-length coding principle becomes explicit from Table 2.5 and from our forthcoming coding efficiency analysis.

2.9.2   Run-Length Coding Compression Ratio [227]

Following Jain's interpretation [227], let us now investigate the RLC efficiency by assuming that a run of r successive logical zeros is followed by a one. Instead of directly transmitting these strings, we represent such a string as an n-bit word giving the length of the zero run between successive logical ones. When a zero run longer than $N = 2^n - 1$ bits occurs, this is signaled as the all-ones codeword, informing the decoder to wait for the next RLC codeword before releasing the decoded sequence. The scheme's operation is characterized by Table 2.5. Clearly, data compression is achieved if the average number of zero data bits per run d is higher than the number of bits, n, required to encode the zero-run length. Let us therefore compute the average number of bits per run without RLC. If a run of r logical

Table 2.5: Run-Length Coding Table © Carlson, 1975 [221]

Length of Zero Run between two Ones	Encoder Output (n-bit codeword)	Decoder Output
0	$00 \cdots 000$	1
1	$00 \cdots 001$	01
2	$00 \cdots 010$	001
3	$00 \cdots 011$	0001
$\vdots$	$\vdots$	$\vdots$
$N-1$	$11 \cdots 110$	$00 \cdots 01$
$\geq N = 2^n - 1$	$11 \cdots 111$	$00 \cdots 00$

zeros is followed by a one, the run length is $(r+1)$. The expected or mean value of $(r+1)$, namely $d = \overline{(r+1)}$, is calculated by weighting each specific $(r+1)$ with its probability of occurrence, that is, with its discrete PDF $c(r)$, and then averaging the weighted components:

$$d = \overline{(r+1)} = \sum_{r=0}^{N-1} (r+1) \cdot c(r) + Nc(N). \qquad (2.41)$$

The PDF of a run of r zeros followed by a one is given by:

$$c(r) = \begin{cases} p^r(1-p) & 0 \leq r \leq N-1 \\ p^N & r = N, \end{cases} \qquad (2.42)$$

since the probability of N consecutive zeros is p^N if $r = N$, while for shorter runs the joint probability of r zeros followed by a one is given by $p^r \cdot (1-p)$. The PDF and CDF of this distribution are shown in Figure 2.15 for $p = 0.9$ and $p = 0.1$, where p represents the probability of a logical zero bit. Substituting Equation (2.42) in Equation (2.41) gives

$$d = N \cdot p^N + \sum_{r=0}^{N-1} (r+1) \cdot p^r \cdot (1-p)$$
$$= N \cdot p^N + 1 \cdot p^0 \cdot (1-p) + 2 \cdot p \cdot (1-p) + \cdots + N \cdot p^{N-1} \cdot (1-p)$$
$$= N \cdot p^N + 1 + 2p + 3p^2 + \cdots + N \cdot p^{N-1} - p - 2p^2 \cdots - N \cdot p^N$$
$$= 1 + p + p^2 + \cdots p^{N-1}. \qquad (2.43)$$

Equation (2.43) is a simple geometric progression, given in closed form as

$$d = \frac{1 - p^N}{1 - p}. \qquad (2.44)$$

RLC Example: *Using a run-length coding memory of $M = 31$ and a zero symbol probability of $p = 0.95$, characterize the RLC efficiency.*

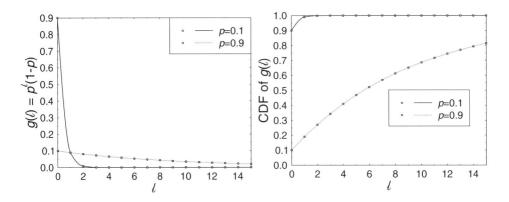

Figure 2.15: CDF and PDF of the geometric distribution of run length l.

Substituting N and p into Equation (2.44) for the average run length we have

$$d = \frac{1 - 0.95^{31}}{1 - 0.95} \approx \frac{1 - 0.204}{0.05} \approx 15.92. \qquad (2.45)$$

The *compression ratio* C achieved by RLC is given by

$$C = \frac{d}{n} = \frac{1 - p^N}{n(1 - p)} \approx \frac{15.92}{5} \approx 3.18. \qquad (2.46)$$

The achieved average bit rate is

$$B = \frac{n}{d} \approx 0.314 \text{ bit/pixel},$$

and the coding efficiency is computed as the ratio of the entropy (i.e., of the lowest possible bit rate and the actual bit rate). The source entropy is given by

$$H \approx -0.95 \cdot 3.322 \cdot \log_{10} 0.95 - 0.05 \cdot 3.322 \cdot \log_{10} 0.05$$
$$\approx 0.286 \text{ bit/symbol}, \qquad (2.47)$$

giving a coding efficiency of

$$E = H/B \approx 0.286/0.314 \approx 91\%.$$

This concludes our RLC example.

2.10   Information Transmission via Discrete Channels

Let us now return to Shannon's classic references [133, 215–217, 225, 226] and assume that both the channel and the source are discrete, and let us evaluate the amount of information transmitted via the channel. We define the channel capacity characterizing the channel and show that according to Shannon nearly error-free information transmission is possible at rates below the channel capacity via the Binary Symmetric Channel (BSC). Let us begin our discourse with a simple introductory example.

2.10.1 Binary Symmetric Channel Example

Let us assume that a binary source is emitting a logical one with a probability of $P(1) = 0.7$ and a logical zero with a probability of $P(0) = 0.3$. The channel's error probability is $p_e = 0.02$. This scenario is characterized by the BSC model of Figure 2.16. The probability of error-free reception is given by that of receiving a logical one when a one is transmitted plus the probability of receiving a zero when a zero is transmitted. This is also plausible from Figure 2.16. For example, the first of these two component probabilities can be computed with the aid of Figure 2.16 as the product of the probability $P(1)$ of a logical one being transmitted and the conditional probability $P(1/1)$ of receiving a one, given the condition that a one was transmitted:

$$P(Y_1, X_1) = P(X_1) \cdot P(Y_1/X_1) \tag{2.48}$$
$$P(1, 1) = P(1) \cdot P(1/1) = 0.7 \cdot 0.98 = 0.686.$$

Similarly, the probability of the error-free reception of a logical zero is given by

$$P(Y_0, X_0) = P(X_0) \cdot P(Y_0/X_0)$$
$$P(0, 0) = P(0) \cdot P(0/0) = 0.3 \cdot 0.98 = 0.294,$$

giving the total probability of error-free reception as

$$P_{correct} = P(1, 1) + P(0, 0) = 0.98.$$

Following similar arguments, the probability of erroneous reception is also given by two components. For example, using Figure 2.16, the probability of receiving a one when a zero was transmitted is computed by multiplying the probability $P(0)$ of a logical zero being transmitted by the conditional probability $P(1/0)$ of receiving a logical one, given the fact that a zero is known to have been transmitted:

$$P(Y_1, X_0) = P(X_0) \cdot P(Y_1/X_0)$$
$$P(1, 0) = P(0) \cdot P(1/0) = 0.3 \cdot 0.02 = 0.006.$$

Conversely,

$$P(Y_0, X_1) = P(X_1) \cdot P(Y_0/X_1)$$
$$P(0, 1) = P(1) \cdot P(0/1) = 0.7 \cdot 0.02 = 0.014,$$

yielding a total error probability of

$$P_{error} = P(1, 0) + P(0, 1) = 0.02,$$

which is constituted by the above two possible error events.

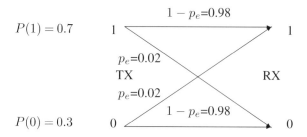

Figure 2.16: The binary symmetric channel. © Shannon [216], BSTJ, 1948.

Viewing events from a different angle, we observe that the total probability of receiving a one is that of receiving a transmitted one correctly plus that of receiving a transmitted zero incorrectly:

$$P_1 = P(1) \cdot (1 - p_e) + P(0) \cdot p_e$$
$$= 0.7 \cdot 0.98 + 0.3 \cdot 0.02 = 0.686 + 0.006 = 0.692. \tag{2.49}$$

On the same note, the probability of receiving a zero is that of receiving a transmitted zero correctly plus that of receiving a transmitted one incorrectly:

$$P_0 = P(0) \cdot (1 - p_e) + P(1) \cdot p_e$$
$$= 0.3 \cdot 0.98 + 0.7 \cdot 0.02 = 0.294 + 0.014 = 0.308. \tag{2.50}$$

In the next example, we further study the performance of the BSC for a range of different parameters in order to gain a deeper insight into its behavior.

Example: *Repeat the above calculations for* $P(1) = 1, 0.9, 0.5,$ *and* $p_e = 0, 0.1, 0.2, 0.5$ *using the BSC model of Figure 2.16. Compute and tabulate the probabilities* $P(1, 1)$, $P(0, 0)$, $P(1, 0)$, $P(0, 1)$, $P_{correct}$, P_{error}, P_1 *and* P_0 *for these parameter combinations, including also their values for the previous example, namely for* $P(1) = 0.7$, $P(0) = 0.3$ *and* $p_e = 0.02$. *Here we neglect the details of the calculations and summarize the results in Table 2.6. Some of the above quantities are plotted for further study in Figure 2.17, which reveals the interdependency of the various probabilities for the interested reader.*

Having studied the performance of the BSC, the next question that arises is how much information can be inferred upon reception of a one and a zero over an imperfect (i.e., error-prone) channel. In order to answer this question, let us first generalize the above intuitive findings in the form of *Bayes's rule*.

2.10.2 Bayes' Rule

Let Y_j represent the received symbols and X_i the transmitted symbols having probabilities of $P(Y_j)$ and $P(X_i)$ respectively. Let us also characterize the forward transition probabilities of the binary symmetric channel as suggested by Figure 2.18.

Then in general, following from the previous introductory example, the joint probability $P(Y_j, X_i)$, of receiving Y_j when the transmitted source symbol was X_i, is computed as the

Table 2.6: BSC Performance Table

p_e	$P(1)$	$P(0)$	$(1-p_e)$ $= P(1/1)$ $= P(0/0)$	$P(1,1)$	$P(0,0)$	p_e $= P(1/0)$ $= P(0/1)$	$P(1,0)$	$P(0,1)$	P_1	P_0
0	1	0	1	1	0	0	0	0	1	0
	0.9	0.1	1	0.9	0.1	0	0	0	0.9	0.1
	0.7	0.3	1	0.7	0.3	0	0	0	0.7	0.3
	0.5	0.5	1	0.5	0.5	0	0	0	0.5	0.5
0.02	1	0	0.98	0.98	0	0.02	0	0.02	0.98	0.02
	0.9	0.1	0.98	0.882	0.098	0.02	0.002	0.018	0.884	0.116
	0.7	0.3	0.98	0.686	0.294	0.02	0.006	0.014	0.692	0.308
	0.5	0.5	0.98	0.49	0.49	0.02	0.01	0.01	0.491	0.509
0.1	1	0	0.9	0.9	0	0.1	0	0.1	0.9	0.1
	0.9	0.1	0.9	0.81	0.09	0.1	0.01	0.09	0.811	0.189
	0.7	0.3	0.9	0.63	0.27	0.1	0.03	0.07	0.723	0.277
	0.5	0.5	0.9	0.45	0.45	0.1	0.05	0.05	0.455	0.545
0.2	1	0	0.8	0.8	0	0.2	0	0.2	0.8	0.2
	0.9	0.1	0.8	0.72	0.08	0.2	0.02	0.18	0.722	0.278
	0.7	0.3	0.8	0.56	0.24	0.2	0.06	0.14	0.566	0.434
	0.5	0.5	0.8	0.40	0.40	0.2	0.1	0.1	0.5	0.5
0.5	1	0	0.5	0.5	0	0.5	0	0.5	0.5	0.5
	0.9	0.1	0.5	0.45	0.05	0.5	0.05	0.45	0.5	0.5
	0.7	0.3	0.5	0.35	0.15	0.5	0.15	0.35	0.5	0.5
	0.5	0.5	0.5	0.25	0.25	0.5	0.25	0.25	0.5	0.5

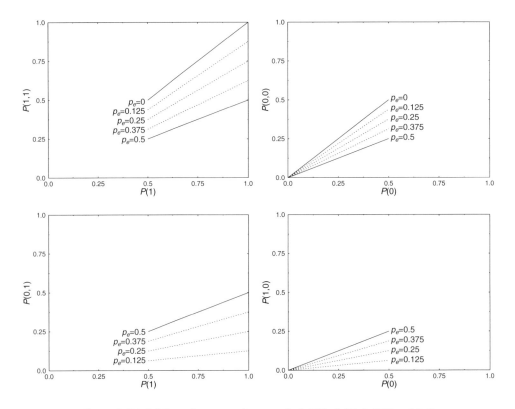

Figure 2.17: BSC performance for $p_e = 0, 0.125, 0.25, 0.375$ and 0.5.

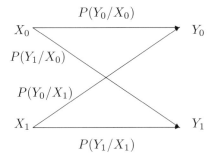

Figure 2.18: Forward transition probabilities of the nonideal binary symmetric channel.

probability $P(X_i)$ of transmitting X_i multiplied by the conditional probability $P(Y_j/X_i)$ of receiving Y_j when X_i is known to have been transmitted:

$$P(Y_j, X_i) = P(X_i) \cdot P(Y_j/X_i), \tag{2.51}$$

a result that we have already intuitively exploited in the previous example. Since for the joint probabilities $P(Y_j, X_i) = P(X_i, Y_j)$ holds, we have

$$\begin{aligned} P(X_i, Y_j) &= P(Y_j) \cdot P(X_i/Y_j) \\ &= P(X_i) \cdot P(Y_j/X_i). \end{aligned} \tag{2.52}$$

Equation (2.52) is often presented in the form

$$\begin{aligned} P(X_i/Y_j) &= \frac{P(X_i, Y_j)}{P(Y_j)} \\ &= \frac{P(Y_j) \cdot P(X_i/Y_j)}{P(Y_j)}, \end{aligned} \tag{2.53}$$

which is referred to as *Bayes's rule*.

Logically, the probability of receiving a particular $Y_j = Y_{j_0}$ is the sum of all joint probabilities $P(X_i, Y_{j_0})$ over the range of X_i. This corresponds to the probability of receiving the transmitted X_i correctly, giving rise to the channel output Y_{j_0} plus the sum of the probabilities of all other possible transmitted symbols giving rise to Y_{j_0}:

$$P(Y_j) = \sum_X P(X_i, Y_j) = \sum_X P(X_i)P(Y_j/X_i). \tag{2.54}$$

Similarly:

$$P(X_i) = \sum_Y P(X_i, Y_j) = \sum_Y P(Y_j)P(X_i/Y_j). \tag{2.55}$$

2.10.3 Mutual Information

In this subsection, we elaborate further on the ramifications of Shannon's information theory [133, 215–217, 225, 226]. Over nonideal channels impairments are introduced, and the received information might be different from the transmitted information. Here we quantify the amount of information that can be inferred from the received symbols over noisy channels. In the spirit of Shannon's fundamental work [133] and Carlson's classic reference [221], let us continue our discourse with the definition of mutual information. We have already used the notation $P(X_i)$ to denote the probability that the source symbol X_i was transmitted and $P(Y_j)$ to denote the probability that the symbol Y_j was received. The joint probability that X_i was transmitted and Y_j was received had been quantified by $P(X_i, Y_j)$, and $P(X_i/Y_j)$ indicated the conditional probability that X_i was transmitted, given that Y_j was received, while $P(Y_j/X_i)$ was used for the conditional probability that Y_j was received given that X_i was transmitted.

In the case of $i = j$, the conditional probabilities $P(Y_j/X_j)j = 1 \cdots q$ represent the error-free transmission probabilities of the source symbols $j = 1 \cdots q$. For example, in

Figure 2.18 the probabilities $P(Y_0/X_0)$ and $P(Y_1/X_1)$ are the probabilities of the error-free reception of a transmitted X_0 and X_1 source symbol, respectively. The probabilities $P(Y_j/X_i)j \neq i$, on the other hand, give the individual error probabilities, which are characteristic of error events that corrupted a transmitted symbol X_i to a received symbol Y_j. The corresponding error probabilities in Figure 2.18 are $P(Y_0/X_1)$ and $P(Y_1/X_0)$.

Let us define the *mutual information* of X_i and Y_j as

$$I(X_i, Y_j) = \log_2 \frac{P(X_i/Y_j)}{P(X_i)} = \log_2 \frac{P(X_i, Y_j)}{P(X_i) \cdot P(Y_j)} = \log_2 \frac{P(Y_j/X_i)}{P(Y_j)} \text{ bits,} \qquad (2.56)$$

which quantifies the amount of information conveyed when X_i is transmitted and Y_j is received. Over a perfect, noiseless channel, each received symbol Y_j uniquely identifies a transmitted symbol X_i with a probability of $P(X_i/Y_j) = 1$. Substituting this probability in Equation (2.56) yields a mutual information of

$$I(X_i, Y_j) = \log_2 \frac{1}{P(X_i)}, \qquad (2.57)$$

which is identical to the self-information of X_i; hence no information is lost over the channel. If the channel is very noisy and the error probability becomes 0.5, then the received symbol Y_j becomes unrelated to the transmitted symbol X_i, since for a binary system upon its reception there is a probability of 0.5 that X_0 was transmitted and the probability of X_1 is also 0.5. Then formally X_i and Y_j are independent, and hence

$$P(X_i/Y_j) = \frac{P(X_i, Y_j)}{P(Y_j)} = \frac{P(X_i) \cdot P(Y_j)}{P(Y_j)} = P(X_i), \qquad (2.58)$$

giving a mutual information of

$$I(X_i, Y_j) = \log_2 \frac{P(X_i)}{P(X_i)} = \log_2 1 = 0, \qquad (2.59)$$

implying that no information is conveyed via the channel. Practical communications channels perform between these extreme values and are usually characterized by the *average mutual information*, defined as

$$I(X, Y) = \sum_{x,y} P(X_i, Y_j) \cdot I(X_i, Y_j)$$

$$= \sum_{x,y} P(X_i, Y_j) \cdot \log_2 \frac{P(X_i/Y_j)}{P(X_i)} \text{ [bits/symbol].} \qquad (2.60)$$

Clearly, the average mutual information in Equation (2.60) is computed by weighting each component $I(X_i, Y_j)$ by its probability of occurrence $P(X_i, Y_j)$ and summing these contributions for all combinations of X_i and Y_j. The average mutual information $I(X, Y)$ defined above gives the average amount of source information acquired per received symbol, as distinguished from that per source symbol, which was given by the entropy $H(X)$. Let us now consolidate these definitions by working through the following numerical example.

2.10.4 Mutual Information Example

Using the same numeric values as in our introductory example of the binary symmetric channel in Section 2.10.1, and exploiting Bayes's rule as given in Equation (2.53), we have

$$P(X_i/Y_j) = \frac{P(X_i, Y_j)}{P(Y_j)}.$$

The following probabilities can be derived, which will be used at a later stage in order to determine the mutual information:

$$P(X_1/Y_1) = P(1/1) = \frac{P(1, 1)}{P_1} = \frac{0.686}{0.692} \approx 0.9913$$

and

$$P(X_0/Y_0) = P(0/0) = \frac{P(0, 0)}{P_0} = \frac{0.294}{0.3080} \approx 0.9545,$$

where $P_1 = 0.692$ and $P_0 = 0.3080$ represent the total probabilities of receiving a one and a zero respectively, which are the unions of the respective events of error-free and erroneous receptions yielding the specific logical value concerned. The mutual information from Equation (2.56) is computed as

$$I(X_1, Y_1) = \log_2 \frac{P(X_1/Y_1)}{P(X_1)}$$

$$\approx \log_2 \frac{0.9913}{0.7} \approx 0.502 \text{ bits} \tag{2.61}$$

$$I(X_0, Y_0) \approx \log_2 \frac{0.9545}{0.3} \approx 1.67 \text{ bits.} \tag{2.62}$$

These figures must be contrasted with the amount of source information conveyed by the source symbols X_0 and X_1:

$$I(0) = \log_2 \frac{1}{0.3} \approx \log_2 3.33 \approx 1.737 \text{ bits/symbol} \tag{2.63}$$

and

$$I(1) = \log_2 \frac{1}{0.7} \approx \log_2 1.43 \approx 0.5146 \text{ bits/symbol.} \tag{2.64}$$

The amount of information 'lost' in the noisy channel is given by the difference between the amount of information carried by the source symbols and the mutual information gained upon inferring a particular symbol at the noisy channel's output. Hence the lost information can be computed from Equations (2.61), (2.62), (2.63) and (2.64), yielding $(1.737 - 1.67) \approx 0.067$ bits and $(0.5146 - 0.502) \approx 0.013$ bits respectively. These values may not seem catastrophic, but in relative terms they are quite substantial and their values rapidly escalate as the channel error probability is increased. For the sake of completeness and for future use, let us compute the remaining mutual information terms, namely $I(X_0, Y_1)$ and $I(X_1, Y_0)$, which necessitate the computation of

$$P(X_0/Y_1) = \frac{P(X_0, Y_1)}{P(Y_1)}$$

$$P(0/1) = \frac{P(0, 1)}{P_1} = \frac{0.3 \cdot 0.02}{0.692} \approx 0.00867$$

$$P(X_1/Y_0) = \frac{P(X_1, Y_0)}{P(Y_0)}$$

$$P(1/0) = \frac{P(1, 0)}{P_0} = \frac{0.7 \cdot 0.02}{0.308} \approx 0.04545$$

$$I(X_0, Y_1) = \log_2 \frac{P(X_0/Y_1)}{P(X_0)} \approx \log_2 \frac{0.00867}{0.3} \approx -5.11 \text{ bits} \qquad (2.65)$$

$$I(X_1, Y_0) = \log_2 \frac{P(X_1/Y_0)}{P(X_1)} \approx \log_2 \frac{0.04545}{0.7} \approx -3.945 \text{ bits}, \qquad (2.66)$$

where the negative sign reflects the amount of 'misinformation' as regards, for example, X_0 upon receiving Y_1. In this example we informally introduced the definition of mutual information. Let us now set out to formally exploit the benefits of our deeper insight into the effects of the noisy channel.

2.10.5   Information Loss via Imperfect Channels

Upon rewriting the definition of mutual information in Equation (2.56), we have

$$I(X_i, Y_j) = \log_2 \frac{P(X_i/Y_j)}{P(X_i)}$$

$$= \log_2 \frac{1}{P(X_i)} - \log_2 \frac{1}{P(X_i/Y_j)}$$

$$= I(X_i) - I(X_i/Y_j). \qquad (2.67)$$

Following Shannon's [133,215–217,225,226] and Ferenczy's [223] approach and rearranging Equation (2.67) yields

$$\underbrace{I(X_i)}_{\text{source info}} - \underbrace{I(X_i, Y_j)}_{\text{info conveyed to receiver}} = \underbrace{I(X_i/Y_j)}_{\text{info loss}}. \qquad (2.68)$$

Briefly returning to Figure 2.18 assists the interpretation of $P(X_i/Y_j)$ as the probability, or the degree of certainty, or in fact uncertainty, associated with the event that X_i was transmitted given that Y_j was received, which justifies the above definition of the information loss. It is useful to observe from this figure that, as was stated before, $P(Y_j/X_i)$ represents the probability of erroneous or error-free reception. Explicitly, if $j = i$, then $P(Y_j/X_i) = P(Y_j/X_j)$ is the probability of error-free reception, while if $j \neq i$, then $P(Y_j/X_i)$ is the probability of erroneous reception.

With the probability $P(Y_j/X_i)$ of erroneous reception in mind, we can actually associate an error information term with it:

$$I(Y_j/X_i) = \log_2 \frac{1}{P(Y_j/X_i)}. \tag{2.69}$$

Let us now concentrate on the expression for average mutual information in Equation (2.60), and expand it as follows:

$$I(X,Y) = \sum_{X,Y} P(X_i, Y_j) \cdot \log_2 \frac{1}{P(X_i)} - \sum_{X,Y} P(X_i, Y_j) \log_2 \frac{1}{P(X_i/Y_j)}. \tag{2.70}$$

Considering the first term on the right-hand side (rhs) of the above equation and invoking Equation (2.55), we have

$$\sum_X \left[\sum_Y P(X_i, Y_j) \right] \log_2 \frac{1}{P(X_i)} = \sum_X P(X_i) \log_2 \frac{1}{P(X_i)} = H(X). \tag{2.71}$$

Then rearranging Equation (2.70) gives

$$H(X) - I(X,Y) = \sum_{X,Y} P(X_i, Y_j) \log_2 \frac{1}{P(X_i/Y_j)}, \tag{2.72}$$

where $H(X)$ is the average source information per symbol and $I(X,Y)$ is the average conveyed information per received symbol.

Consequently, the rhs term must be the average information per symbol lost in the noisy channel. As we have seen in Equations (2.67) and (2.68), the information loss is given by

$$I(X_i/Y_j) = \log_2 \frac{1}{P(X_i/Y_j)}. \tag{2.73}$$

The average information loss $H(X/Y)$, which Shannon [216] terms *equivocation*, is computed as the weighted sum of these components:

$$H(X/Y) = \sum_X \sum_Y P(X_i, Y_j) \cdot \log_2 \frac{1}{P(X_i/Y_j)}. \tag{2.74}$$

Following Shannon, this definition allows us to express Equation (2.72) as:

$$\underbrace{H(X)}_{\text{(average source info/symbol)}} - \underbrace{I(X,Y)}_{\text{(average conveyed info/symbol)}} = \underbrace{H(X/Y)}_{\text{(average lost info/symbol)}}. \tag{2.75}$$

2.10.6   Error Entropy via Imperfect Channels

Similarly to our previous approach and using the probability $P(Y_j/X_i)$ of erroneous reception associated with the information term of

$$I(Y_j/X_i) = \log_2 \frac{1}{P(Y_j/X_i)}, \tag{2.76}$$

we can define the average 'error information' or error entropy. Hence the above error information terms (Equation (2.76)) are weighted using the probabilities $P(X_i, Y_j)$ and averaged for all X and Y values, defining the *error entropy*:

$$H(Y/X) = \sum_X \sum_Y P(X_i, Y_j) \log_2 \frac{1}{P(Y_j/X_i)}. \tag{2.77}$$

Using Bayes's rule from Equation (2.52), we have

$$P(X_i/Y_j) \cdot P(Y_j) = P(Y_j/X_i) \cdot P(X_i)$$
$$\frac{P(X_i/Y_j)}{P(X_i)} = \frac{P(Y_j/X_i)}{P(Y_j)}. \tag{2.78}$$

Following from this, for the average mutual information in Equation (2.56) we have

$$I(X, Y) = I(Y, X), \tag{2.79}$$

which, after interchanging X and Y in Equation (2.75), gives

$$\underbrace{H(Y)}_{\text{destination entropy}} - \underbrace{I(Y, X)}_{\text{conveyed information}} = \underbrace{H(Y/X)}_{\text{error entropy}}. \tag{2.80}$$

Subtracting the conveyed information from the destination entropy gives the error entropy, which is nonzero if the destination entropy and the conveyed information are unequal due to channel errors. Let us now proceed following Ferenczy's approach [223] and summarize the most important definitions for future reference in Table 2.7 before we attempt to augment their physical interpretations using the forthcoming numerical example.

Example: *Using the BSC model of Figure 2.16, as an extension of the worked examples of Subsections 2.10.1 and 2.10.4 and following Ferenczy's interpretation [223] of Shannon's elaborations [133, 215–217, 225, 226], let us compute the following range of system characteristics:*

(a) *The **joint information**, as distinct from the mutual information introduced earlier, for all possible channel input/output combinations.*

(b) *The entropy, i.e., the average information of both the source and the sink.*

(c) *The average joint information $H(X, Y)$.*

(d) *The average mutual information per symbol conveyed.*

(e) *The average information loss and average error entropy.*

With reference to Figure 2.16 and to our introductory example from Section 2.10.1 we commence by computing further parameters of the BSC. Recall that the source information

Table 2.7: Summary of Definitions © Ferenczy [223]

Quantity	Definition
Source information	$I(X_i) = -\log_2 P(X_i)$
Received information	$I(Y_j) = -\log_2 P(Y_j)$
Joint information	$I_{X_i,Y_j} = -\log_2 P(X_i, Y_j)$
Mutual information	$I(X_i, Y_j) = \log_2 \frac{P(X_i/Y_j)}{P(X_i)}$
Average mutual information	$I(X, Y) = \sum_X \sum_Y P(X_iY_j) \log_2 \frac{P(X_i/Y_j)}{P(X_i)}$
Source entropy	$H(X) = -\sum_X P(X_i) \cdot \log_2 P(X_i)$
Destination entropy	$H(Y) = -\sum_Y P(Y_j) \log_2 P(Y_j)$
Equivocation	$H(X/Y) = -\sum_X \sum_Y P(X_i, Y_j) \log_2 P(X_i/Y_j)$
Error entropy	$H(Y/X) = -\sum_X \sum_Y P(X_iY_j) \log_2 P(Y_j/X_i)$

was

$$I(X_0) = \log_2 \frac{1}{0.3} \approx 3.322 \log_{10} 3.333 \approx 1.737 \text{ bits}$$

$$I(X_1) = \log_2 \frac{1}{0.7} \approx 0.515 \text{ bits.}$$

The probability of receiving a logical zero was 0.308 and that of a logical one was 0.692, whether zero or one was transmitted. Hence, the information inferred upon the reception of zero and one, respectively, is given by

$$I(Y_0) = \log_2 \frac{1}{0.308} \approx 3.322 \log_{10} 3.247 \approx 1.699 \text{ bits}$$

$$I(Y_1) = \log_2 \frac{1}{0.692} \approx 0.531 \text{ bits.}$$

Observe that because of the reduced probability of receiving a logical one from $0.7 \rightarrow 0.692$ as a consequence of channel-induced corruption, the probability of receiving a logical zero is increased from $0.3 \rightarrow 0.308$. This is expected to increase the average destination entropy, since the entropy maximum of unity is achieved, when the symbols are equiprobable. We note, however, that this does not give more information about the source symbols, which must be maximized in an efficient communications system. In our example, the information conveyed increases for the reduced probability logical one from 0.515 bits $\rightarrow 0.531$ bits and decreases for the increased probability zero from 1.737 bits $\rightarrow 1.699$ bits. Furthermore, the average information conveyed is reduced, since the reduction from 1.737 to 1.699 bits is more than the increment from 0.515 to 0.531. In the extreme case of an error probability of 0.5 we would have $P(0) = P(1) = 0.5$, and $I(1) = I(0) = 1$ bit, associated with receiving equiprobable random bits, which again would have a maximal destination entropy, but minimal information concerning the source symbols transmitted. Following the above interesting introductory calculations, let us now turn our attention to the computation of the joint information.

(a) The *joint information*, as distinct from the mutual information introduced earlier in Equation (2.56), of all possible channel input/output combinations is computed from Figure 2.16 as follows

$$I_{X_i, Y_j} = -\log_2 P(X_i, Y_j)$$

$$I_{00} = -\log_2(0.3 \cdot 0.98) \approx -3.322 \cdot \log_{10} 0.294 \approx 1.766 \text{ bits}$$

$$I_{01} = -\log_2(0.3 \cdot 0.02) \approx 7.381 \text{ bits}$$

$$I_{10} = -\log_2(0.7 \cdot 0.02) \approx 6.159 \text{ bits}$$

$$I_{11} = -\log_2(0.7 \cdot 0.98) \approx 0.544 \text{ bits}. \tag{2.81}$$

These information terms can be individually interpreted formally as the information carried by the simultaneous occurrence of the given symbol combinations. For example, as it accrues from their computation, I_{00} and I_{11} correspond to the favorable event of error-free reception of a transmitted zero and one respectively, which hence were simply computed by formally evaluating the information terms. By the same token, in the computation of I_{01} and I_{10}, the corresponding source probabilities were weighted by the channel error probability rather than the error-free transmission probability, leading to the corresponding information terms. The latter terms, namely I_{01} and I_{10}, represent low-probability, high-information events due to the low channel error probability of 0.02.

Lastly, a perfect channel with zero error probability would render the probability of the error events zero, which in turn would assign infinite information contents to the corresponding terms of I_{01} and I_{10}, while I_{00} and I_{11} would be identical to the self-information of the zero and one symbols. Then, if under zero error probability we evaluate the effect of the individual symbol probabilities on the remaining joint information terms, the less frequently a symbol is emitted by the source the higher its associated joint information term becomes, and vice versa, which is seen by comparing I_{00} and I_{11}. Their difference can be equalized by assuming an identical probability of 0.5 for both, which would yield $I_{00} = I_{11} = 1$ bit. The unweighted average of I_{00} and I_{11} would then be lower than in the case of the previously used probabilities of 0.3 and 0.7 respectively, since the maximum average would be associated with the case of zero and one, where the associated $\log_2$ terms would be 0 and $-\infty$, respectively. The appropriately weighted average joint information terms are evaluated under paragraph (c) during our later calculations. Let us now move on to evaluate the average information of the source and the sink.

(b) Calculating the *entropy*, that is, the average information for both the source and the sink, is quite straightforward and ensues as follows

$$H(X) = \sum_{i=1}^{2} P(X_i) \cdot \log_2 \frac{1}{P(X_i)}$$

$$\approx 0.3 \cdot \log_2 3.333 + 0.7 \cdot \log_2 1.429$$

$$\approx 0.5211 + 0.3605$$

$$\approx 0.8816 \text{ bits/symbol}. \tag{2.82}$$

For the computation of the sink's entropy, we invoke Equations (2.49) and (2.50), yielding

$$H(Y) = 0.308 \cdot \log_2 \frac{1}{0.308} + 0.692 \log_2 \frac{1}{0.692}$$
$$\approx 0.5233 + 0.3676$$
$$\approx 0.8909 \text{ bits/symbol.} \tag{2.83}$$

Again, the destination entropy $H(Y)$ is higher than the source entropy $H(X)$ due to the more random reception caused by channel errors, approaching $H(Y) = 1$ bit/symbol for a channel bit error rate of 0.5. Note, however, that unfortunately this increased destination entropy does not convey more information about the source itself.

(c) Computing the *average joint information* $H(X, Y)$ gives

$$H(X, Y) = -\sum_{i=1}^{2} \sum_{j=1}^{2} P(X_i, Y_j) \log_2 P(X_i, Y_j)$$
$$= -\sum_{i=1}^{2} \sum_{j=1}^{2} P(X_i, Y_j) I_{X_i, Y_j}. \tag{2.84}$$

Upon substituting the I_{X_i, Y_j} values calculated in Equation (2.81) into Equation (2.84), we have

$$H(X, Y) \approx 0.3 \cdot 0.98 \cdot 1.766 + 0.3 \cdot 0.02 \cdot 7.381$$
$$+ 0.7 \cdot 0.02 \cdot 6.159 + 0.7 \cdot 0.98 \cdot 0.544$$
$$\approx 0.519 + 0.044 + 0.086 + 0.373$$
$$\approx 1.022 \text{ bits/symbol-combination.}$$

In order to interpret $H(X, Y)$, let us again scrutinize the definition given in Equation (2.84), which weights the joint information terms of Equation (2.81) by their probability of occurrence. We have argued before that the joint information terms corresponding to erroneous events are high due to the low error probability of 0.02. Observe, therefore, that these high-information symbol combinations are weighted by their low probability of occurrence, causing $H(X, Y)$ to become relatively low. It is also instructive to consider the above terms in Equation (2.84) for the extreme cases of zero and 0.5 error probabilities and for different source emission probabilities, which are left for the reader to explore. Here we proceed considering the average conveyed mutual information per symbol.

(d) The *average conveyed mutual information per symbol* was defined in Equation (2.60) in order to quantify the average source information acquired per received symbol, and is repeated here for convenience:

$$I(X, Y) = \sum_{X} \sum_{Y} P(X_i, Y_j) \log_2 \frac{P(X_i/Y_j)}{P(X_i)}$$
$$= \sum_{X} \sum_{Y} P(X_i, Y_j) \cdot I(X_i, Y_j).$$

Using the individual mutual information terms from Equations (2.61)–(2.66) in Section 2.10.4, we get the average mutual information representing the average amount of source information acquired from the received symbols:

$$
\begin{aligned}
I(X,Y) &\approx 0.3 \cdot 0.98 \cdot 1.67 + 0.3 \cdot 0.02 \cdot (-5.11) \\
&\quad + 0.7 \cdot 0.02 \cdot (-3.945) + 0.7 \cdot 0.98 \cdot 0.502 \\
&\approx 0.491 - 0.03066 - 0.05523 + 0.3444 \\
&\approx 0.7495 \text{ bits/symbol}.
\end{aligned}
\tag{2.85}
$$

In order to interpret the concept of mutual information, in Section 2.10.4 we noted that the amount of information 'lost' owing to channel errors was given by the difference between the amount of information carried by the source symbols and the mutual information gained upon inferring a particular symbol at the noisy channel's output. These were given in Equations (2.61)–(2.64), yielding $(1.737 - 1.67) \approx 0.067$ bits and $(0.5146 - 0.502) \approx 0.013$ bits, for the transmission of a zero and a one respectively. We also noted that the negative sign of the terms corresponding to the error events reflected the amount of misinformation as regards, for example, X_0 upon receiving Y_1. Over a perfect channel, the cross-coupling transitions of Figure 2.16 are eliminated, since the associated error probabilities are zero, and hence there is no information loss over the channel. Consequently, the error-free mutual information terms become identical to the self-information of the source symbols, since exactly the same amount of information can be inferred upon reception of a symbol as is carried by its appearance at the output of the source.

It is also instructive to study the effect of different error probabilities and source symbol probabilities in the average mutual information definition of Equation (2.84) in order to acquire a better understanding of its physical interpretation and quantitative power as regards the system's performance. It is interesting to note, for example, that assuming an error probability of zero will therefore result in average mutual information, which is identical to the source and destination entropy computed above under paragraph (b). It is also plausible that $I(X,Y)$ will be higher than the previously computed 0.7495 bits/symbol if the symbol probabilities are closer to 0.5, or in general in the case of q-ary sources closer to $1/q$. As expected, for a binary symbol probability of 0.5 and error probability of zero, we have $I(X,Y) = 1$ bit/symbol.

(e) Lastly, let us determine the *average information loss* and *average error entropy*, which were defined in Equations (2.74) and (2.80) and are repeated here for convenience. Again, we will be using some of the previously computed probabilities from Sections 2.10.1 and 2.10.4, beginning with computation of the average information loss of Equation (2.74):

$$
\begin{aligned}
H(X/Y) &= -\sum_X \sum_Y P(X_i, Y_j) \log_2 P(X_i/Y_j) \\
&= -P(X_0, Y_0) \log_2 P(X_0/Y_0) - P(X_0, Y_1) \log_2 P(X_0/Y_1) \\
&\quad - P(X_1, Y_0) \log_2 P(X_1/Y_0) - P(X_1, Y_1) \log_2 P(X_1/Y_1) \\
&= P(0,0) \cdot \log_2 P(0/0) + P(0,1) \cdot \log_2 P(0/1)
\end{aligned}
$$

$$P(1,0) \cdot \log_2 P(1/0) + P(1,1) \cdot \log_2 P(1/1)$$
$$\approx -0.3 \cdot 0.98 \cdot \log_2 0.9545 - 0.3 \cdot 0.02 \cdot \log_2 0.00867$$
$$- 0.7 \cdot 0.02 \cdot \log_2 0.04545 - 0.7 \cdot 0.98 \cdot \log_2 0.9913$$
$$\approx 0.0198 + 0.0411 + 0.0624 + 0.0086$$
$$\approx 0.132 \text{ bits/symbol.}$$

In order to augment the physical interpretation of the above expression for average information loss, let us examine the main contributing factors in it. It is expected to decrease as the error probability decreases. Although it is not straightforward to infer the clear effect of any individual parameter in the equation, experience shows that as the error probability increases the two middle terms corresponding to the error events become more dominant. Again, the reader may find it instructive to alter some of the parameters on a one-by-one basis and study the way each one's influence manifests itself in terms of the overall information loss.

Moving on to the computation of the average error entropy, we find its definition equation repeated below, and on inspecting Figure 2.16 we have

$$H(Y/X) = -\sum_X \sum_Y P(X_i, Y_j) \cdot \log_2 P(Y_j/X_i)$$
$$= -P(X_0, Y_0) \log_2 P(Y_0/X_0) - P(X_0, Y_1) \log_2 P(Y_1/X_0)$$
$$- P(X_1, Y_0) \log_2 P(Y_0/X_1) - P(X_1, Y_1) \log_2 P(Y_1/X_1)$$

$$P(Y_0/X_0) = 0.98$$
$$P(Y_0/X_1) = 0.02$$
$$P(Y_1/X_0) = 0.02$$
$$P(Y_1/X_1) = 0.98$$

$$H(Y/X) = P(0,0) \cdot \log_2 P(0/0) + P(0,1) \cdot \log_2 P(0/1)$$
$$P(1,0) \cdot \log_2 P(1/0) + P(1,1) \cdot \log_2 P(1/1)$$
$$= -0.294 \cdot \log_2 0.98 - 0.014 \cdot \log_2 0.02$$
$$- 0.006 \cdot \log_2 0.02 - 0.686 \cdot \log_2 0.98$$
$$\approx 0.0086 + 0.079 + 0.034 + 0.02$$
$$\approx 0.141 \text{ bits/symbol.}$$

The average error entropy in the above expression is expected to fall as the error probability is reduced and vice versa. Substituting different values into its definition equation further augments its practical interpretation. Using our previous results in this section, we see that the *average loss of information per symbol*, or *equivocation*, denoted by $H(X/Y)$ is given by the difference between the source entropy of Equation (2.82) and the average mutual information of Equation (2.85), yielding:

$$H(X/Y) = H(X) - I(X,Y) \approx 0.8816 - 0.7495 \approx 0.132 \text{ bits/symbol,}$$

which, according to Equation (2.75), is identical to the value of $H(X/Y)$ computed earlier. In harmony with Equation (2.80), the error entropy can also be computed as the difference of the average entropy $H(Y)$ in Equation (2.83) of the received symbols and the mutual information $I(X, Y)$ of Equation (2.85), yielding

$$H(Y) - I(X, Y) \approx 0.8909 - 0.7495 \approx 0.141 \text{ bits/symbol},$$

as seen above for $H(Y/X)$.

Having defined the fundamental parameters summarized in Table 2.7 and used in the information-theoretical characterization of communications systems, let us now embark on the definition of channel capacity. Initially, we consider discrete noiseless channels, leading to a brief discussion of noisy discrete channels, and then we proceed to analog channels before exploring the fundamental message of the Shannon–Hartley law.

2.11   Capacity of Discrete Channels   [216, 223]

Shannon [216] defined the *channel capacity* C of a channel as the maximum achievable information transmission rate at which error-free transmission can be maintained over the channel.

Every practical channel is noisy, but by transmitting at a sufficiently high power the channel error probability p_e can be kept arbitrarily low, providing us with a simple initial channel model for our further elaborations. Following Ferenczy's approach [223], assume that the transmission of symbol X_i requires a time interval of t_i, during which an average of

$$H(X) = \sum_{i=1}^{q} P(X_i) \log_2 \frac{1}{P(X_i)} \frac{\text{bits}}{\text{symbol}} \tag{2.86}$$

information is transmitted, where q is the size of the source alphabet used. This approach assumes that a variable-length coding algorithm, such as the previously described Shannon–Fano or the Huffman coding algorithm, may be used in order to reduce the transmission rate to as low as the source entropy. Then the average time required for the transmission of a source symbol is computed by weighting t_i with the probability of occurrence of symbol $X_i, i = 1 \ldots q$:

$$t_{av} = \sum_{i=1}^{q} P(X_i) t_i \frac{\text{seconds}}{\text{symbol}}. \tag{2.87}$$

Now we can compute the average information transmission rate v by dividing the average information content of a symbol by the average time required for its transmission:

$$v = \frac{H(X)}{t_{av}} \frac{\text{bits}}{\text{second}}. \tag{2.88}$$

The maximum transmission rate v as a function of the symbol probability $P(X_i)$ must be found. This is not always an easy task, but a simple case occurs when the symbol duration is constant; that is, we have $t_i = t_0$ for all symbols. Then the maximum of v is a function of $P(X_i)$ only and we have shown earlier that the entropy $H(X)$ is maximized by equiprobable source symbols, where $P(X_i) = \frac{1}{q}$. Then from Equations (2.86) and (2.87) we have an expression for the channel's maximum capacity:

$$C = v_{\max} = \frac{H(X)}{t_{av}} = \frac{\log_2 q}{t_0} \frac{\text{bits}}{\text{second}}. \tag{2.89}$$

Shannon [216] characterized the capacity of discrete noisy channels using the previously defined mutual information, describing the amount of average conveyed information, given by

$$I(X, Y) = H(Y) - H(Y/X), \tag{2.90}$$

where $H(Y)$ is the average amount of information per symbol at the channel's output, while $H(Y/X)$ is the error entropy. Here a unity symbol-rate was assumed for the sake of simplicity. Hence, useful information is only transmitted via the channel if $H(Y) > H(Y/X)$. Via a channel with $p_e = 0.5$, where communication breaks down we have $H(Y) = H(Y/X)$, and the information conveyed becomes $I(X, Y) = 0$. The amount of information conveyed is maximum if the error entropy $H(Y/X) = 0$. Therefore, Shannon [216] defined the noisy channel's capacity as the maximum value of the conveyed information $I(X, Y)$:

$$C = I(X, Y)_{MAX} = [H(Y) - H(Y/X)]_{MAX}, \tag{2.91}$$

where the maximization of Equation (2.91) is achieved by maximizing the first term and minimizing the second term.

In general, the maximization of Equation (2.91) is an arduous task, but for the BSC seen in Figure 2.19 it becomes fairly simple. Let us consider this simple case and assume that the source probabilities of zero and one are $P(0) = P(1) = 0.5$ and the error probability is p_e. The entropy at the destination is computed as

$$H(Y) = -\frac{1}{2} \log_2 \frac{1}{2} - \frac{1}{2} \log_2 \frac{1}{2} = 1 \text{ bit/symbol},$$

while the error entropy is given by

$$H(Y/X) = -\sum_X \sum_Y P(X_i, Y_j) \cdot \log_2 P(Y_j/X_i). \tag{2.92}$$

In order to be able to compute the capacity of the BSC as a function of the channel's error probability, let us substitute the required joint probabilities

$$P(0, 0) = P(0)(1 - p_e)$$
$$P(0, 1) = P(0)p_e$$
$$P(1, 0) = P(1)p_e$$
$$P(1, 1) = P(1)(1 - p_e), \tag{2.93}$$

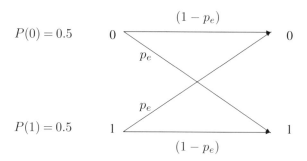

$P(0) = 0.5$

$P(1) = 0.5$

Figure 2.19: BSC model.

and the conditional probabilities

$$P(0/0) = (1 - p_e)$$
$$P(0/1) = p_e$$
$$P(1/0) = p_e$$
$$P(1/1) = (1 - p_e) \tag{2.94}$$

into Equation (2.92), yielding

$$
\begin{aligned}
H(Y/X) &= -[P(0)(1 - p_e) \cdot \log_2(1 - p_e) + P(0) \cdot p_e \log_2 p_e \\
&\quad + P(1) \cdot p_e \log_2 p_e + P(1)(1 - p_e) \log_2(1 - p_e)] \\
&= -[P(0) + P(1)](1 - p_e) \log_2(1 - p_e) \\
&\quad + [P(0) + P(1)]p_e \log_2 p_e \\
&= -(1 - p_e) \cdot \log_2(1 - p_e) - p_e \cdot \log_2 p_e. \tag{2.95}
\end{aligned}
$$

Finally, upon substituting $H(Y)$ and $H(Y/X)$ from above into Equation (2.91), the BSC's channel capacity becomes

$$C = 1 + (1 - p_e) \log_2(1 - p_e) + p_e \log_2 p_e. \tag{2.96}$$

Following Ferenczy's [223] interpretation of Shannon's lessons [133, 215–217, 225, 226], the graphic representation of the BSC's capacity is depicted in Figure 2.20 using various p_e error probabilities.

Observe, for example, that for $p_e = 10^{-2}$ the channel capacity is $C \approx 0.9$ bits/symbol (that is, close to its maximum of $C = 1$ bit/symbol), but for higher p_e values it rapidly decays, falling to $C = 0.5$ bits/symbol around $p_e = 10^{-1}$. If $p_e = 0.5$, we have $C = 0$ bits/symbol; since no useful information transmission takes place, the channel delivers random bits. Notice also that if $P(0) \neq P(1) \neq 0.5$, then $H(Y) < 1$ bit/symbol and hence $C < C_{\max} = 1$ bit/symbol, even if $p_e = 0$.

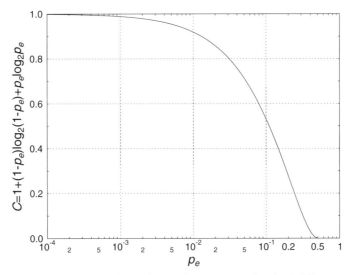

Figure 2.20: Channel capacity versus p_e for the BSC.

2.12   Shannon's Channel Coding Theorem [220, 228]

In the previous section we derived a simple expression for the capacity of the noisy BSC in Equation (2.96), which was depicted in Figure 2.20 as a function of the channel's error probability p_e. In this section we focus on Shannon's *channel coding theorem*, which states that as long as the information transmission rate does not exceed the channel's capacity, the bit error rate can be kept arbitrarily low [225, 226]. In the context of the BSC channel capacity curve of Figure 2.20, this theorem implies that noise over the channel does not preclude the reliable transmission of information; it only limits the rate at which transmission can take place. Implicitly, this theorem prophesies the existence of an appropriate error correction code, which adds redundancy to the original information symbols. This reduces the system's useful information throughput but simultaneously allows error correction coding. Instead of providing a rigorous proof of this theorem, we will follow the approach suggested by Abramson [220], which was also used by Hey and Allen [228] in their compilation of Feynman's lectures, and make it plausible.

The theorem is stated more formally as follows. Let us assume that a message of K useful information symbols is transmitted by assigning it to an N-symbol so-called block code, where the symbols are binary and the error probability is p_e. Then, according to Shannon, upon reducing the *coding rate* $R = \frac{K}{N}$ beyond every limit, the error probability obeys the following relationship:

$$R = \frac{K}{N} \leq C = 1 + (1 - p_e) \log_2(1 - p_e) + p_e \cdot \log_2 p_e. \qquad (2.97)$$

As Figure 2.20 shows, upon increasing the bit error rate p_e, the channel capacity reduces gradually toward zero, which forces the channel coding rate $R = \frac{K}{N}$ to zero in the limit.

This inequality therefore implies that an arbitrarily low BER is possible only when the coding rate R tends to zero, which assumes an infinite-length block code and an infinite coding delay. By scrutinizing Figure 2.20 we can infer that, for example, for a BER of 10^{-1} an approximately $R = \frac{K}{N} \approx \frac{1}{2}$ so-called half-rate code is required in order to achieve asymptotically perfect communications, while for a BER of 10^{-2} an approximately $R \approx 0.9$ code is required.

Shannon's channel coding theorem does not specify how to create error correction codes that can achieve this predicted performance; it merely states their existence. Hence the error correction coding community has endeavored over the years to create such good codes, but until 1993 it had only limited success. Then in that year Berrou *et al.* [229] invented the family of iteratively decoded turbo codes, which are capable of approaching the Shannonian predictions within a fraction of a dB.

Returning to the channel coding theorem, Hey and Feynman [228] offered a witty approach to deepening the physical interpretation of this theorem, which we briefly highlight below. Assuming that the block-coded sequences are long, in each block on the average there are $t = p_e \cdot N$ number of errors. In general, t number of errors can be allocated over the block of N positions in

$$C_N^t = \binom{N}{t} = \frac{N!}{t!(N-t)!}$$

different ways, which are associated with the same number of error patterns. The number of additional parity bits added during the coding process is $P = (N - K)$, which must be sufficiently high for identifying all the C_N^t number of error patterns, in order to allow for inverting (i.e., correcting) the corrupted bits in the required positions. Hence we have [228]

$$\frac{N!}{t!(N-t)!} \leq 2^{(N-K)}. \tag{2.98}$$

Upon exploiting the Stirling formula of

$$N! \approx \sqrt{2\pi N} \cdot \left(\frac{N}{e}\right)^N = \sqrt{2\pi} \cdot \sqrt{N} \cdot N^N \cdot e^{-N}$$

and taking the logarithm of both sides, we have

$$\log_e N! \approx \log_e \sqrt{2\pi} + \frac{1}{2} \log_e N + N \log_e N - N.$$

Furthermore, when N is large, the first and second terms are diminishingly small in comparison with the last two terms. Thus we have

$$\log_e N! \approx N \log_e N - N.$$

Then, after taking the logarithm, the factorial expression on the left-hand side (L) of Equation (2.98) can be written as:

$$L \approx [N \log_e N - N] - [t \log_e t - t] - [(N - t) \log_e (N - t) - (N - t)].$$

Now taking into account that $t \approx p_e \cdot N$, we have [228]

$$
\begin{aligned}
L &\approx [N \log_e N - N] - [p_e N \log_e (p_e N) - p_e N] \\
&\quad - [(N - p_e N) \log_e (N - p_e N) - (N - p_e N)] \\
&\approx [N \log_e N - N] - [p_e N \log_e p_e + p_e N \log_e N - p_e N] \\
&\quad - [N \log_e (N(1 - p_e)) - p_e N \log_e (N(1 - p_e)) - (N - p_e N)] \\
&\approx [N \log_e N - N] - [p_e N \log_e p_e + p_e N \log_e N - p_e N] \\
&\quad - [N \log_e N + N \log_e (1 - p_e) - p_e N \log_e N \\
&\quad - p_e N \log_e (1 - p_e) - (N - p_e N)] \\
&\approx N[\log_e N - 1 - p_e \log_e p_e - p_e \log_e N + p_e \\
&\quad - \log_e N - \log_e (1 - p_e) + p_e \log_e N + p_e \log_e (1 - p_e) + 1 - p_e] \\
&\approx N[-p_e \log_e p_e - \log_e (1 - p_e) + p_e \log_e (1 - p_e)] \\
&\approx N[-p_e \log_e p_e - (1 - p_e) \log_e (1 - p_e)].
\end{aligned}
$$

If we consider that $\log_e a = \log_2 a \cdot \log_e 2$, then we can convert the $\log_e$ terms to $\log_2$ as follows [228]:

$$
L \approx N \log_e 2 [-p_e \log_2 p_e - (1 - p_e) \log_2 (1 - p_e)].
$$

Finally, upon equating this term with the logarithm of the right-hand-side expression of Equation (2.98), we arrive at:

$$
N \log_e 2 [-p_e \log_2 p_e - (1 - p_e) \log_2 (1 - p_e)] \le (N - K) \log_e 2,
$$

which can be simplified to

$$
-p_e \log_2 p_e - (1 - p_e) \log_2 (1 - p_e) \le 1 - \frac{K}{N}
$$

or to a form identical to Equation (2.97):

$$
\frac{K}{N} \le 1 + (1 - p_e) \log_2 (1 - p_e) + p_e \log_2 p_e.
$$

2.13   Capacity of Continuous Channels [217, 223]

During our previous discussions, it was assumed that the source emitted discrete messages with certain finite probabilities, which would be exemplified by an eight-bit analog-to-digital converter emitting one of 256 discrete values with a certain probability. However, after digital source encoding and channel encoding according to the basic schematic of Figure 2.1 the modulator typically converts the digital messages to a finite set of bandlimited analog waveforms, which are chosen for maximum 'transmission convenience.' In this context, transmission convenience can imply a range of issues, depending on the communications channel. Two typical constraints are found with channels that are predominantly either power limited or bandwidth limited, although in many practical scenarios both of these constraints become important. Because of their limited solar power supply, satellite channels tend to be more severely power limited than bandlimited, while typically the reverse situation is experienced in mobile radio systems.

The third part of Shannon's pioneering paper [217] considers many of these issues. Thus, in what follows we define the measure of information for continuous signals, introduce a concept for the continuous channel capacity, and reveal the relationships among channel bandwidth, channel capacity and channel signal-to-noise ratio, as stated by the Shannon–Hartley theorem. Finally, the ideal communications system transpiring from Shannon's pioneering work is characterized, before we conclude with a brief discussion of the ramifications of wireless channels as regards the applicability of Shannon's results.

Let us now assume that the channel's analog input signal $x(t)$ is bandlimited and hence that it is fully characterized by its Nyquist samples and by its Probability Density Function (PDF) $p(x)$. The analogy of this continuous PDF and that of a discrete source are characterized by $P(X_i) \approx p(X_i)\Delta X$, which reflects the practical way of experimentally determining the histogram of a bandlimited analog signal by observing the relative frequency of events, when its amplitude resides in a ΔX wide amplitude bin, centered around X_i. As an analogy to the discrete average information or entropy expression of

$$H(X) = - \sum_i P(X_i) \cdot \log_2 P(X_i), \tag{2.99}$$

Shannon [217] introduced the *entropy of analog sources*, as also noted and exploited, for example, by Ferenczy [223], as

$$H(x) = - \int_{-\infty}^{\infty} p(x) \log_2 p(x) \, dx. \tag{2.100}$$

For our previously used discrete sources, we have shown that the source entropy is maximized for equiprobable messages. The question that arises is whether this is also true for continuous PDFs. Shannon [217] derived the maximum of the analog signal's entropy under the constraints of

$$\int_{-\infty}^{\infty} p(x) \, dx = 1 \tag{2.101}$$

$$\sigma_x^2 = \int_{-\infty}^{\infty} x^2 \cdot p(x) \, dx = \text{constant} \tag{2.102}$$

based on the calculus of variations. He showed that the entropy of a signal $x(t)$ having a constant variance of σ_x^2 is maximum, if $x(t)$ has a Gaussian distribution given by

$$p(x) = \frac{1}{\sqrt{2\pi}\sigma} e^{-(x^2/2\sigma^2)}. \tag{2.103}$$

Then the maximum of the entropy can be derived upon substituting this PDF into the expression of the entropy. Let us first take the natural logarithm of both sides of the PDF and convert it to a base two logarithm by taking into account that $\log_e a = \log_2 a \cdot \log_e 2$, in order to be able to use it in the entropy's $\log_2$ expression. Then the PDF of Equation (2.103) can be written as

$$-\log_2 p(x) = +\log_2 \sqrt{2\pi}\sigma + (x^2/2\sigma^2) \cdot \frac{1}{\log_e 2}, \qquad (2.104)$$

and upon exploiting the fact that $\log_e 2 = 1/\log_2 e$, the entropy is expressed according to Shannon [217] and Ferenczy [223] as

$$H_{\max}(x) = -\int p(x) \cdot \log_2 p(x)\, dx$$

$$= \int p(x) \cdot \log_2 \sqrt{2\pi}\sigma\, dx + \int p(x) \frac{x^2 \cdot \log_2 e}{2\sigma^2}\, dx$$

$$= \log_2 \sqrt{2\pi}\sigma \int p(x)\, dx + \frac{\log_2 e}{2\sigma^2} \underbrace{\int x^2 p(x)\, dx}_{\sigma^2}$$

$$= \log_2 \sqrt{2\pi}\sigma + \frac{\sigma^2}{2\sigma^2} \log_2 e$$

$$= \log_2 \sqrt{2\pi}\sigma + \frac{\log_2 e}{2}$$

$$= \log_2 \sqrt{2\pi}\sigma + \frac{1}{2} \log_2 e$$

$$= \log_2 \sqrt{2\pi e}\sigma. \qquad (2.105)$$

Since the maximum of the entropy is proportional to the logarithm of the signal's average power $S_x = \sigma_x^2$, upon quadrupling the signal's power the entropy is increased by one bit because the range of uncertainty as regards where the signal samples can reside is expanded.

We are now ready to formulate the channel capacity versus channel bandwidth and versus channel SNR relationship of analog channels. Let us assume white, additive, signal-independent noise with a power of N via the channel. Then the received (signal + noise) power is given by

$$\sigma_y^2 = S + N. \qquad (2.106)$$

By the same argument, the channel's output entropy is maximum if its output signal $y(t)$ has a Gaussian PDF, and its value is computed from Equation (2.105) as

$$H_{max}(y) = \frac{1}{2} \log_2(2\pi e \sigma_y^2) = \frac{1}{2} \log_2 2\pi e(S + N). \qquad (2.107)$$

We proceed by taking into account the channel impairments, reducing the amount of information conveyed by the amount of the error entropy $H(y/x)$, giving

$$I(x, y) = H(y) - H(y/x), \qquad (2.108)$$

where again the noise is assumed to be Gaussian, and hence

$$H(y/x) = \frac{1}{2} \log_2(2\pi e N). \qquad (2.109)$$

Upon substituting Equation (2.107) and Equation (2.109) in Equation (2.108), we have

$$I(x, y) = \frac{1}{2} \log_2 \left(\frac{2\pi e (S + N)}{2\pi e N} \right)$$

$$= \frac{1}{2} \log_2 \left(1 + \frac{S}{N} \right), \tag{2.110}$$

where, again, both the channel's output signal and the noise are assumed to have Gaussian distribution.

The analog channel's capacity is then calculated upon multiplying the information conveyed per source sample by the Nyquist sampling rate of $f_s = 2 \cdot f_B$, yielding [225]

$$C = f_B \cdot \log_2 \left(1 + \frac{S}{N} \right) \frac{\text{bits}}{\text{second}}. \tag{2.111}$$

Equation (2.111) is the well-known *Shannon–Hartley law*,[1] establishing the relationship among the channel capacity C, channel bandwidth f_B and channel Signal-to-Noise Ratio (SNR).

Before analyzing the consequences of the Shannon–Hartley law following Shannon's deliberations [225], we make it plausible from a simple practical point of view. As we have seen, the Root Mean Squared (RMS) value of the noise is $\sqrt{N}$, and that of the signal plus noise at the channel's output is $\sqrt{S + N}$. The receiver has to decide from the noisy channel's output signal what signal has been input to the channel, although this has been corrupted by an additive Gaussian noise sample. Over an ideal noiseless channel, the receiver would be able to identify what signal sample was input to the receiver. However, over noisy channels it is of no practical benefit to identify the corrupted received message exactly. It is more beneficial to quantify a discretized version of it using a set of decision threshold values, where the resolution is dependent on how corrupted the samples are. In order to quantify this SNR-dependent receiver dynamic range resolution, let us consider the following argument.

Having very densely spaced receiver detection levels would often yield noise-induced decision errors, while a decision-level spacing of $\sqrt{N}$ according to the RMS noise-amplitude intuitively seems a good compromise between high information resolution and low decision error rate. Then assuming a transmitted sample, which resides at the center of a $\sqrt{N}$ wide decision interval, noise samples larger than $\sqrt{N}/2$ will carry samples across the adjacent decision boundaries. According to this spacing, the number of receiver reconstruction levels is given by

$$q = \frac{\sqrt{S + N}}{\sqrt{N}} = \left(1 + \frac{S}{N} \right)^{\frac{1}{2}}, \tag{2.112}$$

which creates a scenario similar to the transmission of equiprobable q-ary discrete symbols via a discrete noisy channel, each conveying $\log_2 q$ amount of information at the Nyquist

---

[1]The present authors believe that, although the loose definition of capacity is due to Hartley, the underlying relationship is entirely due to Shannon.

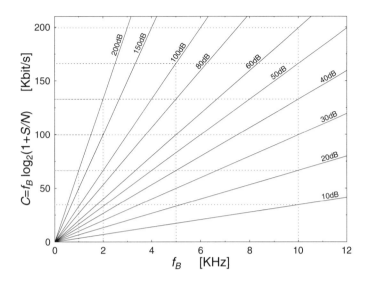

Figure 2.21: Graphical representation of the Shannon–Hartley law: © Ferenczy [223].

sampling rate of $f_s = 2 \cdot f_B$. Therefore, the channel capacity becomes [225]

$$C = 2 \cdot f_B \cdot \log_2 q = f_B \cdot \log_2\left(1 + \frac{S}{N}\right), \tag{2.113}$$

as seen earlier in Equation (2.111).

2.13.1 Practical Evaluation of the Shannon–Hartley Law

The Shannon–Hartley law of Equations (2.111) and (2.113) reveals the fundamental relationship of the SNR, bandwidth and channel capacity. This relationship can be further studied, following Ferenczy's interpretation [223], with reference to Figure 2.21.

Observe from the figure that a constant channel capacity can be maintained, even when the bandwidth is reduced, if a sufficiently high SNR can be guaranteed. For example, from Figure 2.21 we infer that at $f_B = 10$ KHz and SNR $= 30$ dB the channel capacity is as high as $C = 100$ kbits/second. Surprisingly, $C \approx 100$ kbits/second can be achieved even for $f_B = 5$ KHz, if SNR $= 60$ dB is guaranteed.

Figure 2.22 provides an alternative way of viewing the Shannon–Hartley law, where the SNR is plotted as a function of f_B, parameterized with the channel capacity C. It is important to notice how dramatically the SNR must be increased in order to maintain a constant channel capacity C, as the bandwidth f_B is reduced below $0.1 \cdot C$, where C and f_B are expressed in kbits/second and Hz respectively. This is due to the $\log_2(1 + \mathrm{SNR})$ function in Equation (2.111), where a logarithmically increasing SNR value is necessitated to compensate for the linear reduction in terms of f_B.

From our previous discourse, the relationship between the relative channel capacity C/f_B, obtained from Equation (2.113), and the channel SNR now becomes plausible.

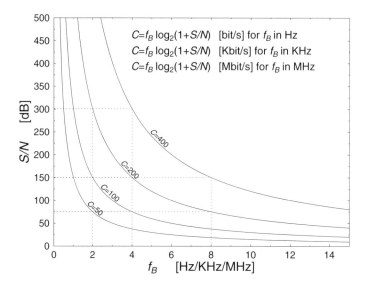

Figure 2.22: SNR versus f_B relations according to the Shannon–Hartley law: © Ferenczy [223].

Table 2.8: Relative channel capacity versus SNR

SNR		C/f_B
Ratio	dB	bps/Hz
1	0	1
3	4.8	2
7	8.5	3
15	11.8	4
31	14.9	5
63	18.0	6
127	21.0	7

This relationship is quantified in Table 2.8 and Figure 2.23 for convenience. Notice that due to the logarithmic SNR scale expressed in dB, the $C/f_B[\frac{\text{bps}}{\text{Hz}}]$ curve becomes near-linear, allowing a near-linearly proportional relative channel capacity improvement upon increasing the channel SNR. A very important consequence of this relationship is that if the channel SNR is sufficiently high to support communications using a high number of modulation levels, the channel is not exploited to its full capacity upon using C/f_B values lower than are afforded by the prevailing SNR. Proposing various techniques in order to exploit this philosophy was the motivation of [230].

The capacity C of a noiseless channel with $SNR = \infty$ is $C = \infty$, although noiseless channels do not exist. In contrast, the capacity of an ideal system with $f_B = \infty$ is finite [221, 224]. Assuming Additive White Gaussian Noise (AWGN) with a double-sided PSD of $\eta/2$,

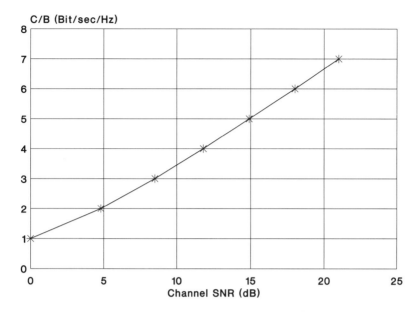

Figure 2.23: Relative channel capacity (C/f_B) versus SNR (dB).

we have $N = \frac{\eta}{2} \cdot 2 \cdot f_B = \eta \cdot f_B$, and applying the Shannon–Hartley law gives [221]

$$C = f_B \cdot \log_2 \left(1 + \frac{S}{\eta f_B} \right)$$
$$= \left(\frac{S}{\eta} \right) \left(\frac{\eta f_B}{S} \right) \log_2 \left(1 + \frac{S}{\eta f_B} \right)$$
$$= \left(\frac{S}{\eta} \right) \log_2 \left(1 + \frac{S}{\eta f_B} \right)^{\frac{\eta f_B}{S}}. \qquad (2.114)$$

Our aim is now to determine $C_\infty = \lim_{f_B \to \infty} C$. Upon exploiting the fact that

$$\lim_{x \to 0} (1 + x)^{\frac{1}{x}} = e, \qquad (2.115)$$

where $x = S/(\eta \cdot f_B)$, we have

$$C_\infty = \lim_{f_B \to \infty} C = \frac{S}{\eta} \log_2 e = 1.45 \cdot \left(\frac{S}{\eta} \right), \qquad (2.116)$$

which is the capacity of the channel with $f_B = \infty$. The practically achievable transmission rate R is typically less than the channel capacity C, although complex turbo-coded modems [229] can approach its value. For example, for a telephone channel with a signal-to-noise ratio of $S/N = 10^3 = 30$ dB and a bandwidth of $B = 3.4$ kHz from Equation (2.113), we have $C = 3.4 \cdot \log_2(1 + 10^3)$ kbit/s $\approx 3.4 \cdot 10 = 34$ kbit/s, which is fairly close to the

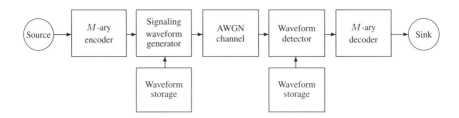

Figure 2.24: Shannon's ideal communications system for AWGN channels.

rate of the V.34 CCITT standard 28.8 kbit/s telephone-channel modem that was recently standardized.

In this chapter we have been concerned with various individual aspects of Shannon's information theory [133, 215–217, 225, 226]. Drawing nearer to concluding our discourse on the foundations of information theory, let us now outline in broad terms the main ramifications of Shannon's work [133, 215–217].

2.13.2 Shannon's Ideal Communications System for Gaussian Channels

The ideal Shannonian communications system shown in Figure 2.24 has the following characteristics. The system's information-carrying capacity is given by the information rate $C = f_B \log_2(1 + S/N)$, while as regards its error rate we have $p_e \to 0$. The transmitted and received signals are bandlimited Gaussian random variables, which facilitate communicating at the highest possible rate over the channel.

Information from the source is observed for T seconds, where T is the symbol duration, and encoded as equiprobable M-ary symbols with a rate of $R = \frac{\log_2 M}{T}$. Accordingly, the signaling waveform generator of Figure 2.24 assigns a bandlimited AWGN representation having a maximum frequency of f_B from the set of $M = 2^{RT}$ possible waveforms to the source message, uniquely representing the signal $x(t)$ to be transmitted for a duration of T. The noisy received signal $y(t) = x(t) + n(t)$ is compared with all $M = 2^{RT}$ prestored waveforms at the receiver, and the most 'similar' is chosen to identify the most likely transmitted source message. The observation intervals at both the encoder and the decoder amount to T, yielding an overall coding delay of $2T$. Signaling at a rate equal to the channel capacity is only possible if the source signal's observation interval is infinitely long, that is, $T \to \infty$.

Before concluding this chapter we offer a brief discussion of the system-architectural ramifications of transmitting over wireless channels rather than over AWGN channels.

2.14 Shannon's Message for Wireless Channels

In wireless communications over power- and bandlimited channels it is always of prime concern to maintain an optimum compromise in terms of the contradictory requirements of low bit rate, high robustness against channel errors, low delay and low complexity. The minimum bit rate at which distortionless communication is possible is determined by the entropy of the speech source message. Note, however, that in practical terms the source rate corresponding to the entropy is only asymptotically achievable as the encoding memory length or delay tends to infinity. Any further compression is associated with information loss or coding distortion. Note that the optimum source encoder generates a perfectly uncorrelated source-coded stream, where all the source redundancy has been removed; therefore, the encoded symbols are independent, and each one has the same significance. Having the same significance implies that the corruption of any of the source-encoded symbols results in identical source signal distortion over imperfect channels.

Under these conditions, according to Shannon's pioneering work [133], which was expanded, for example, by Hagenauer [231] and Viterbi [232], the best protection against transmission errors is achieved if source and channel coding are treated as separate entities. When using a block code of length N channel coded symbols in order to encode K source symbols with a coding rate of $R = K/N$, the symbol error rate can be rendered arbitrarily low, if N tends to infinity and the coding rate to zero. This condition also implies an infinite coding delay. Based on the above considerations and on the assumption of Additive White Gaussian Noise (AWGN) channels, source and channel coding have historically been separately optimized.

Mobile radio channels are subjected to multipath propagation and so constitute a more hostile transmission medium than do AWGN channels, typically exhibiting path loss, log-normal slow fading and Rayleigh fast fading. Furthermore, if the signaling rate used is higher than the channel's coherence bandwidth, over which no spectral-domain linear distortion is experienced, then additional impairments are inflicted by dispersion, which is associated with frequency-domain linear distortions. Under these circumstances the channel's error distribution versus time becomes bursty, and an infinite-memory symbol interleaver is required in Figure 2.1 in order to disperse the bursty errors and hence to render the error distribution random Gaussian-like, such as over AWGN channels. For mobile channels, many of the above mentioned, asymptotically valid ramifications of Shannon's theorems have limited applicability.

A range of practical limitations must be observed when designing mobile radio speech or video links. Although it is often possible to further reduce the prevailing typical bit rate of state-of-art speech or video codecs, in practical terms this is possible only after a concomitant increase in the implementational complexity and encoding delay. A good example of these limitations is the half-rate GSM speech codec, which was required to approximately halve the encoding rate of the 13 kbps full-rate codec, while maintaining less than quadrupled complexity, similar robustness against channel errors, and less than doubled encoding delay. Naturally, the increased algorithmic complexity is typically associated with higher power consumption, while the reduced number of bits used to represent a certain speech segment intuitively implies that each bit will have an increased relative significance. Accordingly, their corruption may inflict increasingly objectionable speech degradations, unless special attention is devoted to this problem.

In a somewhat simplistic approach, one could argue that because of the reduced source rate we could accommodate an increased number of parity symbols using a more powerful, implementationally more complex and lower rate channel codec, while maintaining the same transmission bandwidth. However, the complexity, quality and robustness trade-off of such a scheme may not always be attractive.

A more intelligent approach is required to design better speech or video transceivers for mobile radio channels [231]. Such an intelligent transceiver is portrayed in Figure 2.1. Perfect source encoders operating close to the information-theoretical limits of Shannon's predictions can only be designed for stationary source signals, a condition not satisfied by most source signals. Further previously mentioned limitations are the encoding complexity and delay. As a consequence of these limitations the source-coded stream will inherently contain residual redundancy, and the correlated source symbols will exhibit unequal error sensitivity, requiring unequal error protection. Following Hagenauer [231], we will refer to the additional knowledge as regards the different importance or vulnerability of various speech-coded bits as Source Significance Information (SSI). Furthermore, Hagenauer termed the confidence associated with the channel decoder's decisions 'Decoder Reliability Information' (DRI). These additional links between the source and channel codecs are also indicated in Figure 2.1. A variety of such techniques have successfully been used in robust source-matched source and channel coding.

The role of the interleaver and de-interleaver seen in Figure 2.1 is to rearrange the channel coded bits before transmission. The mobile radio channel typically inflicts bursts of errors during deep channel fades, which often overload the channel decoder's error correction capability in certain speech or video segments. In contrast other segments are not benefiting from the channel codec at all, because they may have been transmitted between fades and hence are error free even without channel coding. This problem can be circumvented by dispersing the bursts of errors more randomly between fades so that the channel codec is always faced with an 'average-quality' channel, rather than the bimodal faded/nonfaded condition. In other words, channel codecs are most efficient if the channel errors are near-uniformly dispersed over consecutive received segments.

In its simplest manifestation, an interleaver is a memory matrix filled with channel-coded symbols on a row-by-row basis, which are then passed on to the modulator on a column-by-column basis. If the transmitted sequence is corrupted by a burst of errors, the de-interleaver maps the received symbols back to their original positions, thereby dispersing the bursty channel errors. An infinite memory channel interleaver is required in order to perfectly randomize the bursty errors and therefore to transform the Rayleigh-fading channel's error statistics to that of a AWGN channel, for which Shannon's information-theoretical predictions apply. Since in interactive video or speech communications the tolerable delay is strictly limited, the interleaver's memory length and efficiency are also limited.

A specific deficiency of these rectangular interleavers is that in case of a constant vehicular speed the Rayleigh-fading mobile channel typically produces periodic fades and error bursts at traveled distances of $\lambda/2$, where λ is the carrier's wavelength, which may be mapped by the rectangular interleaver to another set of periodic bursts of errors. Hence a range of random interleaving algorithms have been proposed in the literature.

Returning to Figure 2.1, the Soft-Decision Information (SDI) or Channel State Information (CSI) link provides a measure of confidence with regard to the likelihood that a specific symbol was transmitted. Then the channel decoder often uses this information in order to invoke Maximum Likelihood Sequence Estimation (MLSE) based on the Viterbi algorithm and thereby improve the system's performance with respect to conventional hard-decision decoding. Following this rudimentary review of Shannon's information theory, let us now turn our attention to the characterization of wireless communications channels.

2.15   Summary and Conclusions

An overview of Shannonian information theory has been given, in order to establish a firm basis for our further discussions throughout the book. Initially we focussed our attention on the basic Shannonian information transmission scheme and highlighted the differences between Shannon's theory, valid for ideal source and channel codecs as well as for Gaussian channels, and its ramifications for Rayleigh channels. We also argued that practical finite-delay source codecs cannot operate at transmission rates as low as the entropy of the source. However, these codecs do not have to operate losslessly, since perceptually unobjectionable distortions can be tolerated. This allows us to reduce the associated bit rate.

Since wireless channels exhibit bursty error statistics, the error bursts can only be randomized with the aid of infinite-length channel interleavers, which are not amenable to real-time communications. Although with the advent of high-delay turbo channel codecs it is possible to operate near the Shannonian performance limits over Gaussian channels, over bursty and dispersive channels different information-theoretical channel capacity limits apply.

We considered the entropy of information sources both with and without memory and highlighted a number of algorithms, such as the Shannon–Fano, the Huffman and run-length coding algorithms, designed for the efficient encoding of sources exhibiting memory. This was followed by considering the transmission of information over noise-contaminated channels, leading to Shannon's channel coding theorem. Our discussions continued by considering the capacity of communications channels in the context of the Shannon–Hartley law. The chapter was concluded by considering the ramifications of Shannon's messages for wireless channels.

In the rest of the book familiarity with the basic concepts outlined in this chapter is assumed.

Part I

Regular Concatenated Codes and Their Design

List of Symbols in Part I

General notation

- The superscript $*$ is used to indicate complex conjugation. Therefore, a^* represents the complex conjugate of the variable a.
- The superscript T is used to denote the matrix transpose operation. Therefore, $\mathbf{a}^T$ represents the transpose of the matrix $\mathbf{a}$.
- The superscript H is used to indicate the complex conjugate transpose operation. Therefore, $\mathbf{a}^H$ represents the complex conjugate transpose of the matrix $\mathbf{a}$.
- The notation $\hat{x}$ represents the estimate of x.

Special symbols

A	A random discrete source.
$\mathscr{A}$	A source alphabet.
$\mathcal{A}$	The area under an EXIT curve.
$\bar{\mathcal{A}}$	The area under the inverse curve of an EXIT function.
$\mathbf{A_c}$	The *a priori* Logarithmic Likelihood Ratios (LLRs) of the encoded bits.
$\mathbf{A_u}$	The *a priori* LLRs of the information bits.
A_k	The forward recursion metric of a MAP decoder in the log domain.
a_i	The legitimate symbols of a source alphabet.
B_k	The backward recursion metric of a MAP decoder in the log domain.
$b_{\min}$	The overall minimum block distance of a VLC.
C	A VLC.
$\mathbf{C}$	The encoded output sequence of a trellis encoder.
C_k	The encoded output symbol of a trellis encoder.
C_{UI}	The channel capacity when using uniformly distributed input.
$c_{\min}$	The minimum convergence distance of the VLC.
$\mathbf{c}_j$	A VLC codeword.
$c(e)$	The output symbol associated with an edge e in a trellis.
d_{b_k}	The minimum block distance for the length L_k of a VLC.
d_c	The convergence distance between two different-length codewords.
d_d	The divergence distance between two different-length codewords.

d_f	The free distance of a VLC.
$d_{\min}$	The minimum divergence distance of a VLC.
d_h	The Hamming distance between two identical-length VLC codewords.
$\mathbf{E_u}$	The *a posteriori* LLRs of the information bits.
$\mathbf{E_c}$	The *a posteriori* LLRs of the encoded bits.
E_b	Bit energy.
E_b/N_0	Ratio of the bit energy to the noise power spectral density.
E_s	Symbol energy.
E_s/N_0	Ratio of the symbol energy to the noise power spectral density.
e	The edge of a trellis.
F_N	The extended code of order N of a VLC.
g_k	The generator polynomials of a convolutional code.
H	The entropy of a source.
h_k	The coefficients of a channel impulse response.
I_i	The information carried by a source symbol.
$\bar{L}$	The average codeword length of a VLC.
$\ell(\mathbf{c}_j)$	The length of a VLC codeword $\mathbf{c}_j$ in terms of code symbols, or bits for binary codes.
N_0	Single-sided power spectral density of white noise.
$\mathbf{n}$	The length distribution vector of a VLC.
P_i	The state probabilities of a discrete Markov model.
p_i	The source symbol probabilities.
p_{ij}	The state transition probabilities of a discrete Markov model.
R	Coding rate.
R_I	The source information rate measured in bits/second.
R_s	The source emission rate measured in symbols/second.
$\mathcal{S}$	The set of valid states of a trellis encoder.
$s^S(e)$	The starting state of an edge e in a trellis.
$s^E(e)$	The ending state of an edge e in a trellis.
$\mathbf{U}$	The input symbol sequence of a trellis encoder.
U_k	The input symbol of a trellis encoder.
$u(e)$	The input symbol associated with an edge e in a trellis.
X_i	The states of a discrete Markov model.
α	The forward recursion metric of a MAP decoder.
β	The backward recursion metric of a MAP decoder.
γ	The branch transition metric of a MAP decoder.
ρ	The correlation coefficient of a first-order Markov model.
η	The efficiency of a VLC.
ν	The code memory.
Γ_k	The branch transition metric of a MAP decoder in the log domain.
Π	Interleaver.
Π^{-1}	Deinterleaver.

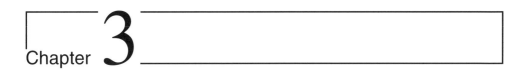

Chapter **3**

Sources and Source Codes

3.1 Introduction

This chapter, and indeed the rest of the book, relies on familiarity with the basic concepts outlined in Chapter 2, where a rudimentary introduction to VLCs, such as entropy codes and Huffman codes, and their benefits, was provided. This chapter primarily investigates the design of VLCs constructed for compressing discrete sources, as well as the choice of soft decoding methods for VLCs. First, in Section 3.2 appropriate source models are introduced for representing both discrete memoryless sources and discrete sources having memory. In order to map discrete source symbols to binary digits, source coding is invoked. This may employ various source codes, such as Huffman codes [9] and RVLCs [12] as well as VLEC codes [30]. Unlike channel codes such as convolutional codes, VLCs constitute nonlinear codes; hence the superposition of legitimate codewords does not necessarily yield a legitimate codeword. The design methods of VLCs are generally heuristic. As with channel codes, the coding rate of VLCs is also an important design criterion in the context of source codes, although some redundancy may be intentionally imposed in order to generate reversible VLCs, where decoding may be commenced from both ends of the codeword, or to incorporate error-correction capability. A novel code construction algorithm is proposed in Section 3.3, which typically constructs more efficient source codes in comparison with the state-of-the-art methods found in the literature.

The soft decoding of VLCs is presented in Section 3.4. In contrast with traditional hard decoding methods, processing binary ones and zeros, the family of soft decoding methods is capable of processing soft input bits, i.e. hard bits associated with reliability information, which are optimal in the sense of maximizing the *a posteriori* probability. Finally, the performance of soft decoding methods is quantified in Section 3.4.3, when communicating over AWGN channels.

3.2 Source Models

Practical information sources can be broadly classified into the families of analog and digital sources. The output of an analog source is a continuous function of time. With the aid

Near-Capacity Variable-Length Coding   Lajos Hanzo, Robert G. Maunder, Jin Wang and Lie-Liang Yang
© 2011 John Wiley & Sons, Ltd

of sampling and quantization an analog source signal can be transformed into its digital representation.

The output of a digital source is constituted by a finite set of ordered, discrete symbols, often referred to as an alphabet. Digital sources are usually described by a range of characteristics, such as the source alphabet, the symbol rate, the individual symbol probabilities and the probabilistic interdependence of symbols.

At the current state of the art, most communication systems are digital, which has a host of advantages, such as their ease of implementation using Very-Large-Scale Integration (VLSI) technology and their superior performance in hostile propagation environments. Digital transmissions invoke a whole range of diverse signal processing techniques, which would otherwise be impossible or difficult to implement in analog forms. In this monograph we mainly consider discrete-time, discrete-valued sources, such as the output of a quantizer.

3.2.1 Quantization

Source symbols with values having unequal probabilities of occurrence may be generated during the scalar quantization [164, 165] of real-valued source samples, for example. In scalar quantization, each real-valued source sample y is quantized separately. More specifically, an approximation $\hat{y}$ of each source sample y is provided by one of N_s number of real-valued quantization levels $\{\hat{y}_i, i = 1, 2, \ldots, N_s\}$. In each case, the selected quantization level $\hat{y}_i$ is that particular one that has the smallest Euclidean distance from the source sample, according to

$$\hat{y} = \underset{\{\hat{y}_i, i=1,2,\ldots,N_s\}}{\mathrm{argmin}} \ (y - \hat{y}_i)^2. \tag{3.1}$$

This selection is indicated using a source symbol having the corresponding value from the alphabet $\mathscr{A} = \{a_1, a_2, \ldots, a_i, \ldots, a_{N_s}\}$. During inverse quantization, each reconstructed source sample $\hat{y}$ approximates the corresponding source sample y using the quantization level $\hat{y}_i$ that is indicated by the source symbol A. Owing to this approximation, quantization noise is imposed upon the reconstructed source samples. It may be reduced by employing a larger number of quantization levels N_s.

Furthermore, in Lloyd–Max quantization [164, 165] typically the K-means algorithm [233] is employed to select the quantization levels $\{\hat{y}_i, i = 1, 2, \ldots, N_s\}$ in order to minimize the imposed quantization noise that results for a given number of quantization levels N_s. This is illustrated in Figure 3.1 for the $N_s = 4$-level Lloyd–Max quantization of Gaussian distributed source samples having a zero mean and unity variance. Observe that the Probability Density Function (PDF) of Figure 3.1 is divided into $N_s = 4$ sections by the decision boundaries, which are located halfway between each pair of adjacent quantization levels, namely $\hat{y}_{i'}$ and $\hat{y}_{i'+1}$ for $i' = 1, 2, \ldots N_s - 1$. Here, each section of the PDF specifies the range of source sample values that are mapped to the quantization level $\hat{y}_i$ at its center of gravity, resulting in the minimum quantization noise.

The varying probabilities of occurrence $\{p_i, i = 1, 2, \ldots, N_s\}$ of the N_s number of source symbol values generated during quantization may be determined by integrating the corresponding sections of the source sample PDF. Figure 3.1 shows the $N_s = 4$ source symbol value probabilities of occurrence $\{p_i, i = 1, 2, \ldots, N_s\}$ that result for the $N_s = 4$-level Lloyd–Max quantization of Gaussian distributed source samples.

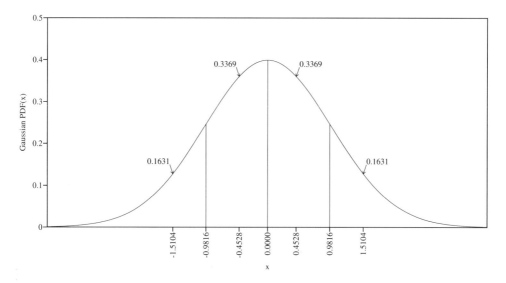

Figure 3.1: Gaussian PDF for unity mean and variance. The x axis is labeled with the $N_s = 4$ Lloyd–Max quantization levels $\{\hat{y}_i, i = 1, 2, \ldots, N_s\}$ and $N_s - 1 = 3$ decision boundaries as provided in [164]. The decision boundaries are employed to decompose the Gaussian PDF into $N_s = 4$ sections. The integral p_i of each PDF section is provided.

In the case where the source samples are correlated, Vector Quantization (VQ) [170] may be employed to generate source symbols with values having unequal probabilities of occurrence. In contrast with scalar quantization, where an individual source sample is mapped to each quantization level, VQ maps a number of correlated source samples to each so-called quantization tile. Quantization tiles that impose a minimum quantization noise may be designed using the Linde–Buzo–Gray (LBG) algorithm [170], which applies the K-means algorithm [233] in multiple dimensions.

3.2.2 Memoryless Sources

Following Shannon's approach [1], let us consider a source A emitting one of N_s possible symbols from the alphabet $\mathscr{A} = \{a_1, a_2, \ldots, a_i, \ldots, a_{N_s}\}$ having symbol probabilities of $\{p_i, i = 1, 2, \ldots, N_s\}$. Here the probability of occurrence for any of the source symbols is assumed to be independent of previously emitted symbols; hence this type of source is referred to as a 'memoryless' source. The information carried by symbol a_i is

$$I_i = -\log_2 p_i \; [\text{bits}].\tag{3.2}$$

The average information per symbol is referred to as the entropy of the source, which is given by

$$H = -\sum_{i=1}^{N_s} p_i \log_2 p_i \; [\text{bits/symbol}],\tag{3.3}$$

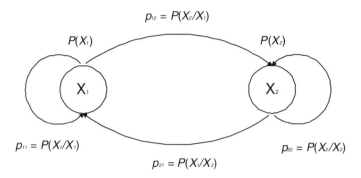

Figure 3.2: A two-state Markov model.

where H is bounded according to

$$0 \leq H \leq \log_2 N_s, \tag{3.4}$$

and we have $H = \log_2 N_s$ when the symbols are equiprobable.

Then the average source information rate can be defined as the product of the information carried by a source symbol, given by the entropy H, and the source emission rate R_s [symbols/second]:

$$R_I = R_s \cdot H \; [\text{bits/second}]. \tag{3.5}$$

3.2.3 Sources with Memory

Most analog source signals, such as speech and video, are inherently correlated, owing to various physical restrictions imposed on the analog sources [2]. Hence all practical analog sources possess some grade of memory, which constitutes a property that is also retained after sampling and quantization. An important feature of sources having memory is that they are predictable to a certain extent; hence, they can usually be more efficiently encoded than unpredictable memoryless sources can.

Predictable sources that exhibit memory can be conveniently modeled by *Markov processes*, as introduced in Section 2.6.1. A source having a memory of one symbol interval directly 'remembers' only the previously emitted source symbol, and, depending on this previous symbol, it emits one of its legitimate symbols with a certain probability, which depends explicitly on the state associated with this previous symbol. A one-symbol-memory model is often referred to as a first-order model.

In general, Markov models are characterized by their state probabilities $P_i = P(X = X_i), i = 1, \ldots, N$, where N is the number of states, as well as by the transition probability $p_{ij} = P(X_j/X_i)$, where p_{ij} explicitly indicates the probability of traversing from state X_i to state X_j. At every state transition, they emit a source symbol. Figure 3.2 shows a two-state first-order Markov model as introduced in Figure 2.9, and repeated here for the reader's convenience.

The entropy of a source having memory is defined as the weighted average of the entropies of the individual symbols emitted from each state, where weighting is carried out by taking into account the probability of occurrence of the individual states, namely P_i. In analytical terms, the symbol entropy for state $X_i, i = 1, \ldots, N$ is given by

$$H_i = -\sum_{j=1}^{N} p_{ij} \cdot \log_2 p_{ij} \quad i = 1, \ldots, N. \tag{3.6}$$

The symbol entropies weighted by their probability of occurrence are summed, which effectively corresponds to averaging over all symbols, hence yielding the source entropy:

$$H = \sum_{i=1}^{N} P_i H_i$$

$$= -\sum_{i=1}^{N} P_i \sum_{j=1}^{N} p_{ij} \log_2 p_{ij}. \tag{3.7}$$

Finally, assuming a source symbol rate of R_s, the average information emission rate R of the source is given by

$$R_I = R_s \cdot H \text{ [bits/second]}. \tag{3.8}$$

A widely used analytical Markov model is the *Autoregressive (AR) model*. A zero-mean random sequence $y(n)$ is referred to as an AR process of order N if it is generated as follows:

$$y(k) = \sum_{n=1}^{N} \rho_n y(k - n) + \varepsilon(k), \quad \forall k, \tag{3.9}$$

where $\varepsilon(k)$ is an uncorrelated zero-mean, random input sequence with variance σ^2. A variety of practical information sources are adequately modeled by the analytically tractable first-order Markov/AR model given by

$$y(k) = \varepsilon(k) + \rho y(k - 1), \tag{3.10}$$

where ρ is the correlation coefficient of the process $y(k)$.

3.3 Source Codes

Source coding is the process by which each source symbol is mapped to a binary codeword; hence the output of an information source is converted to a sequence of binary digits. For a non-uniform Discrete Memoryless Source (DMS) the transmission of a highly probable symbol carries little information, and hence assigning the same number of bits to every symbol in the context of fixed-length encoding is inefficient. Therefore various variable-length coding methods are proposed. The general principle is that the source symbols that occur more frequently are assigned short codewords and those that occur infrequently are assigned long codewords, which generally results in a reduced average codeword length. Hence variable-length coding may also be viewed as a data compression method. Shannon's source coding theorem suggests that by using a *source encoder* before transmission the efficiency of a system transmitting equiprobable source symbols can be arbitrarily approached [1].

Code efficiency can be defined as the ratio of the source information rate and the average output bit rate of the source encoder, which is usually referred to as *coding rate* in the channel coding community [8]. The most efficient VLCs are Huffman codes [9]. Besides code efficiency, there are other VLC design criteria that may be important in the application environment, such as having a high error resilience or a symmetric codeword construction. Thus both RVLCs [12, 20, 21] and VLEC codes [30] are proposed in the literature. We will describe these in the following sections.

Before proceeding further, we now provide some definitions.

3.3.1 Definitions and Notions

Let X be a code alphabet with cardinality[1] q. In what follows, we shall assume that we have $q = 2$, although the results may readily be extended to the general case of $q > 2$. A finite sequence $\mathbf{x} = x_1 x_2 \ldots x_l$ of code symbols (or bits for binary alphabets) is referred to as a codeword over X of length $\ell(\mathbf{x}) = l$, where each code symbol satisfies $x_i \in X$, for all $i = 1, 2, \ldots, l$. A set C of legitimate words is referred to as a code. Let the code C have N_s legitimate codewords $\mathbf{c}_1, \mathbf{c}_2, \ldots, \mathbf{c}_{N_s}$ and let the length of $\mathbf{c}_j$ be given by $l_j = \ell(\mathbf{c}_j)$, $j = 1, 2, \ldots, N_s$. Without loss of generality, assume that we have $l_1 \leq l_2 \leq \cdots \leq l_{N_s}$. Furthermore, let K denote the number of different codeword lengths in the code C and let these lengths be given by $L_1, L_2, \ldots, L_K$, where we have $L_1 < L_2 < \cdots < L_K$, and the corresponding number of codewords having specific lengths is given by $n_1, n_2, \ldots, n_K$, where we have $\sum_{k=1}^{K} n_k = N_s$. Where applicable, we use the VLC $\mathscr{C} = \{00, 11, 010, 101, 0110\}$ as an example.

Definition 1 The *Hamming weight* (or simply *weight*) of a codeword $\mathbf{c}_i$, $w(\mathbf{c}_i)$, is the number of nonzero symbols in $\mathbf{c}_i$. The *Hamming distance* between two equal-length words, $d_h(\mathbf{c}_i, \mathbf{c}_j)$, is the number of positions in which $\mathbf{c}_i$ and $\mathbf{c}_j$ differ. For example, the Hamming distance between the codewords '00' and '11' is $d_h(00, 11) = 2$, while $d_h(010, 101) = 3$.

Definition 2 The *minimum block distance* d_{b_k} at the length of L_k for a VLC C is defined as the minimum Hamming distance between all codewords having the same length L_k, i.e. [29]

$$d_{b_k} = \min\{d_h(\mathbf{c}_i, \mathbf{c}_j) : \mathbf{c}_i, \mathbf{c}_j \in C, i \neq j \quad \text{and} \quad \ell(\mathbf{c}_i) = \ell(\mathbf{c}_j) = L_k\}.$$

There are λ different minimum block distances, one for each different codeword length. However, if for some length L_k there is only one codeword, i.e. we have $n_k = 1$, then in this case the minimum block distance for length L_k is undefined. For example, the minimum block distance at length two for the VLC $\mathscr{C} = \{00, 11, 010, 101, 0110\}$ is $d_{b_2} = 2$, while $d_{b_3} = 3$ and d_{b_4} is undefined.

Definition 3 The *overall minimum block distance*, $b_{\min}$, of a VLC C is defined as the minimum value of d_{b_k} over all possible VLC codeword lengths $k = 1, 2, \ldots, K$, i.e. [29]

$$b_{\min} = \min\{d_{b_k} : k = 1, 2, \ldots, K\}.$$

---

[1]The number of elements in a set.

According to this definition, the overall minimum block distance of the VLC $\mathscr{C} = \{00, 11, 010, 101, 0110\}$ is $b_{\min} = \min(d_{b_2}, d_{b_3}) = 2$.

Definition 4 The Buttigieg–Farrell (B–F) *divergence distance*[2] $d_d(\mathbf{c}_i, \mathbf{c}_j)$ between two different-length codewords $\mathbf{c}_i = c_{i_1} c_{i_2} \cdots c_{i_{l_i}}$ and $\mathbf{c}_j = c_{j_1} c_{j_2} \cdots c_{j_{l_j}}$, where we have $\mathbf{c}_i, \mathbf{c}_j \in C$, $l_i = \ell(\mathbf{c}_i)$ and $l_j = \ell(\mathbf{c}_j)$, is defined as the Hamming distance between the codeword sections spanning the common support of these two codewords when they are left-aligned; i.e., assuming $l_j < l_i$) [29]:

$$d_d(\mathbf{c}_i, \mathbf{c}_j) = d_h(c_{i_1} c_{i_2} \cdots c_{i_{l_j}}, c_{j_1} c_{j_2} \cdots c_{j_{l_j}}).$$

For example, the B–F divergence distance between the codewords '00' and '010' is the Hamming distance between '00' and '01,' i.e. we have a common-support distance of $d_d(00, 010) = 1$. Similarly, we have $d_d(00, 101) = 1$, $d_d(00, 0110) = 1$, $d_d(11, 010) = 1$, $d_d(11, 101) = 1$, $d_d(11, 0110) = 1$, $d_d(010, 0110) = 1$ and $d_d(101, 0110) = 2$.

Definition 5 The *minimum B–F divergence distance* $d_{\min}$ of the VLC C is the minimum value of all the B–F divergence distances between all possible pairs of unequal length codewords of C, i.e. [29]

$$d_{\min} = \min\{d_d(\mathbf{c}_i, \mathbf{c}_j) : \mathbf{c}_i, \mathbf{c}_j \in C, \ell(\mathbf{c}_i) > \ell(\mathbf{c}_j)\}.$$

According to this definition, the minimum divergence distance of the VLC $\mathscr{C} = \{00, 11, 010, 101, 0110\}$ is $d_{\min} = 1$.

Definition 6 The *B–F convergence distance* $d_c(\mathbf{c}_i, \mathbf{c}_j)$ between two different-length codewords $\mathbf{c}_i = c_{i_1} c_{i_2} \cdots c_{i_{l_i}}$ and $\mathbf{c}_j = c_{j_1} c_{j_2} \cdots c_{j_{l_j}}$, where we have $\mathbf{c}_i, \mathbf{c}_j \in C$, $l_i = \ell(\mathbf{c}_i)$ and $l_j = \ell(\mathbf{c}_j)$, is defined as the Hamming distance between the codeword sections spanning the common support of these two codewords when they are right-aligned; i.e., assuming $l_j < l_i$ [29]:

$$d_c(\mathbf{c}_i, \mathbf{c}_j) = d_h(c_{i_{l_i}-l_j+1} c_{i_{l_i}-l_j+2} \cdots c_{i_{l_i}}, c_{j_1} c_{j_2} \cdots c_{j_{l_j}}).$$

For example, the B–F convergence distance between the codewords '00' and '010' is the Hamming distance between '00' and '10', i.e. $d_c(00, 010) = 1$. Similarly, we have $d_c(00, 101) = 1$, $d_c(00, 0110) = 1$, $d_c(11, 010) = 1$, $d_c(11, 101) = 1$, $d_c(11, 0110) = 1$, $d_c(010, 0110) = 1$ and $d_c(101, 0110) = 2$.

Definition 7 The *minimum B–F convergence distance* $c_{\min}$ of the VLC C is the minimum value of all the B–F convergence distances between all possible pairs of unequal-length codewords of C, i.e. [29]

$$c_{\min} = \min\{d_c(\mathbf{c}_i, \mathbf{c}_j) : \mathbf{c}_i, \mathbf{c}_j \in C, \ell(\mathbf{c}_i) > \ell(\mathbf{c}_j)\}.$$

According to this definition, the minimum B–F divergence distance of the VLC $\mathscr{C} = \{00, 11, 010, 101, 0110\}$ is $c_{\min} = 1$.

---

[2]We will also refer to the Buttigieg–Farrell (B–F) divergence distance [29] as the 'left-aligned common-support distance.'

Definition 8 The set of VLC codeword sequences consisting of a total number of N code symbols (or bits for binary codes) is defined as the *extended code* of order N of a VLC. Note that such codeword sequences may consist of an arbitrary number of VLC codewords. Let the set F_N be the extended code of order N of the VLC C and the integer η_i be the number of VLC codewords that constitute an element of the set F_N; then we have [29]

$$F_N = \{\mathbf{f}_i : \mathbf{f}_i = \mathbf{c}_{i_1}\mathbf{c}_{i_2}\cdots\mathbf{c}_{i_{\eta_i}}, \ell(\mathbf{f}_i) = N, \mathbf{c}_{i_j} \in C, \ \forall j = 1, 2, \ldots, \eta_i\}.$$

For example, $F_4 = \{0000, 0011, 1100, 1111, 0110\}$ is the extended code of order four for the VLC $\mathscr{C} = \{00, 11, 010, 101, 0110\}$.

Definition 9 The *free distance*, d_f, of a VLC C is the minimum Hamming distance between any two equal-length VLC codeword sequences, which may consist of an arbitrary number of VLC codewords. With the aid of the definition of extended codes, the free distance of a VLC may be defined as the minimum Hamming distance in the set of all arbitrary extended codes, i.e. [29]

$$d_f = \min\{d_h(\mathbf{f}_i, \mathbf{f}_j) : \mathbf{f}_i, \mathbf{f}_j \in F_n, i \neq j, n = 1, 2, 3, \ldots\},$$

where F_n is the extended code of order n of C. It was found in [29, 30] that d_f is the most important parameter that determines the error-correcting capability of VLEC codes at high channel Signal-to-Noise Ratios (SNRs). The relationship between d_f, $b_{\min}$, $d_{\min}$ and $c_{\min}$ is described by the following theorem.

Theorem 1 *The free distance d_f of a VLC is lower bounded by [29]*

$$d_f = \min(b_{\min}, \ d_{\min} + c_{\min}).$$

For the proof of the theorem please refer to [29]. According to this theorem, the free distance of the VLC $\mathscr{C} = \{00, 11, 010, 101, 0110\}$ is $d_f = \min(2, 1 + 1) = 2$.

Definition 10 Let A be a memoryless data source having an alphabet of $\{a_1, a_2, \ldots, a_{N_s}\}$, where the elements have a probability of occurrence given by $p_i, i = 1, 2, \ldots, N_s$. The source A is encoded using the code C by mapping symbol a_i to codeword c_i for all indices $i = 1, 2, \ldots, N_s$. Then the *average codeword length* $\bar{L}$ is given by

$$\bar{L} = \sum_{i=1}^{N_s} l_i p_i,$$

where $l_i = \ell(\mathbf{c}_i)$ represents the length of the codeword $\mathbf{c}_i$.

Definition 11 A VLC's *efficiency* or *coding rate*, η, is defined as

$$\eta = \frac{H(A)}{\bar{L}},$$

where $H(A)$ is the entropy of source A, while the VLC's *redundancy*, ϱ, is defined as

$$\varrho = 1 - \eta = \frac{\bar{L} - H(A)}{\bar{L}}.$$

3.3.2 Some Properties of Variable-Length Codes

Non-Singular Codes
A code C is said to be non-singular if all the codewords in the code are distinct. Both fixed and variable-length codes must satisfy this property in order to be useful.

Unique Decodability
A code C is said to be uniquely decodable if we can map a string of codewords uniquely and unambiguously back to the original source symbols. It is plausible that all fixed-length codes that are non-singular are also uniquely decodable. However, this is not true in general for variable-length codes.

A necessary and sufficient condition for the existence of a uniquely decodable code is provided by the McMillan inequality [234].

Theorem 2 *A necessary and sufficient condition for the existence of a uniquely decodable code having codeword lengths of $l_1, l_2, \ldots, l_s$ is that we have [234]*

$$\sum_{i=1}^{s} q^{-l_i} \leq 1,$$

where q is the cardinality of the code alphabet.

Instantaneous Decodability
For practical applications, it is required that the decoding delay should be as low as possible in order to facilitate real-time interactive communications. The minimum possible decoding delay is achieved when a codeword becomes immediately decodable as soon as it was completely received. Any delay lower than this would imply that the code is redundant. A code satisfying this property is referred to as an instantaneously decodable code [235]. It is plausible that for a code to satisfy this property none of its codewords may constitute a prefix of another codeword. Therefore, these codes are also known as *prefix codes* [235]. Hence, prefix codes are uniquely decodable and have a decoding delay of at most l_s (maximum codeword length) bit duration.

Interestingly enough, McMillan's inequality, given in [234], is also a necessary and sufficient condition for the existence of a prefix code having codeword lengths of $l_1, l_2, \ldots, l_s$, which was stated as the *Kraft inequality* [236]. Chronologically, the proof of this inequality was provided in [236] before that of McMillan's [234].

3.3.3 Huffman Code

Huffman codes were named after their inventor, D. A. Huffman [9], and because of their high coding efficiency they have been widely adopted in both still image and video coding standards [2]. Huffman codes are optimum in the sense that the average number of binary digits required to represent the source symbols is as low as possible, subject to the constraint

Table 3.1: An example of RVLCs

Symbol	Probability	C_{HUFF}	$C_{\text{sym.RVLC}}$	$C_{\text{asym.RVLC}}$
a	0.33	00	00	00
b	0.30	01	11	01
c	0.18	11	010	10
d	0.10	100	101	111
e	0.09	101	0110	11011
Average code length		2.19	2.46	2.37
Free distance		1	2	1

that the codewords satisfy the above-mentioned prefix condition, which allows the received sequence to be uniquely and instantaneously decodable.

The construction of Huffman codes is fairly straightforward and is described in many textbooks [2, 235], and thus is omitted here. It is noteworthy nonetheless that there are more possible Huffman trees than just one for a given source; i.e., the Huffman code construction is not necessarily unique.

Amongst the diverse variants a meritorious one is the canonical Huffman code, which is defined as the code having the following characteristics [237]:

- Shorter codewords have a numerically higher value than longer codewords, provided that short codewords are filled with zeros to the right so that all codewords have the same length.

- For a specific codeword length, the corresponding numerical values increase monotonically upon the decrease of the source symbol probabilities.

For example, the code $\{11, 10, 01, 001, 000\}$ is the canonical Huffman code for the source given in Table 3.1. It can be seen that we have $(11\underline{0})_2 = (6)_{10} > (10\underline{0})_2 = (4)_{10} > (01\underline{0})_2 = (2)_{10} > (001)_2 = (1)_{10} > (000)_2 = (0)_{10}$, where $\underline{0}$ denotes the extra zeros that are appended to the short codewords. Hence the above-mentioned properties of canonical Huffman codes are satisfied. In general, the canonical Huffman code has the minimum total sum of codeword-lengths $\sum l_i$ and simultaneously the minimum maximum codeword-length $L_{\max}$ in addition to the minimum average codeword-length $\bar{L}$, while allowing fast instantaneous decoding [33]. The construction of this code was described, for example, in [237].

3.3.4 Reversible Variable-Length Codes

Due to the variable-length nature of Huffman codes, they are very sensitive to errors that occur in noisy environments. Even a single-bit error is extremely likely to cause loss of synchronization, which is the term often used to describe the phenomenon of losing track of the beginning and end of VLCs, when a single error renders a VLC codeword unrecognizable, and hence the data received after the bit error position becomes useless until a unique synchronization word is found.

As a potential remedy, RVLCs [12–14, 20–24] have been developed for the sake of mitigating the effects of error propagation. As the terminology suggests, RVLCs often have a symmetric construction, i.e. they have the same bit-pattern both at their beginning and at their end; this facilitates instantaneous decoding in both forward and backward directions, so that we can recover the correct data with a high probability from the received data even when errors occur. To achieve this, a RVLC must satisfy both the so-called prefix and suffix conditions, which implies that no codeword is allowed to be a prefix or a suffix of any other codeword. Thus RVLCs are also known as 'fix-free' codes. The conditions for the existence of RVLCs and their properties were studied in [25–27].

Apart from the above-mentioned symmetric RVLC construction, asymmetric RVLCs also exist. For the same source, a symmetric RVLC generally has a higher average code length and thus becomes less efficient than an asymmetric RVLC. However, a symmetric RVLC has the benefit of requiring a single decoding table for bidirectional decoding, while an asymmetric RVLC employs two. Hence symmetric RVLCs might be preferable for memory-limited applications. Table 3.1 shows an example of both a symmetric and an asymmetric RVLC. In the following sections we discuss the construction of RVLCs for a given source.

3.3.4.1 *Construction of RVLCs*

The construction of RVLCs was studied in [12, 20–24]. Most RVLC constructions commence from a Huffman code designed for the source concerned, and then replace the Huffman codewords by identical-length codewords that satisfy both the above-mentioned prefix and suffix conditions, as necessary. If the number of valid reversible codewords is insufficient, longer reversible codewords have to be generated and assigned, resulting in an increased redundancy. Hence a lower code efficiency is expected. Viewing this process of mapping RVLCs to Huffman codes from a different perspective, it is plausible that at a given codeword length we are likely to run out of legitimate RVLCs owing to their 'fix-free' constraint, typically requiring length extension. In contrast, if there are more RVLC candidate codewords than necessary, different codeword selection mechanisms may be applied. For example, the Maximum Symmetric-Suffix Length (MSSL) scheme [20] may be used for designing symmetric RVLCs, or the Minimum Repetition Gap (MRG) [21] metric and the so-called 'affix index' [23] metric may be invoked for designing asymmetric RVLCs.

The number of codewords of a certain length that satisfy both the prefix and the suffix condition depends on the previously assigned (shorter) RVLC codewords. This dependency was not considered in [12], which hence resulted in reduced-efficiency RVLCs. In [21] it was conjectured that the number of available codewords of any length depends on the above-mentioned MRG metric associated with the shorter RVLC codewords. The codeword assignment policy in [21] was base on MRG and typically produced more efficient codes than the algorithm advocated in [12]. In [23] the formal relationship between the number of available RVLC codewords of any length and the structure of shorter RVLC codewords was established. Based on this relationship, the authors proposed a new codeword selection mechanism, which rendered the 'length-by-length' approach optimal. However, the computational complexity imposed is generally quite high. More details of the MRG-based algorithm [21], which is denoted as Algorithm A.1, and the affix index-based algorithm [23] denoted as Algorithm A.2 can be found in Appendix A.

Both Algorithm A.1 [21] and Algorithm A.2 [23] attempt to match the codeword length distribution of the RVLC to that of the corresponding Huffman code. Since a RVLC has to satisfy the suffix condition in addition to the prefix condition, the desirable codeword length distribution of a Huffman code is usually not matched by that of the RVLC for the same source.

Moreover, both the value of the minimal average codeword length and the optimal codeword length distribution of the RVLC are unknown for a given source. If we find the necessary and sufficient condition for the existence of a RVLC, finding the optimal codeword length distribution would be a constrained minimization problem. More explicitly, given the source symbol probabilities, the problem is to find a codeword length distribution that minimizes the average codeword length, while satisfying the necessary and sufficient condition at the same time. Some results related to this existence condition have been reported in [25–27, 238]; however, even the existence of a general necessary and sufficient condition is uncertain. Therefore, later in this section we propose a heuristic method for finding an improved codeword length distribution.

The proposed algorithm bases its codeword selection procedure on that of the procedures proposed in [21] or [23] and attempts to optimize the codeword length distribution of the resultant RVLC length by length. The construction procedure[3] is given as Algorithm 3.1.

Algorithm 3.1

Step 1: Employ Algorithm A.1 or Algorithm A.2 of Appendix A to construct an initial RVLC, and calculate the length vector, $\mathbf{n}$, which records the numbers of codewords at different lengths, i.e. the histogram of the VLC codewords' lengths.

Step 2: Increase the number of required codewords at length l by one, and decrease the number of required codewords at the maximum codeword length $L_{\max}$ by one, yielding

$$\mathbf{n}(l) := \mathbf{n}(l) + 1, \quad \mathbf{n}(L_{\max}) := \mathbf{n}(L_{\max}) - 1.$$

Use Algorithm A.1 or Algorithm A.2 of Appendix A to construct an RVLC based on the modified length distribution.

Step 3: Recalculate the average codeword length $\bar{L}$; if $\bar{L}$ increased, restore the previous length distribution and then proceed to the next level; otherwise update the length vector $\mathbf{n}$ and go to Step 2.

Step 4: Repeat Steps 2 and 3 until the last level is reached.

3.3.4.2 An Example Construction of a RVLC

In the following, we will demonstrate the construction procedure of our Algorithm 3.1 by using an example. The specific source for which our RVLC example is designed is the English alphabet shown in Table 3.3.

The construction procedure is summarized in Table 3.2. At the first iteration, the length vector is initialized to that of the Huffman code, namely $\mathbf{n} = [0\ 0\ 2\ 7\ 7\ 5\ 1\ 1\ 1\ 2]$, and Tasi's algorithm [21] (Algorithm A.1) is invoked to construct an initial RVLC, resulting in a length distribution of $\mathbf{n} = [0\ 0\ 2\ 7\ 4\ 3\ 5\ 2\ 1\ 2]$ and an average codeword length of $\bar{L} = 4.30675$. Then our length-distribution optimization procedure commences. Firstly, the

---

[3]The proposed algorithm is also applicable to the construction of the symmetric RVLCs considered in [20].

Table 3.2: An example construction of a RVLC designed for the English alphabet

Iteration number	Expected length vector†	Resultant length vector‡	Average length	Control logic§
#1	[0 0 2 7 7 5 1 1 1 2]	[0 0 <u>2</u> 7 <u>4</u> 3 5 2 1 <u>2</u>]	4.30675	start
#2	[0 0 3↑ 7 4 3 5 2 1 1↓]	[0 0 <u>3</u> 6 5 2 2 2 2 3 <u>1</u>]	4.25791	continue
#3	[0 0 4↑ 6 5 2 2 2 2 3 0↓]	[0 0 4 4 4 4 3 2 2 3]	4.26231	restore, next level
#4	[0 0 3 7↑ 5 2 2 2 2 3 0↓]	[0 0 3 <u>6</u> <u>6</u> 2 2 2 2 2 0 <u>1</u>]	4.18785	continue
#5	[0 0 3 7↑ 6 2 2 2 2 0 0↓]	[0 0 3 6 6 2 2 2 2 2 0 0 1]	4.18839	restore, next level
⋮	⋮	⋮	⋮	⋮
#12	[0 0 3 6 6 2 2 2 2 2 1↑ 0↓]	[0 0 3 6 6 2 2 2 2 2 1]	4.18732	end

† The desired RVLC length distribution: the arrows pointing upwards imply that the values at those positions have been increased by one, while arrows pointing downwards imply the opposite.

‡ The actual length distribution of the resultant RVLC after each iteration: the underlines indicate the numbers that will be changed in the next iteration.

§ Decides the next step of the code generator, which depends on the resultant average codeword length.

number of codewords $n(3)$ required at the length of $l = 3$ is increased by one. For the sake of maintaining the total number of codewords, the number of codewords $n(10)$ required at the length of $l = 10$ is decreased by one. Therefore the desired length distribution becomes $n = [0\ 0\ 3\ 7\ 4\ 3\ 5\ 2\ 1\ 1]$, as shown in the second row of Table 3.2. Then Algorithm A.1 is invoked again to construct a RVLC according to the new length distribution, and the resultant average codeword length becomes $\bar{L} = 4.25791$. Since the average codeword length of the new RVLC is lower than that of the old one, the code construction procedure continues to increase the number of codewords required at the length of $l = 3$ during the third iteration. However, the resultant RVLC now has an increased average codeword length of $\bar{L} = 4.26231$; hence the code construction procedure has to restore the length vector to that of the RVLC obtained at the end of the second iteration, which is $n = [0\ 0\ 3\ 6\ 5\ 2\ 2\ 2\ 2\ 3\ 1]$. Then the code construction procedure attempts to increase the number of codewords required at the next level, i.e. at the length of $l = 4$. Hence the desired length distribution at the beginning of the fourth iteration becomes $n = [0\ 0\ 3\ 7\ 5\ 2\ 2\ 2\ 2\ 3\ 0]$. This procedure is repeated until the last level has been reached. As a result, the best RVLC having the minimum average codeword length is found at the end of the 12th iteration.

3.3.4.3 Experimental Results

Table 3.3 compares the RVLCs designed for the English alphabet, constructed using the proposed algorithm (invoking Algorithm A.1 of Appendix A for the initial construction), and the algorithms published in [12, 21, 23]. As a benefit of its improved length distribution, Algorithm 3.1 yields a RVLC construction of the highest efficiency. For example, compared with the algorithms in [12] and [21], the average codeword length is reduced by 2.8% and 1.5% respectively.

Table 3.3: VLCs for the English alphabet

u	$p(u)$	Huffman code	Takishima's [12] RVLC	Tsai's [21] RVLC	Lakovic's [23] RVLC	Proposed RVLC
E	0.14878	111	001	000	000	000
T	0.09351	110	110	111	001	111
A	0.08833	1011	0000	0101	0100	010
O	0.07245	1010	0100	1010	0101	0110
R	0.06872	1001	1000	0010	0110	1001
N	0.06498	1000	1010	1101	1010	1011
H	0.05831	0111	0101	0100	1011	1101
I	0.05644	0110	11100	1011	1100	0011
S	0.05537	0101	01100	0110	1101	1100
D	0.04376	01001	00010	11001	01110	10101
L	0.04124	01000	10010	10011	01111	00100
U	0.02762	00111	01111	01110	10010	01110
P	0.02575	00110	10111	10001	10011	10001
F	0.02455	00101	11111	001100	11110	00101
M	0.02361	00100	111101	011110	11111	10100
C	0.02081	00011	101101	100001	100010	011110
W	0.01868	000101	000111	1001001	100011	100001
G	0.01521	000011	011101	0011100	1000010	0111110
Y	0.01521	000100	100111	1100011	1000011	1000001
B	0.01267	000010	1001101	0111110	1110111	01111110
V	0.01160	000001	01110011	1000001	10000010	10000001
K	0.00867	0000001	00011011	00111100	10000011	011111110
X	0.00146	00000001	000110011	11000011	11100111	100000001
J	0.00080	000000001	0001101011	100101001	100000010	0111111110
Q	0.00080	0000000001	00011010011	0011101001	1000000010	1000000001
Z	0.00053	0000000000	000110100011	1001011100	1000000111	01111111110
Average length		4.15572	4.36068	4.30678	4.25145	4.18732
Max length		10	12	10	10	11

In [24] Lin *et al.* proposed two so-called backtracking-based construction algorithms for asymmetric RVLCs. In comparison with the algorithm of [21], one of the algorithms in [24] is capable of achieving a lower average codeword length at the cost of an equal or higher maximum codeword length, while the other algorithm of [24] is capable of reducing the maximum codeword length at the expense of an increased average codeword length. Table 3.4 compares the achievable performance of the algorithm proposed here with that of various benchmarks, based on the Canterbury Corpus file set (available at http://corpus.canterbury.ac.nz), which was designed specifically for testing new compression algorithms. It shows that our Algorithm 3.1 of Section 3.3.4.1 is capable of reducing both the average codeword length and the maximum codeword length at the same time.

In [22] Lakovic *et al.* proposed a construction algorithm based on the MRG metric for designing RVLCs having a free distance of $d_f = 2$, which significantly outperformed the

Table 3.4: VLCs for Canterbury Corpus[†]

File	Number of codewords	Huffman code		Tsai's [21] RVLC		Lin's [24] RVLC-1		Lin's [24] RVLC-2		Proposed RVLC	
		Average	Max	Average	Max	Average	Max	Average	Max	Average	Max
asyoulik.txt	68	4.84465	15	5.01142	15	5.00954	15	5.13624	11	4.92273	11
alice29.txt	74	4.61244	16	4.80326	17	4.68871	18	4.86762	11	4.70569	11
xargs.1	74	4.92382	12	5.07334	13	5.16087	15	5.27537	11	5.00166	11
grammar.lsp	76	4.66434	12	4.85461	12	4.78581	17	4.96130	11	4.80247	12
plrabn12.txt	81	4.57534	19	4.80659	19	4.64910	17	4.84043	11	4.71036	12
lcet10.txt	84	4.69712	16	4.87868	16	4.74177	17	4.93372	11	4.80024	12
cp.html	86	5.26716	14	5.37113	14	5.77080	16	5.74580	11	5.29342	14
fields.c	90	5.04090	13	5.26987	13	5.20278	13	5.36233	11	5.18341	11
ptt5	159	1.66091	17	1.71814	17	1.70401	15	1.72843	13	1.69580	13
sum	255	5.36504	14	5.49767	13	6.01870	15	5.78572	13	5.47330	13
kennedy.xls	256	3.59337	12	3.89401	13	3.85384	14	3.86296	14	3.79759	14

† Results of Tsai's RVLCs and Lin's RVLCs are from [24].

Table 3.5: Free distance 2 RVLCs for Canterbury Corpus and English alphabet

File	Number of codewords	Lakovic's [22] RVLC ($d_f = 2$)		Proposed RVLC ($d_f = 2$)	
		Average length	Max length	Average length	Max length
asyoulik.txt	68	4.99469	15	4.96686	12
alice29.txt	74	4.75122	16	4.73989	11
xargs.1	74	5.09108	13	5.06624	12
grammar.lsp	76	4.79387	12	4.79118	11
plrabn12.txt	81	4.67692	19	4.65932	12
lcet10.txt	84	4.79175	16	4.78348	12
cp.html	86	5.40999	14	5.28915	14
fields.c	90	5.30430	13	5.24592	12
ptt5	159	1.73990	17	1.71901	13
sum	255	5.54921	14	5.54772	13
kennedy.xls	256	3.94279	13	3.89779	14
English alphabet	26	4.34534	10	4.33957	8

regular RVLCs having a free distance of $d_f = 1$ when using a joint source–channel decoding scheme. Since the RVLC construction of [22] was also based on the codeword length distribution of Huffman codes, the proposed optimization procedure of Algorithm 3.1 may also be invoked in this context for generating more efficient RVLCs having a free distance of $d_f = 2$. The corresponding results are shown in Table 3.5.

3.3.5 Variable-Length Error-Correcting Code

VLEC codes [28, 29] were invented for combing source coding and error correction for employment in specific scenarios, when the assumptions of the Shannonian source and channel coding separation theorem do not apply. In terms of their construction, VLEC codes are similar to classic Huffman codes and RVLCs, in that shorter codewords are assigned to more probable source symbols. Additionally, in order to provide an error-correcting capability, VLEC codes typically have better distance properties, i.e., higher free distances in exchange for their increased average codeword length. The idea of VLEC codes was first discussed in [239], and subsequently in [240–242], where they were considered to be instantaneously decodable, resulting in suboptimal performance [29]. In [28, 29] Buttigieg *et al.* proposed a Maximum Likelihood (ML) decoding algorithm, which significantly improves the decoding performance. It was shown that the free distance is the most important parameter that determines the decoding performance at high E_b/N_0. Detailed treatment of various code properties, decoding of VLEC codes and their performance bounds can be found in [30], for example.

The family of VLECs may be characterized by three different distances, namely the minimum block distance b_{min}, the minimum divergence distance d_{min} and the minimum convergence distance c_{min} [30]. It may be shown that Huffman codes and RVLCs may also be viewed as VLECs. For Huffman codes we have $b_{min} = d_{min} = 1$, $c_{min} = 0$. By contrast, for RVLCs we have $d_{min} = c_{min} = 1$ and $b_{min} = 1$ for RVLCs having a free distance of $d_f = 1$,

Table 3.6: An example of VLEC

Symbol	Probability	C_{HUFF}	C_{VLEC}	C_{VLEC}
0	0.33	00	000	0000
1	0.30	01	0110	1111
2	0.18	11	1011	01010
3	0.10	100	11010	10101
4	0.09	101	110010	001100
Average code length		2.19	3.95	4.46
Free distance		1	3	4

while $b_{\min} = 2$ for RVLCs having a free distance of $d_f = 2$. Table 3.6 shows two VLECs having free distances of $d_f = 3$ and $d_f = 4$ respectively.

3.3.5.1 Construction of VLEC Codes

The construction of an optimal VLEC code requires a variable-length code that satisfies the specific distance requirements, and additionally has a minimum average codeword length, to be found. As for a VLEC code having $d_f > 2$, the code structure becomes substantially different from that of the corresponding Huffman code for the same source. Therefore it is inadequate to use the length distribution of the Huffman code as a starting point for constructing a VLEC code. According to Theorem 1, the free distance of a VLC is determined by the minimum block distance $b_{\min}$ and by the sum of the minimum divergence distance $d_{\min}$ and the minimum convergence distance $c_{\min}$, $(d_{\min} + c_{\min})$, whichever of these is the less. Hence, in order to construct a VLEC code having a given free distance of d_f, it is reasonable to choose $b_{\min} = d_{\min} + c_{\min} = d_f$ and $d_{\min} = \lceil d_f/2 \rceil$ as well as $c_{\min} = d_f - d_{\min}$. In [30] Buttigieg proposed a heuristic VLEC construction method, which is summarized here as Algorithm 3.2.

The procedure of Algorithm 3.2 is repeated until it finds the required number of codewords or has no more options to explore. If no more codewords of the required properties can be found, or if the maximal codeword length is reached, shorter codewords are deleted until one finds an adequate VLEC code structure.

Table 3.7 shows an example of a VLEC code's construction for the English alphabet using Algorithm 3.2. For reasons of brevity, only the results of the first four steps are listed. The target VLEC code has a free distance of $d_f = 3$; hence the minimum block distance should be at least $b_{\min} = 3$ and we choose the minimum B–F divergence distance to be $d_{\min} = 2$ as well as the minimum B–F convergence distance to be $c_{\min} = 1$. Let the minimum codeword length be $L_{\min} = 4$. At the first step, the greedy algorithm is invoked to generate a fixed-length code of length $l = 4$, while having a minimum block distance of $b_{\min} = 3$. This is shown in the first column of Table 3.7. Our target code C is initialized with this fixed-length code. In the first part of Step 2 corresponding to column 2.1 of Table 3.7, all codewords of length 4 are generated, and only those having at least a distance of $d = 2$ from each codeword in the set C are assigned to a temporary codeword set W. Then, in the second part of Step 2, as depicted in column 2.2 of Table 3.7, an extra bit, either '0' or '1,' is affixed at the end of all the codewords

---

Algorithm 3.2

Step 1: Initialize the target VLEC codeword set C to be designed as a fixed-length code of length L_1 having a minimal distance $b_{\min}$. This step and all following steps that result in sets of identical-length codewords having a given distance are carried out with the aid of either the greedy algorithm or the Majority Voting Algorithm (MVA) of Appendix A.3.

Step 2: Create a set W that contains all L_1-tuples having at least a distance $d_{\min}$ from each codeword in C. If the set W is not empty, an extra bit is affixed at the end of all its words. This new set, having twice the number of words, replaces the previous set W.

Step 3: Delete all words in the set W that do not have at least a distance $c_{\min}$ from all codewords of C. At this point, the set W satisfies both the $d_{\min}$ and the $c_{\min}$ minimum distance requirements with respect to the set C.

Step 4: Select the specific codewords from the set W that are at a distance of $b_{\min}$ using the greedy algorithm or the MVA. The selected codewords are then added to set C.

Table 3.7: An example construction of a VLEC code having a free distance of $d_f = 3$ for the English alphabet using Algorithm 3.2

Set C	Set W			Set C
Step 1	Step 2.1	Step 2.2	Step 3	Step 4
0 0 0 0	1 0 0 1	1 0 0 1 0	1 0 0 1 0	0 0 0 0
0 1 1 1	1 0 1 0	1 0 0 1 1	1 0 0 1 1	0 1 1 1
	1 0 1 1	1 0 1 0 0	1 0 1 0 0	1 0 0 1 0
	1 1 0 0	1 0 1 0 1	1 0 1 0 1	1 0 1 0 1
	1 1 0 1	1 0 1 1 0	1 0 1 1 0	
	1 1 1 0	1 0 1 1 1	1 1 0 0 0	
		1 1 0 0 0	1 1 0 0 1	
		1 1 0 0 1	1 1 0 1 0	
		1 1 0 1 0	1 1 0 1 1	
		1 1 0 1 1	1 1 1 0 0	
		1 1 1 0 0	1 1 1 0 1	
		1 1 1 0 1		

in the set W. At Step 3, all codewords in the set W that do not have at least a distance of $d = 1$ from all codewords of the set C are deleted, as shown in column 2.3 of Table 3.7. As a result, all codewords in the set W now satisfy both the minimum B–F divergence distance and the minimum B–F convergence distance requirements with respect to the codewords in the set C. Finally, at Step 4 the greedy algorithm is invoked to select codewords from the set

W that are at a distance of at least $b_{min} = 3$ from each other. The selected codewords are then incorporated in the set C, as shown in Table 3.7.

The basic construction of Algorithm 3.2 attempts to assign as many short codewords as possible in order to minimize the average codeword length without considering the source statistics. However, assigning too many short codewords typically decreases the number of codewords available at a higher length. Therefore the maximum codeword length may have to be increased in order to generate the required number of codewords, which will probably result in an increased average codeword length.

For this reason, Buttigieg [30] proposed an exhaustive search procedure for arriving at a better codeword length distribution. After generating a sufficiently high number of codewords using the basic construction Algorithm 3.2, the optimization procedure starts by deleting the last group of codewords of the same length, and one codeword from the next to last group of codewords. The rest of the codewords are retained and the basic construction Algorithm 3.2 is invoked to add new codewords. After finding the required number of codewords, the same optimization procedure is applied to the new code. The optimization is repeated until it reaches the first group of codewords (i.e., the group of codewords with the minimum length) and there is only one codeword in the first group. At every iteration the average codeword length is calculated, and the code with minimum average codeword length is recorded. In this way, every possible codeword length distribution is tested and the 'best' one is found. This technique is capable of improving the achievable code efficiency; however, its complexity increases exponentially with the size of the source alphabet.

Here, therefore, we propose a codeword length distribution optimization procedure, Algorithm 3.3, similar to that for the construction of RVLCs, as an alternative to the exhaustive search.

Algorithm 3.3

Step 1: Use Algorithm 3.2 to construct an initial VLEC code, and calculate the length vector $\mathbf{n}$.

Step 2: Decrease the number of codewords required at length l, $\mathbf{n}(l)$, by one. The values of $\mathbf{n}(k)$, $k \le l$ are retained, and no restrictions are imposed on the number of codewords at higher lengths. Then use Algorithm 3.2 again to construct a VLEC code based on the modified length distribution.

Step 3: Recalculate the average codeword length $\bar{L}$, and if $\bar{L}$ was increased, restore the previous length distribution, and proceed to the next codeword length; otherwise update the bit length vector $\mathbf{n}$ and go to Step 2.

Step 4: Repeat Steps 2 and 3 until the last codeword length is reached.

3.3.5.2 *An Example Construction of a VLEC Code*

In the following, we will demonstrate the construction procedure of our Algorithm 3.3 using an example. The source for which the VLEC code is designed is the English alphabet as shown in Table 3.9. The construction procedure is summarized in Table 3.8, which is similar to the procedure described in Section 3.3.4.2 and devised for the construction of RVLCs,

Table 3.8: An example construction of a VLEC code having a free distance of $d_f = 3$ for the English alphabet

Iteration number	Expected length vector[†]	Resultant length vector[‡]	Average length	Control logic[§]
#1	NULL	[0 0 0 2 1 2 1 2 5 7 6]	6.94527	start
#2	[0 0 0 1↓]	[0 0 0 1 4 1 2 2 2 8 6]	6.64447	continue
#3	[0 0 0 0↓]	[0 0 0 4 3 1 3 3 6 6]	6.71211	restore, next level
#4	[0 0 0 1 3↓]	[0 0 0 1 3 3 2 3 5 7 2]	6.39466	continue
		⋮		
#9	[0 0 0 1 2 4 4↓]	[0 0 0 1 2 4 4 6 5 2 2]	6.2053	continue
		⋮		
#12	[0 0 0 1 2 4 4 6 5 1↓]	[0 0 0 1 2 4 4 6 4 1 4]	6.21624	end

[†] The desired VLEC code length distribution: the arrows pointing downwards imply that the values at those positions have been decreased by one.

[‡] The actual length distribution of the resultant VLEC code after each iteration: the underlines indicate the numbers that will be changed in the next iteration.

[§] Decides the next step of the code generator, which depends on the resultant average codeword length.

except that no length distribution restrictions are imposed beyond the current level, and that the number of codewords required at the current level is expected to be decreased.

At the first iteration the length vector is initialized to NULL, i.e. no restrictions are imposed, and Algorithm 3.2 is invoked to construct an initial VLEC code, resulting in a length distribution of $\mathbf{n} = [0\ 0\ 0\ 2\ 1\ 2\ 1\ 2\ 5\ 7\ 6]$ and an average codeword length of $\bar{L} = 6.94527$. Then our length-distribution optimization procedure commences. Firstly, the number of codewords $\mathbf{n}(4)$ required at the length of $l = 4$ is decreased by one and the expected length distribution becomes $\mathbf{n} = [0\ 0\ 0\ 1]$, as shown in Table 3.8. Then Algorithm 3.2 is invoked again to construct a VLEC code according to the new length distribution, and the resultant average codeword length becomes $\bar{L} = 6.64447$. Since the average codeword length of the new RVLC is lower than that of the old one, the code generator continues to increase the number of codewords required at the length of $l = 4$ during the third iteration. However, the resultant RVLC has an increased average codeword length of $\bar{L} = 6.71211$; hence the code generator has to restore the length vector to that of the VLEC code obtained at the end of the second iteration, which is $\mathbf{n} = [0\ 0\ 0\ 1\ 4\ 1\ 2\ 2\ 2\ 8\ 6]$. Then the code generator attempts to increase the number of codewords required at the next level, i.e. at the length of $l = 4$. Hence the desired length distribution devised at the beginning of the fourth iteration becomes $\mathbf{n} = [0\ 0\ 0\ 1\ 3]$. This procedure is repeated until the last level has been reached. As seen from Table 3.8, after 12 iterations all levels of the codeword lengths have been optimized. Finally, the best VLEC code having the minimum average codeword length found at the ninth iteration is selected.

Table 3.9: Proposed VLECs for the English alphabet

u	$p(u)$	VLEC $d_f = 1$	VLEC $d_f = 2$	VLEC $d_f = 3$
E	0.14878	000	000	0000
T	0.09351	001	011	00110
A	0.08833	010	101	11000
O	0.07245	0110	110	010100
R	0.06872	0111	0010	100101
N	0.06498	1011	0100	011001
H	0.05831	1100	1001	101000
I	0.05644	1101	1111	0101110
S	0.05537	1110	00111	0110100
D	0.04376	1111	01010	1111010
L	0.04124	10011	10001	1001100
U	0.02762	10100	11100	11110010
P	0.02575	10101	001100	01111110
F	0.02455	100011	010111	11011111
M	0.02361	100100	100001	10111011
C	0.02081	100101	111010	10111100
W	0.01868	1000011	0011010	11101101
G	0.01521	1000100	0101100	011110110
Y	0.01521	1000101	1000001	110110101
B	0.01267	10000011	1110111	111110011
V	0.01160	10000100	00110111	101011110
K	0.00867	10000101	01011010	111011101
X	0.00146	100000011	10000001	1111100001
J	0.00080	100000100	11101100	1110101100
Q	0.00080	100000101	001101100	11101010010
Z	0.00053	1000000011	010110111	11101011111
Average length		4.19959	4.23656	6.20530
Max length		10	9	11

3.3.5.3 Experimental Results

In [30] Buttigieg only considered the construction of VLEC codes having $d_f > 2$. The experimental results of Table 3.9 demonstrate that the proposed Algorithm 3.3 of Section 3.3.5.1 may also be used for constructing VLEC codes having $d_f = 1, 2$, which are the equivalents of the RVLCs having $d_f = 1, 2$ in Section 3.3.4.3. It is worth noting that the VLEC code having $d_f = 2$ seen in Table 3.9 results in an even shorter average codeword length than does the corresponding RVLC having $d_f = 2$ in Table 3.5, which indicates that this is a meritorious way of constructing such codes.

Table 3.10 compares the VLEC codes having $d_f = 3, 5, 7$ constructed by the proposed Algorithm 3.3 of Section 3.3.5.1 with those of [30]. By using the optimization procedure of

Table 3.10: Comparison of VLECs for the English alphabet[†]

	Buttigieg's VLEC [30]		Proposed VLEC	
Free distance	Average length	Max length	Average length	Max length
$d_f = 3$	6.370	13	6.3038	11
$d_f = 5$	8.467	12	8.4752	12
$d_f = 7$	10.70	15	10.7594	15

[†] The probability distribution in [30] was used, which is different from that in Table 3.9.

Algorithm 3.3 instead of the exhaustive search of [30], only a small fraction of all possible codeword length distributions are tested; hence the construction complexity is significantly reduced without compromising the achievable performance. For sources of larger alphabets, the 'accelerated' search techniques of [31] may be invoked for further complexity reduction.

3.4 Soft Decoding of Variable-Length Codes

Classic VLCs, such as Huffman codes and RVLCs, have been widely employed in image and video coding standards, such as the JPEG/JPEG2000 [243, 244], the H.263/H.264 [18, 161] and MPEG-1/2/4 [19, 245, 246], for the sake of source compression. Traditionally, VLCs are instantaneously decoded using hard decisions, implying that, after demodulation and possible channel decoding, the input of the VLC decoder is a stream of binary zeros and ones. The VLC decoder then searches the binary code tree or looks up the code table in order to decode the bit stream. If there are residual errors in the input bit stream, it is very likely that the VLC decoder will lose synchronization and hence have to discard the rest of the bit stream.

On the other hand, the demodulator or channel decoder is often capable of providing more information for the VLC decoder than the hard decision-based bits, such as the probability that a transmitted bit assumes a particular value from the set $\{0, 1\}$. The principle of using probabilities (soft information) rather than hard decisions is often referred to as 'soft decoding.' According to one of Viterbi's three lessons on Shannonian information theory [247], we should '[n]ever discard information prematurely that may be useful in making a decision, until after all decisions related to that information have been completed.' Hence, one may argue that soft decoding could outperform hard decoding. Before describing the soft decoding of VLCs, we now introduce their trellis representations.

3.4.1 Trellis Representation

3.4.1.1 Symbol-Level Trellis

In [40], Bauer and Hagenauer proposed a symbol-level trellis representation of VLCs, where the trellis was constructed based on two types of length information related to a VLC sequence, namely the number of source symbols (or VLC codewords) and the number of bits. Let (n, k) represent a node in the trellis of Figure 3.3 at symbol index k, where the source symbol is represented by n bits of VLC codewords. Assume that we have a VLC sequence consisting of N bits and K symbols. Then the trellis of Figure 3.3 diverges from

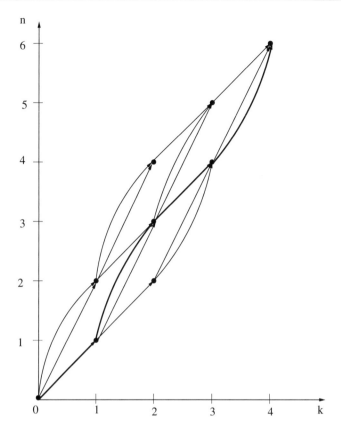

Figure 3.3: Symbol-level trellis for $K = 4$ and $N = 6$ using the VLC alphabet $\mathscr{C} = \{1, 01, 00\}$ [40].

node $(0, 0)$ as the symbol index k increases, and converges to node (N, K) at the end of the sequence. Each path through the trellis represents a legitimate combination of K symbols described by a total of N bits.

For example, consider a VLC having the alphabet $\mathscr{C} = \{1, 01, 00\}$, and a sequence of $K = 4$ source symbols $\mathbf{u} = (0, 2, 0, 1)$. This sequence is mapped to a sequence of the VLC codewords $\mathbf{c} = (1, 00, 1, 01)$. Its length, expressed in terms of bits, is $N = 6$. All sequences associated with $K = 4$ and $N = 6$ can be represented in the trellis diagram shown in Figure 3.3.

In this diagram k denotes the symbol index and n identifies a particular trellis state at symbol index k. The value of n is equivalent to the number of bits of a subsequence ending in the corresponding state. At each node of the trellis, the appropriate match (synchronization) between the received bits and the received symbols is guaranteed.

Note that as the number K of source symbols considered by a symbol-level VLC trellis is increased, its parallelogram shape becomes wider and the number of transitions employed increases exponentially. For this reason, the symbol-level trellis complexity per bit

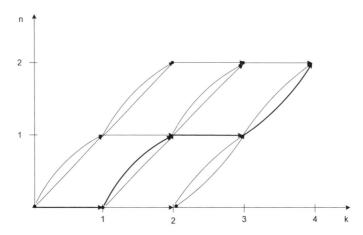

Figure 3.4: Transformed symbol-level trellis for $K = 4$ and $N = 6$ using the VLC alphabet $\mathscr{C} = \{1, 01, 00\}$ [40].

depends on the number K of source symbols considered. In addition to this, the symbol-level trellis complexity depends on the number of entries employed within the corresponding VLC codebook. Furthermore, the difference in length between the longest and shortest VLC codewords influences the symbol-level trellis complexity, since this dictates the size of the acute angles within the symbol-level trellis's parallelogram shape.

Let us now introduce the transformation

$$s = n - kl_{\min}$$

for the sake of obtaining the trellis seen in Figure 3.4, where $l_{\min}$ is the minimum VLC codeword length. The maximal number of states at any symbol index k is

$$s_{\max} = N - Kl_{\min} + 1,$$

which is less than the original trellis of Figure 3.3. Hence the transformed trellis of Figure 3.4 is more convenient for implementation.

3.4.1.2 Bit-Level Trellis

The high complexity of the symbol-level trellis of Section 3.4.1 motivates the use of a bit-level trellis [41], which is similar to that of a binary convolutional code.

The trellis is obtained by assigning its states to the nodes of the VLC tree. The nodes in the tree can be subdivided into a root-node (R) and so-called internal nodes (I) and terminal nodes (T). The root node and all the terminal nodes are assumed to represent the same state, since they all show the start of a new symbol. The internal nodes are assigned one by one to the other states of the trellis. As an example, Figure 3.5 shows the trellis corresponding to the RVLC-2 scheme of $\{00, 11, 010, 101, 0110\}$. Figure 3.6 shows two other examples of RVLC trellises.

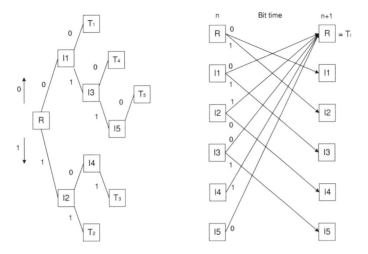

Figure 3.5: Code tree and trellis for the RVLC-2 scheme of {00, 11, 010, 101, 0110} [42].

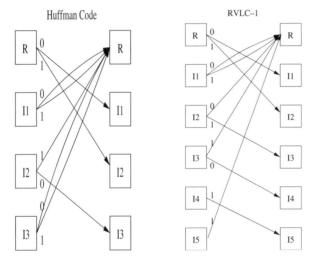

Figure 3.6: Bit-level trellises for the Huffman code {00, 01, 11, 100, 101}, and the RVLC-1 {00 01 10 111 11011}.

The bit-level trellis of a VLC differs from that of a binary convolutional code in many ways. In the bit-level trellis of a VLC the number of trellis states is equal to the number of internal nodes plus one (the root node), which depends on the specific code tree structure. The maximum number of branches leaving a trellis state is two, corresponding to a separate branch for each possible bit value. Merging of several paths only occurs in the root node, where the number of paths converging at the root node corresponds to the total number of codewords in the VLC. If the variable-length coded sequence consists of N bits, the trellis

Table 3.11: Comparison of the trellis of a VLC and that of a binary convolutional code

	VLCs	Convolutional codes
Number of states	Depends on code tree structure	Depends on the size of the code memory
Number of branches Emitting from a state	1 or 2	2
Number of branches Emerging at a state	Equal to the number of VLC codewords for the root state; 1 for other states	2
Parallel transitions	May exist	None

can be terminated after N sections. Table 3.11 summarizes our comparison of VLCs and convolutional codes. Furthermore, in comparison with the symbol-level trellis, the bit-level trellis is relatively simpler and time invariant. More specifically, as the number of source symbols considered is increased, the number of transitions employed by a bit-level trellis increases linearly, rather than exponentially as in a symbol-level trellis. Note, however, that – unlike a symbol-level trellis – a bit-level trellis is unable to exploit knowledge of the number of transmitted symbols K, even if this is available.

Similar to that for convolutional codes, we may define the constraint length of a VLC. Since a constraint-length-K convolutional code has 2^{K-1} states in the trellis, the constraint length of a VLC may be defined as

$$K \triangleq \log_2 N + 1, \tag{3.11}$$

where N is the number of states in the bit-level trellis.

3.4.2 Trellis-Based VLC Decoding

Based on the above trellis representations, either ML/MAP sequence estimation employing the VA [47, 248] or MAP decoding using the BCJR algorithm [48] can be performed. We describe all of these algorithms in the following sections.

3.4.2.1 *Sequence Estimation Using the Viterbi Algorithm*

Sequence estimation can be carried out using either the MAP or the ML criterion. The latter is actually a special case of the former. The VA [47, 248] is invoked in order to search through the trellis for the optimal sequence that meets the MAP/ML criterion. The Euclidean distance is used as the decoding metric for the case of soft decision, while the Hamming distance is used as the decoding metric for the hard-decision scenario.

3.4.2.1.1 **Symbol-Level Trellis-Based Sequence Estimation.** Consider the simple transmission scheme shown in Figure 3.7, where the K-symbol output sequence **u** of a discrete memoryless source is first VLC encoded into a N-bit sequence **b** and then modulated before being transmitted over a channel.

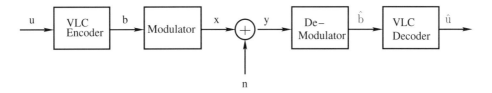

Figure 3.7: A simple transmission scheme of VLC-encoded source information communicating over AWGN channels.

At the receiver side, the MAP decoder selects as its estimate the particular sequence that maximizes the *a posteriori* probability:

$$\hat{\mathbf{u}} = \arg \max_{\mathbf{u}} P(\mathbf{u}|\mathbf{y}). \tag{3.12}$$

Since the source symbol sequence $\mathbf{u}$ is uniquely mapped to the bit sequence $\mathbf{b}$ by the VLC encoder, maximization of the *a posteriori* probability of $P(\mathbf{u}|\mathbf{y})$ is equivalent to the maximization of the *a posteriori* probability of $P(\mathbf{b}|\mathbf{y})$, which is formulated as

$$\hat{\mathbf{b}} = \arg \max_{\mathbf{b}} P(\mathbf{b}|\mathbf{y}). \tag{3.13}$$

Using Bayes's rule, we have

$$P(\mathbf{b}|\mathbf{y}) = \frac{P(\mathbf{y}|\mathbf{b})P(\mathbf{b})}{P(\mathbf{y})}. \tag{3.14}$$

Note that the denominator of (3.14) is constant for all $\mathbf{b}$. Hence, to maximize (3.13) we only have to maximize the numerator of (3.14), and it might be more convenient to maximize its logarithm. Therefore, we have to choose a codeword sequence $\mathbf{b}$ that maximizes

$$\log P(\mathbf{y}|\mathbf{b}) + \log P(\mathbf{b}). \tag{3.15}$$

If we assume that the channel is memoryless, the first term of (3.15) can be rewritten as

$$\log P(\mathbf{y}|\mathbf{b}) = \sum_{i=1}^{N} \log P(y_i|b_i), \tag{3.16}$$

where $\mathbf{b} = b_1 b_2 \cdots b_N$; N is the length of the sequence $\mathbf{b}$ expressed in terms of the number of bits. Furthermore, assuming that the source is memoryless, then the probability of occurrence $P(\mathbf{b})$ for the sequence $\mathbf{b}$ is given by

$$P(\mathbf{b}) = P(\mathbf{c}_1)P(\mathbf{c}_2) \cdots P(\mathbf{c}_K), \tag{3.17}$$

where $\mathbf{c}_1, \mathbf{c}_2, \ldots, \mathbf{c}_K$ are the codewords that constitute the sequence $\mathbf{b}$; i.e., we have $\mathbf{b} = \mathbf{c}_1 \mathbf{c}_2 \cdots \mathbf{c}_K$, and $P(\mathbf{c}_j)$ is determined by the corresponding source symbol probability. Therefore, based on the channel output and source information, the Path Metric (PM) given by (3.15) is transformed to

$$PM_{MAP} = \sum_{i=1}^{N} \log P(y_i|b_i) + \sum_{j=1}^{K} \log P(\mathbf{c}_j). \tag{3.18}$$

Let $l_j \triangleq \ell(\mathbf{c}_j)$ be the length of the codeword $\mathbf{c}_j$ in terms of bits. Then the branch metric (BM) calculated for the jth branch of the path is formulated as

$$BM_{MAP} = \sum_{m=1}^{l_j} \underbrace{\log P(y_{j_m}|c_{j_m})}_{\text{channel info}} + \underbrace{\log P(\mathbf{c}_j)}_{\text{source info}}, \tag{3.19}$$

where we have $\mathbf{c}_j = (c_{j_1} c_{j_2} \cdots c_{j_{l_j}})$, and $(y_{j_1} y_{j_2} \cdots y_{j_{l_j}})$ represents the corresponding received bits.

Let us detail this algorithm further with the aid of an example. Assume that the source described in Figure 3.7 is a DMS emitting three possible symbols, namely $A = \{a_0, a_1, a_2\}$, which have different probabilities of occurrence. Furthermore, the source symbol sequence $\mathbf{u}$ is encoded by the VLC encoder of Figure 3.7, which employs the VLC table of $C = \{1, 01, 00\}$. Assume that a sequence of $K = 4$ source symbols $\mathbf{u} = (0, 2, 0, 1)$ is generated and encoded to a sequence of VLC codewords $\mathbf{c} = (1, 00, 1, 01)$, which has a length of $N = 6$ in terms of bits. This bit sequence is then modulated and transmitted over the channel. At the VLC decoder of Figure 3.7, a symbol-level trellis may be constructed using the method described in Section 3.4.1.1, which is shown in Figure 3.8. Essentially, the trellis seen in Figure 3.8 is the same as that of Figure 3.3, since we are using the same VLC and source symbol sequence for the sake of comparability. According to Equation (3.19), the branch metric of the MAP Sequence Estimation (SE) algorithm can be readily exemplified, as seen in Figure 3.8 for a specific path corresponding to the VLC codeword sequence $\mathbf{c} = (1, 00, 1, 01)$.

For Binary Phase-Shift Keying (BPSK) transmission over a memoryless AWGN channel, the channel-induced transition probability is [8]

$$P(y_i|b_i) = \frac{1}{\sqrt{\pi N_0}} \exp\left\{ -\frac{[y_i - \sqrt{E_s}(2b_i - 1)]^2}{N_0} \right\}, \tag{3.20}$$

where $N_0 = 2\sigma^2$ is the single-sided power spectral density of the AWGN, σ^2 is the noise variance and E_s is the transmitted signal energy per channel symbol. Consequently, upon substituting Equation (3.20) into Equation (3.18), the path metric to be maximized is

$$PM_{MAP} = -\sum_{i=1}^{N} \frac{[y_i - \sqrt{E_s}(2b_i - 1)]^2}{N_0} + \sum_{j=1}^{K} \log P(\mathbf{c}_j). \tag{3.21}$$

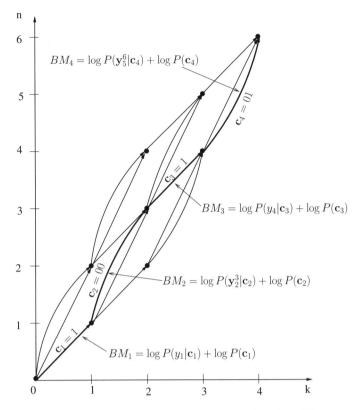

Figure 3.8: An example of the branch metric calculation for the MAP SE algorithm employed in the VLC decoder of Figure 3.7. The VLC table employed is $C = \{1, 01, 00\}$. The transmitted VLC codeword sequence has a length of $N = 6$ bits, which represents $K = 4$ source symbols. The received channel observation is $\mathbf{y} = y_1 \cdots y_6$.

Upon removing the common components in Equation (3.21), which may be neglected during the optimization, it can be further simplified to

$$PM_{MAP} = \sum_{i=1}^{N} \frac{L_c}{2} y_i (2b_i - 1) + \sum_{j=1}^{K} \log P(\mathbf{c}_j), \qquad (3.22)$$

where

$$L_c = \frac{4\sqrt{E_s}}{N_0} \qquad (3.23)$$

is the channel reliability value.

Correspondingly, the branch metric of Equation (3.19) can be expressed as

$$BM_{MAP} = \sum_{m=1}^{l_j} \frac{L_c}{2} y_{j_m} (2c_{j_m} - 1) + \log P(\mathbf{c}_j). \qquad (3.24)$$

If we assume that all paths in the trellis are equiprobable, i.e. the probability of occurrence $P(\mathbf{b})$ of sequence $\mathbf{b}$ is equiprobable, then the MAP decoder becomes a ML decoder, and the path metric to be maximized is reduced from the expression seen in Equation (3.15) to

$$PM_{ML} = \log P(\mathbf{y}|\mathbf{b}). \tag{3.25}$$

In particular, for BPSK transmission over a memoryless AWGN channel the path metric formulated for the ML decoder is

$$PM_{ML} = \sum_{i=1}^{N} \frac{L_c}{2} y_i (2b_i - 1), \tag{3.26}$$

and the corresponding branch metric is given by

$$BM_{ML} = \sum_{m=1}^{l_j} \frac{L_c}{2} y_{j_m} (2c_{j_m} - 1). \tag{3.27}$$

As to hard-decision decoding, the input of the MAP/ML VLC decoder is a bit sequence $\hat{\mathbf{b}}$. The path metric of the MAP decoder is

$$PM_{MAP} = \log P(\hat{\mathbf{b}}|\mathbf{b}) + \log P(\mathbf{b}). \tag{3.28}$$

In the case of hard decisions and BPSK modulation, a memoryless AWGN channel is reduced to a Binary Symmetric Channel (BSC) having a crossover probability p. Therefore, we have

$$P(\hat{\mathbf{b}}|\mathbf{b}) = p^h (1-p)^{N-h}, \tag{3.29}$$

where $h = H(\mathbf{b}, \hat{\mathbf{b}})$ is the Hamming distance between the transmitted bit sequence $\mathbf{b}$ and the received bit sequence $\hat{\mathbf{b}}$.

Hence, the path metric is formulated as

$$PM_{MAP} = h(\log p - \log(1-p)) + \sum_{j=1}^{K} \log P(\mathbf{c}_j), \tag{3.30}$$

while the branch metric is given by

$$BM_{MAP} = (\log p - \log(1-p))H(\mathbf{c}_j, \hat{b}_{j_1}\hat{b}_{j_2} \cdots \hat{b}_{j_{l_j}}) + \log P(\mathbf{c}_j). \tag{3.31}$$

As for the ML decoding, assuming that we have $0 \leq p < 0.5$ and all transmitted sequences are equiprobable, the branch metric is simply the Hamming distance between the transmitted codeword and the received codeword, namely $H(\mathbf{c}_j, \hat{b}_{j_1}\hat{b}_{j_2} \cdots \hat{b}_{j_{l_j}})$.

3.4.2.1.2 Bit-Level Trellis-Based Sequence Estimation.

Consider the same transmission scheme as in the symbol-level trellis scenario of Figure 3.7. It is readily seen that the path metric of MAP/ML decoding in the bit-level trellis scenario is the same as that in the symbol-level trellis case, e.g. the path metrics derived in Equations (3.18), (3.22) and (3.26).

By contrast, since there is only a single bit associated with each transition in the bit-level trellis, the branch metric is different from that derived for the symbol-level trellis as in Equations (3.19), (3.24) and (3.27). Generally, there are two convenient ways of calculating the branch metric of a bit-level VLC trellis.

For a bit-level trellis each transition ending in the root node represents the termination or completion of a legitimate source codeword, and the merging only occurs at the root node. Hence, for transitions that are not ending at the root node, the source symbol probabilities $P(\mathbf{c}_j)$ need not to be included in the branch metric. Instead, they are added only when calculating branch metrics for the transitions that do end at the root node.

To be specific, for the soft MAP decoding the branch metrics of the transitions that do not end at the root node of Figure 3.5 are

$$BM_{MAP}^{non\text{-}root} = \log P(y_i|b_i), \tag{3.32}$$

which may be expressed as

$$BM_{MAP}^{non\text{-}root} = \frac{L_c}{2} y_i(2b_i - 1), \tag{3.33}$$

in the case of BPSK transmission over a memoryless AWGN channel. By contrast, the branch metrics of the transitions ending at the root node are

$$BM_{MAP}^{root} = \log P(y_i|b_i) + \log P(\mathbf{c}_j), \tag{3.34}$$

or

$$BM_{MAP}^{root} = \frac{L_c}{2} y_i(2b_i - 1) + \log P(\mathbf{c}_j), \tag{3.35}$$

in the case of BPSK transmission over a memoryless AWGN channel, where $\mathbf{c}_j$ is the source symbol corresponding to the transition that ends at the root node.

The second way of calculating the branch metric of a bit-level trellis is to align the source information with each of the trellis transitions. For example, consider the VLC $\mathscr{C} = \{00, 11, 101, 010, 0110\}$ used in Figure 3.5 and assume that the corresponding codeword probabilities are $\mathscr{P} = \{p_1, p_2, p_3, p_4, p_5\}$. The probability of the codeword $'010'$ being transmitted can be expressed as

$$P('010') = P('0')P('1'/'0')P('0'/'01'). \tag{3.36}$$

Furthermore, $P('0')$, $P('1'/'0')$, $P('0'/'01')$ can be calculated as the State Transition Probabilities (STPs) of the trellis as defined in [249]. In this example, as seen from the code tree of Figure 3.5, we have

$$P('0') = P(s_{k+1} = \text{I}1, b_k = 0|s_k = \text{R}) = p_1 + p_4 + p_5, \tag{3.37}$$

$$P('1'/'0') = P(s_{k+1} = \text{I}3, b_k = 1|s_k = \text{I}1) = \frac{p_4 + p_5}{p_1 + p_4 + p_5}, \tag{3.38}$$

$$P('0'/'01') = P(s_{k+1} = \text{R(T4)}, b_k = 0|s_k = \text{I}3) = \frac{p_4}{p_4 + p_5}. \tag{3.39}$$

Hence we arrive at

$$P('010') = (p_1 + p_4 + p_5) \cdot \frac{p_4 + p_5}{p_1 + p_4 + p_5} \cdot \frac{p_4}{p_4 + p_5} = p_4. \qquad (3.40)$$

For a generic VLC trellis, the state transition probabilities between two adjacent states associated with an input bit i can be formulated as

$$P(s_{k+1} = n, b_k = i | s_k = m) = \frac{\sum_{\alpha \in g(s_{k+1})} p_\alpha}{\sum_{\beta \in g(s_k)} p_\beta}, \qquad (3.41)$$

where $g(s_k)$ represents all the codeword indices associated with node s_k in the code tree.

Hence, the branch metric of soft MAP decoding may be expressed as

$$BM_{MAP} = \log P(y_i | b_i) + \log P(s_{i+1}, b_i | s_i), \qquad (3.42)$$

or

$$BM_{MAP} = \frac{L_c}{2} y_i (2b_i - 1) + \log P(s_{i+1}, b_i | s_i), \qquad (3.43)$$

when considering BPSK transmission over a memoryless AWGN channel.

More explicitly, we exemplify this algorithm in Figure 3.9. Consider again the transmission scheme seen in Figure 3.8, except that the VLC decoder employs a bit-level trellis constructed by the method described in Section 3.4.1.2. As seen from Figure 3.9, the branch metrics corresponding to the different state transitions can be readily computed according to Equation (3.42).

For ML soft decoding, source information is not considered, for the sake of reducing complexity. Hence the branch metric is simply expressed as

$$BM_{ML} = \log P(y_i | b_i), \qquad (3.44)$$

and we have

$$BM_{ML} = \frac{L_c}{2} y_i (2b_i - 1), \qquad (3.45)$$

for BPSK transmission over a memoryless AWGN channel.

Similarly, for hard-decision decoding, given the crossover probability p, the branch metric of MAP decoding is expressed as

$$BM_{MAP} = (\log p - \log(1-p)H(b_i, \hat{b}_i) + \log P(s_{i+1}, b_i | s_i), \qquad (3.46)$$

while it is

$$BM_{ML} = -H(b_i, \hat{b}_i) \qquad (3.47)$$

for ML decoding.

3.4.2.2 Symbol-by-Symbol MAP Decoding

3.4.2.2.1 Symbol-Level Trellis-Based MAP Decoding. In [40] Bauer and Hagenauer applied the BCJR algorithm [48] to the symbol-level trellis in order to perform symbol-by-symbol MAP decoding and hence minimize the SER. The objective of the decoder then is to

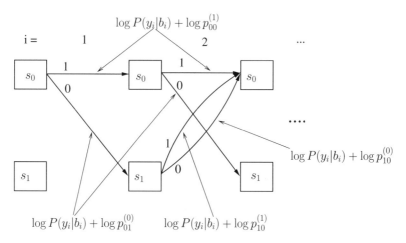

Figure 3.9: An example of the branch metric calculation for the bit-level trellis-based MAP SE algorithm employed in the VLC decoder of Figure 3.7. The VLC table employed is $C = \{1, 01, 00\}$. The probability $P(y_i|b_i)$ is the channel-induced transition probability for the ith transmitted bit and the state transition probabilities are defined as $p_{00}^{(1)} \triangleq P(s_0, b_i = 1|s_0)$, $p_{01}^{(0)} \triangleq P(s_1, b_i = 0|s_0)$, $p_{10}^{(1)} \triangleq P(s_0, b_i = 1|s_1)$, $p_{10}^{(0)} \triangleq P(s_0, b_i = 0|s_1)$, which can be calculated according to Equation (3.41), given a specific source symbol distribution.

determine the APPs of the transmitted symbols u_k, $(1 \leq k \leq K)$ from the observed sequence $\mathbf{y}$ according to

$$u_k = \arg\max_{u_k} P(u_k|\mathbf{y}). \tag{3.48}$$

Using Bayes's rule, we can write

$$P(u_k = i|\mathbf{y}) = \frac{P(u_k = i, \mathbf{y})}{P(\mathbf{y})}$$

$$= \frac{P(u_k = i, \mathbf{y})}{\displaystyle\sum_{i=0}^{M-1} P(u_k = i, \mathbf{y})}, \tag{3.49}$$

where M is the cardinality of the source alphabet. The numerator of (3.49) can be rewritten as

$$P(u_k = i, \mathbf{y}) = \sum_{n' \in \mathcal{N}_{k-1}} \sum_{n \in \mathcal{N}_k} P(s_{k-1} = n', s_k = n, u_k = i, \mathbf{y}), \tag{3.50}$$

where s_{k-1} and s_k are the states of the source encoder trellis at symbol indices $k-1$ and k respectively, and $\mathcal{N}_{k-1}$ and $\mathcal{N}_k$ are the sets of possible states at symbol indices $(k-1)$ and k respectively. The state transition of n' to n is encountered when the input symbol is $u_k = i$. Furthermore, the received sequence $\mathbf{y}$ can be split into three sections, namely the received codeword $y_{n'+1}^n$ associated with the present transition, the received sequence $y_1^{n'}$ prior to the

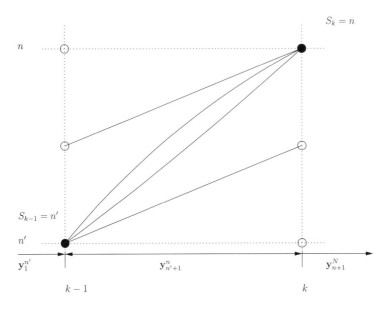

Figure 3.10: A section of transitions from the trellis in Figure 3.3.

present transition, and the received sequence y_{n+1}^N after the present transition. The received sequence after the split can be seen in Figure 3.10.

Consequently, the probability $P(s_{k-1} = n', s_k = n, u_k = i, \mathbf{y})$ in Equation (3.50) can be written as

$$P(s_{k-1} = n', s_k = n, u_k = i, \mathbf{y}) = P(s_{k-1} = n', s_k = n, u_k = i, \mathbf{y}_1^{n'}, \mathbf{y}_{n'+1}^n, \mathbf{y}_{n+1}^N).$$
(3.51)

Again, using Bayes's rule and the assumption that the channel is memoryless, we can expand Equation (3.51) as follows:

$$
\begin{aligned}
P(s_{k-1} = n', s_k = n, u_k = i, \mathbf{y}) &= P(\mathbf{y}_{n+1}^N | s_k = n) \\
&\quad \cdot P(s_{k-1} = n', s_k = n, u_k = i, \mathbf{y}_1^{n'}, \mathbf{y}_{n'+1}^n) \\
&= P(\mathbf{y}_{n+1}^N | s_k = n) \\
&\quad \cdot P(s_k = n, u_k = i, \mathbf{y}_{n'+1}^n | s_{k-1} = n') \\
&\quad \cdot P(s_{k-1} = n', \mathbf{y}_1^{n'}) \\
&= \beta_k(n) \gamma_k(i, n', n) \alpha_{k-1}(n').
\end{aligned}
$$
(3.52)

The basic operations of the decoder are the classic forward recursion [70] required for determining the values

$$
\begin{aligned}
\alpha_k(n) &= P(s_k = n, \mathbf{y}_1^n) \\
&= \sum_{n' \in \mathcal{N}_{k-1}} \sum_{i=0}^{M-1} P(s_k = n, \mathbf{y}_1^n, u_k = i, s_{k-1} = n')
\end{aligned}
$$

$$= \sum_{n' \in \mathcal{N}_{k-1}} \sum_{i=0}^{M-1} P(s_k = n, u_k = i, \mathbf{y}_{n'+1}^n | s_{k-1} = n') P(s_{k-1} = n', \mathbf{y}_1^{n'})$$

$$= \sum_{n' \in \mathcal{N}_{k-1}} \sum_{i=0}^{M-1} \gamma_k(i, n', n) \cdot \alpha_{k-1}(n'), \tag{3.53}$$

and the backward recursion [70] invoked for obtaining the values

$$\beta_k(n) = P(\mathbf{y}_{n+1}^N | s_k = n)$$

$$= \sum_{n' \in \mathcal{N}_{k+1}} \sum_{i=0}^{M-1} P(\mathbf{y}_{n+1}^N, u_{k+1} = i, s_{k+1} = n' | s_k = n)$$

$$= \sum_{n' \in \mathcal{N}_{k+1}} \sum_{i=0}^{M-1} P(\mathbf{y}_{n'+1}^N | s_{k+1} = n') P(s_{k+1} = n', u_{k+1} = i, \mathbf{y}_{n+1}^{n'} | s_k = n)$$

$$= \sum_{n' \in \mathcal{N}_{k+1}} \sum_{i=0}^{M-1} \gamma_{k+1}(i, n, n') \cdot \beta_{k+1}(n'). \tag{3.54}$$

Both quantities have to be calculated for all symbol indices k and for all possible states $n \in \mathcal{N}_k$ at symbol index k. In order to carry out the recursions of Equations (3.53) and (3.54) we also need the probability

$$\gamma_k(i, n', n) = P(u_k = i, s_k = n, \mathbf{y}_{n'+1}^n | s_{k-1} = n')$$

$$= P(\mathbf{y}_{n'+1}^n | u_k = i) \cdot P(u_k = i, s_k = n | s_{k-1} = n'), \tag{3.55}$$

which includes the source *a priori* information in terms of the state transition probability $P(u_k = i, s_k = n | s_{k-1} = n')$ and the channel characteristics quantified in terms of the channel transition probability $P(\mathbf{y}_{n'+1}^n | u_k = i)$.

Furthermore, assuming that the source is memoryless, we have

$$P(u_k = i, s_k = n | s_{k-1} = n') = \begin{cases} P(u_k = i) & n' \to n \text{ is a valid transition,} \\ 0 & \text{otherwise.} \end{cases} \tag{3.56}$$

Additionally, for a memoryless channel we can express the channel-induced transition probability seen in Equation (3.55) as the product of the bitwise transition probabilities:

$$P(\mathbf{y}_{n'+1}^n | u_k = i) = \prod_{j=1}^{|\mathbf{c}_i|} P(y_{n'+j} | c_{i_j}). \tag{3.57}$$

For BPSK transmission over an AWGN channel, the bitwise transition probability is given by [8]

$$P(y | x = \pm 1) = \frac{1}{\sqrt{\pi N_0}} \exp\left(-\frac{(y - \sqrt{E_s} x)^2}{N_0}\right), \tag{3.58}$$

where E_s is the transmitted energy per channel symbol and N_0 is the single-sided power spectral density of the AWGN.

So far we have obtained all the quantities that we need for computing the APPs of the information symbols u_k, which are summarized as follows:

$$P(u_k = m|\mathbf{y}) = \frac{\sum_{n \in \mathcal{N}_k} \sum_{n' \in \mathcal{N}_{k-1}} \gamma_k(m, n', n) \cdot \tilde{\alpha}_{k-1}(n') \cdot \tilde{\beta}_k(n)}{\sum_{n \in \mathcal{N}_k} \sum_{n' \in \mathcal{N}_{k-1}} \sum_{i=0}^{M-1} \gamma_k(i, n', n) \cdot \tilde{\alpha}_{k-1}(n') \cdot \tilde{\beta}_k(n)}, \qquad (3.59)$$

where instead of using $\alpha_k(n)$ and $\beta_k(n)$ we used their normalized versions $\tilde{\alpha}_k(n)$ and $\tilde{\beta}_k(n)$ to avoid numerical problems such as overflows and underflows. This is because both $\alpha_k(n)$ and $\beta_k(n)$ drop toward zero exponentially. For a sufficiently large value of K, the dynamic range of these quantities will exceed the number-representation range of conventional processors. To obtain a numerically stable algorithm, these quantities must be scaled as their computation proceeds. Consequently, the forward recursion $\tilde{\alpha}_k(n)$ can be written as

$$\tilde{\alpha}_k(n) = \frac{\sum_{n' \in \mathcal{N}_{k-1}} \sum_{i=0}^{M-1} \gamma_k(i, n', n) \cdot \tilde{\alpha}_{k-1}(n')}{\sum_{n \in \mathcal{N}_k} \sum_{n' \in \mathcal{N}_{k-1}} \sum_{i=0}^{M-1} \gamma_k(i, n', n) \cdot \tilde{\alpha}_{k-1}(n')}, \qquad (3.60)$$

with $\tilde{\alpha}_0(0) = 1$ and $k = 1, \ldots, K$, while the backward recursion $\tilde{\beta}_k(n)$ can be expressed as

$$\tilde{\beta}_k(n) = \frac{\sum_{n' \in \mathcal{N}_{k+1}} \sum_{i=0}^{M-1} \gamma_k(i, n, n') \cdot \tilde{\beta}_{k+1}(n')}{\sum_{n \in \mathcal{N}_k} \sum_{n' \in \mathcal{N}_{k+1}} \sum_{i=0}^{M-1} \gamma_k(i, n, n') \cdot \tilde{\alpha}_k(n)} \qquad (3.61)$$

with $\tilde{\beta}_K(N) = 1$ and $k = K - 1, \ldots, 1$.

Furthermore, the term in the form of $\gamma(i, n, n')$ in Equations (3.59), (3.60) and (3.61) can be computed using Equation (3.55). Note that the scaling or normalization operations used in Equations (3.60) and (3.61) do not affect the result of the APP computation, since they are canceled out in Equation (3.59).

Further simplification of Equation (3.59) can be achieved by using the logarithms of the above quantities instead of the quantities themselves [70]. Let $A_k(n)$, $B_k(n)$ and $\Gamma_k(i, n', n)$ be defined as follows:

$$A_k(n) \triangleq \ln(\tilde{\alpha}_k(n)), \qquad (3.62)$$

$$B_k(n) \triangleq \ln(\tilde{\beta}_k(n)), \qquad (3.63)$$

and

$$\Gamma_k(i, n', n) \triangleq \ln(\gamma_k(i, n', n)). \qquad (3.64)$$

Then, we have

$$A_k(n) = \ln\left(\frac{\sum_{n' \in \mathcal{N}_{k-1}} \sum_{i=0}^{M-1} \gamma_k(i, n', n) \cdot \tilde{\alpha}_{k-1}(n')}{\sum_{n \in \mathcal{N}_k} \sum_{n' \in \mathcal{N}_{k-1}} \sum_{i=0}^{M-1} \gamma_k(i, n', n) \cdot \tilde{\alpha}_{k-1}(n')} \right)$$

$$= \ln\left(\sum_{n' \in \mathcal{N}_{k-1}} \sum_{i=0}^{M-1} \exp[\Gamma_k(i, n', n) + A_{k-1}(n')] \right)$$

$$-\ln\left(\sum_{n\in\mathcal{N}_k}\sum_{n'\in\mathcal{N}_{k-1}}\sum_{i=0}^{M-1}\exp[\Gamma_k(i,n',n)+A_{k-1}(n')]\right),\qquad(3.65)$$

$$B_k(n)=\ln\left(\frac{\sum_{n'\in\mathcal{N}_{k+1}}\sum_{i=0}^{M-1}\gamma_{k+1}(i,n,n')\cdot\tilde{\beta}_{k+1}(n')}{\sum_{n\in\mathcal{N}_k}\sum_{n'\in\mathcal{N}_{k+1}}\sum_{i=0}^{M-1}\gamma_{k+1}(i,n,n')\cdot\tilde{\beta}_{k+1}(n')}\right)$$

$$=\ln\left(\sum_{n'\in\mathcal{N}_{k+1}}\sum_{i=0}^{M-1}\exp[\Gamma_{k+1}(i,n,n')+B_{k+1}(n')]\right)$$

$$-\ln\left(\sum_{n\in\mathcal{N}_k}\sum_{n'\in\mathcal{N}_{k+1}}\sum_{i=0}^{M-1}\exp[\Gamma_{k+1}(i,n,n')+B_{k+1}(n')]\right),\qquad(3.66)$$

and

$$\Gamma_k(i,n',n)=\ln(P(\mathbf{y}_{n'+1}^n|u_k=i)\cdot P(u_k=i))$$
$$=\ln P(\mathbf{y}_{n'+1}^n|u_k=i)+\ln P(u_k=i)$$
$$=-\sum_{j=1}^{l(\mathbf{c}_i)}\left(\frac{1}{2}\ln\pi N_0+\frac{(y_{n'+j}-\sqrt{E_s}(2c_{i_j}-1))^2}{N_0}\right)+\ln P(u_k=i).\quad(3.67)$$

Finally, the logarithm of the *a posteriori* probability $P(u_k=m|\mathbf{y})$ can be written as

$$\ln P(u_k=m|\mathbf{y})=\ln\left(\frac{\sum_{n\in\mathcal{N}_k}\sum_{n'\in\mathcal{N}_{k-1}}\gamma_k(m,n',n)\cdot\tilde{\alpha}_{k-1}(n')\cdot\tilde{\beta}_k(n)}{\sum_{n\in\mathcal{N}_k}\sum_{n'\in\mathcal{N}_{k-1}}\sum_{i=0}^{M-1}\gamma_k(i,n',n)\cdot\tilde{\alpha}_{k-1}(n')\cdot\tilde{\beta}_k(n)}\right)$$

$$=\ln\left(\sum_{n\in\mathcal{N}_k}\sum_{n'\in\mathcal{N}_{k-1}}\exp[\Gamma_k(m,n',n)+A_{k-1}(n')+B_k(n)]\right)$$

$$-\ln\left(\sum_{n\in\mathcal{N}_k}\sum_{n'\in\mathcal{N}_{k-1}}\sum_{i=0}^{M-1}\exp[\Gamma_k(i,n',n)+A_{k-1}(n')+B_k(n)]\right).$$

$$(3.68)$$

The logarithms in these quantities can be calculated using either the Max-Log approximation [250]:

$$\ln\left(\sum_i e^{x_i}\right)\approx\max_i(x_i),\qquad(3.69)$$

or the Jacobian logarithm [250]:

$$\ln(e^{x_1}+e^{x_2})=\max(x_1,x_2)+\ln(1+e^{-|x_1-x_2|})$$
$$=\max(x_1,x_2)+f_c(|x_1-x_2|)$$
$$=g(x_1,x_2),\qquad(3.70)$$

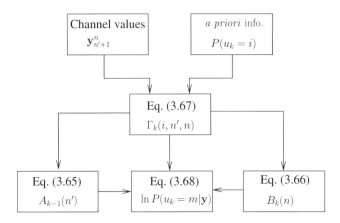

Figure 3.11: Summary of the symbol-level trellis-based MAP decoding operations.

where the correction function $f_c(x)$ can be approximated by [250]

$$f_c(x) = \begin{cases} 0.65 & 0 < x \le 0.2 \\ 0.55 & 0.2 < x \le 0.43 \\ 0.45 & 0.43 < x \le 0.7 \\ 0.35 & 0.7 < x \le 1.05 \\ 0.25 & 1.05 < x \le 1.5 \\ 0.15 & 1.5 < x \le 2.25 \\ 0.05 & 2.25 < x \le 3.7 \\ 0 & 3.7 < x, \end{cases} \tag{3.71}$$

and, upon recursively using $g(x_1, x_2)$, we have

$$\ln \left(\sum_{i=1}^{N} e^{x_i} \right) = g(x_N, g(x_{N-1}, \ldots, g(x_3, g(x_2, x_1)))). \tag{3.72}$$

Let us summarize the operations of the MAP decoding algorithm using Figure 3.11. The MAP decoder processes the channel observation sequence $\mathbf{y}$ and the source symbol probabilities $P(u_k)$ as its input, and calculates the transition probability $\Gamma_k(i, n', n)$ for each of the trellis branches using Equation (3.67). Then both forward- and backward-recursions are carried out in order to calculate the quantities of $A_k(n)$ and $B_k(n)$ for each of the trellis states, using Equations (3.65) and (3.66) respectively. Finally, the logarithm of the a *posteriori* probability $P(u_k = m|\mathbf{y})$ is computed using Equation (3.68).

3.4.2.2.2 Bit-Level Trellis-Based MAP Decoding. Like the MAP decoding algorithm based on the symbol-level trellis, as shown for example in Figure 3.3, the MAP decoding algorithm based on a bit-level trellis such as that in Figure 3.5 can also be formulated.

The forward and backward recursions are defined as follows [42]:

$$\alpha_n(s) = P(s_n = s, \mathbf{y}_1^n)$$

$$= \sum_{\text{all } s'} \sum_{i=0}^{1} \gamma_n(i, s', s) \cdot \alpha_{n-1}(s'), \tag{3.73}$$

with $\alpha_0(0) = 1$ and $\alpha(s \neq 0) = 0$; and

$$\beta_n(s) = P(\mathbf{y}_{n+1}^N | s_n = s)$$

$$= \sum_{\text{all } s'} \sum_{i=0}^{1} \gamma_{n+1}(i, s, s') \cdot \beta_{n+1}(s'), \tag{3.74}$$

with $\beta_N(0) = 1$ and $\beta_N(s \neq 0) = 0$. Furthermore, we have

$$\gamma_n(i, s', s) = P(y_n, b_n = i, s_n = s | s_{n-1} = s')$$

$$= P(y_n | b_n = i) \cdot P(b_n = i, s_n = s | s_{n-1} = s'), \tag{3.75}$$

where b_n is the input bit necessary to trigger the transition from state $s_{n-1} = s'$ to state $s_n = s$. The first term in Equation (3.75) is the channel-induced transition probability. For BPSK transmission over a memoryless AWGN channel, this term can be expressed as [70]

$$P(y_n | b_n) = \frac{1}{\sqrt{\pi N_0}} \exp\left(-\frac{(y_n - \sqrt{E_s}(2b_n - 1))^2}{N_0}\right)$$

$$= \frac{1}{\sqrt{\pi N_0}} \cdot \exp\left(-\frac{E_s}{N_0}\right) \cdot \exp\left(-\frac{y_n^2}{N_0}\right) \cdot \exp\left(\frac{2\sqrt{E_s}}{N_0} y_n (2b_n - 1)\right)$$

$$= C \cdot \exp\left(\frac{L_c}{2} y_n (2b_n - 1)\right), \tag{3.76}$$

where

$$C = \frac{1}{\sqrt{\pi N_0}} \cdot \exp\left(-\frac{E_s}{N_0}\right) \cdot \exp\left(-\frac{y_n^2}{N_0}\right) \tag{3.77}$$

is the same for both $b_n = 0$ and $b_n = 1$.

The second term in Equation (3.75) is the *a priori* source information, which can be calculated using Equation (3.41) of Section 3.4.2.1.2.

According to Equations (3.75) and (3.76), we have

$$\gamma_k(i, s', s) = C \cdot \exp\left(\frac{L_c}{2} y_n (2b_n - 1)\right) \cdot P(b_n = i, s_n = s | s_{n-1} = s'). \tag{3.78}$$

Finally, from Equations (3.73) to (3.78) the conditional Logarithmic Likelihood Ratio (LLR) of b_n for a given received sequence $\mathbf{y}$ can be expressed as

$$L(b_n|\mathbf{y}) = \ln\left(\frac{\sum_{(s',s)\Rightarrow b_n=0}\gamma_n(0, s', s)\alpha_{n-1}(s')\beta_n(s)}{\sum_{(s',s)\Rightarrow b_n=1}\gamma_n(1, s', s)\alpha_{n-1}(s')\beta_n(s)}\right)$$

$$= L_c y_k + \ln\left(\frac{\sum_{(s',s)\Rightarrow b_n=0}\alpha_{n-1}(s')\beta_n(s)P(b_n = 0, s|s')}{\sum_{(s',s)\Rightarrow b_n=1}\alpha_{n-1}(s')\beta_n(s)P(b_n = 1, s|s')}\right). \qquad (3.79)$$

It is worth noting that the source *a priori* information and the extrinsic information in Equation (3.79) cannot be separated. A further simplification of Equation (3.79) can be achieved by using the logarithms of the above quantities, as in Section 3.4.2.2.1.

3.4.2.3 Concatenation of MAP Decoding and Sequence Estimation

There are several reasons for concatenating MAP decoding with sequence estimation.

1. For bit-by-bit MAP decoding, the decoder maximizes the *a posteriori* probability of each bit, which results in achieving the minimum BER. However, the bit-by-bit MAP decoding is unable to guarantee that the output bit sequence will form a valid symbol-based path through the trellis, which implies that there may be invalid codewords in the decoded sequence. Hence MLSE may be performed after the bit-by-bit MAP decoding so as to find a valid VLC codeword sequence. Specifically, given the APPs $P(b_n|\mathbf{y})$ provided by the MAP decoder, the sequence estimator selects the specific sequence $\hat{\mathbf{b}}$ that maximizes the probability $P(\mathbf{b}|\mathbf{y})$, which is formulated as

$$\hat{\mathbf{b}} = \arg\max_{\mathbf{b}}\left\{P(\mathbf{b}|\mathbf{y}) \approx \prod_n P(b_n|\mathbf{y})\right\}. \qquad (3.80)$$

2. For symbol-by-symbol MAP decoding, the same problem as outlined above exists. The MAP decoder is capable of guaranteeing that the number of decoded symbols is equal to the number of transmitted symbols and that the corresponding SER is minimized. However, the number of decoded bits may still deviate from that of the transmitted bits. This might be a problem when converting the symbol-by-symbol reliability information to bit-by-bit reliability information. In this case, the MLSE may be performed in order to find a valid path in the trellis. Specifically, given the APPs $P(u_k|\mathbf{y})$ generated by the MAP decoder, the sequence estimator selects the specific sequence $\hat{\mathbf{u}}$ that maximizes the probability $P(\mathbf{u}|\mathbf{y})$, which is formulated as

$$\hat{\mathbf{u}} = \arg\max_{\mathbf{u}}\left\{P(\mathbf{u}|\mathbf{y}) = \prod_k P(u_k|\mathbf{y})\right\}. \qquad (3.81)$$

3.4.3 Simulation Results

In this section we characterize the achievable performance of the decoding schemes described in the previous sections.

For all simulations BPSK is used as the baseband modulation scheme and an AWGN channel is assumed. Furthermore, assume that the source is memoryless and it is encoded by the various VLCs [12] as shown in Table 3.12. To be more specific, HUFF represents a

Table 3.12: VLC codes used in the simulations

Symbol	Probability	HUFF	RVLC-1	RVLC-2
0	0.33	00	00	00
1	0.30	01	01	11
2	0.18	11	10	010
3	0.10	100	111	101
4	0.09	101	11011	0110
Average length		2.19	2.37	2.46
Code rate/efficiency		0.98	0.90	0.87
Constraint length		3	3.58	3.58
Free distance		1	1	2

Huffman code, RVLC-1 is a asymmetric RVLC having a free distance of 1 and RVLC-2 is a symmetric RVLC having a free distance of 2.

As we can see, the different source codes have different average codeword lengths; hence for encoding the same source message the resultant bit sequences will have different lengths, which implies that they require different total transmitted signal power. For the sake of fair comparison, we introduce the source code rate R_s, which is defined as

$$R_s \triangleq \frac{H(u)}{\bar{L}}, \tag{3.82}$$

where $H(u)$ is the entropy of the symbol source U, and $\bar{L}$ is the average codeword length. Consequently, in our simulations E_b/N_0 is calculated as

$$E_b/N_0 = E_s/(R_s \cdot \log_2 M)/N_0, \tag{3.83}$$

where E_s is the transmitted energy per channel symbol, while M is the number of modulation levels in the signal space of M-ary modulation. For BPSK modulation we have $M = 2$.

As for our performance measure, we use the so-called Levenshtein SER [40], since, owing to the self-synchronization properties of VLCs, a simple symbol-by-symbol comparison of the decoded sequence with the original sequence is not fair in the context of numerous applications. If a decoding error occurs, there might be VLC symbol insertions or deletions in the decoded sequence. After resynchronization the remaining sequence therefore might be a shifted version of the original one, although a simple symbol comparison would identify the entire remaining part of the sequence as corrupted. A detection metric that is more appropriate in this case is the Levenshtein distance [251], which is defined as the minimal number of symbol insertions, deletions or substitutions that transform one symbol sequence into another. In our simulations we evaluate the SER in terms of the Levenshtein distance.

3.4.3.1 Performance of Symbol-Level Trellis-Based Decoding

In our simulations, six different decoding methods are used, namely MAP/ML sequence estimation using hard decisions (MAP/MLSE-HARD), MAP/ML sequence estimation using soft decisions (MAP/MLSE-SOFT), symbol-by-symbol MAP decoding (SMAP) and

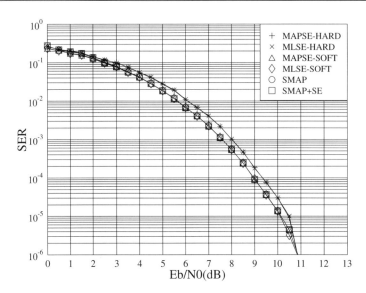

Figure 3.12: SER versus E_b/N_0 results for the *Huffman code HUFF* of Table 3.12, using *symbol-level trellis* decoding and BPSK transmission over AWGN channels.

symbol-by-symbol MAP decoding concatenated with sequence estimation (SMAP+SE). For every decoding scheme, we use source symbol packets of length $K = 100$ and transmit 1000 different realizations of the source packets. Furthermore, each of these 1000 packets is transmitted over 50 channel realizations. The simulations are curtailed when all the packets have been transmitted or the number of source symbol errors reaches 5000.

Our results for the Huffman code HUFF are shown in Figure 3.12. As we can see, for the Huffman code the performance of soft-decoding schemes is slightly better than that of the corresponding hard-decoding schemes (gain < 0.5 dB). Furthermore, the performance of the MAP decoding scheme is similar to that of the ML decoding scheme, both in the case of soft decoding and for hard decoding. Moreover, the performance of sequence estimation using soft decision is similar to that of the symbol-by-symbol MAP decoding, although the former has a significantly lower complexity. The results of the concatenated MAP decoding and sequence estimation are also shown in Figure 3.12. The concatenation of these two schemes does not affect the achievable SER performance, but it is necessary for solving the previously mentioned length inconsistency problem, namely that the number of decoded bits may be different from the number of transmitted bits.

Our results recorded for the RVLCs RVLC-1 and RVLC-2 scheme of Table 3.12 are shown in Figures 3.13 and 3.14 respectively. In this case it can be seen that the soft-decoding schemes significantly outperform the hard-decoding schemes. More explicitly, RVLC-2 achieves a gain up to 3 dB, as shown in Figure 3.14. By contrast, observe in Figure 3.14 that in the case of hard decision the MAP decoding scheme performs slightly better than ML decoding, while they perform similarly in the case of soft decisions. As in the Huffman coded scenario, the performance of soft sequence estimation is almost the same as that of symbol-by-symbol MAP decoding, as evidenced by Figures 3.13 and 3.14.

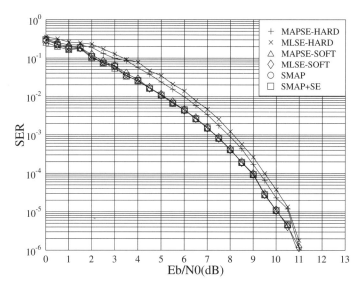

Figure 3.13: SER versus E_b/N_0 results for the *RVLC-1* of Table 3.12, using *symbol-level trellis* decoding and BPSK transmission over AWGN channels.

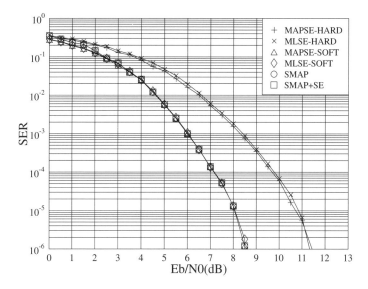

Figure 3.14: SER versus E_b/N_0 results for the *RVLC-2* of Table 3.12, using *symbol-level trellis* decoding and BPSK transmission over AWGN channels.

The performances of the HUFF, RVLC-1 and RVLC-2 schemes are compared in Figure 3.15. In the case of hard decoding, the RVLC-1 arrangement performs similarly to Huffman coding and slightly better than the RVLC-2 scheme. In the case of soft decoding,

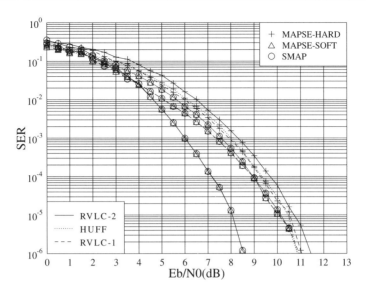

Figure 3.15: SER versus E_b/N_0 *performance comparison* for HUFF, RVLC-1 and RVLC-2 of Table 3.12, using *symbol-level trellis* decoding and BPSK transmission over AWGN channels.

the RVLC-2 significantly outperforms both the Huffman code and the RVLC-1 at medium to high SNRs in excess of $E_b/N_0 > 4$ dB. The achievable E_b/N_0 gain is as high as about 2 dB at a SER of 10^{-5}.

3.4.3.2 *Performance of Bit-Level Trellis-Based Decoding*

In these investigations five different decoding methods are used, namely MAP/ML sequence estimation using hard decisions (MAP/MLSE-HARD), MAP/ML sequence estimation using soft decisions (MAP/MLSE-SOFT) and bit-by-bit MAP decoding concatenated with sequence estimation (BMAP+SE). For each of the decoding schemes, we use the same simulation parameters as those set up for the symbol-level trellis-based decoding described in Section 3.4.3.1.

The results recorded for the Huffman code of Table 3.12 are shown in Figure 3.16. As we can see, for the Huffman code the achievable performance of all decoding schemes is similar.

The corresponding results recorded for RVLC-1 and RVLC-2 of Table 3.12 are shown in Figures 3.17 and 3.18 respectively. This time, the soft-decoding schemes significantly outperform the hard-decoding-aided schemes, especially for the code RVLC-2. The achievable E_b/N_0 gain is approximately 2.5 dB at SER $= 10^{-5}$. In the case of hard decisions, the MAP decoding scheme performs slightly better than the ML decoding scheme, while they perform similarly in the case of soft decisions.

The performances of the Huffman code, RVLC-1 and RVLC-2 are compared in Figure 3.19. In the case of hard decoding the Huffman code performs the best, while in the case of soft decoding the RVLC-1 and RVLC-2 schemes outperform the Huffman code at medium to high channel SNRs in excess of $E_b/N_0 > 2$ dB, and RVLC-2 performs the best. Its gain is as high as about 2 dB at SER $= 10^{-5}$.

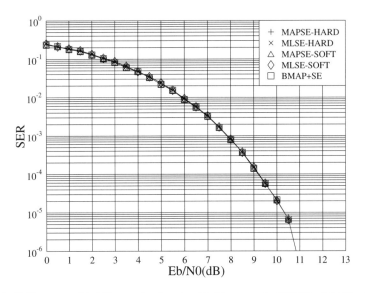

Figure 3.16: SER versus E_b/N_0 results for the *Huffman code HUFF* of Table 3.12, using *bit-level trellis* decoding and BPSK transmission over AWGN channels.

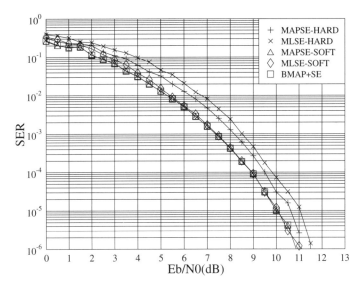

Figure 3.17: SER versus E_b/N_0 results for the *RVLC-1* of Table 3.12, using *bit-level trellis* decoding and BPSK transmission over AWGN channels.

3.4.3.3 *Comparison of Symbol-Level Trellis and Bit-Level Trellis-Based Decoding*

The achievable performance of various symbol-level trellis-based decoding schemes and those of their bit-level counterparts are compared in Figures 3.20, 3.21 and 3.22 for the Huffman code and the RVLC RVLC-1 and RVLC-2 schemes respectively.

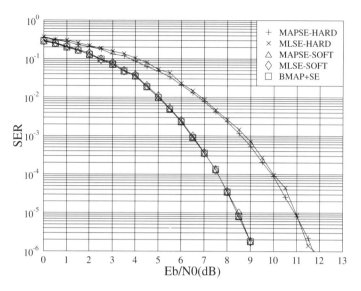

Figure 3.18: SER versus E_b/N_0 results for the *RVLC-2* of Table 3.12, using *bit-level trellis* decoding and BPSK transmission over AWGN channels.

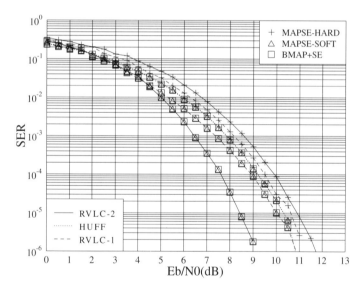

Figure 3.19: SER versus E_b/N_0 *performance comparison* for HUFF, RVLC-1 and RVLC-2 of Table 3.12, using *bit-level trellis* decoding and BPSK transmission over AWGN channels.

In all of the decoding schemes considered, we assume that the length of the transmitted packets, expressed in terms of bits or symbols as applicable, is known at the decoders. More specifically, the length information includes the number of transmitted bits for the bit-level

Table 3.13: Comparison of various VLC construction algorithms when using the English alphabet as the source

$d_f = 1$ RVLC

	Takishima's Algorithm [12]	Tsai's Algorithm [21]	Lakovic's Algorithm [23]	Algorithm 3.1 of Table 3.3	Algorithm 3.3 of Table 3.9
Average Length	4.36068	4.30678	4.25145	4.18732	4.19959
Max Length	12	10	10	11	10
Efficiency	0.95	0.96	0.97	0.98	0.98

$d_f = 2$ RVLC

	Lakovic's Algorithm [22]	Algorithm 3.1 of Table 3.5	Algorithm 3.3 of Table 3.9
Average Length	4.34534	4.33957	4.23656
Max Length	10	8	9
Efficiency	0.95	0.95	0.97

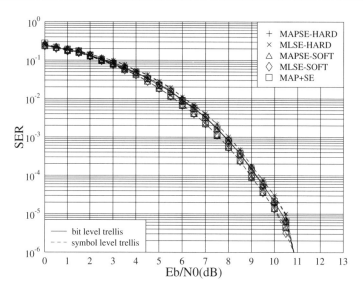

Figure 3.20: SER versus E_b/N_0 *performance comparison* of symbol-level trellis-based decoding and bit-level trellis-based decoding for the *Huffman code HUFF* of Table 3.12, using BPSK transmission over AWGN channels.

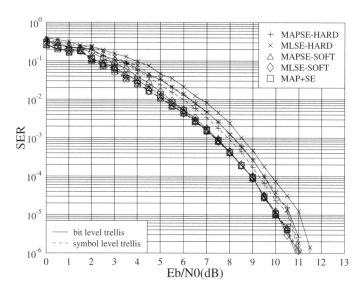

Figure 3.21: SER versus E_b/N_0 *performance comparison* of symbol-level trellis-based decoding and bit-level trellis-based decoding for *RVLC-1* of Table 3.12, using BPSK transmission over AWGN channels.

trellis-based decoding schemes, while both the number of transmitted bits and the number of source symbols has to be known for the symbol-level trellis-based decoding schemes.

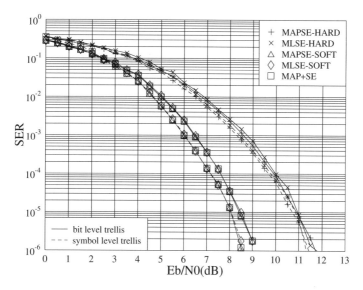

Figure 3.22: SER versus E_b/N_0 *performance comparison* of symbol-level trellis-based decoding and bit-level trellis-based decoding for *RVLC-2* of Table 3.12, using BPSK transmission over AWGN channels.

As shown in Figure 3.20, for the Huffman code the symbol-level trellis-based decoding schemes perform slightly better than the bit-level trellis-based decoding schemes, when using soft decoding. However, the above observations are reversed in Figure 3.20 in the case of hard decoding. For the RVLC-1, the symbol-level trellis-based decoding schemes perform slightly better than the corresponding bit-level trellis-based decoding schemes when using hard decoding, while they perform similarly in the case of soft decoding, as seen in Figure 3.21. Finally, for RVLC-2 the symbol-level trellis-based decoding schemes perform consistently better (gain < 0.5 dB) than the corresponding bit-level trellis-based decoding schemes, for both hard decoding and soft decoding.

3.5 Summary and Conclusions

In this chapter we have investigated the design of efficient VLC source codes and their soft decoding methods. After reviewing their construction methods in Section 3.3, we proposed a novel algorithm for the design of efficient RVLCs and VLEC codes in Sections 3.3.4 and 3.3.5 respectively. This optimizes the VLC codeword length distribution on a length-by-length basis. For RVLCs it results in codes of higher efficiency and/or shorter maximum codeword length than the algorithms previously disseminated in the literature, as shown in Tables 3.3, 3.4, 3.5 and 3.9. For VLEC codes, the techniques proposed in Section 3.3.5 significantly reduce the associated construction complexity, as shown in Table 3.10. The optimization procedure is independent of the codeword selection mechanism; therefore, it can be combined with all existing algorithms aiming for efficient codeword design. Our main results are summarized in Table 3.13.

Table 3.14: Summary of ML VLC decoding performance

	Symbol-level trellis			Bit-level trellis		
	Soft Dec.[†]	Hard Dec.[†]	Gain[‡]	Soft Dec.[†]	Hard Dec.[†]	Gain[‡]
HUFF	10.1 dB	10.5 dB	0.4 dB	10.3 dB	10.3 dB	0.0 dB
RVLC-1	10.0 dB	10.5 dB	0.5 dB	10.0 dB	11.0 dB	1.0 dB
RVLC-2	8.0 dB	10.9 dB	2.9 dB	8.5 dB	11.0 dB	2.5 dB

[†] The numerical values in this column are the minimum E_b/N_0 values required for achieving a SER of 10^{-5} for the different decoding algorithms.

[‡] The numerical values in this column are the E_b/N_0 gains of soft-decision decoding over the corresponding hard-decision decoding.

In Section 3.4 we considered the soft decoding of VLCs. Both symbol-level trellises and bit-level trellises were described. Based on these trellis representations, two different types of decoding methods – namely sequence estimation using the VA, and MAP decoding employing the BCJR algorithm – were discussed with the aid of a host of simulation results, as shown in Section 3.4.3. Table 3.14 lists some of the ML VLC decoding results.

Based on the simulation results in Table 3.14 and those in Section 3.4.3, we can reach the following conclusions:

- **Soft decoding versus hard decoding**
  At the beginning of Section 3.4 we conjectured that soft decoding may outperform hard decoding according to one of Viterbi's three lessons [247]. Whether this conjecture is true depends on both the source code and the bit- versus symbol-based trellis structure used. For the Huffman code, soft decoding performs similar to hard decoding in the case of using a bit-level trellis, while soft decoding performs better in the case of symbol-level trellis as evidenced in Figure 3.20. Furthermore, observe in Figures 3.21 and 3.22 that for RVLC-1 and RVLC-2 schemes soft decoding outperforms hard decoding in the case of both trellises. It can be shown that the larger the free distance and the higher the constraint length of the source code, the larger the gain that soft decoding achieves over hard decoding.

- **MAP versus ML sequence estimation**
  Our simulation results in Section 3.4.3 show that we cannot attain any performance gain for MAP sequence estimation over ML sequence estimation in the case of soft decoding, although the former exploits the source symbol-probability information. However, in the case of hard decoding MAP sequence estimation achieves some gain over ML sequence estimation for both RVLC-1 and RVLC-2 schemes, as shown in Figures 3.13, 3.14, 3.17 and 3.18.

In addition to the effects of different source codes, Buttigieg [30] investigated the effects of different source statistics on the achievable performance and proposed the so-called MAP factor performance metric. It has been shown that the larger the MAP factor, the larger the attainable E_b/N_0 gain [30].

- **Bit-level trellis versus symbol-level trellis**

  Since the symbol-level trellis is capable of exploiting more source information than the bit-level trellis can, the decoding schemes based on a symbol-level trellis are typically capable of performing slightly better than their bit-level counterparts, which can be seen from Figures 3.20, 3.21 and 3.22, as well as Table 3.14. However, according to the construction methods described in Section 3.4.1 the bit-level trellis structure only depends on the specific VLC source code, which is time invariant. By contrast, the symbol-level trellis construction depends on both the specific VLC source code and the specific length of the frame to be decoded; hence it is data dependent. Therefore, the decoding scheme based on the bit-level trellis is typically of lower complexity than that based on the symbol-level trellis.

- **SE versus symbol-by-symbol MAP decoding**

  We observed in our investigations that the symbol-by-symbol or bit-by-bit MAP decoding schemes perform similarly to the corresponding SE schemes using soft decisions, as shown in Figures 3.12–3.19. However, the MAP decoding schemes are capable both of exploiting the soft inputs generated by the previous decoding stage and of providing soft outputs, which can be used in a successive decoding stage. These issues will be explored in more detail in the next chapter.

- **Effect of source codes**

  In the investigations of this chapter, various source codes have been used, including a $d_f = 1$ Huffman code, a $d_f = 1$ RVLC and a $d_f = 2$ RVLC scheme. It is clear that the free distance of the VLC plays an important role in determining the MAP/ML decoding performance when using soft-decision-based decoding. Our results, seen in Figures 3.15 and 3.19, demonstrate that, as expected, the $d_f = 2$ RVLC significantly outperforms the $d_f = 1$ RVLC and the Huffman code.

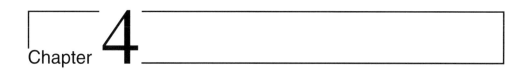

Chapter 4

Iterative Source–Channel Decoding[1]

Code concatenation constitutes a convenient technique for constructing powerful codes capable of achieving huge coding gains, while keeping the decoding complexity manageable. The concept of Serially Concatenated Codes (SCCs) having an 'outer' and an 'inner' code used in cascade was proposed in [125]. The discovery of Parallel Concatenated 'turbo' Codes (PCCs) [69] considerably improved the achievable performance gains by separating component codes through interleavers and using iterative decoding in order to further reduce the BER. The intensive research efforts of the ensuing era demonstrated [70] that the employment of iterative processing techniques is not limited to traditional concatenated coding schemes. In other words, the 'turbo principle' [53] is applicable to numerous other algorithms that can be found in digital communications, for example in turbo equalization [82], spectrally efficient modulation [83, 84], turbo multiuser detection [87, 88] and channel-coded code-division multiple-access [90].

Previously, substantial research efforts have been focussed on optimizing concatenated coding schemes in order to improve the asymptotic slopes of their error probability curves, especially at moderate to high SNRs. Recently, researchers have focussed their efforts on investigating the convergence behavior of iterative decoding. In [77] the authors proposed a so-called density evolution algorithm for calculating the convergence thresholds of randomly constructed irregular LDPC codes designed for the AWGN channel. This method was used to study the convergence of iterative decoding for turbo-like codes in [104], where a SNR measure was proposed for characterizing the inputs and outputs of the component decoders. The mutual information between the data bits at the transmitter and the soft values at the receiver was used to describe the flow of extrinsic information between two soft-in soft-out decoders in [106]; the exchange of extrinsic information between constituent codes may be visualized as a staircase-shaped gradually increasing decoding trajectory within the EXIT charts of both parallel [98, 111] and serially concatenated codes [112]. A comprehensive

---

[1]Parts of this chapter are based on the collaborative research outlined in [209] © IEEE (2005), and [210] © IEEE (2006).

Near-Capacity Variable-Length Coding   Lajos Hanzo, Robert G. Maunder, Jin Wang and Lie-Liang Yang

study of the convergence behavior of iterative decoding schemes for bit-interleaved coded modulation, equalization and serially concatenated binary codes was presented in [115].

In this chapter we will investigate an intriguing application of the turbo principle [53], namely ISCD [42, 45, 252], where the channel decoder and the source decoder can be viewed as a pair of serial concatenated codes. By applying the turbo principle, the residual redundancy in the source after source encoding may be efficiently utilized for improving the system's performance in comparison with that of separate source–channel decoding. Furthermore, the convergence behavior of various ISCD schemes will be analyzed with the aid of EXIT charts.

This chapter is organized as follows. Section 4.1 provides a brief introduction to various concatenated schemes and that of the iterative decoding principle: the turbo principle. The kernel of the turbo principle – the SISO APP decoding algorithm – and the semi-analytical tool of EXIT charts are described in Section 4.2. The achievable performance of the ISCD schemes is investigated in Sections 4.3 and 4.4, when communicating over AWGN channels and dispersive channels respectively. Section 4.5 concludes the chapter.

4.1   Concatenated Coding and the Turbo Principle

In this section we briefly describe several commonly used concatenated schemes, namely parallel concatenation [69,126] serial concatenation [71,130] and hybrid concatenation [253–255] of various constituent codes. Furthermore, we demonstrate that the turbo principle may be applied for detecting any of the above schemes.

4.1.1   Classification of Concatenated Schemes

4.1.1.1   Parallel Concatenated Schemes

The classic turbo codes [69] employ a parallel concatenated structure. Figure 4.1 shows two examples of the family of turbo codes. A so-called systematic code is shown in Figure 4.1a, where the input information bits are simply copied to the output, interspersed with the parity bits. By contrast a non-systematic code is shown in Figure 4.1b, where the original information bits do not explicitly appear at the output. The component encoders are often convolutional encoders, but binary Bose–Chaudhuri–Hocquenghem (BCH) codes [70] have also been used. At the encoder, the input information bits are encoded by the first constituent encoder (Encoder I) and an interleaved version of the input information bits is encoded by the second constituent encoder (Encoder II). The encoded output bits may be punctured, and hence arbitrary coding rates may be obtained. Furthermore, this structure can be extended to a parallel concatenation of more than two component codes, leading to multiple-stage turbo codes [126, 256].

4.1.1.2   Serially Concatenated Schemes

The basic structure of a SCC [71] is shown in Figure 4.2. The SCC encoder consists of an outer encoder (Encoder I) and an inner encoder (Encoder II), interconnected by an interleaver. The introduction of the interleaver scrambles the bits before they are passed to the other constituent encoder, which ensures that even if a specific bit has been gravely contaminated by the channel, the chances are that the other constituent decoder is capable of providing more reliable information concerning this bit. This is a practical manifestation of time diversity. Iterative processing is used in the SCC decoder, and a performance similar to that of a classic PCC may be achieved [71]. In fact, the serial concatenation constitutes quite a general

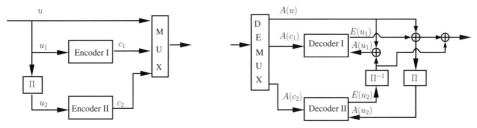

(a) A systematic parallel concatenated code: the encoder and the decoder.

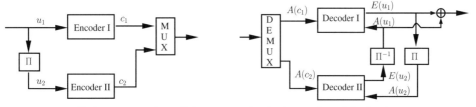

(b) A non-systematic parallel concatenated code: the encoder and the decoder.

Figure 4.1: Two examples of parallel concatenated codes. The constituent codes are non-systematic codes having arbitrary code rates.

Figure 4.2: The schematic of a serially concatenated code: the encoder and the decoder.

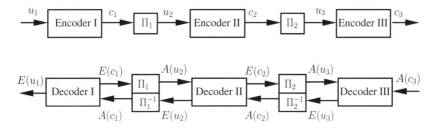

Figure 4.3: The schematic of a three-stage serially concatenated code: the encoder and the decoder.

structure and many decoding and detection schemes can be described as serially concatenated structures, such as those used in turbo equalization [82], coded modulation [83, 84], turbo multiuser detection [87, 88], joint source–channel decoding [42, 45, 252], LDPC decoding [77] and others [53, 81]. Similarly, a serially concatenated scheme may contain more than two components, as in [120–122, 130]. Figure 4.3 shows the schematic of a three-stage serially concatenated code [122, 130].

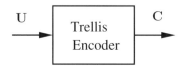

Figure 4.4: A trellis encoder, which processes the input information symbol sequence $\mathbf{U}$ and outputs encoded symbol sequence $\mathbf{C}$. The encoder's state transitions may be described by certain trellis diagrams.

4.1.1.3 Hybrid Concatenated Schemes

A natural extension of PCCs and SCCs is the family of hybrid concatenated codes [253–255], which combine the two principles. More specifically, some of the components are concatenated in parallel and some are serially concatenated. Again, the employment of this scheme is not restricted to channel coding. For example, combined turbo decoding and turbo equalization [257, 258] constitutes a popular application.

4.1.2 Iterative Decoder Activation Order

Since the invention of turbo codes [69], iterative (turbo) decoding has been successfully applied to diverse detection and decoding problems, which may inherit any of the three concatenated structures described above. Hence this technique has been referred to as the 'Turbo Principle' [53]. The crucial point at the receiver is that the component detectors or decoders have to be soft-in soft-out decoders that are capable of accepting and delivering probabilities or soft values, and the extrinsic part of the soft output of one of the decoders is passed on to the other decoder to be used as *a priori* input. If there are only two component decoders, the decoding alternates between them. When there are more component decoders, the activation order of the decoders is an important issue, which is sometimes referred to as 'scheduling' in the related literature – borrowing the terminology from the field of resource allocation. The specific activation order may substantially affect the decoding complexity, delay and performance of the system [121, 122].

4.2 SISO APP Decoders and their EXIT Characteristics

The key point of iterative decoding of various concatenated schemes is that each component decoder employs a SISO algorithm. Benedetto *et al.* [254] proposed a SISO APP module, which implements the APP algorithm [48] in its basic form for any trellis-based decoder/detector. This SISO APP module is employed in our ISCD schemes in the rest of this treatise unless stated otherwise. Moreover, the capability of processing *a priori* information and generating extrinsic information is an inherent characteristic of a SISO APP module, which may be visualized by EXIT charts. Let us first introduce the SISO APP module.

4.2.1 A Soft-Input Soft-Output APP Module

4.2.1.1 The Encoder

Figure 4.4 shows a trellis encoder, which processes the input information symbol sequence $\mathbf{U}$ and outputs encoded symbol sequence $\mathbf{C}$. The code trellis can be time invariant or time varying. For simplicity of exposition, we will refer to the case of binary time-invariant convolutional codes having code rate of $R = k_0/n_0$.

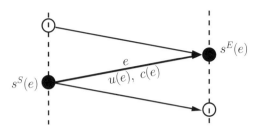

Figure 4.5: An edge of the trellis section.

The input symbol sequence $\mathbf{U} = (U_k)_{k \in \mathcal{K}}$ is defined over a symbol index set $\mathcal{K}$ and drawn from the alphabet $\mathcal{U} = \{u_1, \ldots, u_{N_I}\}$. Each input symbol U_k consists of k_0 bits $U_k^j, j = 1, 2, \ldots, k_0$ with realization $u^j \in \{0, 1\}$. The encoded output sequence $\mathbf{C} = (C_k)_{k \in \mathcal{K}}$ of Figure 4.4 is defined over the same symbol index set $\mathcal{K}$, and drawn from the alphabet $\mathcal{C} = \{c_1, \ldots, c_{N_o}\}$. Each output symbol C_k consists of n_0 bits $C_k^j, j = 1, 2, \ldots, n_0$ with realization $c^j \in \{0, 1\}$.

The legitimate state transitions of a time-invariant convolutional code are completely specified by a single trellis section, which describes the transitions ('edges') between the states of the trellis at time instants k and $k + 1$. A trellis section is characterized by:

- a set of N states $\mathcal{S} = \{s_1, \ldots, s_N\}$ (the state of the trellis at time k is $S_k = s$, with $s \in \mathcal{S}$);
- a set of $N \cdot N_I$ edges obtained by the Cartesian product $\mathcal{E} = \mathcal{S} \times \mathcal{U} = \{e_1, \ldots, e_{N \cdot N_I}\}$, which represent all legitimate transitions between the trellis states.

As seen in Figure 4.5, the following functions are associated with each edge $e \in \mathcal{E}$:

- the starting state $s^S(e)$;
- the ending state $s^E(e)$;
- the input symbol $u(e)$;
- the output symbol $c(e)$.

The relationship between these functions depends on the particular encoder, while the assumption that the pair $[s^S(e), u(e)]$ uniquely identifies the ending state $s^E(e)$ should always be fulfilled, since it is equivalent to stating that, given the initial trellis state, there is a one-to-one correspondence between the input sequences and the state transitions.

4.2.1.2   The Decoder

The APP algorithm designed for the SISO decoder operates on the basis of the code's trellis representation. It accepts as inputs the probabilities of each of the original information symbols and of the encoded symbols, labels the edges of the code trellis and generates as its outputs an updated version of these probabilities based upon the code constraints. Hence the SISO module may be viewed as a four-terminal processing block, as shown in Figure 4.6.

More specifically, we associate the sequence of information symbols $\mathbf{u}$ with the sequence of *a priori* probabilities $\mathbf{P}(\mathbf{u}; I) = (P_k(u; I))_{k \in \mathcal{K}}$. Furthermore, we associate the sequence of encoded symbols $\mathbf{c}$ with the sequence of *a priori* probabilities $\mathbf{P}(\mathbf{c}; I) = (P_k(c; I))_{k \in K}$,

Figure 4.6: A SISO module that processes soft information input of $\mathbf{P}(\mathbf{u}; I)$ as well as $\mathbf{P}(\mathbf{c}; I)$ and generates soft information output of $\mathbf{P}(\mathbf{u}; O)$ as well as $\mathbf{P}(\mathbf{c}; O)$.

Algorithm 4.1

---

1. The *a posteriori* probabilities $\tilde{P}_k(u^j : O)$ and $\tilde{P}_k(c^j : O)$ associated with the jth bit within each symbol at time k are computed as [254]

$$\tilde{P}_k(u : O) = \tilde{H}_{u^j} \sum_{e:U_k^j(e)=u^j} \alpha_{k-1}(s^S(e))\beta_k(s^E(e))\gamma_k(e) \tag{4.1}$$

and

$$\tilde{P}_k(c : O) = \tilde{H}_{c^j} \sum_{e:C_k^j(e)=c^j} \alpha_{k-1}(s^S(e))\beta_k(s^E(e))\gamma_k(e) \tag{4.2}$$

respectively, where $\tilde{H}_{u^j}$ and $\tilde{H}_{c^j}$ are normalization factors ensuring that $\sum_{u^j} \tilde{P}_k(u : O) = 1$ and $\sum_{c^j} \tilde{P}_k(c : O) = 1$, respectively.

2. $\gamma_k(e) \triangleq P(s^E(e), e|s^S(e))$ is the state transition probability associated with each edge at the trellis section k of Figure 4.5, which is computed as

$$\gamma_k(e) = P_k[u(e); I] \cdot P_k[c(e); I]. \tag{4.3}$$

3. The quantity $\alpha_k(s)$ is associated with each state s, which can be obtained with the aid of the *forward* recursions originally proposed in [48] and detailed in [70] as

$$\alpha_k(s) = \sum_{e:s^E(e)=s} \alpha_{k-1}[s^S(e)]\gamma_k(e), \quad k = 1, \ldots, n, \tag{4.4}$$

with the initial values of $\alpha_0(s) = 1$ if $s = S_0$, and $\alpha_0(s) = 0$ otherwise.

4. The quantity $\beta_k(s)$ is also associated with each state s, which can be obtained through the *backward* recursions as [254]

$$\beta_k(s) = \sum_{e:s^S(e)=s} \beta_{k+1}[s^S(e)]\gamma_{k+1}(e), \quad k = n - 1, \ldots, 0, \tag{4.5}$$

with the initial values of $\beta_n(s) = 1$ if $s = S_n$, and $\beta_n(s) = 0$ otherwise.

---

where we have $P_k(c; I) = \prod_{j=1}^{n_0} P_k(c^j; I)$. The assumption that the *a priori* probabilities of specific symbols can be represented as the product of the probabilities of its constituent bits may be deemed valid when sufficiently long independent bit interleavers, rather than symbol interleavers, are used in the iterative decoding scheme of a concatenated code. The two outputs of the SISO module are the updated probabilities $\mathbf{P}(\mathbf{u}; O)$ and $\mathbf{P}(\mathbf{c}; O)$. Here, the letters I and O refer to the input and the output of the SISO module of Figure 4.6 respectively.

Let the symbol index set $\mathcal{K} = \{1, \ldots, n\}$ be finite. The SISO module operates by evaluating the output probabilities, as shown in Algorithm 4.1.

The extrinsic probabilities $P_k(u^j; O)$ and $P_k(c^j; O)$ are obtained by excluding the *a priori* probabilities $P_k(u^j; I)$ and $P_k(c^j; I)$ of the individual bits concerned from the *a posteriori* probabilities, yielding [254]

$$P_k(u^j; O) = H_{u^j} \frac{\tilde{P}_k(u:O)}{P_k(u^j; I)} \tag{4.6}$$

and [254]

$$P_k(c^j; O) = H_{c^j} \frac{\tilde{P}_k(c:O)}{P_k(c^j; I)}, \tag{4.7}$$

where, again, H_{u^j} and H_{c^j} are normalization factors ensuring that $\sum_{u^j} P_k(u:O) = 1$ and $\sum_{c^j} P_k(c:O) = 1$ respectively.

Note that in order to avoid the reuse of already-exploited information, it is necessary to remove the *a priori* information from the *a posteriori* information before feeding it back to the concatenated decoder. More specifically, since the *a priori* information originated from the concatenated decoder it is unnecessary, and in fact detrimental, to feed this information back. Indeed, the extrinsic information that remains after the *a priori* information has been removed from the *a posteriori* information represents the *new* information that is unknown, and therefore useful, to the concatenated decoder.

For the sake of avoiding any numerical instability and reducing the computational complexity, the above algorithm relying on multiplications can be converted to an additive form in the log domain. More details on the released operations can be found in Appendix A. Furthermore, we may use Logarithmic Likelihood Ratios (LLRs) instead of probabilities, where we define

$$L(U_k^j; I) \triangleq \ln[P_k(U_k^j = 0; I)] - \ln[P_k(U_k^j = 1; I)], \tag{4.8}$$

$$L(C_k^j; I) \triangleq \ln[P_k(C_k^j = 0; I)] - \ln[P_k(C_k^j = 1; I)], \tag{4.9}$$

$$L(U_k^j; O) \triangleq \ln[P_k(U_k^j = 0; O)] - \ln[P_k(U_k^j = 1; O)], \tag{4.10}$$

$$L(C_k^j; O) \triangleq \ln[P_k(C_k^j = 0; O)] - \ln[P_k(C_k^j = 1; O)]. \tag{4.11}$$

Consequently, in the log domain, the inputs of the four-terminal SISO module are $L(\mathbf{u}; \mathbf{I})$ and $L(\mathbf{c}; \mathbf{I})$, while its outputs are $L(\mathbf{u}; \mathbf{O})$ and $L(\mathbf{c}; \mathbf{O})$, as shown in Figure 4.7.

Figure 4.7: A SISO module operating in the log domain.

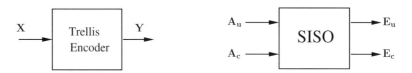

Figure 4.8: A trellis encoder and the corresponding SISO Logarithmic *A Posteriori* Probability (log-APP) decoder.

4.2.2 EXIT Chart

4.2.2.1 *Mutual Information*

For convenience, we redraw the previously introduced trellis encoder of Figure 4.4 and the corresponding SISO APP module of Figure 4.6 as seen in Figure 4.8. We will use the following notations:

- $\mathbf{X} = (X_n)_{n \in \{0,\dots,N_u-1\}}$ represents the sequence of information bits of length N_u, which is input to the trellis encoder, while X_n is a binary random variable denoting the information bits having realizations $x_n \in \{\pm1\}$;
- $\mathbf{Y} = (Y_n)_{n \in \{0,\dots,N_c-1\}}$ is the sequence of encoded bits of length N_c output by the trellis encoder, while Y_n is a binary random variable, denoting the encoded bits with assumed values of $y_n \in \{\pm1\}$;
- $\mathbf{A_u} = (A_{u,n})_{n \in \{0,\dots,N_u-1\}}$ is the sequence of the *a priori* LLR input $\mathbf{L}(\mathbf{u}; I)$ and $A_{u,n} \in \mathbb{R}$ denotes the *a priori* LLR of the information bit X_n;
- $\mathbf{A_c} = (A_{c,n})_{n \in \{0,\dots,N_c-1\}}$ is the sequence of the *a priori* LLR inputs $\mathbf{L}(\mathbf{c}; I)$, where $A_{c,n} \in \mathbb{R}$ denotes the *a priori* LLR of the code bit Y_n;
- $\mathbf{E_u} = (E_{u,n})_{n \in \{0,\dots,N_u-1\}}$ is the sequence of extrinsic LLR outputs $\mathbf{L}(\mathbf{u}; O)$ and $E_{u,n} \in \mathbb{R}$ denotes the extrinsic LLR of the information bit X_n;
- $\mathbf{E_c} = (E_{c,n})_{n \in \{0,\dots,N_c-1\}}$ is the sequence of extrinsic LLR outputs $\mathbf{L}(\mathbf{c}; O)$, while $E_{c,n} \in \mathbb{R}$ denotes the extrinsic LLR for the code bit Y_n.

We assume that the statistical properties of $X_n, Y_n, A_{u,n}, A_{c,n}, E_{u,n}$ and $E_{c,n}$ are independent of the index n; hence the subscript n is omitted in our forthcoming derivations. Then the Shannonian mutual information between the *a priori* LLR A_u and the information bit X is defined as [98]

$$I(X; A_u) = \frac{1}{2} \sum_{x=\pm1} \int_{-\infty}^{+\infty} f_{A_u}(\xi|X=x) \log_2 \frac{2 f_{A_u}(\xi|X=x)}{f_{A_u}(\xi|X=-1) + f_{A_u}(\xi|X=+1)} \, d\xi,$$

$$(4.12)$$

where $f_{A_u}(\xi|X=x)$ is the conditional probability density function associated with the *a priori* LLR A_u, and the information bits X are assumed to be equiprobable; i.e., we have $P(X=+1)=P(X=-1)=1/2$.

In order to compute the mutual information $I(X;A_u)$ of Equation (4.12), the conditional PDF $f_{A_u}(\xi|X=x)$ of the LLR A_u has to be known. If we model the *a priori* LLR A_u as an independent Gaussian random variable n_{A_u} having a variance of $\sigma^2_{A_u}$ and a mean of zero in conjunction with the information bit X, then we have

$$A_u = \mu_{A_u} X + n_{A_u}, \qquad (4.13)$$

where μ_{A_u} must fulfill

$$\mu_{A_u} = \sigma^2_{A_u}/2. \qquad (4.14)$$

According to the consistency condition [77] of LLR distribution, $f_{A_u}(\xi|X=x) = f_{A_u}(-\xi|X=x) \cdot e^{x\xi}$ and the conditional PDF $f_{A_u}(\xi|X=x)$ becomes

$$f_{A_u}(\xi|X=x) = \frac{1}{\sqrt{2\pi}\sigma_{A_u}} \exp\left\{ \frac{-(\xi - (\sigma^2_{A_u}/2)x)^2}{2\sigma^2_{A_u}} \right\}. \qquad (4.15)$$

Consequently, the mutual information $I_{A_u} \triangleq I(X;A_u)$ defined in Equation (4.12) can be expressed as

$$I_{A_u} = 1 - \int_{-\infty}^{+\infty} \frac{1}{\sqrt{2\pi}\sigma_{A_u}} \exp\left\{ \frac{-(\xi - \sigma^2_{A_u}/2)^2}{2\sigma^2_{A_u}} \right\} \log_2(1+e^{-\xi})\,d\xi. \qquad (4.16)$$

For the sake of notational compactness we define

$$J(\sigma) \triangleq I_{A_u}(\sigma_{A_u}=\sigma), \quad \sigma > 0. \qquad (4.17)$$

Similarly, the same Gaussian model can be applied to the *a priori* LLR A_c, yielding

$$A_c = \mu_{A_c} Y + n_{A_c}, \qquad (4.18)$$

where $\mu_{A_c} = \sigma^2_{A_c}/2$ [98], and n_{A_c} is a Gaussian random variable with a variance of $\sigma^2_{A_c}$ and a mean of zero. Then, the mutual information between the encoded bit Y and the *a priori* LLR A_c can be expressed as

$$I_{A_c} = 1 - \int_{-\infty}^{+\infty} \frac{1}{\sqrt{2\pi}\sigma_{A_c}} \exp\left\{ \frac{-(\xi - \sigma^2_{A_c}/2)^2}{2\sigma^2_{A_c}} \right\} \log_2(1+e^{-\xi})\,d\xi, \qquad (4.19)$$

or in terms of the function $J(\cdot)$ as

$$I_{A_c} = J(\sigma_{A_c}). \qquad (4.20)$$

The mutual information is also used to quantify the relationship of the extrinsic output $I_{E_u} \triangleq I(X;E_u)$ and $I_{E_c} \triangleq I(Y;E_c)$ with the original information bits and the encoded bits as follows:

$$I_{E_u} = \frac{1}{2} \sum_{x=\pm 1} \int_{-\infty}^{+\infty} f_{E_u}(\xi|Y=x) \log_2 \frac{2f_{E_u}(\xi|X=x)}{f_{E_u}(\xi|X=-1)+f_{E_u}(\xi|X=+1)} \, d\xi, \quad (4.21)$$

$$I_{E_c} = \frac{1}{2} \sum_{y=\pm 1} \int_{-\infty}^{+\infty} f_{E_c}(\xi|Y=y) \log_2 \frac{2f_{E_c}(\xi|Y=y)}{f_{E_c}(\xi|Y=-1)+f_{E_c}(\xi|Y=+1)} \, d\xi. \quad (4.22)$$

4.2.2.2 Properties of the $J(\cdot)$ Function

The function $J(\cdot)$ defined in Equation (4.17) is useful for determining the mutual information. Hence we list some of its properties below.

- It is monotonically increasing, and thus it has a unique and unambiguous inverse function [98]

$$\sigma = J^{-1}(I). \quad (4.23)$$

- Its range is $[0 \ldots 1]$ and it satisfies [98]

$$\lim_{\sigma \to 0} J(\sigma) = 0, \quad \lim_{\sigma \to \infty} J(\sigma) = 1. \quad (4.24)$$

- The capacity C_G of a binary input/continuous output AWGN channel is given by [98]

$$C_G = J(2/\sigma_n), \quad (4.25)$$

where σ_n^2 is the variance of the Gaussian noise.
- It is infeasible to express $J(\cdot)$ or its inverse in closed form. However, they can be closely approximated by [122]

$$J(\sigma) \approx \left(1 - 2^{-H_1 \sigma^{2H_2}}\right)^{H_3} \quad (4.26)$$

$$J^{-1}(I) \approx \left(-\frac{1}{H_1} \log_2(1 - I^{\frac{1}{H_3}})\right)^{\frac{1}{2H_2}}. \quad (4.27)$$

Numerical optimization to minimize the total squared difference between Equation (4.17) and Equation (4.26) results in the parameter values of $H_1 = 0.3073$, $H_2 = 0.8935$ and $H_3 = 1.1064$. The solid curve in Figure 4.9 shows the values of Equation (4.17) and its visibly indistinguishable approximation in Equation (4.26).

4.2.2.3 Evaluation of the EXIT Characteristics of a SISO APP Module

According to the SISO APP algorithm outlined in Section 4.2.1, the extrinsic outputs E_u and E_c of a SISO APP module depend entirely on the a priori input A_u and A_c. Hence the mutual information I_{E_u} or I_{E_c} quantifying the extrinsic output is a function of the mutual information I_{A_u} and I_{A_c} quantifying the a priori input, which can be defined as

$$I_{E_u} = T_u(I_{A_u}, I_{A_c}), \quad (4.28)$$

$$I_{E_c} = T_c(I_{A_u}, I_{A_c}). \quad (4.29)$$

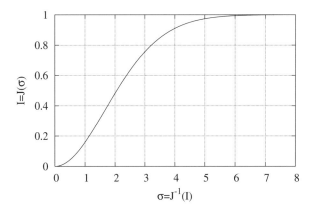

Figure 4.9: The $J(\cdot)$ function of Equation (4.17) and its visibly indistinguishable approximation by Equation (4.26).

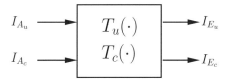

Figure 4.10: An EXIT module defined by two EXIT functions (4.28) and (4.29) of the corresponding SISO APP module.

These two functions are often referred to as the EXIT characteristics of a SISO module, since they characterize the SISO module's capability of enhancing our confidence in its input information. Consequently, we can define an EXIT module, which is the counterpart of the corresponding SISO APP module in terms of mutual information transfer. Therefore the EXIT module has two inputs, namely, I_{A_u} and I_{A_c} as well as two outputs, namely, I_{E_u} and I_{E_c}, as depicted in Figure 4.10. The relationships between the inputs and the outputs are determined by the EXIT functions of Equations (4.28) and (4.29).

For simple codes such as repetition codes or single parity check codes, analytical forms of the associated EXIT characteristics may be derived. However, for the SISO modules of sophisticated codecs, their EXIT characteristics have to be obtained by simulation. In other words, for given values of I_{A_u} and I_{A_c} we artificially generate the *a priori* inputs A_u and A_c, which are fed to the SISO module. Then the SISO APP algorithm of Section 4.2.1 is invoked, resulting in the extrinsic outputs E_u and E_c. Finally, we evaluate the mutual information I_{E_u} and I_{E_c} of Equations (4.21) and (4.22). In this way, the EXIT characteristics can be determined once a sufficiently high number of data points are accumulated. The detailed procedure is shown in Algorithm 4.2.

This is the original method proposed in [98] for the evaluation of the EXIT characteristics. Actually, Step 4 in Algorithm 4.2 may be significantly simplified [132, 259], as long as the outputs $\mathbf{E_u}$ and $\mathbf{E_c}$ of the SISO module are true *a posteriori* LLRs. In fact, the mutual

Algorithm 4.2

1) For given *a priori* input (I_{A_u}, I_{A_c}), compute the noise variance $\sigma_{A_u}^2$ and $\sigma_{A_c}^2$ of the Gaussian model of Equation (4.13) using the inverse of the $J(\cdot)$ function, i.e. $\sigma_{A_u} = J^{-1}(I_{A_u})$ and $\sigma_{A_c} = J^{-1}(I_{A_c})$.
2) Generate a random binary sequence $\mathbf{X}$ with equally distributed values of $\{+1, -1\}$ and encode it, resulting in the encoded bit sequence $\mathbf{Y}$. Then generate the *a priori* inputs $\mathbf{A_u}$ and $\mathbf{A_c}$ using the Gaussian model of Equation (4.13), i.e. by scaling the input and adding Gaussian noise, where the noise variances are those obtained in Step 1.
3) Invoke the SISO APP algorithm to compute the extrinsic output $\mathbf{E_u}$ and $\mathbf{E_c}$.
4) Use a histogram-based estimate of the conditional PDFs of $f_{E_u}(\xi|Y=y)$ and $f_{E_c}(\xi|Y=y)$ in Equation (4.21) and Equation (4.22) respectively. Then compute the mutual information I_{E_u} and I_{E_c} according to Equation (4.21) and Equation (4.22) respectively.
5) Set $I_{A_u} = I_{A_u} + \triangle I_{A_u}$, $0 \leq I_{A_u} \leq 1$ and $I_{A_c} = I_{A_c} + \triangle I_{A_c}$, $0 \leq I_{A_c} \leq 1$. Repeat Steps 1–4 until a sufficiently high number of data points are accumulated.

information can be expressed as the expectation of a function of solely the absolute values of the *a posteriori* LLRs [132]. This result provides a simple method for computing the mutual information by simulation. Unlike the original method [98], it does not require explicit measurements of histograms of the soft outputs. More details are given below.

4.2.2.4 Simplified Computation of Mutual Information

Let us first consider the mutual information $I(X_n; E_{u,n})$ between the *a posteriori* LLR $E_{u,n}$ and the transmitted bit X_n, which can be expressed as

$$I(X_n; E_{u,n}) = H(X_n) - H(X_n|E_{u,n}), \tag{4.30}$$
$$= H(X_n) - \mathrm{E}\{H(X_n|E_{u,n} = \lambda_n)\}, \tag{4.31}$$

where $E\{x\}$ denotes the expected value of x. In tangible physical terms Equation (4.30) quantifies the extra information gleaned with the advent of the knowledge of $E_{u,n}$. Note that, according to the definition of LLRs, the *a posteriori* probability $P(X_n = \pm1|\lambda_n)$ can be computed from the *a posteriori* LLR $E_{u,n} = \lambda_n$ as $P(X_n = \pm1|\lambda_n) = \frac{1}{1+e^{\mp\lambda_n}}$ [70,97]. Hence the conditional entropy $H(X_n|E_{u,n} = \lambda_n)$ can be obtained as

$$H(X_n|E_{u,n} = \lambda_n) = -\frac{1}{1+e^{\lambda_n}} \log_2 \frac{1}{1+e^{\lambda_n}} - \frac{1}{1+e^{-\lambda_n}} \log_2 \frac{1}{1+e^{-\lambda_n}}, \tag{4.32}$$
$$\triangleq h(\lambda_n). \tag{4.33}$$

Therefore, the symbol-wise mutual information $I(X_n; E_{u,n})$ of Equation (4.30) may be expressed as

$$I(X_n; E_{u,n}) = H(X_n) - \mathrm{E}\{h(\lambda_n)\}. \tag{4.34}$$

Similarly, we obtain the symbol-wise mutual information $I(Y_n; E_{c,n})$ as

$$I(Y_n; E_{c,n}) = H(Y_n) - E\{h(\lambda_n)\}. \tag{4.35}$$

Consequently, the average mutual information I_{E_u} and I_{E_c} can be expressed as

$$I_{E_u} = \frac{1}{N_u} \sum_{n=0}^{N_u-1} (H(X_n) - E\{h(\lambda_n)\}), \tag{4.36}$$

$$I_{E_c} = \frac{1}{N_c} \sum_{n=0}^{N_c-1} (H(Y_n) - E\{h(\lambda_n)\}). \tag{4.37}$$

In general, both the input and the output bit values X_n and Y_n are equiprobable; hence $H(X_n) = H(Y_n) = 1$. Then Equations (4.36) and (4.37) may be simplified to

$$I_{E_u} = \frac{1}{N_u} \sum_{n=0}^{N_u-1} (1 - E\{h(E_{u,n} = \lambda_n)\})$$

$$= \frac{1}{N_u} \sum_{n=0}^{N_u-1} E\{f_I(E_{u,n} = \lambda_n)\}, \tag{4.38}$$

$$I_{E_c} = \frac{1}{N_c} \sum_{n=0}^{N_c-1} E\{f_I(E_{c,n} = \lambda_n)\}, \tag{4.39}$$

where $f_I(\cdot)$ is

$$f_I(\lambda) \triangleq 1 - h(\lambda)$$

$$= \frac{1}{1+e^\lambda} \log_2 \frac{2}{1+e^\lambda} + \frac{1}{1+e^{-\lambda}} \log_2 \frac{2}{1+e^{-\lambda}}. \tag{4.40}$$

Assume furthermore that the *a posteriori* LLRs are ergodic and the block length is sufficiently high. Then the expectation values in Equations (4.38) and (4.39) can be replaced by the time averages [132, 259]

$$I_{E_u} \approx \frac{1}{N_u} \sum_{n=0}^{N_u-1} f_I(E_{u,n} = \lambda_n), \tag{4.41}$$

$$I_{E_c} \approx \frac{1}{N_c} \sum_{n=0}^{N_c-1} f_I(E_{c,n} = \lambda_n). \tag{4.42}$$

The method outlined in Equations (4.41) and (4.42) has the following advantages.

- No LLR histograms have to be recorded; hence the computational complexity is significantly reduced. This method can operate 'on-line,' because as soon as a new *a posteriori* LLR becomes available it can be used to update the current estimate of the mutual information.

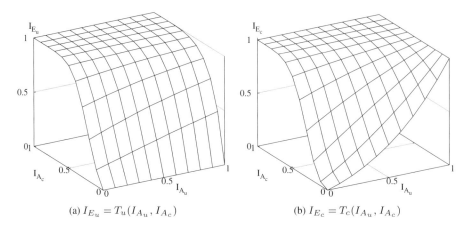

(a) $I_{E_u} = T_u(I_{A_u}, I_{A_c})$ (b) $I_{E_c} = T_c(I_{A_u}, I_{A_c})$

Figure 4.11: EXIT functions for CC(7,5) used as Decoder II in Figure 4.3.

- The results are unbiased even for codes with time-varying trellises such as punctured convolutional codes, and bits having different statistical properties do not have to be treated differently [132].

4.2.2.5 *Examples*

In this section we will show examples of various EXIT functions that belong to different SISO modules. They are evaluated using the procedure outlined in Section 4.2.2.3. In general, a SISO module needs two EXIT functions (Equations (4.28) and (4.29)) in order to characterize its information transfer behavior in iterative processing, for example when it is used as the intermediate decoder in multiple serial concatenation of component decoders. However, some of the input/output terminals of a SISO module may not be actively exploited in the iterative processing; hence, the employment of a single EXIT function may be sufficient on some occasions. We will discuss these scenarios separately in our forthcoming discourse.

4.2.2.5.1 The Intermediate Code. This is the general case. As seen from Figure 4.3, the intermediate SISO decoder receives its *a priori* input of I_{A_c} from the preceding SISO decoder, while its *a priori* input of I_{A_u} comes from the successive decoder. Correspondingly, it feeds back the extrinsic outputs of I_{E_c} and I_{E_u} respectively to the other two constituent decoders, as seen in Figure 4.3. Therefore, the pair of 2D EXIT functions described by Equations (4.28) and (4.29) are needed; these may be depicted in two 3D charts. For example, we show the two 2D EXIT functions of a half-rate convolutional code having the octal generator polynomials of $(7, 5)$ in Figure 4.11.

4.2.2.5.2 The Inner Code. Some SISO modules, such as soft demappers (used, for example, in [84]) and SISO equalizers operate directly on channel observations. These SISO modules receive the *a priori* input I_{A_u} from the successive SISO module and feed back the extrinsic output I_{E_u} as seen in Figure 4.2. The input terminal I_{A_c} and the output terminal I_{E_c} are not used. Hence only a single EXIT function is needed. Moreover, the extrinsic output I_{E_u} also depends on the channel input; hence, I_{E_u} is usually expressed as a function of I_{A_u}

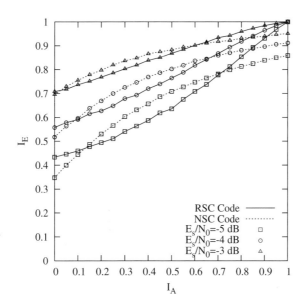

Figure 4.12: The EXIT functions of inner codes exemplified in Figure 4.2. The half-rate RSC code having octal generator polynomials of $(5/7)$ and the half-rate NSC code having the octal generator polynomials of $(7, 5)$ are used.

combined with E_s/N_0 as a parameter, i.e. we have

$$I_{E_u} = T_u(I_{A_u}, E_s/N_0). \tag{4.43}$$

When using BPSK modulation, the EXIT function of the SISO decoder that is connected to the soft demapper, as exemplified in Figure 4.2, may also be characterized by Equation (4.43), since no iterative processing between the demapper and the decoder is needed. For example, we depict the EXIT functions of a half-rate Recursive Systematic Convolutional (RSC) code having octal generator polynomials of $(5/7)$ and those of a half-rate Non-Systematic Convolutional (NSC) code using the same generator polynomials at several E_s/N_0 values in Figure 4.12. It can be seen that the EXIT functions of the RSC code always satisfy the condition of $T_u(I_{A_u} = 1, E_s/N_0) = 1$, while the EXIT functions of the NSC code do not.

4.2.2.5.3 The Outer Code.   For SISO modules which are in the 'outer' positions of the concatenated scheme exemplified in Figure 4.2, the resultant EXIT functions may also be simplified to One-Dimensional (1D) functions. Since the input terminal I_{A_u} and the output terminal I_{E_u} are not used in the associated iterative processing, the EXIT function of an outer SISO module can be expressed as

$$I_{E_c} = T_c(I_{A_c}). \tag{4.44}$$

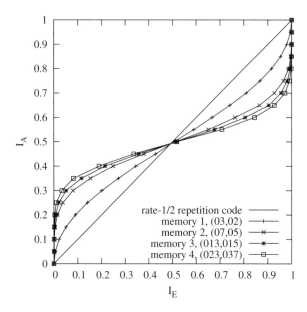

Figure 4.13: The EXIT functions of various outer rate-1/2 non-recursive NSC decoders, where (g_0, g_1) represent the octal generator polynomials.

Figure 4.13 shows the EXIT functions of various non-recursive NSC codes used as outer codes. According to our informal observations, the RSC codes using the same generator polynomials have the same EXIT functions as the corresponding NSC codes.

4.3   Iterative Source–Channel Decoding Over AWGN Channels

This section investigates the performance of iterative source–channel decoding techniques. In general, channel coding is invoked after source coding for the sake of protecting the source information against the noise and other sources of impairments imposed by the transmission channel. In an approach inspired by the iterative decoding philosophy of turbo codes [69], the source code and channel code together may be viewed as a serially concatenated code [71], which can be decoded jointly and iteratively, provided that both the source decoder and the channel decoder are soft-in and soft-out components [42]. The performance of the iterative receiver is evaluated for transmission over AWGN channels, and the EXIT chart technique [98] is employed for analyzing the performance of iterative decoding and to assist in the design of efficient systems. Let us first introduce our system model.

4.3.1   System Model

The schematic diagram of the transmission system considered is shown in Figure 4.14. At the transmitter side of Figure 4.14, the non-binary source $\{u_k\}$ represents a typical source in a video, image or text compression system. It is modeled as a discrete memoryless source using a finite alphabet of $\{a_1, a_2, \ldots, a_s\}$. The output symbols of the source are encoded by a source encoder, which outputs a binary sequence $\{b_k\}$. We assume that the source encoder is a simple entropy encoder, which employs a Huffman code [9], RVLC [12] or VLEC [29]

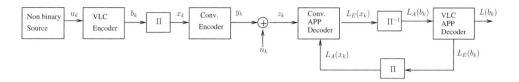

Figure 4.14: Iteratively detected source and channel coding scheme.

code. As shown in Figure 4.14, after interleaving the source-encoded bit stream is protected by an inner channel code against channel impairments. To be more specific, the inner code we employ is a convolutional code. Note that the interleaver seen in Figure 4.14 is used for mitigating the dependencies among the consecutive extrinsic LLRs in order to produce independent *a priori* LLRs. This is necessary because the independence of the *a priori* LLRs is assumed when the SISO APP algorithm multiplies the corresponding bit value probabilities in order to obtain their joint probabilities. The interleaver's ability to achieve this goal is commensurate with its length and depends on how well its design facilitates 'randomize' the order of the LLRs.

At the receiver side of Figure 4.14, an iterative source–channel decoding structure is invoked, as in [42]. The iterative receiver consists of an inner APP channel decoder and an outer APP source decoder. Both of them employ the SISO APP module described in Section 4.2.1 and exchange LLRs as their reliability information.

The first stage of an iteration is constituted by the APP channel decoder. The input of the channel decoder is the channel output z_k and the *a priori* information $L_A(x_k)$ provided by the outer VLC decoder, which is $L_A(x_k) = 0$ for the first iteration. Based on these two inputs, the channel decoder computes the *a posteriori* information $L(x_k)$ and only the extrinsic information $L_E(x_k) = L(x_k) - L_a(x_k)$ is forwarded to the outer VLC decoder.

The second stage of an iteration is constituted by the bit-level trellis-based APP VLC decoder, which accepts as its input the *a priori* information $L_A(b_k)$ from the inner channel decoder and generates the *a posteriori* information $L(b_k)$. Similarly, only the extrinsic information $L_E(b_k) = L(b_k) - L_A(b_k)$ is fed back to the inner channel decoder. After the last iteration the source sequence estimation based on the same trellis, namely on that of the APP VLC decoder, is invoked in order to obtain the decoded source symbol sequence. Generally, the symbols of the non-binary source have a non-uniform distribution, which may be interpreted as *a priori* information. In order to be able to exploit this *a priori* source information, the APP algorithm of the SISO module has to be appropriately adapted, as detailed in Appendix B.

4.3.2 EXIT Characteristics of VLCs

We evaluated the associated EXIT characteristics for the various VLCs listed in Table 4.1. The corresponding results are depicted in Figure 4.15.

It can be seen in the figure that for the Huffman code the SISO VLC decoder only marginally benefits from having extrinsic information, even if exact *a priori* information is provided in the form of $I_A = 1$. Upon increasing the code's redundancy, i.e. decreasing the code rate, the SISO VLC decoder benefits from more extrinsic information, as observed for the RVLC-1, RVLC-2 and VLEC-3 schemes of Table 4.1; this indicates that higher iteration

Table 4.1: VLCs used in the simulations

Symbol	Probability	HUFF	RVLC-1	RVLC-2	VLEC-3
0	0.33	00	00	00	000
1	0.30	01	01	11	0110
2	0.18	11	10	010	1011
3	0.10	100	111	101	11010
4	0.09	101	11011	0110	110010
Average length		2.19	2.37	2.46	3.95
Code rate		0.98	0.90	0.87	0.54
Constraint length		3	3.58	3.58	4.70
Free distance		1	1	2	3

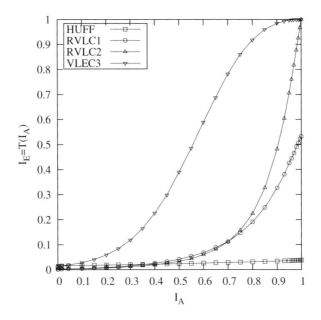

Figure 4.15: EXIT characteristics of VLCs listed in Table 4.1.

gain will be obtained when using iterative decoding. It is also worth noting that for VLCs having a free distance of $d_f \geq 2$, such as the RVLC-2 and VLEC-3 schemes, the condition of $T(I_A = 1) = 1$ is satisfied. In contrast, for those having $d_f = 1$ this condition is not satisfied, which indicates that a considerable number of residual errors will persist in the decoder's output even in the high SNR areas. Note also that the starting point of the VLC EXIT function at the abscissa value of $I_A = 0$ is given by $T(I_A = 0) = 1 - H_b$, where H_b is the entropy of the VLC-encoded bits, which typically approaches one, giving a very low EXIT function starting point.

4.3.3 Simulation Results for AWGN Channels

In this section we present some simulation results for the proposed transmission scheme. All simulations were performed for BPSK modulation when communicating over an AWGN channel. The channel code used is a terminated memory-2, rate-$1/2$ RSC code having the generator polynomial of $(g_1/g_0)_8 = (5/7)_8$, where g_0 is the feedback polynomial. For every simulation, 1000 different frames of length $K = 1000$ source symbols were encoded and transmitted. For the iterative receiver, six decoding iterations were executed, except for the Huffman code, where only three iterations were performed. The simulations were terminated either when all the frames were transmitted or when the number of detected source symbol errors reached 1000.

For the source encoder we use the VLCs as listed in Table 4.1. Since these codes have different rates, for the sake of fair comparison the associated E_b/N_0 values have to be scaled by both the source code rate R_s and the channel code rate R_c as follows:

$$E_b/N_0 = (E_s/(R_s \cdot R_c \cdot \log_2 M))/N_0, \tag{4.45}$$

or in terms of decibels

$$E_b/N_0 \, [\text{dB}] = E_s/N_0 \, [\text{dB}] - 10 \log_{10}(R_s \cdot R_c \cdot \log_2 M), \tag{4.46}$$

where E_s is the transmitted energy per modulated symbol, and M is the number of signals in the signal space of M-ary modulation. For example, $M = 2$ for BPSK modulation. The effect of the tail bits of the channel code is ignored here.

For every VLC, the transfer functions of the outer VLC decoder and the inner channel decoder are plotted in the form of EXIT charts in order to predict the performances of the iterative receivers.

The EXIT chart of the iterative receiver using the Huffman code of Table 4.1 and the corresponding SER performance are depicted in Figures 4.16 and 4.17 respectively. In the EXIT chart the transfer functions of the inner RSC decoder recorded at $E_b/N_0 = 0$ dB, 1 dB and 2 dB are plotted. Moreover, the extrinsic information outputs, I_E^{VLC} and I_E^{RSC}, of both decoders were recorded during our simulations and the average values at each iteration were plotted in the EXIT chart, which form the so-called *decoding trajectories*. The iterative decoding algorithm may possibly converge to a near-zero BER, when the decoding trajectory approaches the point ($I_A = 1$, $I_E = 1$). However, it can be readily seen that since the outer Huffman code has a flat transfer characteristic, it provides only limited extrinsic information for the inner channel decoder; hence, the overall performance can hardly be improved by iterative decoding. This is confirmed by the SER performance seen in Figure 4.17, where only a modest iteration gain is attained.

For the transmission scheme employing the RVLC-1 arrangement of Table 4.1, the iterative receiver did exhibit some iteration gain, as shown in Figure 4.19. However, the achievable improvements saturated after $I = 2$ or 3 iterations. This is because the two EXIT curves intersect, before reaching the ($I_A = 1$, $I_E = 1$) point, as shown in the EXIT chart of Figure 4.18, and the trajectories get trapped at these intersect points; hence, no more iteration gain can be obtained.

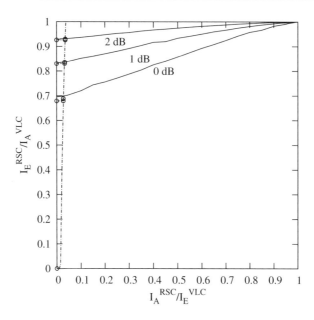

Figure 4.16: EXIT chart for the iterative receiver of Figure 4.14 using a *Huffman code* having $d_f = 1$ of Table 4.1 as the outer code and the RSC$(2, 1, 2)$ code as the inner code for transmission over AWGN channels.

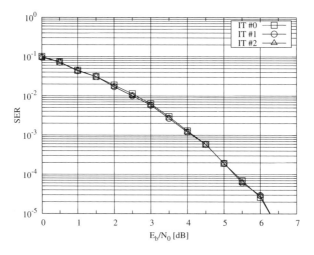

Figure 4.17: SER performance of the iterative receiver of Figure 4.14 using a *Huffman code* of Table 4.1 as the outer code and the RSC$(2, 1, 2)$ code as the inner code for transmission over AWGN channels.

As already stated, the VLCs having $d_f \geq 2$ exhibit better transfer characteristics, and hence also better convergence properties. This is confirmed in Figure 4.21, where the SER of

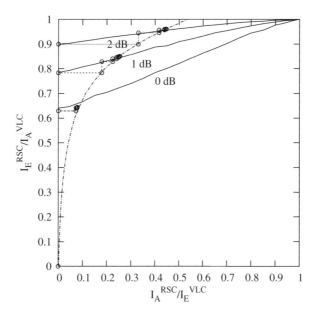

Figure 4.18: EXIT chart for the iterative receiver of Figure 4.14 using a *RVLC* having $d_f = 1$ of Table 4.1 as the outer code and the RSC$(2, 1, 2)$ code as the inner code for transmission over AWGN channels.

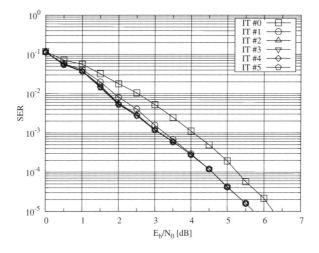

Figure 4.19: SER performance of the iterative receiver of Figure 4.14 using a *RVLC* having $d_f = 1$ of Table 4.1 as the outer code and the RSC$(2, 1, 2)$ code as the inner code for transmission over AWGN channels.

the system employing the RVLC-2 scheme of Table 4.1 was continuously reduced for up to six iterations. At $\mathrm{SER} = 10^{-4}$, the iteration gain is as high as about 2.5 dB. As seen from the

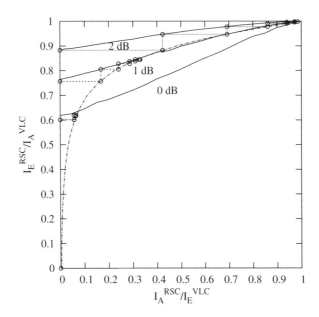

Figure 4.20: EXIT chart for the iterative receiver of Figure 4.14 using a *RVLC* having $d_f = \mathbf{2}$ of Table 4.1 as the outer code and the RSC$(2, 1, 2)$ code as the inner code for transmission over AWGN channels.

EXIT chart of Figure 4.20, for $E_b/N_0 > 1$ dB there will be an open EXIT 'tunnel' between the curves of the two transfer functions. This implies that if enough decoding iterations are performed, the iterative decoding trajectory will reach the $(I_A = 1, I_E = 1)$ point of the EXIT chart, which is associated with achieving a near-zero BER. Note, however, that this will be only be possible if the iterative decoding trajectory follows the EXIT functions sufficiently well. This requires that the interleaver is sufficiently long and well designed in order for it to successfully mitigate the dependencies between the consecutive *a priori* LLRs, as described in Section 4.3.1.

In the corresponding SER chart of Figure 4.21, the SER curve approaches the so-called 'waterfall' region or 'turbo cliff' region beyond this point. Hence, this E_b/N_0 value is also referred to as the *convergence threshold*. It can also be seen from the EXIT chart of Figure 4.20 that, beyond the convergence threshold, the EXIT 'tunnel' between the two curves becomes narrow towards the top right corner near the point $I_A = 1, I_E = 1$. This indicates that the iteration gains will become lower during the course of iterative decoding. In particular, in the case of a limited interleaver size and for a low number of iterations a considerable number of residual errors will persist at the decoder's output, and an error floor will appear even in the high channel SNR region.

Figures 4.22 and 4.23 show the EXIT chart and the SER performance of the system respectively, when employing the VLEC-3 scheme of Table 4.1. The convergence threshold of this system is about $E_b/N_0 = 0$ dB. Hence, beyond this point a significant iteration gain was obtained, as a benefit of the excellent transfer characteristics of the VLEC-3 code.

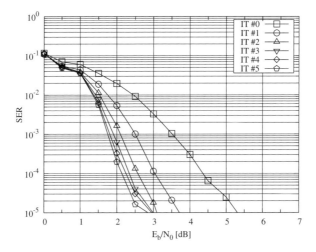

Figure 4.21: SER performance of the iterative receiver of Figure 4.14 using a *RVLC* having $d_f = 2$ of Table 4.1 as the outer code and the RSC$(2, 1, 2)$ code as the inner code for transmission over AWGN channels.

At SER $= 10^{-4}$, the achievable E_b/N_0 gain is about 4 dB after six iterations and no error floor is observed. A performance comparison of the systems employing different VLCs is depicted in Figure 4.24. Clearly, in the context of iterative decoding, when the inner code is fixed, the achievable system performance is determined by the free distance of the outer code. The VLEC code having a free distance of $d_f = 3$ performs best after $I = 6$ iterations, while it has the lowest code rate and hence results in the lowest achievable system throughput. Similarly, the RVLC-2 scheme having $d_f = 2$ also outperforms the RVLC-1 having $d_f = 1$ and the Huffman code.

4.4 Iterative Channel Equalization, Channel Decoding and Source Decoding

In practical communications environments, the same signal may reach the receiver via multiple paths, as a result of scattering, diffraction and reflections. In simple terms, if all the significant multipath signals reach the receiver within a fraction of the duration of a single transmitted symbol, they result in non-dispersive fading, whereby occasionally they add destructively, resulting in a low received signal amplitude. However, if the arrival time difference among the multipath signals is higher than the symbol duration, then the received signal inflicts dispersive time-domain fading, yielding Inter-Symbol Interference (ISI).

The fading channel hence typically inflicts bursts of transmission errors, and the effects of these errors may be mitigated by adding redundancy to the transmitted signal in the form of error-correction coding, such as convolutional coding, as discussed in the previous section. However, adding redundancy reduces the effective throughput perceived by the end user. Furthermore, if the multipath signals are spread over multiple symbols they tend to further increase the error rate experienced, unless a channel equalizer is used. The channel equalizer collects all the delayed and faded multipath replicas, and linearly or nonlinearly combines

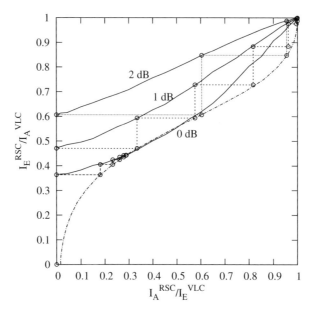

Figure 4.22: EXIT chart for the iterative receiver of Figure 4.14 using a *VLEC code* having $d_f = 3$ of Table 4.1 as the outer code and the $\text{RSC}(2, 1, 2)$ code as the inner code for transmission over AWGN channels.

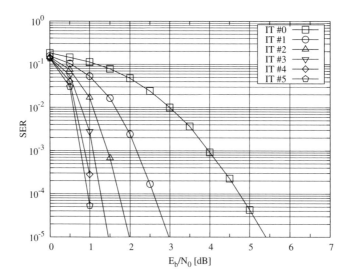

Figure 4.23: SER performance of the iterative receiver of Figure 4.14 using a *VLEC code* having $d_f = 3$ of Table 4.1 as the outer code and the $\text{RSC}(2, 1, 2)$ code as the inner code for transmission over AWGN channels.

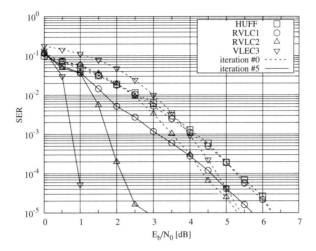

Figure 4.24: Performance comparison of the schemes employing different VLCs including Huffman code, RVLC-1 having $d_f = 1$, RVLC-2 having $d_f = 2$ and VLEC-3 having $d_f = 3$ of Table 4.1, after the first and the sixth iterations.

them [260]. The equalizers typically achieve a diversity gain, because, even if one of the signal paths becomes severely faded, the receiver may still be able to extract sufficient energy from the other independently faded paths.

Capitalizing on the impressive performance gains of turbo codes, turbo equalization [82, 113, 261] has been proposed as a combined iterative channel equalization and channel decoding technique that has the potential of achieving equally impressive performance gains when communicating over ISI-contaminated channels. The turbo-equalization scheme introduced by Douillard *et al.* [82] may be viewed as an extension of the turbo decoding algorithm [69], which interprets the information spreading effects of the ISI as another form of error protection, reminiscent of that imposed by a rate-1 convolutional code. To elaborate a little further, the impulse response of the dispersive channel has the effect of smearing each transmitted symbol over a number of consecutive symbols. This action is similar to that of a convolutional encoder.

By applying the turbo detection principle of exchanging extrinsic information between channel equalization and source decoding, the receiver becomes capable of efficiently combatting the effects of ISI. In the following sections we describe such a scheme, where the redundancy inherent in the source is exploited for the sake of eliminating the channel-induced ISI. Transmission schemes both with and without channel coding are considered.

4.4.1 Channel Model

We assume having a coherent symbol-spaced receiver benefiting from the perfect knowledge of the signal's phase and symbol timing, so that the transmit filter, the channel and the receiver filter can be jointly represented by a discrete-time linear filter having a finite-length impulse

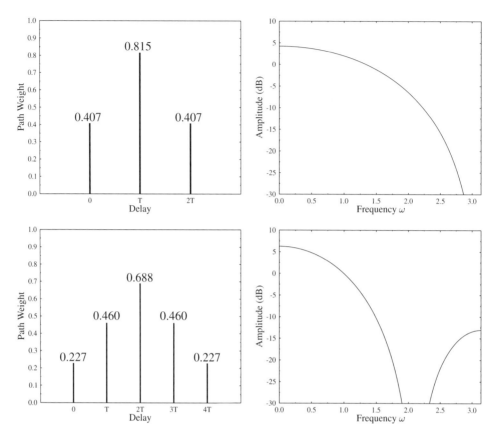

Figure 4.25: Two dispersive Channel Impulse Responses (CIRs) and their frequency-domain representations.

response expressed as

$$h_k = \sum_{m=0}^{M} h_m \delta(k - m), \tag{4.47}$$

where the real-valued channel coefficients $\{h_m\}$ are assumed to be time-invariant and known to the receiver. Figure 4.25 shows two examples of ISI-contaminated dispersive channels and their frequency domain responses introduced in [8]. The channel characterized in the upper two graphs is a three-path channel having moderate ISI, while the channel depicted in the lower two is a five-path channel suffering from severe ISI.

Given the Channel Impulse Response (CIR) of Equation (4.47) and a BPSK modulator, the channel output y_k, which is also the equalizer's input, is given by

$$y_k = \sum_{m=0}^{M} h_m d_{k-m} + n_k, \quad k = 1, \ldots, N_c, \tag{4.48}$$

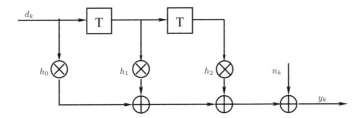

Figure 4.26: Tapped delay line circuit of the channel model Equation (4.48) for $M = 2$.

where n_k is the zero-mean AWGN having a variance of σ^2. We assume that we have $N_c = N_b + N_h$, where N_h represents the so-called tailing bits concatenated by the transmitter, in order to take into account the effect of ISI-induced delay. More specifically, $N_h = M$ number of binary zeros may be appended at the tail of an N_b-bit message, so that the discrete-time linear filter's buffer content converges to the all-zero state. Explicitly, the channel's output samples y_k obey a conditional Gaussian distribution having a PDF expressed as

$$p_{y_k}(y|d_k, d_{k-1}, \ldots, d_{k-M}) = \frac{1}{\sqrt{2\pi}\sigma} \exp\left(\frac{[y - \sum_{m=0}^{M} h_m d_{k-m}]^2}{2\sigma^2}\right). \qquad (4.49)$$

Figure 4.26 shows a tapped delay line model of the equivalent discrete-time channel of Equation (4.48) for $M = 2$, associated with the channel coefficients of $h_0 = 0.407$, $h_1 = 0.815$ and $h_2 = 0.407$. Assuming an impulse response length of $(M + 1)$, the tapped delay line model contains M delay elements. Hence, given a binary input alphabet of $\{+1, -1\}$, the channel can be in one of 2^M states q_i, $i = 1, 2, \ldots, 2^M$, corresponding to the 2^M different possible binary contents in the delay elements. Let us denote the set of possible states by $\mathcal{S} = \{q_1, q_2, \ldots, q_{2^M}\}$. At each time instance of $k = 1, 2, \ldots, N_c$ the state of the channel is a random variable $s_k \in \mathcal{S}$. More explicitly, we define $s_k = (d_k, d_{k-1}, \ldots, d_{k-M+1})$. Note that, given s_k, the next state s_{k+1} can only assume one of the two values determined by a $+1$ or a -1 being fed into the tapped delay line model at time $k + 1$. The possible evolution of states can thus be described in the form of trellis diagram. Any path through the trellis corresponds to a sequence of input and output symbols read from the trellis branch labels, where the channel's output symbol ν_k at time instance k is the noise-free output of the channel model of Equation (4.48) given by

$$\nu_k = \sum_{m=0}^{M} h_m d_{k-m}. \qquad (4.50)$$

The trellis constructed for the channel of Figure 4.26 is depicted in Figure 4.27. A branch of the trellis is a four-tuple $(i, j, d_{i,j}, \nu_{i,j})$, where the state $s_{k+1} = q_j$ at time $k + 1$ can be reached from state $s_k = q_i$ at time k, when the channel encounters the input $d_k = d_{i,j}$ and produces the output $\nu_k = \nu_{i,j}$, where $d_{i,j}$ and $\nu_{i,j}$ are uniquely identified by the index pair (i, j). The set of all index pairs (i, j) corresponding to valid branches is denoted as $\mathcal{B}$. For the

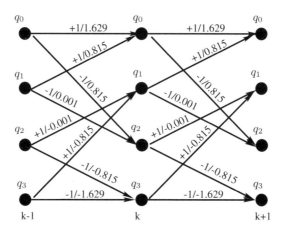

Figure 4.27: Trellis representation of the channel seen in Figure 4.26. The states $q_0 = (1, 1), q_1 = (-1, 1), q_2 = (1, -1), q_3 = (-1, -1)$ are the possible contents of the delay elements in Figure 4.26. The transitions from a state $s_k = q_i$ at time k to a state $s_{k+1} = q_j$ at time $k + 1$ are labeled with the input/output pair $d_{i,j}/\nu_{i,j}$.

trellis seen in Figure 4.27, the set $\mathcal{B}$ is as follows:

$$\mathcal{B} = \{(0, 0), (0, 1), (1, 2)(1, 3), (2, 0), (2, 1), (3, 3), (3, 2)\}. \qquad (4.51)$$

4.4.2   Iterative Channel Equalization and Source Decoding

In this section we investigate the principles of iterative equalization as well as source decoding, and derive the extrinsic information exchanged between the equalizer and the source decoder. We assume that the channel is an ISI-contaminated Gaussian channel, as expressed in Equation (4.48).

4.4.2.1   System Model

At the transmitter side, the output symbols of the non-binary source $\{x_n\}$ seen in Figure 4.28 are forwarded to an entropy encoder, which outputs a binary sequence $\{b_k\}$, $k = 1, 2, \ldots, N_b$. The source code used in the entropy encoder may be a Huffman code, RVLC [12] or VLEC code [29]. Then the resultant binary sequence is interleaved and transmitted over an ISI-contaminated channel also imposing AWGN, as shown in Figure 4.28.

The receiver structure performing iterative equalization and source decoding is shown in Figure 4.29. The ultimate objective of the receiver is to provide estimates of the source symbols $\{x_n\}$, given the channel output samples $\mathbf{y} \triangleq y_1^{N_c}$, where we have $y_1^{N_c} = \{y_1, y_2, \ldots, y_{N_c}\}$. More specifically, in the receiver structure of Figure 4.29 the channel equalizer's feed-forward information generated in the context of each received data bit is applied to the source decoder through a deinterleaver described by the turbo deinterleaver function Π^{-1}. The information conveyed along the feed-forward path is described by $L_f(\ )$. In contrast, the source decoder feeds back the *a priori* information associated with each data bit to the channel equalizer through a turbo interleaver having the interleaving function

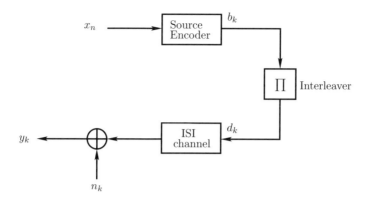

Figure 4.28: Representation of a data transmission system including source coding and a turbo interleaver communicating over an ISI-channel inflicting additive white Gaussian noise.

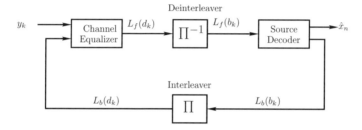

Figure 4.29: Receiver schematic to perform iterative channel equalization and source decoding.

described by Π. The information conveyed along the feedback path is described by $L_b(\ )$. This process is repeated for a number of times, until finally the source decoder computes the estimates $\{\hat{x}_n\}$ for the source symbols $\{x_n\}$, with the aid of the information provided by $\{L_f(b_m)\}$. Let us first investigate the equalization process in a little more detail.

4.4.2.2 EXIT Characteristics of Channel Equalizer

The channel equalization considered here is based on the APP algorithm, which makes the equalizer another manifestation of the SISO APP module described in Section 4.2.1. More details of the APP channel equalization can be found in Appendix C. Let us now study the transfer characteristics of the channel equalizer. Consider two ISI channels, namely the three-path and five-path CIRs depicted in Figure 4.25, which are represented by their CIR taps as

$$\mathbf{H}_1 = [0.407\ 0.815\ 0.407]^T; \tag{4.52}$$

$$\mathbf{H}_2 = [0.227\ 0.46\ 0.688\ 0.46\ 0.227]^T. \tag{4.53}$$

Figure 4.30 shows the EXIT functions of the channel equalizer for the dispersive AWGN channels of Equations (4.52) and (4.53) at different channel SNRs. It can be seen that the channel equalizer has the following transfer characteristics.

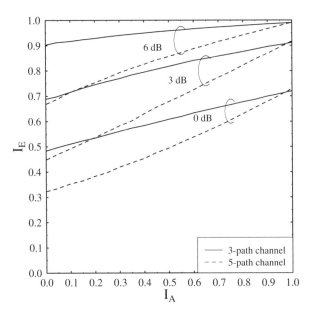

Figure 4.30: EXIT functions of the channel equalizer for transmission over the dispersive AWGN channels of Equations (4.52) and (4.53).

- The EXIT curves are almost straight lines, and those recorded for the same channel at different SNRs are almost parallel to each other.
- At low SNRs, $I_E = T(I_A = 1)$ can be significantly less than 1, and $I_E = T(I_A = 1)$ will be close to 1 only at high SNRs. This indicates that a considerable number of residual errors will persist at the output of the channel equalizer at low to medium SNRs.
- Observe in Figure 4.30 that the channel equalizer generally provides less extrinsic information I_E for the five-path channel $\mathbf{H}_2$ than for the three-path channel $\mathbf{H}_1$ at the same input *a priori* information I_A, which predicts potentially worse BER performance. Interestingly, the transfer functions of the two equalizers almost merge at $I_A = 1$, i.e. they have a similar value of $T(I_A = 1)$. This indicates that they would have the same BER performance bound provided that all ISI were eliminated.

4.4.2.3 Simulation Results

In this section we evaluate the SER performance of the iterative receiver. In the simulations we used the three-path channel of Equation (4.52) and the five-path channel of Equation (4.53). For the source encoder we employed two different VLCs, namely the RVLC-2 and VLEC-3 schemes listed in Table 4.1. The encoded data is permutated by a random bit interleaver of size $L = 4096$ bits.

The SER was calculated by using Levenshtein distance [251] for the combined transceiver using a RVLC, which was depicted in Figure 4.31, when transmitting over the three-path ISI channel $\mathbf{H}_1$ of Equation (4.52) and over the five-path ISI channel $\mathbf{H}_2$ of Equation (4.53). The performance of the same system over a non-dispersive AWGN channel is also depicted

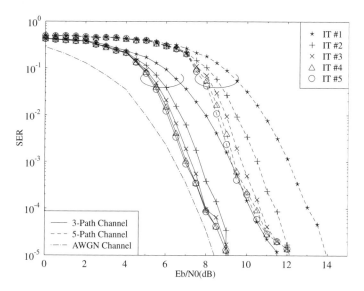

Figure 4.31: SER performance of the RVLC-2 of Table 4.1 for transmission over the dispersive AWGN channels of Equations (4.52) and (4.53).

as a best-case 'bound.' It can be seen in Figure 4.31 that the iterative equalization and source decoding procedure converges as the number of iterations increases. This behavior is similar to that of a turbo equalizer [82] and effectively reduces the detrimental effects of ISI after $I = 4$ iterations, potentially approaching the non-dispersive AWGN channel's performance.

The SER performance of the system using a VLEC code of Table 4.1 is depicted in Figure 4.32, when transmitting over the three-path ISI channel $\mathbf{H}_1$ of Equation (4.52) and over the five-path ISI channel $\mathbf{H}_2$ of Equation (4.53). Since the VLEC code has a larger free distance, and hence a stronger error-correction capability, the iterative equalization and source decoding procedure succeeds in effectively eliminating the ISI, and approaches the non-dispersive AWGN performance bound after $I = 4$ iterations.

4.4.2.4 Performance Analysis Using EXIT Charts

In order to analyze the simulation results obtained, we measured the EXIT characteristics of the channel equalizer as well as those of the VLC decoder, and plotted their EXIT charts in Figure 4.33. Specifically, both the EXIT chart and the simulated decoding trajectory of the system using RVLC-2 of Table 4.1 for communicating over the three-path channel $\mathbf{H}_1$ are shown in Figure 4.33a. At $E_b/N_0 = 4$ dB, the transfer function of the channel equalizer and that of the VLC decoder intersect in the middle, and after three iterations the trajectory becomes trapped, which implies that little iteration gain can be obtained, even if more iterations are executed. Upon increasing E_b/N_0, the two transfer functions intersect at higher (I_A, I_E) points and the tunnel between the two EXIT curves becomes wider, which now indicates that higher iteration gains can be achieved, and hence the resultant SER becomes lower. Similar observations can be made from the EXIT chart of the same system communicating over the five-path channel $\mathbf{H}_2$ of Figure 4.33b, except that the channel

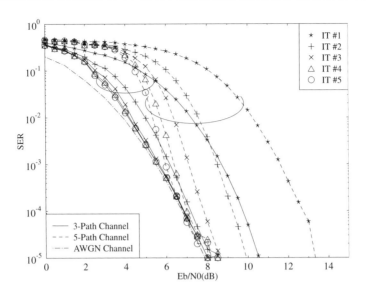

Figure 4.32: SER performance of the VLEC code of Table 4.1 for transmission over the dispersive AWGN channels of Equations (4.52) and (4.53).

equalizer needs an almost 3 dB higher E_b/N_0 to obtain a similar transfer characteristic. Hence the SER curves are shifted to the right by almost 3 dB, as shown in Figure 4.31.

Figure 4.33c shows the EXIT chart and the simulated decoding trajectory of the system using VLEC-3 of Table 4.1 for communicating over the three-path channel $\mathbf{H_1}$. Compared with the system using RVLC-2 (Figure 4.33a), there is a significantly wider tunnel between the two EXIT curves at the same E_b/N_0; hence higher iteration gains can be obtained at a lower number of iterations. This is mainly the benefit of the significantly lower coding rate of the VLEC code used ($R_{VLEC-3} = 0.54 < R_{RVLC-2} = 0.87$). However, the system still suffers from a significant number of residual errors, since the two EXIT curves intersect in the middle and the resultant decoding trajectory gets trapped before reaching the top right corner of ($I_A = 1, I_E = 1$).

Figure 4.33d shows the EXIT chart and the actual decoding trajectory of the same system when communicating over the five-path channel $\mathbf{H_2}$. For the same E_b/N_0, the equalizer communicating over the five-path channel has a steeper EXIT curve than for transmission over the three-path channel, and its initial point at $(I_E^{EQ} = T(I_A^{EQ} = 0))$ is significantly lower. After the first iteration, the extrinsic output of the VLEC decoder remains smaller. Hence the SER performance recorded after the first iteration is always worse than that of the system communicating over the three-path channel. At low E_b/N_0 values (e.g. <4 dB), the EXIT tunnel between the inner code's curve and the outer code's curve remains rather narrow. By contrast, at high E_b/N_0 values (e.g. ≥6 dB), the tunnel becomes significantly wide and has almost the same extrinsic output as that recorded for the three-path channel (Figure 4.33c) after five iterations. Hence the SER performances of the system recorded over the three-path channel and the five-path channel converge at high E_b/N_0 values, as shown in Figure 4.32.

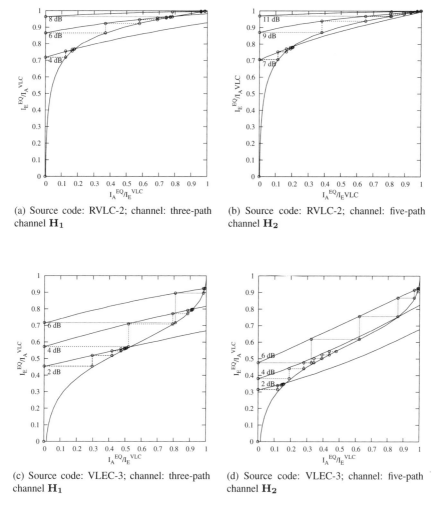

(a) Source code: RVLC-2; channel: three-path channel $\mathbf{H}_1$

(b) Source code: RVLC-2; channel: five-path channel $\mathbf{H}_2$

(c) Source code: VLEC-3; channel: three-path channel $\mathbf{H}_1$

(d) Source code: VLEC-3; channel: five-path channel $\mathbf{H}_2$

Figure 4.33: EXIT charts of the channel equalizer as well as the VLC decoder and the actual decoding trajectories at various E_b/N_0 values.

4.4.3 Precoding for Dispersive Channels

When the channel is non-recursive, which is the case for all wireless channels, then the gain of the iterative equalization remains limited. More explicitly, the performance of the outer channel code for transmission over AWGN channels represents an upper bound on the achievable receiver performance. In order to achieve a performance better than this, either the outer code can be replaced with a turbo code, as suggested in [262], or the channel can be made to appear recursive to the receiver, thus enabling us to achieve a higher interleaver gain. In the family of the latter techniques, recursive rate-one precoders, which effectively render the channel recursive, have been shown to yield significant performance gains in iterative equalization systems [263]. The concept was first presented in [105, 264, 265] for

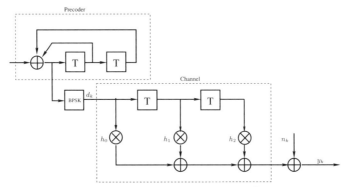

(a) Shift register representation of the precoder $1 + D + D^2$ concatenated with a BPSK modulator and the three-path channel of Equation (4.52)

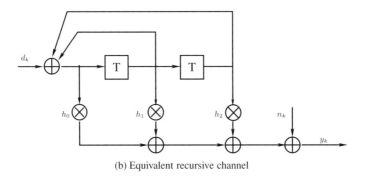

(b) Equivalent recursive channel

Figure 4.34: An example of precoding for BPSK modulation.

magnetic recording channels, and has subsequently been investigated in [263, 266] using BPSK modulations and in [212] using higher order modulations for general dispersive channels. A further advantage of the precoder is that there is no substantial increase in receiver complexity associated with its inclusion, as long as the memory of the precoder is no higher than that of the dispersive channel.

The precoder is generally described in terms of a shift register having feedback connections, described by a feedback polynomial. The structure is essentially the same as that of a recursive convolutional encoder. Figure 4.34 shows an example of a precoded channel.

4.4.3.1 EXIT Characteristics of Precoded Channel Equalizers

In this section we investigate the EXIT characteristics of precoded channel equalizers. Again, the three-path channel of Equation (4.52) and the five-path channel of Equation (4.53) are used as examples. For the three-path channel, the generator polynomials of the precoders having a memory of $m \leq 2$ are $1 + D$, $1 + D + D^2$ and $1 + D^2$. The EXIT functions of the operating channel equalizer both with and without these precoders for transmission over the three-path channel are shown in Figure 4.35a. It can be seen that, when using a precoder, the EXIT function can reach the $(I_A = 1, I_E = 1)$ point, while this is not the case for the non-precoded system. Hence the channel equalizer is capable of achieving near error-free detection for transmission over precoded channels, while it encounters an error rate floor when communicating over non-precoded channels. Furthermore, all the EXIT functions recorded for precoded channels have lower initial values at $(T(I_A = 0))$ in comparison with the non-precoded case, and this initial value decreases upon increasing the polynomial's order.

Similarly, we study the EXIT function of the channel equalizer for transmission over the five-path channel. As the five-path channel has a memory of 4, there are a total of 15 different precoders having a memory of $m \leq 4$. Only eight of them were selected for quantitative characterization in Figure 4.35b. Similar observations are obtained as in the case of the three-path channel, such as that the transfer functions arrive at the $(I_A = 1, I_E = 1)$ point when using precoders as well as having lower start values.

4.4.3.2 Performance Analysis

The performance analysis of the joint channel equalization and source decoding system communicating over precoded channels is presented in this section. As observed in Section 4.4.3.1, precoders have similar effects to the channel equalizer's EXIT characteristics in the case of both the three-path and the five-path channels. Hence, we only consider the three-path channel here, and the precoders used are $(1 + D)$ and $(1 + D^2)$. For source coding we used the RVLC-2 and VLEC-3 schemes of Table 4.1. In all simulations, $L = 1000$ source symbols were encoded and transmitted in each frame.

First, the EXIT charts of the system using the RVLC-2 of Table 4.1 and precoder $(1 + D)$ as well as $(1 + D^2)$ are depicted in Figures 4.36a and 4.36b respectively. As seen from Figure 4.36a, the transfer function of the inner channel equalizer and that of the outer source decoder intersect before reaching the $(I_A = 1, I_E = 1)$ point. The coordinates of the intersection point increase upon increasing the channel SNR, until the $(I_A = 1, I_E = 1)$ point is reached. Hence we can predict that for the system using the precoder $(1 + D)$ the iteration gains will saturate after a certain number of iterations, and the achievable error rate is limited, which is similar to the case of non-precoded channels. In contrast, for the case of the precoder $(1 + D^2)$ different observations may be made. As seen from Figure 4.36b, $E_b/N_0 = 5.2$ dB is the system's convergence threshold. Above this E_b/N_0 value, the EXIT functions of the inner channel equalizer and the outer source decoder intersect only at the $(I_A = 1, I_E = 1)$ point. Hence it is possible to achieve error-free detection provided that the interleaver length and the number of iterations are sufficiently high. Also shown in Figure 4.36 are the simulated decoding trajectories, which follow the EXIT chart predictions quite well. Further confirmations accrue from the simulated SER performances shown in Figure 4.37. For the system using the precoder $(1 + D)$, the achievable SER is bounded by the performance when transmitting over an AWGN channel, as seen from Figure 4.37a. In contrast, for the system using the precoder $(1 + D^2)$, the SER can in fact outperform the bound and attain a lower value at high channel SNRs, as seen from Figure 4.37b, although it performs slightly worse in the low SNR region in comparison with the system using the precoder $(1 + D)$, due to the lower extrinsic output of the equalizer previously observed in Figure 4.36.

Similarly, we analyze the achievable performances of the various systems using VLEC-3 of Table 4.1. The corresponding EXIT charts are shown in Figure 4.38. Both systems using

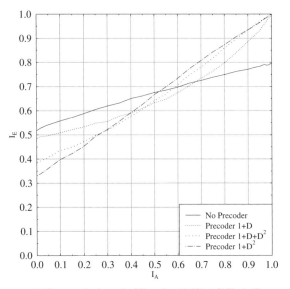

(a) Three-path channel of Equation (4.52) at SNR=1 dB

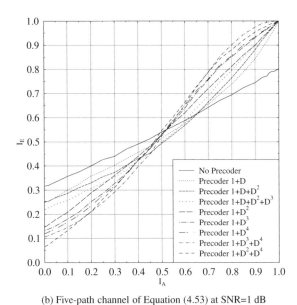

(b) Five-path channel of Equation (4.53) at SNR=1 dB

Figure 4.35: EXIT functions of the APP equalizer for transmission over precoded dispersive AWGN channels.

the precoders $(1 + D)$ and $(1 + D^2)$ show meritorious EXIT characteristics, where an open tunnel can be obtained between the two curves until the $(I_A = 1, I_E = 1)$ point is reached.

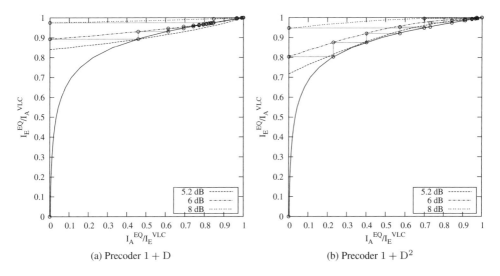

(a) Precoder $1 + D$ (b) Precoder $1 + D^2$

Figure 4.36: EXIT charts of the channel equalizer and the VLC decoder using the RVLC-2 scheme of Table 4.1 over the precoded three-path channel of Equation (4.52).

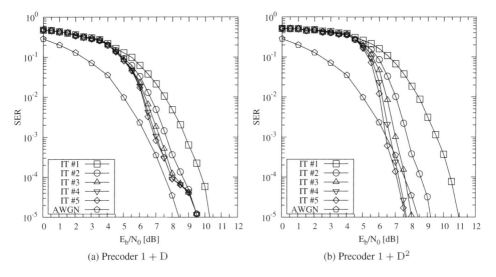

(a) Precoder $1 + D$ (b) Precoder $1 + D^2$

Figure 4.37: SER performance of the iterative receiver using the RVLC-2 scheme of Table 4.1 for transmission over the precoded dispersive AWGN channel of Equation (4.52).

Furthermore, the system using the precoder $(1 + D)$ has a lower convergence threshold of $E_b/N_0 = 2.1$ dB than that using the precoder $(1 + D^2)$, which attains convergence at $E_b/N_0 = 2.8$ dB. The corresponding SER performances are shown in Figure 4.39. Both systems can outperform the performance bounds of non-precoded systems. In particular, the system using the precoder $(1 + D)$ benefits from earlier convergence and attains a

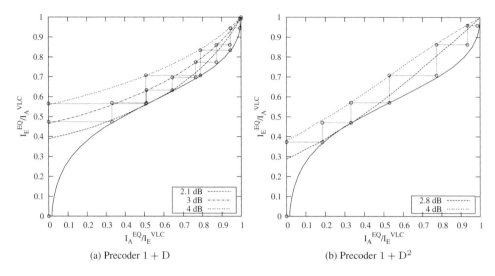

Figure 4.38: EXIT charts of the channel equalizer and the VLC decoder using the VLEC-3 scheme of Table 4.1 for transmission over the precoded three-path channel of Equation (4.52).

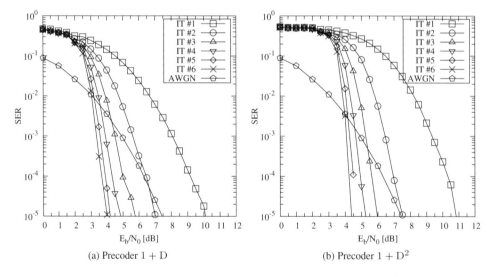

Figure 4.39: SER performance of the iterative receiver using the VLEC-3 scheme of Table 4.1 for transmission over the precoded three-path channel of Equation (4.52).

SER of 10^{-4} at $E_b/N_0 = 3.6$ dB, while the system using the precoder $(1 + D^2)$ needs $E_b/N_0 = 4.2$ dB to achieve the same SER.

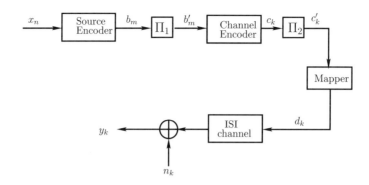

Figure 4.40: Transmitter schematic of the joint source–channel coding scheme for transmission over dispersive channels.

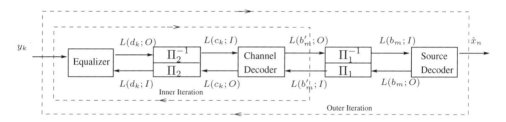

Figure 4.41: Receiver schematic of iterative channel equalization, channel decoding and source decoding.

4.4.4 Joint Turbo Equalization and Source Decoding

In this section we will investigate the achievable performance of iterative equalization, when both channel decoding and source decoding are considered, which constitutes a three-stage serially concatenated scheme. Let us first introduce our system model.

4.4.4.1 System Model

We assume that the source-encoded data is protected by a channel code before its transmission over an ISI-contaminated channel, as shown in Figure 4.40. At the receiver channel equalization, channel decoding and source decoding are performed iteratively. Generally, the iterative channel equalization and channel decoding processes, carried out in a two-stage structure, are referred to together as turbo equalization. However, when using SISO source decoding as well, the joint channel equalization, channel decoding and source decoding scheme of Figure 4.41 also becomes capable of exploiting the residual redundancy inherent in the source.

More explicitly, the corresponding receiver consists of three SISO modules, the channel equalizer, the channel decoder and the source decoder, as shown in Figure 4.41. All of them employ similar APP decoding algorithms and generate soft output for the benefit of the next processing stage. The channel equalization and channel decoding form the inner iterations. After a certain number of inner iterations, the source decoder is invoked and the extrinsic

Table 4.2: System parameters used in the simulations

Channel Impulse Response (CIR)	$\mathbf{H}_1 = [0.407\ 0.815\ 0.407]^T$
Channel Code	RSC Code
	$K = 5,\ R = 0.5$
	$G = (023, 035)$
Source Code	RVLC-2
Interleaver #1	Random Sequence Interleaver
	size $= 2048$ bits
Interleaver #2	Random Sequence Interleaver
	size $= 4096$ bits

information of the source is fed back for the next inner iteration. During our experiments, we found that the activation of one outer iteration after two inner iterations offered a good trade-off between the complexity imposed and the achievable performance. At the last outer iteration the source decoder generates the estimate $\{\hat{x}_n\}$ of the source symbol sequence by employing a sequence detection on the basis of the classic Viterbi algorithm using the same bit-level trellis.

4.4.4.2 Simulation Results

The attainable performance of the amalgamated turbo receiver depicted in Figure 4.41 has been evaluated when communicating over the three-path ISI channel $\mathbf{H}_1$ of Equation (4.52). The channel code used here is a half-rate, constraint-length $K = 5$ RSC code, using the octally represented generator polynomial of $G = (035/023)$. The source code used is the RVLC-2 scheme of Table 4.1. The two interleavers are random bit-interleavers having a memory of 2048 bits and 4096 bits respectively. These system parameters are also listed in Table 4.2.

The corresponding simulation results are depicted in Figure 4.42. The performances of both the joint and the separate source-decoding and turbo-equalization schemes are depicted. No outer iterations were executed in the separate source-decoding-based scheme. The performance of the same system for transmission over the non-dispersive AWGN channel after $I = 6$ iterations is also depicted as the best-case performance 'bound.' As we can observe in Figure 4.42, the joint three-stage turbo-detection scheme outperforms the separate source-decoding-based scheme by about 2 dB at SER $= 10^{-4}$ after $I = 6$ iterations.

4.5 Summary and Conclusions

In this chapter the achievable performances of iterative source–channel decoding schemes were investigated for both non-dispersive and dispersive AWGN channels. Since a source decoder combined with a channel decoder or a channel equalizer may be viewed as a serially

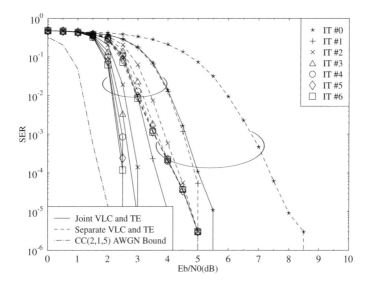

Figure 4.42: SER performance of joint RVLC decoding, channel decoding and channel equalization over the three-path dispersive AWGN channel of Equation (4.52).

concatenated system, the design principle of serial concatenated codes [71] applies, where the free distance of the outer code predetermines the achievable interleaver gain.

To be more specific, in the case of non-dispersive AWGN channels our simulation results show that the RVLC-2 of Table 4.1 having $d_f = 2$ outperforms the Huffman code and RVLC-1 having $d_f = 1$, while VLEC-3 having $d_f = 3$ outperforms the RVLC-2 arrangement. When the source codes have the same free distance, as with the Huffman code and RVLC-1 of Table 4.1, the constraint length determines the achievable performance. The larger the constraint length, the better the attainable performance. Our EXIT chart analysis seen in Figure 4.33 confirms the simulation results of Figures 4.37 and 4.39. It was found that VLCs having $d_f \geq 2$, such as RVLC-2 and VLEC-3, have better convergence properties than their lower-free-distance counterparts.

Unlike those of convolutional codes, the free distance and the code rate of a source code cannot be adjusted separately. Increasing the free distance usually requires a decrease in the code rate. For example, as the free distance increases from 1 to 3, the code rate of the Huffman code, RVLC-1, RVLC-2 and VLEC-3 schemes decreases from 0.98 to 0.54. Increasing the free distance generally improves the attainable decoding performance. However, decreasing the code rate reduces the system's throughput. For the design of the entire system, we have to find an attractive balance between the code rate and the free distance. Our main results are summarized in Table 4.3.

Furthermore, in the case of dispersive AWGN channels, the performance of iterative channel equalization and source decoding is evaluated for transmission over a three-path channel having mild ISI and a five-path channel having severe ISI. For both channels, the system using either the RVLC-2 or the VLEC-3 code of Table 4.1 exhibits significant iteration gains. Indeed, the EXIT chart analysis of Figure 4.30 shows that the EXIT function of a

Table 4.3: Summary of the simulation results for the iterative source–channel decoding scheme investigated in Section 4.3: various source codes listed in Table 4.1 are evaluated

	HUFF	RVLC-1	RVLC-2	VLEC-3
VLC d_f	1	1	2	3
Throughput (bpc)[†]	0.48	0.45	0.44	0.27
E_b/N_0 Gain (dB)[‡]	0	0.8	3.1	4.4

[†] The system throughput is measured in bits per channel use (bpc).

[‡] The E_b/N_0 gain is based on the SNR value required to achieve a SER of 10^{-4} after six iterations, and the benchmark is the scheme using the Huffman code.

Table 4.4: Summary of the simulation results of the iterative source decoding and channel equalization scheme for transmission over the three-path channel H_1 investigated in Section 4.4 – the RVLC-2 and the VLEC-3 code listed in Table 4.1 are employed, resulting in a system throughput of 0.87 bpc and 0.54 bpc respectively

	No precoding	Using Precoder $1 + D$	Using Precoder $1 + D^2$
RVLC-2			
Iteration Gain (dB)[†]	2.0	1.7	3.5
E_b/N_0 (dB)[‡]	8.0	8.0	7.0
VLEC-3			
Iteration Gain (dB)[†]	3.0	5.6	6.2
E_b/N_0 (dB)[‡]	6.8	3.6	4.2

[†] The iteration gain is measured after five iterations.

[‡] The E_b/N_0 value is the SNR value required for achieving a SER of 10^{-4} after five iterations.

channel equalizer cannot reach the ($I_A = 1$, $I_E = 1$) point, except for high SNRs, which implies that residual errors persist at the channel equalizer's output, regardless how high the interleaver length is and how many iterations are executed. This is further confirmed by the simulation results of Figures 4.31 and 4.32, where the performance of the iterative receiver is bounded by the performance of the same system transmitting over an AWGN channel exhibiting no ISI, resulting in 'error shoulders' in the SER curves. The system using the VLEC-3 code performs closer to the non-dispersive AWGN bound than that using the RVLC-2 due to the better EXIT characteristics of the VLEC-3 code.

The effect of precoding on the iterative channel equalization and source decoding was also investigated. A rate-one precoder renders the channel recursive for the receiver, and thus enables the attainment of higher interleaver gains. This effect is more obvious in the EXIT chart analysis of Figure 4.35, where the EXIT function of a channel equalizer recorded for a precoded channel can reach the $(I_A = 1, I_E = 1)$ point. It was found in Section 4.4.3 that for different source codes different precoders should be used. For example, when using RVLC-2, the system employing the precoder $(1 + D)$ inflicts an error floor, while the system employing the precoder $(1 + D^2)$ does not. By contrast, when using VLEC-3, the system employing the precoder $(1 + D)$ achieves earlier convergence than that employing the precoder $(1 + D^2)$, while both systems avoid having error floors. Therefore it is concluded that the source code and the precoder should be designed jointly. Our main findings are summarized in Table 4.4.

Finally, when a channel code is employed at the transmitter, a three-stage iterative receiver is formed consisting of a channel equalizer, a channel decoder and a source decoder. It was found in Figure 4.42 that up to 2 dB gain in terms of E_b/N_0 values required for achieving a SER of 10^{-5} can be obtained, when compared with separate turbo equalization and source decoding.

Chapter 5

Three-Stage Serially Concatenated Turbo Equalization[1]

5.1 Introduction

Turbo equalization [82] is an effective means of eliminating the channel-induced ISI imposed on the received signal. Hence, the achievable performance may approach that recorded over the non-dispersive AWGN channel, as detailed in [70] in the context of diverse turbo equalizers. When a simple rate-1 precoder is applied before the modulator, which renders the channel to appear recursive to the receiver, the attainable performance may be further improved [105, 263].

The SISO Minimum Mean Square Error (MMSE) equalizer [131], which is capable of utilizing *a priori* information from other SISO modules such as a SISO channel decoder, and generating extrinsic information, forms an attractive design alternative to the MAP equalizer owing to its lower computational complexity. This is particularly so for channels having long CIRs [113, 131]. The precoder can be readily integrated into the shift register model of the ISI channel [263], and may be modeled by combining its trellis with the trellis of a MAP/Soft-Output Viterbi Algorithm (SOVA)-based equalizer. However, the precoder's trellis description cannot be directly combined with the model of a MMSE equalizer. Hence, the achievable performance of MMSE turbo equalization is potentially limited [113, 131].

EXIT [98] charts have been proposed for analyzing the convergence behavior of iterative decoding schemes, and indicate that an infinitesimally low BER may only be achieved by an iterative receiver if an open tunnel exists between the EXIT curves of the two SISO components. Recently, both the convergence analysis and the best activation order of the component codes has been studied in the context of multiple-stage concatenated codes [120, 122], which generally require the employment of three-dimensional (3D) EXIT charts. For the sake of simplifying the associated analysis, a 3D to 2D EXIT chart projection

---

[1]Parts of this chapter are based on the collaborative research outlined in [211] © IEEE (2006), [213] © IEEE (2006) and [214] © IEEE (2009).

technique was proposed in [120, 122]. It has been shown [113] that the EXIT curve of a MMSE equalizer intersects with that of the channel decoder before reaching the decoding convergence point of $(1, 1)$; hence, residual errors persist after turbo equalization. However, it is natural to conjecture that there might exist an open tunnel leading to the convergence point in the 3D EXIT chart of a well-designed three-stage SISO system.

Against this backdrop, we propose a combined serially concatenated channel coding and MMSE equalization scheme, which is capable of achieving a precoding-aided convergence acceleration effect for a MAP/SOVA equalizer. Moreover, the convergence behavior of the proposed scheme is investigated with the aid of the 3D to 2D EXIT chart projection technique developed in [120, 122], and further design guidelines are derived from an EXIT-chart perspective. For illustration and comparison, let us commence with the family of traditional two-stage turbo equalization schemes.

5.2   Soft-In Soft-Out MMSE Equalization

We assume a coherent, symbol-spaced receiver front end as well as perfect knowledge of the signal phase and symbol timing, where the transmitter filter, the channel and the receiver filter are represented by a discrete-time linear filter having the Finite-length Impulse Response (FIR)

$$h[n] = \sum_{k=0}^{M-1} h_k \delta[n-k] \tag{5.1}$$

of length M. The coefficients h_k are assumed to be time invariant and known to the receiver. For simplicity, we assume Binary Phase-Shift Keying (BPSK) and that the channel impulse response coefficients h_k and the noise samples w_n are real valued. For higher-order constellations and complex-valued h_k and w_n, please refer to [131] for the details.

Let $x_n, n = 1, \ldots, K_c, x_n \in \{+1, -1\}$ denote the transmitted symbols and w_n represent the AWGN samples, which are independent and identically distributed (i.i.d.). Given Equation (5.1), the receiver's input z_n is given by

$$z_n = \sum_{k=0}^{M-1} h_k x_{n-k} + w_n. \tag{5.2}$$

Then the MMSE equalizer computes the estimates $\hat{x}_n$ of the transmitted symbols x_n from the received symbols z_n by minimizing the cost function $E[|x_n - \hat{x}_n|^2]$.

In contrast with conventional MMSE-based equalization methods, in the SISO equalizer advocated the mean squared error is averaged over both the distribution of the noise and the distribution of the transmitted symbols. This is because, in contrast to classic MMSE-based non-iterative equalization, in the context of turbo equalization the symbol distribution is no longer i.i.d., as is typically assumed, due to the information fed back to the equalizer from the error-correction decoder. Let $L_A(x_n) = \ln \frac{P(x_n=+1)}{P(x_n=-1)}$ denote the *a priori* LLR provided by the channel decoder. Then the SISO equalizer's output $L_E(x_n)$ is obtained using the estimate $\hat{x}_n$:

$$L_E(x_n) \triangleq \ln \frac{P(x_n = +1|\hat{x}_n)}{P(x_n = -1|\hat{x}_n)} - \ln \frac{P(x_n = +1)}{P(x_n = -1)}$$

$$= \ln \frac{p(\hat{x}_n|x_n = +1)}{p(\hat{x}_n|x_n = -1)}, \tag{5.3}$$

$n = 1, \ldots, K_c$, which requires the knowledge of the distribution $p(\hat{x}_n|x_n = x)$.

In order to perform MMSE estimation, the statistics $\bar{x}_n \triangleq \mathrm{E}[x_n]$ and $v_n \triangleq \mathrm{Cov}(x_n, x_n)$ of the symbols x_n are required. Usually, the symbols x_n are assumed to be equiprobable and i.i.d., which corresponds to $L_A(x_n) = 0, \forall n$, and yields $\bar{x}_n = 0$ as well as $v_n = 1$. For the general case of $L_A(x_n) \in \mathbb{R}$, when the symbols x_n are not equiprobable, $\bar{x}_n$ and v_n are obtained as

$$\bar{x}_n = \sum_{x = \pm 1} x \cdot P(x_n = x)$$

$$= \frac{e^{L_A(x_n)}}{1 + e^{L_A(x_n)}} - \frac{1}{1 + e^{L_A(x_n)}}$$

$$= \tanh\left(\frac{1}{2}L_A(x_n)\right), \tag{5.4}$$

$$v_n = \sum_{x = \pm 1} |x - \mathrm{E}(x_n)|^2 \cdot P(x_n = x)$$

$$= 1 - |\bar{x}_n|^2. \tag{5.5}$$

After MMSE estimation, we assume that the PDFs $p(\hat{x}_n|x_n = x)$ are Gaussian, having the parameters of $\mu_{n,x} \triangleq \mathrm{E}[\hat{x}_n|x_n = x]$ and $\sigma_{n,x}^2 \triangleq \mathrm{Cov}(\hat{x}_n, \hat{x}_n|x_n = x)$ [90]:

$$p(\hat{x}_n|x_n = x) \approx \frac{1}{\sqrt{2\pi}\sigma_{n,x}} \exp\left\{-\frac{(\hat{x}_n - \mu_{n,x})^2}{2\sigma_{n,x}^2}\right\}, \tag{5.6}$$

and the corresponding output LLRs $L_E(x_n)$ are formulated as

$$L_E(x_n) = \ln \frac{\frac{1}{\sqrt{2\pi}\sigma_{n,+1}} \exp\left\{-\frac{(\hat{x}_n - \mu_{n,+1})^2}{2\sigma_{n,+1}^2}\right\}}{\frac{1}{\sqrt{2\pi}\sigma_{n,-1}} \exp\left\{-\frac{(\hat{x}_n - \mu_{n,-1})^2}{2\sigma_{n,-1}^2}\right\}}$$

$$= \frac{2\hat{x}_n \mu_{n,+1}}{\sigma_{n,+1}^2}. \tag{5.7}$$

The employment of the Gaussian assumption tremendously simplifies the computation of the SISO equalizer output LLRs $L_E(x_n)$. We emphasize that $L_E(x_n)$ should not depend on the particular *a priori* LLR $L_A(x_n)$. Therefore, we require that $\hat{x}_n$ does not depend on $L_A(x_n)$, which affects the derivation of the MMSE equalization algorithms. For more details we refer to [113, 131].

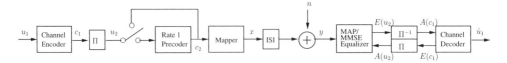

Figure 5.1: Turbo equalization system using MAP/MMSE equalizer both with and without precoding. The system parameters are summarized in Table 5.1.

5.3 Turbo Equalization Using MAP/MMSE Equalizers

5.3.1 System Model

Figure 5.1 shows the system model of a classic turbo equalization scheme. At the transmitter, a block of length L information data bits u_1 is first encoded by a channel encoder. After channel coding the coded bits c_1 are interleaved, yielding the data bits u_2, and they are either directly fed to the bit-to-modulated-symbol mapper or first fed through a rate-1 precoder and encoded to produce the coded bits c_2, as seen in Figure 5.1. After mapping, the modulated signal x is transmitted over a dispersive channel contaminated by AWGN n. At the receiver of Figure 5.1, an iterative detection/decoding structure is employed, where extrinsic information is exchanged between the channel equalizer and the channel decoder in a number of consecutive iterations. To be specific, the channel equalizer processes two inputs, namely the received signal y and the *a priori* information $A(u_2)$ fed back by the channel decoder. Then the channel equalizer of Figure 5.1 generates the extrinsic information $E(u_2)$, which is deinterleaved and forwarded as the *a priori* information to the channel decoder. Furthermore, the channel decoder capitalizes on the *a priori* information $A(c_1)$ provided by the channel equalizer and generates the extrinsic information $E(c_1)$, which is interleaved and fed back to the channel equalizer as the *a priori* information. Following the last iteration, the estimates $\hat{u}_1$ of the original bits are generated by the channel decoder, as seen in Figure 5.1.

In our forthcoming EXIT chart analysis and Monte Carlo simulations, we assume that the channel is time invariant and that the CIR is known at the receiver. To be specific, the three-path CIR of [8], described by

$$h[n] = 0.407\delta[n] + 0.815\delta[n-1] + 0.407\delta[n-2], \tag{5.8}$$

is used. We employ a constraint-length 3, half-rate RSC code RSC$(2, 1, 3)$ having the octally represented generator polynomials of $(5/7)$, where 7 is the feedback polynomial and 5 is the feed-forward polynomial. Then we use a simple rate-1 precoder described by the generator polynomials of $1/(1 + D)$. Either MAP or MMSE equalization is invoked, but precoding is only combined with the MAP equalizer. For the sake of simplicity, BPSK modulation is used. Our system parameters are summarized in Table 5.1.

5.3.2 EXIT Chart Analysis

Figure 5.2 depicts the EXIT functions of both the MAP/MMSE equalizers and the outer convolutional decoder. It is clear that the EXIT curves of both the MMSE equalizer and the MAP equalizer (without precoding) intersect with that of the outer RSC$(2, 1, 3)$ decoder, before reaching the convergence point of $(I_A^{EQ} = 1, I_E^{EQ} = 1)$. Hence residual errors may

Table 5.1: System parameters used in this section

	RSC(2,1,3)
Channel encoder	Generator polynomials $(5/7)$
Precoder	Generator polynomials $1/(1 + D)$
Modulation	BPSK
CIR	$[0.407\ 0.815\ 0.407]^T$
Block length	$L = 4096$ bits

persist, regardless of both the number of iterations used and the size of the interleaver. Furthermore, the MMSE equalizer generally outputs less extrinsic information than the MAP equalizer does, resulting in a poorer performance. On the other hand, with the advent of precoding, the EXIT curve of the MAP equalizer becomes capable of reaching the convergence point, as seen in Figure 5.2. We note, however, that there is a crossover between the EXIT curves of the precoded and the non-precoded MAP equalizer, which implies that the non-precoded MAP equalizer would perform better in the low-SNR region, while the precoded MAP equalizer is capable of achieving an near error-free performance, provided that enough iterations are performed. The convergence threshold of the precoded MAP equalizer is about 2.3 dB.

5.3.3 Simulation Results

In order to verify the convergence prediction of the EXIT chart analysis outlined in Section 5.3.2, Monte Carlo simulations were also performed and the corresponding BER results are depicted in Figure 5.3. It can be seen that the BER performance of the precoded system becomes better than that of the non-precoded system at an E_b/N_0 of about 2.7 dB.

5.4 Three-Stage Serially Concatenated Coding and MMSE Equalization

5.4.1 System Model

Simply incorporating an interleaver between the precoder and the signal mapper in the transmitter of Figure 5.1 enables the receiver to perform iterative equalization/decoding by exchanging extrinsic information between three SISO modules, namely the MMSE equalizer, Decoder II and Decoder I of Figure 5.4. Here, we would prefer not to refer to Encoder II as a precoder, since it cannot be directly combined with the equalizer at the receiver as in Figure 5.1. The same three-path channel of Equation (5.8) is used as in Section 5.3. The length of the non-causal and the causal parts of the MMSE filter are $N_1 = 5$ and $N_2 = 3$ respectively, resulting in an overall filter length of $N = N_1 + N_2 + 1 = 9$.

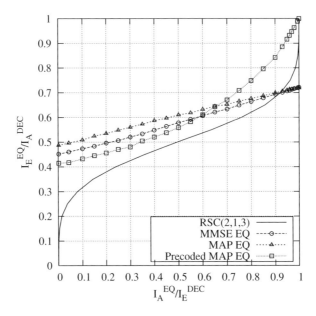

Figure 5.2: EXIT charts for the iterative receiver using either a MMSE equalizer or a MAP equalizer, where the latter is investigated for both a non-precoded and a precoded channel at $E_b/N_0 = 3$ dB.

5.4.2 EXIT Chart Analysis

In the following, we will carry out the EXIT chart analysis of the three-stage system of Figure 5.4. As with the two-stage system, the convergence SNR threshold of the three-stage system can be determined. At the same time, the outer code is optimized to give the lowest convergence SNR threshold. Finally, the activation order of the three SISO modules is optimized.

5.4.2.1 Determination of the Convergence Threshold

As seen in Figure 5.4, Decoder II exploits two a *priori* inputs, namely $A(c_2)$ and $A(u_2)$. At the same time, it generates two extrinsic outputs, $E(c_2)$ and $E(u_2)$. Hence, in order to describe the EXIT characteristics of Decoder II, we need the following two 2D EXIT functions [98, 122]:

$$I_{E(u_2)} = T_{u_2}[I_{A(u_2)}, I_{A(c_2)}], \tag{5.9}$$

$$I_{E(c_2)} = T_{c_2}[I_{A(u_2)}, I_{A(c_2)}]. \tag{5.10}$$

In contrast, for the MMSE equalizer and Decoder I only one a *priori* input is available in Figure 5.4, and the corresponding EXIT functions are

$$I_{E(u_3)} = T_{u_3}[I_{A(u_3)}, E_b/N_0] \tag{5.11}$$

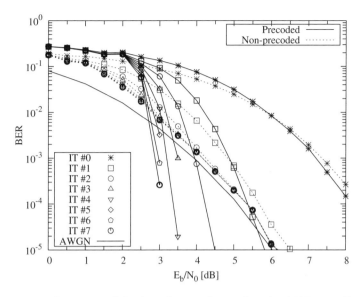

Figure 5.3: BER performance of the iterative receiver using the MAP equalizer both with and without precoding, for transmission over the dispersive AWGN channel having a CIR of Equation (5.8). The system parameters are outlined in Table 5.1.

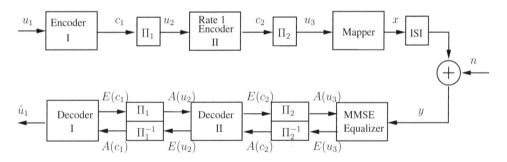

Figure 5.4: System diagram of two serially concatenated codes and MMSE equalization.

for the equalizer and

$$I_{E(c_1)} = T_{c_1}[I_{A(c_1)}] \tag{5.12}$$

for Decoder I, where the second parameter, E_b/N_0, in Equation (5.11) indicates that the extrinsic information also depends on the channel SNR. Therefore, with the aid of the EXIT module concept as described in Section 4.2.2, the three-stage system of Figure 5.4 may be viewed as a serial concatenation of three EXIT modules, which is shown in Figure 5.5.

In order to plot all of the EXIT functions, two 3D EXIT charts are required, namely one for the EXIT functions of both Equations (5.10) and (5.11) as shown in Figure 5.6a, and another for the EXIT functions of both Equations (5.9) and (5.12) as shown in Figure 5.6b. Note that $I_{E(u_3)}$ of Equation (5.11) is independent of $I_{A(u_2)}$; hence the MMSE equalizer's

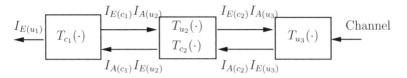

Figure 5.5: Concatenation of the EXIT modules corresponding to the concatenation of the SISO modules in Figure 5.4.

EXIT surface seen in Figure 5.6a is generated by sliding its EXIT curve in the 2D EXIT chart of Figure 5.2 along the $I_{A(u_2)}$ axis. The EXIT surface of Decoder I was generated similarly, as shown in Figure 5.6b, where $I_{E(c_1)}$ of Equation (5.12) is independent of $I_{A(c_2)}$.

Let us first consider the extrinsic information exchange between the MMSE equalizer and Decoder II. Let l be the time index. Only one SISO module is invoked each time. Note that we have $I_{A(c_2)}^{(l)} = I_{E(u_3)}^{(l-1)}$ and $I_{A(u_3)}^{(l)} = I_{E(c_2)}^{(l-1)}$. Considering Equations (5.9), (5.10) and (5.11), for a given *a priori* information $I_{A(u_2)}$, we have

$$I_{A(c_2)}^{(l)} = T_{u_3}[T_{c_2}[I_{A(u_2)}, I_{A(c_2)}^{(l-2)}], E_b/N_0], \tag{5.13}$$

with $I_{A(c_2)}^{(0)} = 0$ and

$$I_{E(u_2)}^{(l)} = T_{u_2}[I_{A(u_2)}, I_{A(c_2)}^{(l)}]. \tag{5.14}$$

The recursive equation of (5.13) actually represents an iteration, including the activation of both Decoder II and the MMSE equalizer. After a sufficiently high number of iterations, $I_{A(c_2)}^{(l)}$ and $I_{E(u_2)}^{(l)}$ will converge to a value between 0 and 1 that depends on the channel SNR and on the *a priori* input $I_{A(u_2)}$ only; i.e., we have

$$I_{A(c_2)} = \lim_{l\to\infty} I_{A(c_2)}^{(l)}, \tag{5.15}$$

$$I_{E(u_2)} = \lim_{l\to\infty} I_{E(u_2)}^{(l)}$$

$$= T_{u_2}\left[I_{A(u_2)}, \lim_{l\to\infty} I_{A(c_2)}^{(l)}\right]. \tag{5.16}$$

Hence the overall EXIT function of the combined module of the MMSE equalizer and Decoder II is a 1D function of $I_{A(u_2)}$ with E_b/N_0 as a parameter, which can be expressed as

$$I_{E(u_2)} = T'_{u_2}[I_{A(u_2)}, E_b/N_0]. \tag{5.17}$$

Figure 5.7 shows the EXIT function of Equation (5.14) for the combined module of the equalizer and Decoder II. It can be seen that the gain obtained by the second iteration ($l = 4$) is considerable, while the gain obtained by the third iteration is very limited ($l = 6$). Moreover, the extreme values of $I_{A(c_2)}$ in Equation (5.15), which correspond to different $I_{A(u_2)}$ abscissa values, can be visualized as the intersection of the two EXIT surfaces seen in Figure 5.6a,

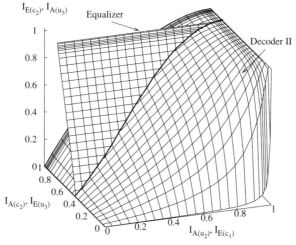

(a) Decoder II and the equalizer at $E_b/N_0 = 4$ dB

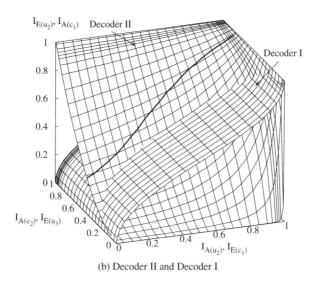

(b) Decoder II and Decoder I

Figure 5.6: 3D EXIT charts for the three-stage SISO system.

which is shown as a thick solid line. Furthermore, the EXIT function of Equation (5.16) corresponding to the extreme values of $I_{A(c_2)}$ is shown as a solid line in Figure 5.6b.

The EXIT function of Equation (5.17) plotted for the combined module is shown in a 2D EXIT chart in Figure 5.8. From this 2D EXIT chart, the convergence threshold of the three-stage system can be readily determined. When using a RSC(2, 1, 3) code having octal generator polynomials of 5/7 as the outer code, the EXIT curve of the outer code intersects

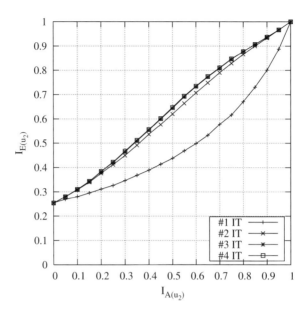

Figure 5.7: EXIT chart for the combined module of the equalizer and Decoder II at $E_b/N_0 = 4$ dB.

with the combined EXIT curve of the equalizer and Decoder II at $E_b/N_0 = 4$ dB. Hence the convergence threshold of this system is around 4.1 dB. Note that it has been shown in [122] that the convergence point of a multiple-stage concatenated system is independent of the activation order of the component decoders. Hence, the convergence threshold determined in this way is the true convergence threshold, regardless of the activation order.

5.4.2.2 Optimization of the Outer Code

After obtaining the EXIT function of the combined module of the equalizer and Decoder II, we can optimize the outer code to provide an open tunnel between the EXIT curve of the outer code and that of the combined module at the lowest possible SNR, and hence approach the channel capacity. We carried out a code search for different generator polynomials having constraint lengths up to $K = 5$. Interestingly, we found that the relatively weak code RSC$(2, 1, 2)$ having generator polynomials of $2/3$ yields the lowest convergence threshold of about 2.8 dB. The EXIT function of this code and that of the combined module are also shown in Figure 5.8 at $E_b/N_0 = 2.8$ dB.

5.4.2.3 Optimization of the Activation Order

Unlike that in the two-stage system of Section 5.3, the activation order of the decoders in the three-stage system is an important issue. Although different activation orders will result in the same final convergence point [122], they incur different decoding complexities and delays. A trellis-based search algorithm was proposed in [122] in order to find the optimal activation order of multiple concatenated codes according to certain criteria, such as, for example, minimizing the decoding complexity. However, for the relatively simple case of the

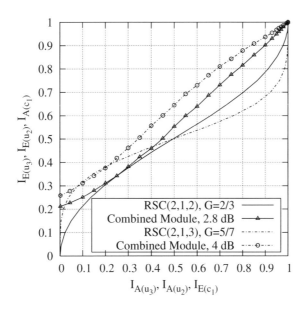

Figure 5.8: 2D EXIT chart for Decoder I and the combined module of the equalizer and Decoder II.

three-stage system, we resorted to a heuristic search by invoking the EXIT functions of the three SISO modules according to different activation orders. In these investigations no actual decoding or detection is invoked; hence this design procedure is of low complexity.

In our initial investigations the associated decoding complexity is not considered. Our sole target is that of determining the minimum number of decoder activations required for reaching the convergence point of $I_{E(u_1)} \approx 1$. We tested a number of different activation orders at several SNRs. For example, the natural decoder activation order is $[3\,2\,1\,3\,2\,1\ldots]$, where the integers represent the Index (I) of the various SISO decoder modules. Specifically, $I = 3$ denotes the MMSE equalizer, $I = 2$ represents Decoder II and $I = 1$ corresponds to Decoder I. Hence, according to this activation order, the MMSE equalizer is invoked first, then Decoder II and then Decoder I, and this pattern is repeated for a number of iterations.

Additionally, we can increase the number of the iterations either between the equalizer and Decoder II or between Decoder II and Decoder I. The former scenario includes the activation orders of $[3, 2, 3, 2, 1, \ldots]$ and $[3, 2, 3, 2, 3, 2, 1, \ldots]$. By contrast, the activation orders of $[3, 2, 1, 2, \ldots]$, $[3, 2, 1, 2, 1, \ldots]$, $[3, 2, 1, 2, 1, 2, \ldots]$ and $[3, 2, 1, 2, 1, 2, 1, \ldots]$ correspond to the latter scenario. Finally, the activation order of $[3, 2, 3, 2, 1, 2, 1, \ldots]$ increases both. The associated convergence test results are listed in Table 5.2. Note that for those activation orders that do not invoke Decoder I in the end, an additional activation of Decoder I is arranged at the end of the final iteration so that $I_{E(u_1)}$ is updated.

We can observe in Table 5.2 that, upon increasing the number of iterations between the equalizer and Decoder II, the total number of activations required for convergence is increased. For example, at $E_b/N_0 = 3$ dB, the number of activations needed for the

Table 5.2: Convergence test results (A/I) for different activation orders

Activation Order	2.8 dB	3 dB	3.5 dB	4 dB	5 dB
$[3, 2, 1, \ldots]$	108/36[†]	66/22	39/13	30/10	21/7
$[3, 2, 3, 2, 1, \ldots]$	130/26	80/16	50/10	40/8	25/5
$[3, 2, 3, 2, 3, 2, 1, \ldots]$	175/25	112/16	70/10	49/7	35/5
$[3, 2, 1, 2, \ldots]$	101/25	61/15	37/9	29/7	19/5
$[3, 2, 1, 2, 1, \ldots]$	100/20	**60/12**[‡]	35/7	28/6	20/4
$[3, 2, 1, 2, 1, 2, \ldots]$	**99**/17	61/**10**	**35/6**	**27/5**	**19/3**
$[3, 2, 1, 2, 1, 2, 1, \ldots]$	108/**16**	66/10	38/6	28/**4**	19/3
$[3, 2, 3, 2, 1, 2, 1, \ldots]$	112/16	70/10	42/6	33/5	26/4

[†] The first number, A, denotes the number of activations required to reach the convergence point of $I_{E(u_1)} \approx 0.9999$. The second number, I, is the corresponding number of iterations.

[‡] Bold numbers highlight the minimum number of activations/iterations by using different activation orders for each E_b/N_0 value.

natural activation order of $[3, 2, 1, \ldots]$ to fully converge is $A = 66$, while that required for the activation order of $[3, 2, 3, 2, 1, \ldots]$ is $A = 80$, and $A = 112$ for the activation order of $[3, 2, 3, 2, 3, 2, 1, \ldots]$. By contrast, when increasing the number of iterations between Decoder II and Decoder I, the number of activations required for full convergence is decreased. For example, again at $E_b/N_0 = 3$ dB, the numbers of activations needed for the activation orders of $[3, 2, 1, 2, \ldots]$, $[3, 2, 1, 2, 1, \ldots]$ and $[3, 2, 1, 2, 1, 2, \ldots]$ to reach convergence are $A = 61, 60$ and 61, respectively. Further increasing the number of iterations between Decoder II and Decoder I, as in the scenario of $[3, 2, 1, 2, 1, 2, 1, \ldots]$ will also increase the total number of activations imposed. Among the three best activation orders, $[3, 2, 1, 2, 1, 2, \ldots]$ invokes the equalizer the smallest number of times, namely $A_{EQ} = 10$ times. Note that the MMSE equalizer is of the highest computational complexity among the three SISO modules; hence, the less frequently the MMSE equalizer is activated, the lower the total decoding complexity. On the whole, the activation order of $[3, 2, 1, 2, 1, 2, \ldots]$ constitutes an attractive choice in terms of both fast convergence and low decoding complexity.

In general, we conclude that by invoking more iterations between Decoder II and Decoder I, while activating the MMSE equalizer from time to time, the three-stage system converges at a lower total number of activations. This is not unexpected, since the error-correction capability is mainly provided by the serial concatenation of Decoder II and Decoder I. Figure 5.9 shows the EXIT function recorded for the combined module of

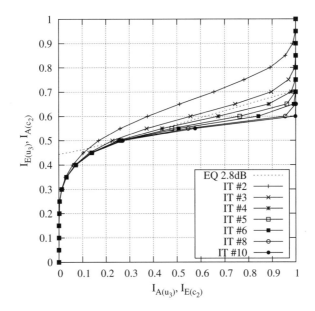

Figure 5.9: EXIT chart for the combined module of Decoder II and Decoder I.

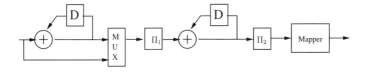

Figure 5.10: A specific manifestation of the generic transmitter schematic seen in Figure 5.4.

Decoder II and Decoder I. It can be seen that significant gains can be obtained by iteratively activating Decoder II and Decoder I for up to 10 iterations, especially in the medium *a priori* input information range of $0.5 < I_{A(c_2)} < 0.75$. In contrast, the EXIT function remains similar after three iterations when iterating between the equalizer and Decoder II, as seen from Figure 5.7. Hence no further iterations are necessary.

5.4.3 BER Performance

In our BER investigations, the RSC$(2, 1, 2)$ code having the octal generator polynomials of $(2/3)$ was invoked in Encoder/Decoder I. For the rate-1 Encoder II, again, the generator polynomial of $1/(1 + D)$ was used. Figure 5.10 shows a more specific manifestation of the generic transmitter schematic seen in Figure 5.4. The block length is $L = 10^5$ bits. The activation order of the three SISO modules is [3 2 1 2 1 2] and this pattern is repeated 12 times.

Our BER results are depicted in Figure 5.11. In addition to the three-stage SISO system, the BER performance of the two-stage SISO system seen in Figure 5.1 and using a MMSE equalizer is also shown. Observe in Figure 5.11 that the two-stage SISO system performs

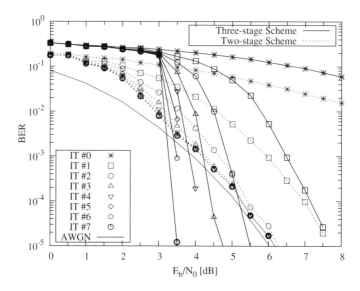

Figure 5.11: BER performance of the three-stage concatenation scheme of Figure 5.10 and the two-stage concatenation scheme of Figure 5.1 using a MMSE equalizer for transmission over the dispersive AWGN channel having a CIR of Equation (5.8). Note that for the three-stage concatenation scheme we used a specially selected activation order of the three SISO modules, namely one iteration is equivalent to an activation period of [3 2 1 2 1 2], where the numbers denote the indices of the SISO modules.

better in the low-SNR region, while the three-stage SISO system has the edge in the medium to high SNR region, achieving an infinitesimally low BER for E_b/N_0 values above 3 dB and hence closely matching the convergence threshold of $E_b/N_0 = 2.8$ dB predicted by the EXIT chart analysis of Figure 5.8. Although the two-stage scheme is of lower decoding complexity, it cannot outperform the AWGN BER bound, regardless of the number of iterations. To achieve a near-zero BER in the medium SNR range, the three-stage scheme has to be used. Additionally, as seen from Figure 5.11, the actual numbers of iterations required for achieving near-zero BER at $E_b/N_0 = 3$ dB, 3.5 dB, 4 dB and 5 dB are $I = 12, 7, 5$ and 4 respectively, which match the predictions of $I = 10, 6, 5$ and 3 seen in Table 5.2 quite well.

5.4.4 Decoding Trajectories

The mutual information vector of $[I_{E(u_3)}, I_{E(c_2)}, I_{E(u_2)}, I_{E(c_1)}]$ recorded during the simulations may be used to describe the evolution of the system status in the course of iterative decoding. They can be visualized in two 3D EXIT charts, resulting in the so-called decoding trajectories. More explicitly, Figure 5.12 shows the decoding trajectory associated with the evolution of $I_{E(c_1)}$, $I_{E(u_3)}$ and $I_{E(c_2)}$. Similarly Figure 5.13 depicts the decoding trajectory associated with the evolution of $I_{E(c_1)}$, $I_{E(u_3)}$ and $I_{E(u_2)}$. Note that each graph depicts one and only one extrinsic output of each SISO module. Hence, each time a SISO module is invoked, the decoding trajectory evolves in one dimension. For example, in Figure 5.12, the activation of the equalizer allows the trajectory to evolve along the $I_{A(c_2)}$, $I_{E(u_3)}$ axis, while

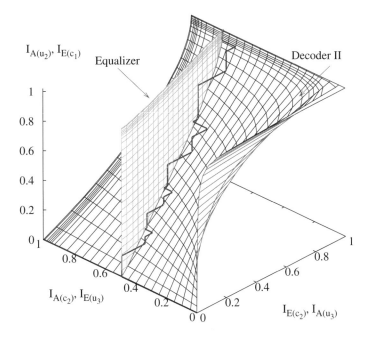

Figure 5.12: The decoding trajectories recorded at Decoder II's output $I_{E(c_2)}$, the equalizer's output $I_{E(u_3)}$ and Decoder I's output $I_{E(c_1)}$ for $E_b/N_0 = 4$ dB. The EXIT function of the equalizer and that of Decoder II are also shown. Each time a SISO module is invoked, the decoding trajectory evolves in the dimension associated with the corresponding extrinsic information output.

the activation of Decoder II evolves the trajectory along the $I_{E(c_2)}$, $I_{A(u_3)}$ axis. Finally the activation of Decoder I triggers the evolution of the trajectory along the $I_{A(u_2)}$, $I_{E(c_1)}$ axis.

Alternatively, the decoding process can also be visualized using two 2D graphs, namely one trajectory associated with the extrinsic outputs $I_{E(u_2)}$ and $I_{E(c_1)}$ of Figure 5.4, while the other is associated with the extrinsic outputs $I_{E(u_3)}$ and $I_{E(c_2)}$. These characterize the constituent decoder pairs of Decoder I and Decoder II, and of Decoder II and the equalizer of Figure 5.4, respectively. The former is depicted in Figure 5.14, which again characterizes the exchange of mutual information between Decoder I and the combined module of the equalizer and Decoder II. The vertical segments of the trajectory represent the activation of the combined equalizer and Decoder II module exchanging extrinsic information a certain number of times, while the horizontal segments represent a single activation of Decoder I.

For example, in the simulations we used an activation order of $[3\ 2\ 1\ 2\ 1\ 2\ldots]$. At the beginning of the iterations the equalizer was activated, but given the architecture of Figure 5.4 and the activation regime assumed, neither the value of $I_{E(u_2)}$ nor that of $I_{A(c_1)}$ was changed; hence, the trajectory stayed at the point A0 in Figure 5.14. Then Decoder II was activated, resulting in an increased value of $I_{E(u_2)}$, and hence the trajectory reached the point A1. Subsequently, Decoder I was activated, which increased the value of $I_{E(c_1)}$, and hence the

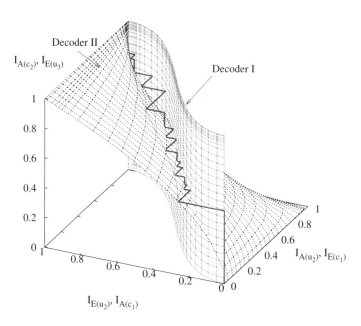

Figure 5.13: The decoding trajectories recorded at Decoder I's output $I_{E(c_1)}$, Decoder II's output $I_{E(u_2)}$ and the equalizer's output $I_{E(u_3)}$ for $E_b/N_0 = 4$ dB. The EXIT function of Decoder II and that of Decoder I are also shown. Each time a SISO module is invoked, the decoding trajectory evolves in the dimension associated with the corresponding extrinsic information output.

trajectory traversed to the point A2. Similarly, the segment between A2 and A3 represents the activation of Decoder II, the segment between A3 and A4 denotes the activation of Decoder I, and the segment between A4 and A5 corresponds to the activation of Decoder II. The segment between A5 and A6 represents the beginning of a new combined iteration cycle associated with similar decoding activations.

Figure 5.15 shows the trajectory associated with the extrinsic outputs $I_{E(u_3)}$ and $I_{E(c_2)}$, which characterizes the exchange of mutual information between the equalizer and the combined module of Decoder II and Decoder I. The vertical segments of the trajectory represent a single activation of the equalizer, while the horizontal segments represent the activation of Decoder II and Decoder I at a certain number of times. Specifically, the activation schedule of $[3\ 2\ 1\ 2\ 1\ 2\ \ldots]$ is visualized as follows. After an initial activation of the equalizer, the trajectory traverses from B0 to B1. Then it remains at B1 after the activation of Decoder II and Decoder I. This is because the output $I_{E(c_2)}$ of Decoder II remained zero after the first activation. However, these two activations are visualized in Figure 5.14 by the segment traversing from A0 to A1 while representing the activation of Decoder II. Similarly the segment spanning from A1 to A2 denotes the activation of Decoder I. Then the second activation of Decoder II triggers the evolution of the trajectory to the point B2 in Figure 5.15. The following activation of Decoder I does not change the trajectory in Figure 5.15, but is visualized by the segment spanning from A3 to A4 in Figure 5.14. Then the activation of

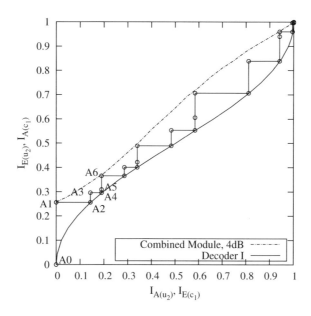

Figure 5.14: The decoding trajectories measured at Decoder II's output $I_{E(u_2)}$ and Decoder I's output $I_{E(c_1)}$ for $E_b/N_0 = 4$ dB. The activations of Decoder II cause the trajectory to grow along the y-dimension, while the activations of Decoder I cause the trajectory to grow along the x-dimension. The activation of the equalizer is not visualized. More explicitly, the trajectories can be interpreted as [EQ: A0, DEC2: A0$\rightarrow$A1, DEC1: A1$\rightarrow$A2, DEC2: A2$\rightarrow$A3, DEC1: A3$\rightarrow$A4, DEC2: A4$\rightarrow$A5], where EQ denotes the activation of the equalizer and DEC1/DEC2 represent the activations of Decoders I and II respectively.

Decoder II triggers a trajectory transition to B3. A new iteration cycle is indicated by the transition from B3 to B4. In conclusion, we surmise that the convergence behavior of a three-stage SISO system can be adequately visualized using two 2D EXIT charts.

5.4.5 Effects of Interleaver Block Length

As described in Section 4.2.2.3, EXIT functions are usually obtained by recording the extrinsic outputs of the SISO decoders, given certain *a priori* inputs. Since the *a priori* LLRs are artificially generated using the Gaussian model, they are independent of each other. However, this assumption is not entirely valid in real systems, where the *a priori* inputs are generated by the interleaved extrinsic outputs of the adjacent SISO decoders. The independence assumption only holds for sufficiently large interleaver block lengths and for the first few iterations. Hence it transpires that the analysis using EXIT functions becomes more and more inaccurate upon increasing the number of iterations, and further aggravated when decreasing the interleaver's block size.

We evaluated the performance of the three-stage system described in Section 5.4.3 for block length $L = 10^5$, 10^4, 10^3 and 10^2 bits using simulations, and the resultant BER performances are listed in Table 5.3. Since the outer code has a rate of $R = 1/2$ and the

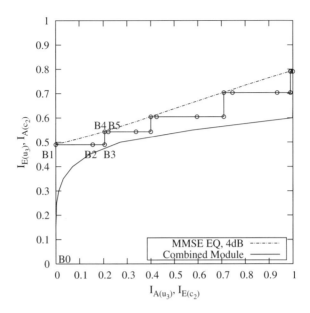

Figure 5.15: The decoding trajectories measured at the equalizer's output $I_{E(u_3)}$ and Decoder II's output $I_{E(c_2)}$ for $E_b/N_0 = 4$ dB. The activations of the equalizer cause the trajectory to grow along the y-dimension, while the activations of Decoder II cause the trajectory to grow along the x-dimension. The activation of Decoder I is not visualized. More explicitly, the trajectories can be interpreted as [EQ: B0→B1, DEC2: B1, DEC1: B1, DEC2: B1→B2, DEC1: B2, DEC2: B2→B3].

Table 5.3: BER performances for different block lengths after 12 iterations

E_b/N_0	$L = 10^5$	$L = 10^4$	$L = 10^3$	$L = 10^2$
3 dB	2.0×10^{-6}	3.2×10^{-2}	9.4×10^{-2}	1.6×10^{-1}
3.5 dB	0	0	2.0×10^{-2}	1.2×10^{-1}
4 dB	0	0	1.8×10^{-3}	7.2×10^{-2}
4.5 dB	0	0	2.4×10^{-4}	4.8×10^{-2}
5 dB	0	0	0	2.9×10^{-2}

intermediate code has a unity rate, the length of both of the interleavers seen in Figure 5.4 is $2L$. According to the EXIT chart analysis in Section 5.4.2.2, the convergence threshold of the three-stage system is 2.8 dB. For a block length of $L = 10^5$ bits, the system achieves convergence at about 3 dB, while the convergence is achieved at around 3.5 dB for $L = 10^4$ bits, at 5 dB for $L = 10^3$ bits and at around 11 dB for $L = 10^2$ bits.

Furthermore, for a given E_b/N_0 value and fixed number of iterations, the BER increases upon decreasing the block length. For example, at $E_b/N_0 = 3$ dB, the BER increases from 2.0×10^{-6} to 1.6×10^{-1} when the block length decreases from 10^5 to 10^2. This effect can also be visualized in the EXIT chart of Figure 5.16. For instance, the decoding trajectories of the systems using block lengths of $L = 10^5$, 10^4, 10^3 and 10^2 at $E_b/N_0 = 3$ dB are depicted in Figure 5.16. It can be seen that upon decreasing the block length the decoding trajectory drifts away from the EXIT curves more and more severely. In other words, the extrinsic information outputs of each SISO decoder become less than the EXIT chart predictions, resulting in increasing BERs.

Additionally, we show the achievable coding gain of the three-stage system at BER=10^{-4} in Figure 5.17. Here, the coding gain is defined as the reduction of the required E_b/N_0 when using the three-stage system as opposed to the uncoded system (i.e. when only MMSE equalization is employed without channel coding, using a block length of $L = 10^5$ bits). It can be seen that the maximum achievable coding gain decreases upon decreasing the block length. In comparison with the system employing the block length of $L = 10^5$ bits, the maximum coding gain of the system using a block length of $L = 10^4$ bits is decreased by less than 1 dB, while it is decreased by about 2 dB for the system using a block length of $L = 10^3$ bits and around 8 dB for $L = 10^2$ bits. Moreover, for $L = 10^5$ and $L = 10^4$, less than 0.5 dB iteration gain is achieved after $I = 8$ iterations, while for $L = 10^3$ the iteration gain diminishes after $I = 6$ iterations, and for $L = 10^2$ only a marginal iteration gain is obtained after $I = 2$ iterations.

5.5 Approaching the Channel Capacity Using EXIT-Chart Matching and IRCCs

In Section 5.4.2.2, we optimized the outer code of the three-stage system for achieving the lowest possible convergence threshold. The optimization was carried out by searching for different generator polynomials for the convolutional codes used. The lowest achievable convergence threshold was about 2.8 dB, when using a RSC(2, 1, 2) code having octally represented generator polynomials of $2/3$. In this section we will investigate whether we can reduce this threshold further down by considering the area properties of EXIT charts.

5.5.1 Area Properties of EXIT Charts

Let us now discuss the relevance of the EXIT chart area $\bar{A}_I$ beneath the inverted outer EXIT function and A_{II} beneath the inner EXIT function. Various proofs relating to these areas were provided in [204, 267] for the case when the a priori LLRs are provided for the respective APP SISO decoder over a Binary Erasure Channel (BEC) having either zero magnitudes or infinite magnitudes (and the correct sign). However, it has been shown that the shapes of the EXIT functions do not significantly depend on the particular channel considered [268], and hence the results discussed in this section hold approximately for more general channels.

In [204, 267], the area $\bar{A}_I$ beneath the inverted EXIT function of an optimal outer APP SISO decoder having a coding rate of R_I was shown to be given by

$$\bar{A}_I = R_I. \tag{5.18}$$

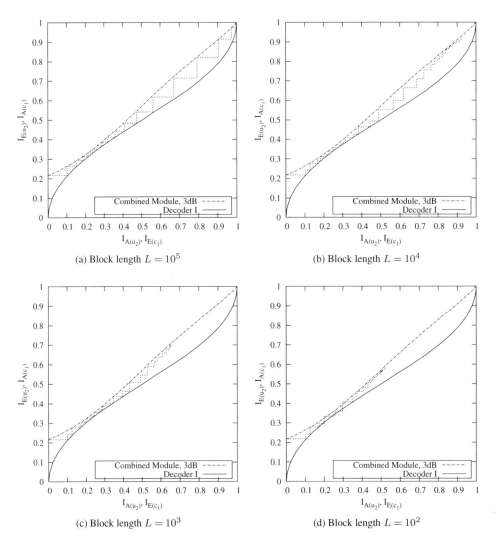

(a) Block length $L = 10^5$

(b) Block length $L = 10^4$

(c) Block length $L = 10^3$

(d) Block length $L = 10^2$

Figure 5.16: Decoding trajectories of the system of Figure 5.4 using different block lengths at $E_b/N_0 = 3$ dB.

In the case where this outer code is serially concatenated with a R_{II}-rate inner code that is employed for protecting M_{mod}-ary modulated transmissions, the effective throughput η in bits of source information per channel use is given by

$$\eta = R_I \cdot R_{II} \cdot \log_2(M_{\mathrm{mod}}) = \bar{\mathcal{A}}_I \cdot R_{II} \cdot \log_2(M_{\mathrm{mod}}). \tag{5.19}$$

As described in Section 4.3, maintaining an open EXIT chart tunnel is a necessary condition for achieving iterative decoding convergence to an infinitesimally low probability

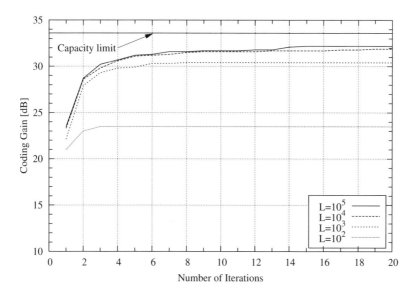

Figure 5.17: Achievable coding gain of the three-stage system of Figure 5.4 at BER=10^{-4}, when employing various block lengths and different numbers of iterations.

of error. Since an open EXIT chart tunnel can only be created if the EXIT chart area beneath the inverted outer EXIT function $\bar{\mathcal{A}}_I$ is less than that beneath the inner code's EXIT function $\mathcal{A}_{II}$, we have $\bar{\mathcal{A}}_I < \mathcal{A}_{II}$ and hence maintaining

$$\eta < \mathcal{A}_{II} \cdot R_{II} \cdot \log_2(M_{\mathrm{mod}}) \tag{5.20}$$

constitutes a necessary condition for the support of iterative decoding convergence to an infinitesimally low probability of error.

In the case where we have $R_{II} = 1$ and an optimal inner APP SISO decoder is employed, [204, 267] showed that

$$\mathcal{A}_{II} \cdot R_{II} \cdot \log_2(M_{\mathrm{mod}}) = C, \tag{5.21}$$

where C is the Discrete-Input Continuous-Output Memoryless Channel's (DCMC's) capacity [269] expressed in bits of source information per channel use. Hence, in the case where $R_{II} = 1$ and an optimal inner APP SISO decoder is employed,

$$\eta < C \tag{5.22}$$

constitutes a necessary condition for the support of iterative decoding convergence to an infinitesimally low probability of error. Note that this is as Shannon stated in his seminal publication of 1948 [133].

However, in the case where we have $R_{II} < 1$ or a suboptimal inner APP SISO decoder is employed, [204, 267] showed that

$$\mathcal{A}_{II} \cdot R_{II} \cdot \log_2(M_{\text{mod}}) = \bar{C} \leq C, \tag{5.23}$$

where $\bar{C}$ is the *attainable* DCMC capacity. More explicitly, in this case some capacity loss occurs, since the necessary condition for iterative decoding convergence to an infinitesimally low probability of error to be supported becomes

$$\eta < \bar{C}. \tag{5.24}$$

Note that the EXIT chart area within an open EXIT chart tunnel $\mathcal{A}_{II} - \bar{\mathcal{A}}_I$ is proportional to the discrepancy between the effective throughput and the (attainable) DCMC capacity. Hence, we may conclude that near-capacity transmissions are facilitated when a narrow, marginally open EXIT chart tunnel can be created for facilitating convergence to an infinitesimally low probability of error. This motivates the employment of irregular coding for EXIT chart matching, as will be discussed in the following sections.

5.5.2 Analysis of the Three-Stage System

For convenience, we first redraw some of the previously used EXIT functions in Figure 5.18. According to the above-described area properties [116] of EXIT charts, the area under the EXIT curve of the inner code is approximately proportional to the channel capacity attained. Furthermore, the area under the inverted EXIT curve of the outer code is approximately equal to R_I, where R_I is the outer code's rate. More explicitly, let $\mathcal{A}_I$ and $\bar{\mathcal{A}}_I$ be the areas under the outer code's EXIT curve $T_{c_1}(i)$ and its inverse $T_{c_1}^{-1}(i)$, $i \in [0, 1]$ respectively, which are expressed as

$$\mathcal{A}_I = \int_0^1 T_{c_1}(i)\, di, \quad \bar{\mathcal{A}}_I = \int_0^1 T_{c_1}^{-1}(i)\, di = 1 - \mathcal{A}_I. \tag{5.25}$$

Similarly, we define the areas $\mathcal{A}_{II}$ and $\bar{\mathcal{A}}_{II}$ for the EXIT curve $T_{u_2}'(i)$ of the combined module of the equalizer and Decoder II, as well as $\mathcal{A}_{III}$ and $\bar{\mathcal{A}}_{III}$ for the EXIT curve $T_{u_3}(i)$ of the inner equalizer. Then we have

$$\bar{\mathcal{A}}_I \approx R_I, \tag{5.26}$$

and for BPSK modulation and MAP equalization

$$\mathcal{A}_{III} \approx C_{UI}, \tag{5.27}$$

where C_{UI} is the uniform-input capacity of the communication channel seen in the schematic of Figure 5.4.

Specifically, in Figure 5.18 the area under the EXIT curve of the MAP equalizer is $\mathcal{A}_{III} \approx 0.59$ and the area under the EXIT curve of the MMSE equalizer is $\mathcal{A}_{III}' \approx 0.57$. When expressed in terms of bits per channel use, these values approximate the channel's capacity. The slight throughput loss of the latter is due to the employment of the MMSE criterion, which is inferior to the MAP criterion. Since the intermediate code has a unity rate, the area $\mathcal{A}_{II}$ under the EXIT curve of the combined module is equal to $\mathcal{A}_{III}'$. In order to achieve convergence, there has to be an open tunnel between the EXIT curve of the

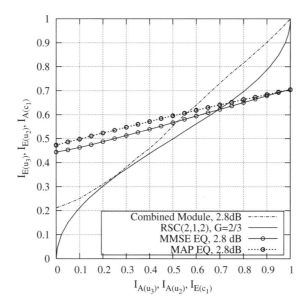

Figure 5.18: EXIT functions of the MMSE equalizer, the MAP equalizer and the combined module of the MMSE equalizer and Decoder II at $E_b/N_0 = 2.8$ dB, as well as the EXIT function of the outer RSC$(2, 1, 2)$ code.

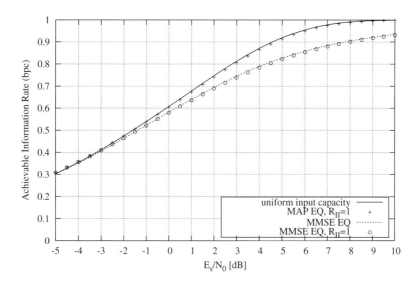

Figure 5.19: The achievable information rate for systems employing MAP/MMSE equalization for transmission over the three-path channel of Equation (5.8).

combined module and that of the outer code. Accordingly we have $\mathcal{A}'_{III} > \bar{A}_I$ or $\mathcal{A}_{II} > \bar{A}_I$. Hence the area under the EXIT curve of the MMSE equalizer is equal to the maximum achievable information rate when using MMSE equalization, which is achieved when the EXIT curve of the outer code is perfectly fitted to the EXIT curve of the combined module. According to these area properties, the channel capacity derived for a uniformly distributed input and the maximum information rate achieved when using MAP/MMSE equalization may be readily obtained for the three-path channel of Equation (5.8), which is depicted as a function of E_s/N_0 in Figure 5.19. It can be seen that, compared with MAP equalization, there is an inherent information rate loss when employing the MMSE equalization, especially at high channel SNRs. The achievable information rates of the systems employing rate-1 intermediate codes are also plotted, and these show negligible information rate loss.

As seen from Figure 5.18, there is an open tunnel between the EXIT curve of the combined module and that of the outer RSC code at the convergence threshold of $E_b/N_0 = 2.8$ dB; i.e., we have $\mathcal{A}_{II} - \bar{A}_I > 0$, which implies that the channel capacity is not reached. On the other hand, if we fix the system's throughput to 0.5 bits per channel use, the convergence threshold may be further lowered, provided that a matching outer code can be found. More explicitly, we found that the area under the EXIT curve of the equalizer at $E_b/N_0 = 1.6$ dB (or equivalently at $E_s/N_0 = -1.4$ dB) is about 0.50. Hence our task now is to search for an outer code whose EXIT curve is perfectly matched to that of the combined module of the equalizer and Decoder II, so that the maximum information rate is achieved.

Irregular codes, such as irregular LDPC codes, irregular RA codes [270] and IRCCs [80, 81], are reported to have flexible EXIT characteristics, which can be optimized to match the EXIT curve of the combined equalizer and Decoder II module, in order to create an appropriately shaped EXIT-chart convergence tunnel. For the sake of simplicity, IRCCs are considered in the forthcoming section.

5.5.3 Design of Irregular Convolutional Codes

The family of IRCCs was proposed in [80, 81]. It consists of a set of convolutional codes having different code rates. They were specifically designed with the aid of EXIT charts [98], for the sake of improving the convergence behavior of iteratively decoded systems. To be more specific, an IRCC is constructed from a family of P subcodes. First, a rate-r convolutional parent code C_1 is selected and the $(P - 1)$ remaining subcodes C_k of rate $r_k > r$ are obtained by puncturing. Let L denote the total number of encoded bits generated from the K uncoded information bits. Each subcode encodes a fraction $\alpha_k r_k L$ of the original uncoded information bits and generates $\alpha_k L$ encoded bits. Given our overall average code rate target of $R \in [0, 1]$, the weighting coefficient α_k has to satisfy

$$1 = \sum_{k=1}^{P} \alpha_k, \quad R = \sum_{k=1}^{P} \alpha_k r_k, \quad \text{and} \quad \alpha_k \in [0, 1], \; \forall k. \tag{5.28}$$

Clearly, the individual code rates r_k and the weighting coefficients α_k play a crucial role in shaping the EXIT function of the resultant IRCC. For example, in [81] a family of $P = 17$ subcodes were constructed from a systematic, rate-1/2, memory-4 parent code defined by the generator polynomial $(1, g_1/g_0)$, where $g_0 = 1 + D + D^4$ is the feedback polynomial and $g_1 = 1 + D^2 + D^3 + D^4$ is the feed-forward one. Higher code rates may be obtained by

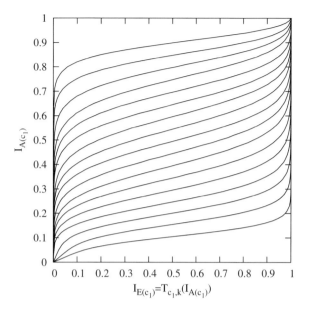

Figure 5.20: EXIT functions of the 17 subcodes in [81].

puncturing, while lower rates are created by adding more generators and by puncturing under the constraint of maximizing the achievable free distance. The two additional generators used are $g_2 = 1 + D + D^2 + D^4$ and $g_3 = 1 + D + D^3 + D^4$. The resultant 17 subcodes of [81] have coding rates spanning from $0.1, 0.15, 0.2, \ldots$ to 0.9.

The IRCC constructed using this procedure has the advantage that the decoding of all subcodes may be performed using the same parent code trellis, except that at the beginning of each block of $\alpha_k r_k L$ number of trellis transitions/sections corresponding to the subcode C_k the puncturing pattern has to be restarted. Trellis termination is necessary only after all of the K uncoded information bits have been encoded.

The EXIT function of an IRCC can be obtained from those of its subcodes. Denote the EXIT function of the subcode k as $T_{c_1,k}(I_{A(c_1)})$. Assuming that the trellis segments of the subcodes do not significantly interfere with each other, which might change the associated transfer characteristics, the EXIT function $T_{c_1}(I_{A(c_1)})$ of the target IRCC is the weighted superposition of the EXIT function $T_{c_1,k}(I_{A(c_1)})$ [81], yielding

$$T_{c_1}(I_{A(c_1)}) = \sum_{k=1}^{P} \alpha_k T_{c_1,k}(I_{A(c_1)}). \tag{5.29}$$

For example, the EXIT functions of the 17 subcodes used in [81] are shown in Figure 5.20. We now optimize the weighting coefficients, $\{\alpha_k\}$, so that the IRCC's EXIT curve of Equation (5.29) matches the combined EXIT curve of Equation (5.17). The area under the combined EXIT curve of Equation (5.17) at $E_b/N_0 = 1.8$ dB is $\mathcal{A}_{II} \approx 0.51$, which indicates that this E_b/N_0 value is close to the lowest possible convergence threshold for a system having an outer coding rate of $R_I = 0.5$. We optimize $T_{c_1}(I_{A(c_1)})$ of Equation (5.29) at $E_b/N_0 = 1.8$ dB by minimizing the square of the error function

$$e(i) = [T'_{u_2}(i, E_b/N_0) - T_{c_1}^{-1}(i)] \tag{5.30}$$

subject to the constraints of (5.28) and $e(i) > 0$ over all i:

$$J(\alpha_1, \ldots, \alpha_P) = \int_0^1 e^2(i)\, di, \quad e(i) > 0, \forall i \in [0, 1]. \tag{5.31}$$

In more tangible physical terms, we minimize the area between the projected EXIT curve and the EXIT curve of the outer code, as seen in Figure 5.8. This results in a good match between the two curves, ultimately leading to a narrow EXIT-chart tunnel, which implies a near-capacity operation attained at the cost of a potentially high number of decoding iterations. The associated high complexity is characteristic of schemes operating beyond the Shannonian cut-off rate. The curve-fitting problem portrayed in Equation (5.31) is a quadratic programming problem, which can be solved by the gradient-descent-based iterative solution proposed in [80]. With the aid of this algorithm [80], the optimized weighting coefficients are obtained as

$$\begin{aligned}[\alpha_1, \ldots, \alpha_{17}] = [&0, \ 0.01374, \ 0, \ 0.28419, \ 0.09306, \ 0, \\ &0, \ 0.10614, \ 0.09578, \ 0.06183, \ 0, \ 0, \\ &0.19685, \ 0, \ 0.02059, \ 0, \ 0.12784]. \end{aligned} \tag{5.32}$$

The resultant EXIT curve of the optimized IRCC is shown in Figure 5.21.

Note that the ability of EXIT chart matching to create narrow but still open EXIT chart tunnels depends on the availability of a suite of component codes having a wide variety of EXIT function shapes. However, in general it is challenging to design component codes having arbitrary EXIT function shapes, motivating the irregular code design process depicted in Figure 5.22. This advocates the design of diverse candidate component codes, the characterization of their EXIT functions and the selection of a specific suite having a wide variety of EXIT function shapes, potentially involving a significant amount of 'trial-and-error'-based human interaction. Following this, EXIT chart matching may be achieved by selecting the component fractions, as described above. Throughout Part II of this book we shall describe a number of modifications to this process.

5.5.4 Simulation Results

Monte Carlo simulations were performed for characterizing both the IRCC design and the convergence predictions of Section 5.5.3. As our benchmark scheme, the RSC(2, 1, 2) code having octal generator polynomials of $2/3$ was employed as Encoder I in the schematic of Figure 5.4. As our proposed scheme, the IRCC designed in Section 5.5.3 was used as Encoder I. The rest of the system components were the same for both schemes, and all of the system parameters are listed in Table 5.4.

Figure 5.23 depicts the BER performance of both schemes. It can be seen that after 12 iterations, the three-stage scheme employing an IRCC achieves a BER lower than 10^{-5} at

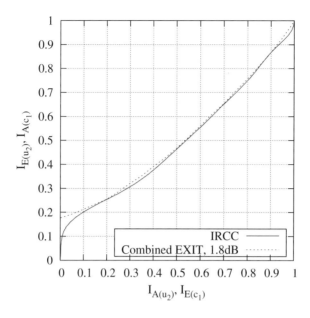

Figure 5.21: The combined EXIT function of Equation (5.17) and the EXIT function of the optimized IRCC at $E_b/N_0 = 1.8$ dB.

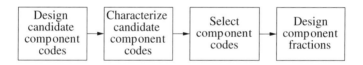

Figure 5.22: Conventional irregular coding design process.

Table 5.4: System parameters used in the simulations of Section 5.5.4

Encoder I	Half-rate RSC$(2, 1, 2)$ code, $G = 2/3$
	or Half-rate IRCC of Equation (5.32)
Encoder II	Generator polynomials $1/3$
Modulation	BPSK
CIR	$[0.407\ 0.815\ 0.407]^T$
Block length	$L = 100\ 000$ bits
Overall coding rate	0.5

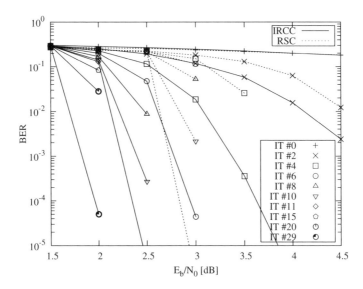

Figure 5.23: BER performances for the three-stage serial concatenation scheme of Figure 5.4 using both regular RSC code and IRCC for transmission over the dispersive AWGN channel having a CIR of Equation (5.8). The system parameters are outlined in Table 5.4.

about $E_b/N_0 = 2.4$ dB, which outperforms the three-stage scheme using a regular RSC code by about 0.5 dB. Moreover, if more iterations are invoked, for example $I = 30$ iterations, the three-stage scheme employing an IRCC is capable of achieving convergence at $E_b/N_0 = 2$ dB, which is very close to the convergence threshold of $E_b/N_0 = 1.8$ dB predicted by the EXIT chart analysis. Hence, a total additional gain of about 1 dB is achievable by using the optimized IRCC.

The decoding trajectory recorded during our Monte Carlo simulations using the optimized IRCC is depicted in Figure 5.24. It can be seen that at $E_b/N_0 = 2$ dB a narrow tunnel exists between the EXIT curve of the outer IRCC code and that of the combined module of the equalizer and Decoder II. Observe furthermore that the simulation-based recorded trajectory matches these EXIT curves quite closely, except for a few iterations close to the point of convergence. Furthermore, since the tunnel between the two EXIT curves is narrow, numerous iterations are required in order to enable the iterative receiver to converge to the point of $(1, 1)$. When aiming to approach the channel capacity even more closely, the tunnel becomes even narrower, and therefore more iterations would be needed, resulting in a drastically increased complexity. At the same time, even longer interleavers will be required in order to ensure that the soft information exchanged between the SISO modules remains uncorrelated, so that the recorded decoding trajectory follows the EXIT curves. This is another manifestation of Shannon's channel capacity theorem [1].

5.6 Rate Optimization of Serially Concatenated Codes

In Section 5.4.2 we have shown that by using a unity-rate intermediate code and a half-rate $RSC(2, 1, 2)$ outer code, the three-stage system of Figure 5.4 became capable

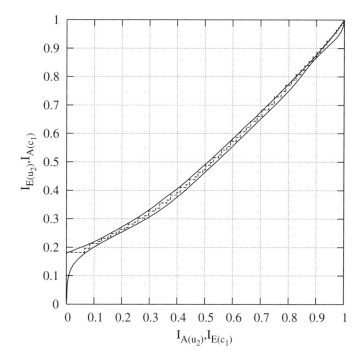

Figure 5.24: The decoding trajectory of the three-stage system using IRCC at $E_b/N_0 = 2$ dB. The EXIT function of the IRCC, $I_{E(c_1)} = T_{c_1}(I_{A(c_1)})$, and that of the combined module of the equalizer and Decoder II, $I_{E(u_2)} = T'_{u_2}(I_{A(u_2)})$, are also shown.

of achieving convergence at $E_b/N_0 = 2.8$ dB, which is 1.4 dB away from the channel capacity achievable in case of uniformly distributed inputs, and 1.2 dB away from the maximum achievable information rate of MMSE equalization. Below we investigate the performance of the three-stage system, when employing an intermediate code having a rate of $R_{II} < 1$. In other words, we fix the overall coding rate $R_{tot} = 0.5$ (i.e., we fix the system throughput to be 0.5 bits per channel use assuming BPSK), and optimize the rate allocation between the intermediate code and the outer code as well as their generator polynomials and puncturing matrix, so that convergence is achieved at the lowest possible E_b/N_0 value.

For simplicity, we limit our search space to regular convolutional codes having constraint lengths of $K < 4$. For the rate allocation between the intermediate code and the outer code, we consider two cases: $R^1_{II} = 3/4$, $R^1_I = 2/3$, and $R^2_{II} = 2/3$, $R^2_I = 3/4$. The rate-2/3 and rate-3/4 codes are generated by puncturing the rate-1/2 parent code, and the puncturing matrices used are listed in Table 5.5. An exhaustive search was carried out, and the combinations that exhibit good convergence properties are listed in Table 5.6. Furthermore, we depict the EXIT functions of the SCCs in Figure 5.25. Note that the EXIT function of each SCC was obtained by using the EXIT module of each component code, rather than by simulations. This incurs a substantially lower complexity.

Table 5.5: Puncturing matrices for generating rate-$2/3$ and rate-$3/4$ codes from rate-$1/2$ parent codes

<table>
<tr><td colspan="2" align="center">Rate-$2/3$</td></tr>
<tr>
<td align="center">$P_{11} = \begin{bmatrix} 1 & 0 \\ 1 & 1 \end{bmatrix}$</td>
<td align="center">$P_{12} = \begin{bmatrix} 1 & 1 \\ 1 & 0 \end{bmatrix}$</td>
</tr>
</table>

<table>
<tr><td colspan="4" align="center">Rate-$3/4$</td></tr>
<tr>
<td>$P_{21} = \begin{bmatrix} 1 & 0 & 1 \\ 1 & 1 & 0 \end{bmatrix}$</td>
<td>$P_{22} = \begin{bmatrix} 1 & 1 & 0 \\ 1 & 0 & 1 \end{bmatrix}$</td>
<td>$P_{23} = \begin{bmatrix} 1 & 0 & 0 \\ 1 & 1 & 1 \end{bmatrix}$</td>
<td>$P_{24} = \begin{bmatrix} 1 & 1 & 1 \\ 1 & 0 & 0 \end{bmatrix}$</td>
</tr>
</table>

Table 5.6: Optimized serially concatenated codes and the convergence thresholds of the corresponding three-stage systems

	Inter rate-$3/4$ code	Outer rate-$2/3$ code	$E_b/N_0|_{th}$
SCC-A1	RSC$(3,4,2)$, $G = 2/3$, $P = P_{24}$	RSC$(2,3,2)$, $G = 2/3$, $P = P_{12}$	2.2 dB
SCC-A2	RSC$(3,4,2)$, $G = 2/3$, $P = P_{24}$	RSC$(2,3,3)$, $G = 6/7$, $P = P_{12}$	2.2 dB
SCC-A3	RSC$(3,4,2)$, $G = 2/3$, $P = P_{24}$	RSC$(2,3,3)$, $G = 5/7$, $P = P_{12}$	2.5 dB
SCC-A4	RSC$(3,4,3)$, $G = 7/5$, $P = P_{24}$	RSC$(2,3,2)$, $G = 2/3$, $P = P_{12}$	2.3 dB
SCC-A5	RSC$(3,4,3)$, $G = 7/5$, $P = P_{24}$	RSC$(2,3,3)$, $G = 5/7$, $P = P_{11}$	2.4 dB
	Inter rate-$2/3$ code	Outer rate-$3/4$ code	$E_b/N_0|_{th}$
SCC-B1	RSC$(2,3,2)$, $G = 2/3$, $P = P_{12}$	RSC$(3,4,3)$, $G = 5/7$, $P = P_{21}$	2.5 dB
SCC-B2	RSC$(2,3,2)$, $G = 2/3$, $P = P_{12}$	RSC$(3,4,3)$, $G = 5/7$, $P = P_{22}$	2.5 dB
SCC-B3	RSC$(2,3,2)$, $G = 2/3$, $P = P_{12}$	RSC$(3,4,3)$, $G = 5/7$, $P = P_{23}$	2.4 dB
SCC-B4	RSC$(2,3,2)$, $G = 2/3$, $P = P_{12}$	RSC$(3,4,3)$, $G = 5/7$, $P = P_{24}$	2.2 dB
SCC-B5	RSC$(2,3,3)$, $G = 5/7$, $P = P_{12}$	RSC$(3,4,2)$, $G = 2/3$, $P = P_{21}$	2.3 dB
SCC-B6	RSC$(2,3,3)$, $G = 5/7$, $P = P_{12}$	RSC$(3,4,2)$, $G = 2/3$, $P = P_{24}$	2.5 dB

It can be seen from Table 5.6 that the lowest achievable convergence threshold of 2.2 dB is achievable by the SCC-A1/A2 and SCC-B4 schemes. This value is 0.6 dB lower than that of the system using rate-1 intermediate codes as discussed in Section 5.4. In order to verify our EXIT chart analysis, Monte Carlo simulations were also conducted, and the BER performance of the corresponding three-stage system using, for instance, the SCC-A2 scheme, is depicted in Figure 5.26. It can be seen that a BER of 10^{-5} is achieved at around $E_b/N_0 = 2.4$ dB, which is close to our EXIT-chart prediction.

Although the three-stage system using an intermediate code having rate less than unity may achieve a lower convergence threshold than the system using a rate-1 intermediate code, as expected, the maximum achievable information rate is reduced. Let $\mathcal{A}_{II}$ be the area under the EXIT curve of the combined module of the inner equalizer and the intermediate decoder. According to the area properties of EXIT charts [116], the coding rate of the outer code R_I

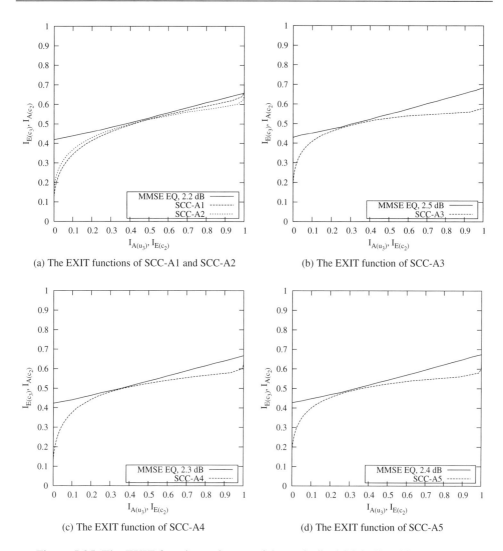

(a) The EXIT functions of SCC-A1 and SCC-A2

(b) The EXIT function of SCC-A3

(c) The EXIT function of SCC-A4

(d) The EXIT function of SCC-A5

Figure 5.25: The EXIT functions of some of the optimized SCCs listed in Table 5.6.

should satisfy $R_I < \mathcal{A}_{II}$ so that convergence may be achieved. Therefore the total coding rate $R_{tot} \triangleq R_I R_{II}$ should satisfy $R_{tot} < R_{II} \mathcal{A}_{II}$. Since $\mathcal{A}_{II} \leq 1$, it is evident that the system's throughput is bounded by R_{II}. For example, the achievable information rates of the systems using intermediate codes of rates of $R_{II} = \{1, 3/4, 2/3\}$ are depicted in Figure 5.27. It can also be seen that the maximum achievable information rate is affected by the constraint length of the intermediate code. Additional information rate loss is incurred by using a relatively low constraint length code.

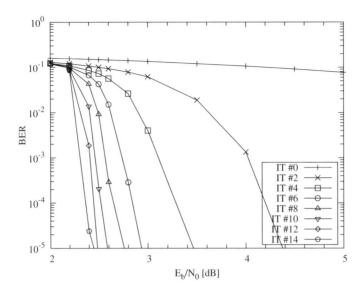

Figure 5.26: BER performance of the three-stage system of Figure 5.4 when employing the SCC of SCC-A2 listed in Table 5.6 for transmission over the dispersive AWGN channel having a CIR of Equation (5.8). The block length of the input data is $L = 10^5$ bits.

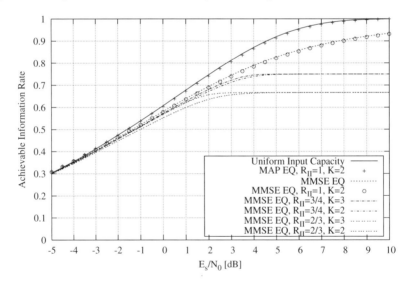

Figure 5.27: Achievable information rates for systems employing MAP/MMSE equalization and various intermediate codes for transmission over the three-path channel of Equation (5.8).

5.7 Joint Source-Channel Turbo Equalization Revisited

In Section 4.4.4 we introduced a three-stage joint source and channel decoding as well as turbo equalization scheme, as depicted in Figures 4.40 and 4.41. Moreover, it was shown

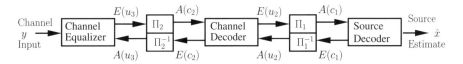

Figure 5.28: The receiver schematic of the joint source–channel turbo equalization scheme described in Section 4.4.4.

in Figure 4.42 that the iteratively detected amalgamated scheme significantly outperformed the separate source and channel decoding as well as the turbo equalization scheme. In our forthcoming discourse, we investigate the convergence behavior of the iterative-detection-aided amalgamated source–channel decoding and turbo equalization scheme with the aid of the EXIT-module-based convergence analysis technique described in Section 5.4.

For the sake of convenience, we redraw the iterative receiver structure in Figure 5.28, which is similar to the three-stage turbo-equalization-aided receiver of Figure 5.4. However, the channel equalizer considered here employs the APP equalization algorithm described in Section 4.4 instead of the MMSE equalization algorithm of Section 5.2. Furthermore, the intermediate decoder is constituted by a half-rate memory-4 RSC decoder and the outer decoder is constituted by the SISO VLC decoder characterized in Section 4.3. The rest of the system parameters are the same as in Table 4.2. As shown in Figure 5.28, the input of the iterative receiver is constituted by the ISI-contaminated channel observations, and after a number of joint channel equalization, channel decoding and source decoding iterations, which are performed according to a specific activation order, the source symbol estimates are generated.

Let us first consider the EXIT function of the combined module constituted by the channel equalizer and the channel decoder, which is shown in Figure 5.29 and indicated by a solid line with '+' symbols. The EXIT function of the outer VLC decoder is also depicted using a solid line with '×' symbols. It can be seen that at $E_b/N_0 = 1.5$ dB an open tunnel is formed between the two EXIT functions, which implies that the amalgamated source–channel decoding and turbo equalization scheme is capable of converging to the point C_2 of ($I_{A(u_2)}^{CH} = I_{E(c_1)}^{VLC} \approx 1$, $I_{E(u_2)}^{CH} = I_{A(c_1)}^{VLC} \approx 1$), and hence an infinitesimally SER may be achieved. By contrast, if a separate source–channel decoding and turbo equalization scheme is employed, which implies that the VLC decoder is invoked once after a certain number of channel equalization and channel decoding iterations, the receiver can only converge to the relatively low mutual information point of C_1, as indicated in Figure 5.29.

On the other hand, let us consider the EXIT function of the combined module constituted by the channel decoder and the VLC decoder, which is depicted in Figure 5.29 using a dense dashed line. The EXIT function of the outer channel equalizer is also shown, as a solid line. As expected, at $E_b/N_0 = 1.5$ dB an open tunnel is formed between the two EXIT functions, and the receiver becomes capable of converging to the point C_4 of ($I_{A(u_3)}^{EQ} = I_{E(c_2)}^{CH} \approx 1$, $I_{E(u_3)}^{EQ} = I_{A(c_2)}^{CH} \approx 0.55$), which implies that residual errors may persist at the output of the channel equalizer, but, nonetheless, a near-zero probability of error can be achieved at the output of the channel decoder and the VLC decoder. By contrast, a conventional two-stage turbo equalization scheme consisting only of channel equalization

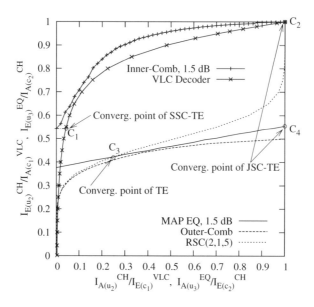

Figure 5.29: EXIT chart of the joint source–channel turbo equalization receiver of Figure 5.28, where SSC-TE denotes the separate source–channel turbo equalization scheme, JSC-TE represents the joint source–channel turbo equalization scheme and TE represents the two-stage turbo equalization scheme.

and channel decoding without source decoding can only converge to the relatively low mutual information point of C_3, as indicated in Figure 5.29.

In summary, when incorporating the outer VLC decoder into the iterative decoding process, the three-stage amalgamated source–channel turbo equalization scheme has a convergence threshold of about $E_b/N_0 = 1.5$ dB, which is significantly lower than that of the separate source–channel turbo equalization scheme.

5.8   Summary and Conclusions

In this chapter we have investigated several three-stage serially concatenated turbo schemes from an EXIT-chart-aided perspective. As a result a number of system design guidelines were obtained, which are summarized below.

- **The three-stage concatenation principle**

  The advocated system design principles are exemplified by an iterative three-stage MMSE equalization scheme, where the iterative receiver is constituted by three SISO modules, namely the inner MMSE equalizer, the intermediate channel decoder and the outer channel decoder. From the point of view of precoding, the inner equalizer and the intermediate decoder of Figure 5.4 may be regarded as a combined module and the EXIT function of this combined module is shown in Figure 5.7. Although the EXIT function of the stand-alone equalizer cannot reach the point of $(1, 1)$ (i.e., $T_{c_1}(1) \neq 1$ c. f. Equation (5.12)), it can be seen that the EXIT function of the combined module can (i.e., $T'_{u_2}(I_{A(u_2)} = 1, E_b/N_0) = 1$, c.f. Equation (5.17)). Therefore the three-stage system becomes capable of achieving the convergence point of $(1, 1)$, which corresponds to an infinitesimally low BER.

Alternatively, the intermediate decoder and the outer decoder can be viewed as a combined module, which results in a serially concatenated turbo-like code. The EXIT function of this combined module is depicted in Figure 5.9. It can be seen that the combined module has steep EXIT characteristics. For example, after $I = 10$ iterations between the intermediate decoder and the outer decoder, the EXIT function of the combined module reaches the upper bound value of 1 for any *a priori* information input larger than 0.6; i.e., we have $T'_{c_2}(I_{A(c_2)}) = 1, \forall I_{A(c_2)} > 0.6$. This characteristic matches that of the inner equalizer quite well and the corresponding three-stage system is capable of achieving $I_{E(c_2)} = 1$, which implies that we have $I_{E(c_1)} = I_{A(u_2)} = 1$ according to the EXIT characteristics of Decoder II, as seen in Figure 5.6a. Again, an infinitesimally low BER can be achieved.

- **EXIT-chart matching-based optimization and the use of irregular codes**
  In the context of three-stage concatenations, the outer code is optimized, so that its EXIT function is matched to that of the combined module of the inner equalizer and the intermediate decoder, yielding the lowest E_b/N_0 convergence threshold. Interestingly, we found in the context of Figure 5.8 that, as far as conventional convolutional codes are concerned, relatively weak codes having a short memory may result in the lowest convergence thresholds in the investigated three-stage context. This is due to the low initial value $(I_{E(u_2)} = T'_{u_2}(0, E_b/N_0))$ of the EXIT function of the combined module, which results in an undesirable EXIT function intersection with that of a strong code earlier than a weak code, as seen in Figure 5.18. Furthermore, if a well-designed IRCC is invoked as the outer code, the convergence threshold of the three-stage system may become even lower. For example, in our system of Section 5.5 employing an IRCC, a near-zero BER is achieved at $E_b/N_0 = 2$ dB, which is only about 0.3 dB away from the lowest achievable convergence threshold constituted by the capacity.

- **Rate optimization of SCCs**
  In addition to rate-1 codes, lower-rate intermediate codes may also be used in three-stage systems. Such systems have been shown in Figures 5.25 and 5.26 to exhibit lower convergence thresholds and better BER performances than those using rate-1 intermediate codes and regular outer codes. Even lower convergence thresholds can be achieved if irregular outer codes are used; however, the decoding complexity typically becomes significantly higher. The disadvantage of using intermediate codes having $R_{II} < 1$ is the reduction of the maximum achievable information rate as seen in Figure 5.27; however, several SCCs having different intermediate coding rates can be designed and activated in order to render the channel coding scheme adaptive to the time-variant channel conditions.

- **Optimization of activation orders**
  For multiple concatenation of more than two SISO modules, the activation order of the component decoders is an important issue, which may affect the computational complexity, delay and performance. With the aid of 1D and 2D EXIT functions, we are able to predict the number of three-stage activations required in order to

achieve converge. It was observed in Table 5.2 that the component module that has the strongest error correction capability, for example the outer code, should be invoked more frequently, in order that the total number of three-stage activations is minimized. Moreover, considering the computational complexities of the component modules, their activation order should be tuned to achieve the target trade-off between fast convergence and low complexity. The activation order thus obtained is rather heuristic and might not be the optimal one. However, the optimal activation order would vary as a function of the channel SNR, the interleaver block length, the affordable total complexity etc. Hence, we surmise that the proposed heuristic techniques are adequate.

Additionally, like those of two-stage systems, the decoding/detection processes of the three-stage systems may be visualized using EXIT charts as decoding trajectories. We have shown in Section 5.4.4 that either two 3D EXIT charts or two 2D EXIT charts can be used to characterize the decoding process in terms of the associated mutual information exchange. Since 2D charts are easier to manipulate and interpret, they are preferable.

Finally, the analysis and design procedure advocated here may be applied in the context of diverse iterative receivers employing multiple SISO modules, such as the jointly designed source coding and space–time coded modulation schemes of [271, 272]. The MMSE equalizer used in the three-stage system of Figure 5.4 discussed here may also be replaced by an M-ary demapper or a MIMO detector, and the resultant system can be optimized with the aid of the methods proposed here. The design of further near-capacity systems constitutes our future research.

Part II

Irregular Concatenated VLCs and Their Design

List of Symbols in Part II

Schematics

$\mathbf{f}_n$	Current video frame.
$\hat{\mathbf{f}}_n$	Quantized current video frame.
$\tilde{\mathbf{f}}_n$	Reconstructed current video frame.
$\hat{\mathbf{f}}_{n-1}$	Quantized previous video frame.
$\tilde{\mathbf{f}}_{n-1}$	Reconstructed previous video frame.
$\mathbf{e}$	Source sample frame.
$\hat{\mathbf{e}}$	Quantized source sample frame.
$\tilde{\mathbf{e}}$	Reconstructed source sample frame.
$\mathbf{s}$	Source symbol frame.
$\tilde{\mathbf{s}}$	Reconstructed source symbol frame.
$\mathbf{u}$	Transmission frame.
$\hat{\mathbf{u}}$	Received transmission frame.
$\tilde{\mathbf{u}}$	Reconstructed transmission frame.
π	Interleaving.
π^{-1}	De-interleaving.
$\mathbf{u}'$	Interleaved transmission frame.
$\mathbf{v}$	Encoded frame.
$\mathbf{v}'$	Interleaved encoded frame.
$L_a(\cdot)$	*A priori* Logarithmic Likelihood Ratios (LLRs) pertaining to the specified bits.
$L_p(\cdot)$	*A posteriori* LLRs/logarithmic *A Posteriori* Probabilities (log-APPs) pertaining to the specified bits/symbols.
$L_e(\cdot)$	Extrinsic LLRs pertaining to the specified bits/symbols.
$\mathbf{x}$	Channel's input symbols.
$\mathbf{y}$	Channel's output symbols.

Channel

η	Effective throughput.
E_c/N_0	Channel Signal-to-Noise Ratio (SNR).
E_b/N_0	Channel SNR per bit of source information.

Near-Capacity Variable-Length Coding   Lajos Hanzo, Robert G. Maunder, Jin Wang and Lie-Liang Yang
© 2011 John Wiley & Sons, Ltd

Video blocks (VBs)

J_x^{MB} Number of VB columns in each Macro-Block (MB).
J_y^{MB} Number of VB rows in each MB.
J^{MB} Number of VBs in each MB.

Sub-frames

M Number of sub-frames.
m Sub-frame index.
$\mathbf{e}^m$ Source sample sub-frame.
$\hat{\mathbf{e}}^m$ Quantized source sample sub-frame.
$\tilde{\mathbf{e}}^m$ Reconstructed source sample sub-frame.
$\mathbf{u}^m$ Transmission sub-frame.
$\tilde{\mathbf{u}}^m$ Reconstructed transmission sub-frame.
$\mathbf{s}^m$ Source symbol sub-frame.
$\tilde{\mathbf{s}}^m$ Reconstructed source symbol sub-frame.

Source sample sub-frames

J Number of source samples comprising each source sample sub-frame.
J^{sum} Number of source samples comprising each source sample frame.
j Source sample index.
e_j^m Source sample.
$\hat{e}_j^m$ Quantized source sample.
$\tilde{e}_j^m$ Reconstructed source sample.

Transmission sub-frames

I Number of bits comprising each transmission sub-frame.
I^{sum} Number of bits comprising each transmission frame.
$I_{\min}$ Minimum number of bits that may comprise each transmission sub-frame.
$I_{\max}$ Maximum number of bits that may comprise each transmission sub-frame.
i Transmission sub-frame bit index.
u_i^m Transmission sub-frame bit.
b Binary value.

Codebooks

K Number of entries in the codebook.
k Codebook entry index.

VQ codebook

$\mathbf{VQ}$ Vector Quantization (VQ) codebook.
$\mathbf{VQ}^k$ VQ tile.
J^k Number of VBs that comprise the VQ tile $\mathbf{VQ}^k$.
j^k VQ tile VB index.
$VQ_{j^k}^k$ VQ tile VB.

VL-coding codebook

VLC   Variable-length coding (VL-coding) codebook.
$\mathbf{VLC}^k$ Variable-length codeword.
I^k Number of bits that comprise the variable-length codeword $\mathbf{VLC}^k$.
I_b^k Number of bits in the variable-length codeword $\mathbf{VLC}^k$ assuming a value $b \in \{0, 1\}$.
i^k Variable-length codeword bit index.
$VLC_{i^k}^k$ Variable-length codeword bit.

VL-coding codebook parameters

E Entropy.
$L(\mathbf{VLC})$ Variable-length codebook average codeword length.
$R(\mathbf{VLC})$ Variable-length coding rate.
$E(\mathbf{VLC})$ Variable-length-encoded bit entropy.
$T(\mathbf{VLC})$ VL-coding trellis complexity.
$O^{\mathrm{APP}}(\mathbf{VLC})$ Average number of Add, Compare and Select (ACS) operations performed per source symbol during *A Posteriori* Probability (APP) Soft-Input Soft-Output (SISO) variable-length decoding.
$O^{\mathrm{MAP}}(\mathbf{VLC})$ Average number of Add, Compare and Select (ACS) operations performed per source symbol during Maximum *A posteriori* Probability (MAP) VL-coding sequence estimation.
$d_{\mathrm{free}}(\mathbf{VLC})$ Variable-length codebook free distance.
$d_{b_{\min}}(\mathbf{VLC})$ Variable-length codebook minimum block distance.
$d_{d_{\min}}(\mathbf{VLC})$ Variable-length codebook minimum divergence distance.
$d_{c_{\min}}(\mathbf{VLC})$ Variable-length codebook minimum convergence distance.
$\bar{d}_{\mathrm{free}}(\mathbf{VLC})$ Variable-length codebook free distance lower bound.
$D(\mathbf{VLC})$ Variable-length codebook Real-Valued Free Distance Metric (RV-FDM)

Irregular Variable-Length Codes (IrVLCs)

N Component variable-length codebook count.
n Component variable-length codebook index.
$\mathbf{u}^n$ Transmission sub-frame.
$\mathbf{s}^n$ Source symbol sub-frame.
C^n Component variable-length codebook source symbol frame fraction.
α^n Component variable-length codebook transmission frame fraction.

IrVLC codebooks

$\mathbf{VLC}^n$ Component variable-length codebook.
$\mathbf{VLC}^{n,k}$ Component variable-length codeword.
$I^{n,k}$ Number of bits that comprise the component variable-length codeword $\mathbf{VLC}^{n,k}$.
$i^{n,k}$ Component variable-length codeword bit index.
$VLC_{i^{n,k}}^k$ Component variable-length codeword bit.

Irregular Unity-Rate Codes (IrURCs)

R Component Unity-Rate Code (URC) count.
r Component URC index.
$\mathbf{u'}^r$ Interleaved transmission sub-frame.
$\mathbf{v}^r$ Encoded sub-frame.
$\mathbf{URC}^r$ Component URC.

Code parameters

$R_{(\cdot)}$ Coding rate.
$M_{(\cdot)}$ Number of modulation constellation points.
$L_{(\cdot)}$ Coding memory.

EXtrinsic Information Transfer (EXIT) chart

I_a A *priori* mutual information.
I_e Extrinsic mutual information.
y Importance of seeking a reduced computational complexity during EXIT-chart matching.

Trellises

$\ddot{i}$ Bit state index.
$\ddot{j}$ Symbol state index.
$\ddot{n}$ Node state index.
$S_{(\ddot{i},\ddot{j})}$ Symbol-based trellis state.
$S_{(\ddot{i},\ddot{n})}$ Bit-based trellis state.

Trellis transitions

T Trellis transition.
k^T Codebook entry index associated with the symbol-based trellis transition T.
b^T Bit value represented by the bit-based trellis transition T.
i^T Index of bit considered by the bit-based trellis transition T.
$\ddot{i}^T$ Bit state index of the trellis state from which the symbol-based trellis transition T emerges.
$\ddot{j}^T$ Symbol state index of the trellis state from which the symbol-based trellis transition T emerges.
$\ddot{n}^T$ Node state index of the trellis state from which the bit-based trellis transition T emerges.

Trellis sets

$\mathrm{en}(u_i^m)$ The set of all trellis transitions that encompasses the transmission sub-frame bit u_i^m.
$\mathrm{en}(u_i^m = b)$ The subset of $\mathrm{en}(u_i^m)$ that maps the binary value b to the transmission sub-frame bit u_i^m.
$\mathrm{en}(\hat{e}_j^m)$ The set of all trellis transitions that encompasses the VB $\hat{e}_j^m$.
$\mathrm{en}(\hat{e}_j^m = VQ_{j^k}^k)$ The subset of $\mathrm{en}(\hat{e}_j^m)$ that maps the VQ tile $VQ_{j^k}^k$ to the VB $\hat{e}_j^m$.

$\mathrm{fr}(S)$	The set of all transitions that emerge from the trellis state S.
$\mathrm{to}(S)$	The set of all transitions that merge to the trellis state S.
$\mathrm{fr}(T)$	The state from which the transition T emerges.
$\mathrm{to}(T)$	The state to which the transition T merges.
$\mathrm{nr}(\hat{e}_j^m)$	The set of all VBs that immediately surround the VB $\hat{e}_j^m$.

Viterbi algorithm

$d(T)$	The distortion of the trellis transition T.
$D(T)$	The minimum cumulative distortion of all trellis paths between the trellis state $S_{(0,0)}$ and the trellis transition T.
$D(S)$	The minimum cumulative distortion of all trellis paths to the state S.
$m(T)$	The Viterbi algorithm metric of the trellis transition T.
$M(T)$	The maximum cumulative Viterbi algorithm metric of all trellis paths between the trellis state $S_{(0,0)}$ and the trellis transition T.
$M(S)$	The maximum cumulative Viterbi algorithm metric of all trellis paths to the state S.

Bahl–Cocke–Jelinek–Raviv (BCJR) algorithm

$P_a(u_i^m = b)$	A *priori* probability of the transmission sub-frame bit u_i^m taking the value b.
$P(k)$	Probability of occurrence of the codebook entry with index k.
$P(S)$	Probability of occurrence of the trellis state S.
$P(T|\mathrm{fr}(T))$	Conditional probability of the occurrence of the trellis transition T given the occurrence of the trellis state from which it emerges.
$P_p(T)$	A *posteriori* trellis transition probability.
C_1	A *posteriori* trellis transition probability normalization factor.
$\gamma(T)$	A *priori* trellis transition probability.
$\gamma'(T)$	Weighted a *priori* trellis transition probability.
$C_2(S)$	A *priori* trellis transition probability normalization factor used for all trellis transitions that emerge from the trellis state S.
$\alpha(S)$	Alpha value obtained for the trellis state S.
$\beta(S)$	Beta value obtained for the trellis state S.
C_{L_a}	BCJR algorithm LLR pruning threshold.
C_γ	BCJR algorithm a *priori* probability pruning threshold.
C_α	BCJR algorithm forward recursion pruning threshold.
C_β	BCJR algorithm backwards recursion pruning threshold.

Genetic Algorithm (GA) for VL-coding codebook design

$\mathbf{L}$	List of candidate VL-coding codebooks.
L^{tar}	Target GA list length.
$M(\mathbf{VLC})$	GA VL-coding quality metric.
D^{lim}	GA VL-coding RV-FDM limit.
R^{lim}	GA VL-coding coding rate limit.
α^D	GA VL-coding RV-FDM importance.
α^R	GA VL-coding coding rate importance.
α^E	GA VL-coding bit entropy importance.
α^T	GA VL-coding trellis complexity importance.

β^D	GA VL-coding RV-FDM increase/decrease constant.
β^R	GA VL-coding coding rate increase/decrease constant.
D^{best}	Most desirable RV-FDM of VL-coding codebooks admitted to the GA list.
R^{best}	Most desirable coding rate of VL-coding codebooks admitted to the GA list.
E^{best}	Most desirable bit entropy of VL-coding codebooks admitted to the GA list.
T^{best}	Most desirable trellis complexity of VL-coding codebooks admitted to the GA list.
P^{max}	Maximum number of GA mutations.

Chapter 6

Irregular Variable-Length Codes for Joint Source and Channel Coding[1]

6.1 Introduction

As described in Chapter 2, Shannon's source and channel coding separation theorem [133] states that, under certain idealized conditions, source and channel coding can be performed in isolation without any loss in performance. This motivated the design of the Vector Quantization (VQ) [170] based video transmission scheme of [273], in addition to the MPEG-4 [158] based scheme of [274]. Following video encoding in the transmitters of these schemes, the resultant bit sequences were protected using a serial concatenation [275] of Reversible Variable-Length Codes (RVLCs) [276] and Trellis-Coded Modulation (TCM) [277]. In the corresponding receivers, iterative *A Posteriori* Probability (APP) Soft-Input Soft-Output (SISO) TCM and RVLC decoding [278] was employed, and the resultant reconstructed video-encoded bit sequence was decoded. As described in Chapter 3, RVLC encoders typically fulfill a joint source and channel coding role. However, in the schemes of [273] and [274] the RVLC encoder was unable to achieve any source compression, but played a beneficial channel coding role, similar to that of the TCM encoder. This was because only a negligible amount of residual redundancy was apparent in the video-encoded bit sequences, owing to the high level of video compression achieved. The schemes of [273] and [274] may therefore be deemed to employ a separate source and channel coding philosophy.

However, Shannon's findings are only valid under a number of idealized assumptions, namely that the information is transmitted over an uncorrelated non-dispersive narrowband Additive White Gaussian Noise (AWGN) channel, while potentially encountering an infinite decoding complexity and buffering latency, as discussed in Chapter 2. These assumptions clearly have limited validity for practical transmissions over realistic wireless channels [151]. In addition to this, Shannon assumed that the source is stationary and is losslessly encoded

---

[1]Part of this chapter is based on [194] © IEEE (2007).

Near-Capacity Variable-Length Coding   Lajos Hanzo, Robert G. Maunder, Jin Wang and Lie-Liang Yang
© 2011 John Wiley & Sons, Ltd

with the aid of symbols having equal significance and error sensitivity. These assumptions are invalid in the case of video transmission, since video information is typically non-stationary, having characteristics that vary in both time and space [151]. Furthermore, typically lossy video coding [151] is employed in order to achieve a high level of compression and a low bandwidth requirement, while exploiting the psycho-visual properties of the human eye. Finally, the video-encoded bits typically have a varying significance and error sensitivity, because these bits are generated using a number of diverse video encoding techniques, such as the Discrete Cosine Transform (DCT) [153], VQ [170], Motion Compensation (MC) [154] and Huffman coding [155].

As discussed in Chapter 2, the application of joint source and channel coding techniques to wireless video transmissions is therefore motivated. In this chapter we consider the iterative joint source and channel decoding of video information, as described in Chapter 4. However, let us commence by considering the current state of the art in this field. The trellis-based Maximum *A posteriori* Probability (MAP) sequence estimation of the Context-Adaptive Variable-Length Codes (CAVLC) of the H.264 video standard [161] was proposed in [279], facilitating the decoding of soft information in a manner similar to that advocated in [180]. Similar techniques were proposed for Variable-Length Coded MPEG-4 streams in [280–282]. This approach was extended in [283], where the MPEG-4 stream was encoded using RVLCs, providing a joint source and channel coding capability, as discussed in Chapter 3. Further joint source and channel coding techniques were proposed for H.264 and MPEG-4 video streams in [284, 285] by employing MAP sequence estimation of jointly trellis-decoded Variable-Length Codes (VLCs) and Convolutional Codes (CCs) [286]. Additionally, the serial concatenation [275] and iterative decoding [278] of video information was demonstrated in [287], where an APP SISO decoder was employed for the Context-Adaptive Binary Arithmetic Codes (CABACs) of H.264 [161].

However, as in the separate source- and channel-coding-aided VQ- and MPEG-4-based video transmission schemes of [273] and [274], a hard decision is made before video decoding commences in each of the above-mentioned joint source and channel coding approaches. If any errors are present in the resultant hard information, then often significant distortion may be observed in the reconstructed video signal owing to the complex interdependencies that are typical in video streams [151]. In the worst-case scenario, a single error may render an entire video frame invalid and inflict significant distortion, which may propagate to all subsequent differentially encoded video frames [151]. Hence, in [288] a VLC MAP sequence estimator was proposed that additionally considers any underlying code constraints that restrict the legitimate sequences of VLC codewords, ensuring that entire video frames can never be rendered invalid by decoding errors. In the alternative approach of [289], the soft reconstruction of Channel-Optimized Vector Quantization (COVQ) [168, 169] was employed, eliminating the requirement for hard decisions. In this chapter we combine the approaches of [288] and [289] in the design of a novel iterative joint source- and channel-decoding video transmission scheme.

More specifically, a novel video codec is serially concatenated [275] and iteratively decoded [278] with a TCM codec [277], in a manner similar to that of [185]. Our novel video codec employs Variable-Dimension Vector Quantization (VDVQ) [290] in order to represent the video frames at a high reconstruction quality, while maintaining a high level of compression. As in the context of the conventional VQ-based video codecs, VDVQ represents video frames with the aid of a tessellation of VQ tiles, employing a K-entry VQ codebook. However, in VDVQ these K number of VQ tiles have a variety of dimensions, facilitating a greater degree of VQ codebook design freedom and, hence, a lower degree of lossy-compression-induced distortion in the reconstructed video frames. Furthermore, our novel video codec employs RVLC [276] in order to achieve joint source and channel coding, as described in Chapter 3. Note that there is a duality between VDVQ and RVLC, since the former employs VQ tiles with various dimensions, while the latter employs binary codewords having various lengths.

In the video codec of our amalgamated scheme, termed the VDVQ/RVLC-TCM scheme, VDVQ and RVLC are employed in a manner that maintains a simple transmission frame structure. Here we refer to the rules governing the formation of the legitimate bit-sequences of a transmission frame as the *VDVQ/RVLC-induced code constraints*. These VDVQ/RVLC-induced code constraints specify which VQ tiles and which RVLC codewords may be employed at which stages of the video coding process. Note that since the set of RVLC codewords that may be employed at a particular stage of the video coding process depends on the particular stage considered, our novel video codec may be deemed to employ Irregular Variable-Length Codes (IrVLCs), following the principles of Irregular Convolutional Codes (IrCCs), as described in Section 5.5.3. This chapter therefore exemplifies the application of IrVLCs to joint source and channel coding.

The VDVQ/RVLC-induced code constraints were specially selected so that they may be completely described by a novel VDVQ/RVLC trellis structure, which is reminiscent of the RVLC symbol-level trellis structure [181] described in Chapter 3, and may be employed to jointly perform VDVQ and RVLC encoding or APP SISO decoding. As we outline in Section 6.3.4, the employment of the proposed VDVQ/RVLC trellis structure represents the consideration of all legitimate transmission frame permutations. This allows us to perform optimal Minimum Mean Square Error (MMSE) VDVQ/RVLC encoding of the video sequence in a novel manner that is reminiscent of [291]. Additionally, the employment of the proposed VDVQ/RVLC trellis structure during VDVQ/RVLC decoding guarantees the recovery of legitimate – although not necessarily error-free – video information, in a manner similar to that outlined in [288]. Hence, unlike in the VQ- and MPEG-4-based video transmission schemes of [273] and [274] respectively, the proposed video decoder is never forced to discard information. During APP SISO VDVQ/RVLC decoding, a novel modification of the Bahl–Cocke–Jelinek–Raviv (BCJR) [48] algorithm provided in [174] is performed on the basis of the proposed VDVQ/RVLC trellis structure. This allows the unconventional soft APP-based [292] MMSE reconstruction of the transmitted video frames, in a manner similar to that of [289]. Additionally, this allows the exploitation of all redundancy associated with the VDVQ/RVLC-induced code constraints during joint iterative VDVQ/RVLC and TCM decoding. Hence we may expect a better error resilience than with the schemes of [273] and [274], in which none of the video redundancy is exploited during iterative decoding.

This chapter is organized as follows. In Section 6.2, the proposed VDVQ/RVLC-TCM scheme is introduced. Section 6.3 describes the VDVQ/RVLC-TCM scheme's transmission frame structure and the VDVQ/RVLC-induced code constraints. Additionally, the complete description of the VDVQ/RVLC-induced code constraints by the proposed VDVQ/RVLC

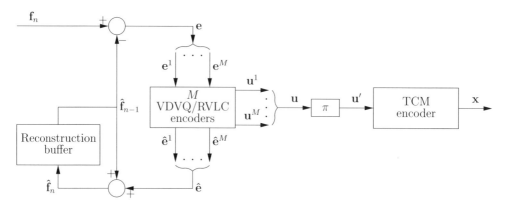

Figure 6.1: The proposed VDVQ/RVLC-TCM scheme's transmitter. ©IEEE [194] Maunder *et al.* 2007.

trellis structure is discussed in Section 6.3. The employment of the VDVQ/RVLC trellis structure in VDVQ/RVLC encoding and APP SISO decoding is described in Sections 6.4 and 6.5 respectively. In Section 6.6, both low- and high-latency implementations of the proposed VDVQ/RVLC-TCM scheme are introduced and their video reconstruction quality is assessed. Additionally, the performance of the proposed joint source and channel coding approach is compared with that of the VQ- and MPEG-4-based schemes of [273] and [274] respectively, which employ a separated source and channel coding philosophy. Our conclusions are offered in Section 6.7.

6.2   Overview of Proposed Scheme

In this section we introduce the proposed VDVQ/RVLC-TCM video transmission scheme, which employs a joint source and channel coding philosophy in order to convey grayscale video sequences over a narrowband uncorrelated Rayleigh fading channel. The video codec of this scheme achieves compression by employing low-complexity frame differencing [151], VDVQ [290] and RVLC [276]. These are chosen in favor of the more advanced but higher-complexity compression techniques, such as MC [154] or the DCT [153]. This ensures that a simple transmission frame structure is maintained. Rather than performing an excessive-complexity single VDVQ/RVLC encoding or APP SISO decoding operation for the entire video frame as usual, a decomposition into a number of smaller operations is performed. This approach is associated with a significant computational complexity reduction. The proposed video codec is serially concatenated with the TCM codec of [293], and joint iterative decoding is employed in the receiver. These issues are further detailed in the following subsections with reference to the proposed VDVQ/RVLC-TCM scheme's transmitter and receiver schematics, seen in Figures 6.1 and 6.2 respectively.

6.2.1   Compression

Video sequences are typically correlated in both space and time [151], with both corresponding pixels in consecutive video frames and adjacent pixels within individual video frames having similar luminance values. This time-domain and spatial-domain correlation

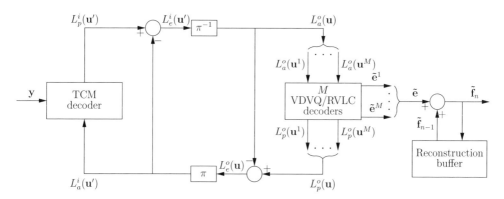

Figure 6.2: The proposed VDVQ/RVLC-TCM scheme's receiver. ©IEEE [194] Maunder *et al.* 2007.

induces redundancy within the video sequence. The redundant part of a video sequence is defined as that which may theoretically be precisely predicted by considering only the remaining information-bearing part of the video sequence. Hence, the redundancy inherent in the video sequence may be removed without jeopardizing our ability to precisely reconstruct it. Indeed, this is the basis on which lossless video compression operates. Note that the characteristic correlation between the corresponding pixels in consecutive video frames induces so-called *inter-frame redundancy* within the video sequence. Similarly, so-called *intra-frame redundancy* is induced by the characteristic correlation between adjacent pixels within individual video frames. A significant amount of further so-called lossy compression can be achieved by additionally discarding some of the information-bearing part of the video sequence, although this jeopardizes our ability to precisely reconstruct it. Therefore successful lossy coding techniques should exploit the psycho-visual inadequacies of human vision in order to discard video sequence information in a manner that is imperceivable to the human eye [151].

In the proposed VDVQ/RVLC-TCM scheme, the characteristic inter-frame temporal redundancy of video sequences is exploited by the employment of low-complexity frame differencing [151]. This differential coding technique employs the previous reconstructed video frame $\hat{\mathbf{f}}_{n-1}$ for predicting the current video frame $\mathbf{f}_n$, as outlined below. In the VDVQ/RVLC-TCM scheme's transmitter of Figure 6.1, the Frame Difference (FD) $\mathbf{e} = \mathbf{f}_n - \hat{\mathbf{f}}_{n-1}$ is calculated as the resultant prediction error. This can be conveyed to the receiver, where the current reconstructed video frame estimate $\tilde{\mathbf{f}}_n = \tilde{\mathbf{e}} + \tilde{\mathbf{f}}_{n-1}$ may be obtained by adding the reconstructed FD estimate $\tilde{\mathbf{e}}$ to the previous reconstructed video frame estimate $\tilde{\mathbf{f}}_{n-1}$, as shown in Figure 6.2. Note that there is no video frame previous to the first frame $\mathbf{f}_1$. Hence, a synthetic video frame having a consistent mid-gray luminance is employed as the reference video frame $\hat{\mathbf{f}}_0$ during the calculation of the FD $\mathbf{e}$ for the first video frame $\mathbf{f}_1$.

Since the characteristic inter-frame temporal redundancy is manifested as correlation between consecutive video frames, the FD $\mathbf{e}$ has a lower variance than the current video frame $\mathbf{f}_n$ [151]. This is illustrated in Figure 6.3, which provides plots of the distribution of the luminance values within the frames of the grayscale 'Lab' video sequence, as well as the

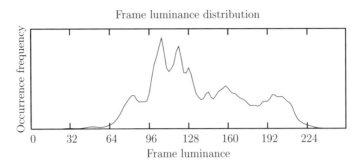

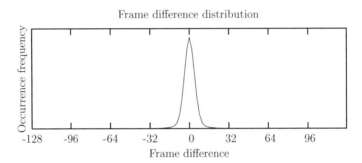

Figure 6.3: The distribution of the luminance values within the frames of the grayscale 'Lab' video sequence and the distribution of the differences between the corresponding luminance values of consecutive frames of the 'Lab' video sequence.

distribution of the differences between the corresponding luminance values of consecutive frames [151]. Indeed, the standard deviation of the luminance values is 39.4, whilst that of the frame differences is just 5.9. It therefore requires less information to convey the FD e instead of the current video frame $\mathbf{f}_n$, and compression is achieved as a result.

However, the employment of frame differencing is associated with a cost. More specifically, any unexpected distortion that was induced within the receiver's previous video frame estimate $\tilde{\mathbf{f}}_{n-1}$ by the effects of transmission errors will propagate into the receiver's current video frame estimate $\tilde{\mathbf{f}}_n$. This is because the receiver's current reconstructed video frame estimate $\tilde{\mathbf{f}}_n = \tilde{\mathbf{e}} + \tilde{\mathbf{f}}_{n-1}$ is obtained by adding the reconstructed FD estimate $\tilde{\mathbf{e}}$ to the *distorted* previous reconstructed video frame estimate $\tilde{\mathbf{f}}_{n-1}$, as described above. Unless measures are taken to counteract the propagation of unexpected distortion, it will continue indefinitely into all subsequent differentially encoded video frames. However, as will be shown in Section 6.6, the objective of the proposed VDVQ/RVLC-TCM scheme is to mitigate all transmission errors at channel Signal-to-Noise Ratios (SNRs) within our region of interest. As a result, all unexpected distortion will be mitigated before it has the opportunity to propagate into subsequent differentially encoded video frames. Therefore, no measures are employed to counteract this propagation in the proposed VDVQ/RVLC-TCM scheme.

The characteristic intra-frame spatial redundancy of video sequences is exploited by the employment of VDVQ [290]. As discussed in Section 6.3, this allows the representation of the FD e in a compact bit-sequence form, namely the transmission frame u. Hence, the VDVQ offers compression. This compression is lossy because the VDVQ approximates the FD e by the reconstructed FD $\hat{e}$, as shown in Figure 6.1. Hence, the VDVQ's high level of compression is achieved at the cost of inducing distortion within the reconstructed FD $\hat{e}$. Note that, unlike distortion that is induced by the effects of transmission errors, the distortion that is induced by lossy coding *is* expected by the receiver and hence does not cause propagation between consecutive video frames in the manner described above. This is due to the employment of the distorted previous reconstructed video frame $\hat{f}_{n-1}$ as the reference for frame differencing in the transmitter, as described above. In the absence of transmission errors, the transmitter's frame differencing reference will be identical to that of the receiver, namely the previous reconstructed video frame estimate $\tilde{f}_{n-1}$, as shown in Figure 6.2. Hence, synchronization between the transmitter and the receiver will be maintained. Note in Figure 6.1 that local decoding is employed in the transmitter in order to obtain the current reconstructed video frame $\hat{f}_n = \hat{e} + \hat{f}_{n-1}$, which is stored for the sake of providing the prediction of the next video frame. This is identical to the generation of the current reconstructed video frame estimate $\tilde{f}_n = \tilde{e} + \tilde{f}_{n-1}$ at the receiver.

6.2.2 VDVQ/RVLC Decomposition

In the proposed VDVQ/RVLC-TCM scheme's transmitter and receiver, VDVQ/RVLC encoding and APP SISO decoding are performed, respectively. As detailed in Sections 6.3.1 and 6.3.4, we decompose each FD e into M smaller sub-frames. This decomposition implies the corresponding decomposition of both the reconstructed FD $\hat{e}$ and the transmission frame u, as indicated by the braces { and } in Figure 6.1. In the transmitter, M number of separate VDVQ/RVLC operations are employed. Each VDVQ/RVLC encoder approximates the FD sub-frame e^m by the reconstructed FD sub-frame $\hat{e}^m$, where $m \in [1 \ldots M]$ denotes the sub-frame index. Additionally, the transmission sub-frame u^m is generated, as detailed in Section 6.4. The bit-based transmission frame u is constituted by the concatenation of the set of these M transmission sub-frames.

6.2.3 Serial Concatenation and Iterative Decoding

In the proposed VDVQ/RVLC-TCM scheme, the video codec is serially concatenated with the TCM codec of [293]. In the transmitter, the current video frame f_n is conveyed by the reconstructed FD $\hat{e}$, where the latter is generated using VDVQ/RVLC encoding and is represented by the transmission frame u, as discussed in Section 6.2.1. As shown in Figure 6.1, the interleaver or scrambling function [275] π is applied to the transmission frame u in order to obtain the interleaved transmission frame u' and TCM encoding is employed to generate the channel's input symbols x. These are transmitted over the channel and are received as the channel's output symbols y in the proposed VDVQ/RVLC-TCM scheme's receiver, as shown in Figure 6.2. Finally, TCM and VDVQ/RVLC decoding are employed to recover the reconstructed FD estimate $\tilde{e}$. This allows the recovery of the current video frame estimate $\tilde{f}_n = \tilde{e} + \tilde{f}_{n-1}$, as shown in Figure 6.2.

In the receiver, APP SISO TCM and VDVQ/RVLC decoding are performed iteratively, as shown in Figure 6.2. Soft information, represented in the form of Logarithmic Likelihood-Ratios (LLRs) [292], is exchanged between the TCM and VDVQ/RVLC iterative decoding stages for the sake of assisting each other's operation, as described in Chapter 4. With each successive decoding iteration, the reliability of this soft information increases, until iterative decoding convergence is achieved [294]. In Figure 6.2 and throughout our following discussions, $L(\cdot)$ is an operator that denotes the LLRs that pertain to the bit-value probabilities of the specified bits. These LLRs may carry a superscript to identify a specific iterative decoding stage, with i signifying inner APP SISO TCM decoding and o representing outer APP SISO VDVQ/RVLC decoding. Additionally, a subscript denotes the role of the LLRs, with a, p and e indicating *a priori*, *a posteriori* and extrinsic information respectively.

During each decoding iteration, *a priori* LLRs $L_a^o(\mathbf{u})$ pertaining to the transmission frame $\mathbf{u}$ are provided for the outer APP SISO VDVQ/RVLC decoding stage, as shown in Figure 6.2. Similarly, *a priori* LLRs $L_a^i(\mathbf{u}')$ pertaining to the interleaved transmission frame $\mathbf{u}'$ are provided for the inner APP SISO TCM decoding stage. Naturally, in the case of APP SISO TCM decoding the channel's output symbols $\mathbf{y}$ are also exploited. Note that the *a priori* LLRs $L_a(\mathbf{u})$ and $L_a(\mathbf{u}')$ are obtained from the most recent operation of the other decoding stage, as will be highlighted below. In the case of the first decoding iteration no previous APP SISO VDVQ/RVLC decoding has been performed. In this case, the *a priori* LLRs $L_a^i(\mathbf{u}')$ provided for APP SISO TCM decoding are all zero, corresponding to a probability of 0.5 for both '0' and '1.' Each iterative decoding stage applies the BCJR [48] algorithm, as described in [293] and Section 6.5 for APP SISO TCM and for VDVQ/RVLC decoding respectively. The result is the generation of the *a posteriori* LLRs $L_p(\mathbf{u})$ and $L_p(\mathbf{u}')$, as shown in Figure 6.2.

During iterative decoding, it is necessary to prevent the reuse of already-exploited information, since this would limit the attainable iteration gain [278]. This is achieved following each decoding stage by the subtraction of $L_a^o(\mathbf{u})$ from $L_p^o(\mathbf{u})$ and $L_a^i(\mathbf{u}')$ from $L_p^i(\mathbf{u}')$, as shown in Figure 6.2. Following APP SISO VDVQ/RVLC decoding, we arrive at the extrinsic LLRs $L_e^o(\mathbf{u})$. In the case of APP SISO TCM decoding, the LLRs $L_e^i(\mathbf{u}')$ additionally contain information extracted from the channel's output symbols $\mathbf{y}$. It is these sets of LLRs that provide the *a priori* LLRs for the next iteration of the other decoding stage. De-interleaving, indicated by the block π^{-1} in Figure 6.2, is applied to $L_e^i(\mathbf{u}')$ in order to generate $L_a^o(\mathbf{u})$. Similarly, interleaving is applied to $L_e^o(\mathbf{u})$ for the sake of providing $L_a^i(\mathbf{u}')$, as shown in Figure 6.2. These interleaving and de-interleaving operations are necessary for the sake of mitigating the correlation of consecutive LLRs, before forwarding them to the next iterative decoding stage, since the BCJR algorithm assumes that consecutive *a priori* LLRs are uncorrelated [48]. As always, the interleaver's ability to maintain this desirable statistical independence is related to its length, as described in Chapter 4.

As stated in Section 6.2.2, M separate APP SISO VDVQ/RVLC decoding steps are employed in the proposed VDVQ/RVLC-TCM scheme's receiver. As in the decomposition of the bit-based transmission frame $\mathbf{u}$, the *a priori* LLRs $L_a^o(\mathbf{u})$ are decomposed into M sub-frames, as shown in Figure 6.2. This decomposition is accompanied by the corresponding decomposition of the *a posteriori* LLRs $L_p^o(\mathbf{u})$ and the reconstructed FD estimate $\tilde{\mathbf{e}}$. Each APP SISO VDVQ/RVLC decoder is provided with the *a priori* LLR sub-frame $L_a^o(\mathbf{u}^m)$ and generates the *a posteriori* LLR sub-frame $L_p^o(\mathbf{u}^m)$. Additionally, each APP SISO VDVQ/RVLC decoder recovers the reconstructed FD sub-frame estimate $\tilde{\mathbf{e}}^m$, as detailed in Section 6.5.

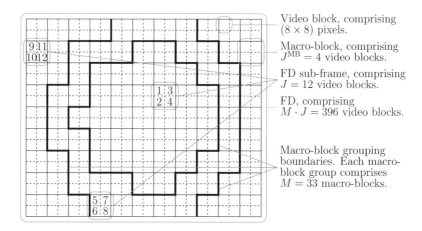

Video block, comprising (8×8) pixels.

Macro-block, comprising $J^{\mathrm{MB}} = 4$ video blocks.

FD sub-frame, comprising $J = 12$ video blocks.

FD, comprising $M \cdot J = 396$ video blocks.

Macro-block grouping boundaries. Each macro-block group comprises $M = 33$ macro-blocks.

Figure 6.4: Example of selecting $J = 12$ (8×8)-pixel VBs from a (176×144)-pixel FD to provide one of the $M = 33$ FD sub-frames. ©IEEE [194] Maunder *et al.* 2007.

6.3 Transmission Frame Structure

As stated in Section 6.2, each video frame $\mathbf{f}_n$ is conveyed between the proposed VDVQ/-RVLC-TCM scheme's transmitter and receiver by means of a single bit-based transmission frame $\mathbf{u}$. This comprises the concatenation of M transmission sub-frames. Again, the formation of legitimate bit-sequences within these transmission sub-frames is governed by simple VDVQ/RVLC-induced code constraints. These are imposed by the decomposition of FDs into sub-frames and by the specific nature of the VDVQ/RVLC codebook employed in the proposed video codec. Owing to their simplicity, the complete set of VDVQ/RVLC-induced code constraints may be described by a novel VDVQ/RVLC trellis structure. These issues are discussed in the following subsections and are described with the aid of an example that continues through Figures 6.4, 6.5, 6.6 and 6.7. Throughout the following subsections, the parameter values of the continuing example are provided during our discussions. It should be noted that generalization is straightforward.

6.3.1 Frame Difference Decomposition

As stated in Section 6.2.2, the FD $\mathbf{e}$, as well as the reconstructed FD $\hat{\mathbf{e}}$ and the reconstructed FD estimate $\tilde{\mathbf{e}}$, are decomposed into M sub-frames. This decomposition is detailed with the aid of the example provided in Figure 6.4.

For the sake of implementational simplicity, the FD decomposition is designed to yield FD sub-frames with statistical properties that are similar to each other. This allows the allocation of an equal number of bits, namely I, to each transmission sub-frame $\mathbf{u}^m = \{u_i^m\}_{i=1}^I$, where $u_i^m \in \{0, 1\}$ and $i \in [1 \ldots I]$ is the bit index. Additionally, identical codebook designs may be employed for the M VDVQ/RVLC encoding steps that constitute the proposed VDVQ/RVLC-TCM scheme's transmitter as well as for the M APP SISO VDVQ/RVLC decoding operations of the receiver.

The proposed video codec operates on a block-based philosophy. In this contribution, the *Video Block* (VB) has the dimensions of (8×8) pixels and is defined as the smallest unit of video information that is considered in isolation. Each of the M FD sub-frames $\mathbf{e}^m$ comprises a unique combination of J (8×8)-pixel VBs of the FD $\mathbf{e}$. Hence, the FD $\mathbf{e}$ comprises $M \cdot J$ VBs in total. In the example of Figure 6.4, each of the $M = 33$ FD sub-frames comprises $J = 12$ of the $M \cdot J = 396$ (8×8)-pixel VBs that are shown with dashed boundaries.

We proceed by decomposing the FD $\mathbf{e}$ into a set of perfectly tiling *Macro-Blocks* (MBs). Each MB has the dimensions of $(J_x^{\mathrm{MB}} \times J_y^{\mathrm{MB}}) = J^{\mathrm{MB}}$ number of VBs, where $J = 12$ is an integer multiple of J^{MB}. In the example of Figure 6.4, MBs are shown with solid boundaries and have the dimensions of $(J_x^{\mathrm{MB}} \times J_y^{\mathrm{MB}}) = (2 \times 2)$ VBs, giving $J^{\mathrm{MB}} = 4$. The MBs are then assigned to $J/J^{\mathrm{MB}} = 3$ different groups on the basis of the distance between their center and the FD center, with each group comprising $M = 33$ MBs. This results in $J/J^{\mathrm{MB}} = 3$ quasi-concentric MB groups, as indicated by the thick boundaries in Figure 6.4. A pseudorandom selection of one MB from each of the $J/J^{\mathrm{MB}} = 3$ groups is performed for each FD sub-frame $\mathbf{e}^m$. It is the $J = 12$ (8×8)-pixel VBs identified by this pseudorandom MB selection that constitute the FD sub-frame $\mathbf{e}^m$. We note that a predetermined fixed seed is employed for the sake of allowing identical pseudorandom MB selections to be made independently by both the video encoder and the decoder.

Each of the $J = 12$ (8×8)-pixel VBs e_j^m, which constitute the FD sub-frame $\mathbf{e}^m = \{e_j^m\}_{j=1}^J$, is allocated a VB index $j \in [1 \ldots J]$. These $J = 12$ VB indices are allocated amongst the $J/J^{\mathrm{MB}} = 3$ pseudorandomly selected MBs in a quasi-radial ordering. In this way, the $J^{\mathrm{MB}} = 4$ lowest-valued VB indices are allocated to the MB nearest to the FD center, as seen in Figure 6.4. Similarly, the MB nearest to the FD perimeter is assigned the $J^{\mathrm{MB}} = 4$ highest-valued VB indices. The allocation of VB indices within MBs should be made with specific consideration of the VDVQ codebook employed, as will be detailed in Section 6.3.3. Again, Figure 6.4 exemplifies the indices allocated to the $J = 12$ VBs of a FD sub-frame.

As mentioned, the FD sub-frames have statistical properties that are similar to each other; however, it should be noted that each of the $J/J^{\mathrm{MB}} = 3$ constituent MBs is likely to exhibit different statistical properties. For example, the MB that is allocated the lowest-valued VB indices in each FD sub-frame can be expected to exhibit a high level of video activity, since this MB is located near the center of the FD. Similarly, a low level of video activity can be expected for the MB that is assigned the highest-valued VB indices.

The reconstructed FD sub-frame $\hat{\mathbf{e}}^m = \{\hat{e}_j^m\}_{j=1}^J$ employs the same VB selection and indexing as the corresponding FD sub-frame $\mathbf{e}^m = \{e_j^m\}_{j=1}^J$. The same applies to the reconstructed FD sub-frame estimate $\tilde{\mathbf{e}}^m = \{\tilde{e}_j^m\}_{j=1}^J$.

In contrast to the FD decomposition described above, an alternative approach could comprise a selection of $J = 12$ *adjacent* VBs for each FD sub-frame. Although this alternative would permit the exploitation of correlation between adjacent VBs, each FD sub-frame would have different statistical properties. As a result, the corresponding transmission sub-frames would require different individual bit-allocations on a demand basis. Additionally, each transmission sub-frame's different bit-allocation would require a separate VDVQ/RVLC encoder and APP SISO decoder design. Finally, the transmission of side information would be required for the sake of signaling this allocation between the transmitter and the receiver. Hence, this alternative approach was discarded for the sake of maintaining simplicity.

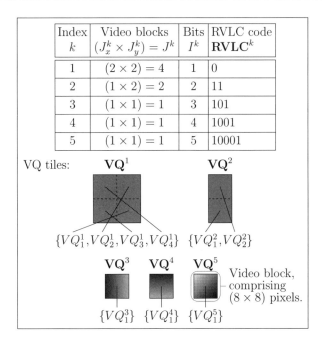

Index k	Video blocks $(J_x^k \times J_y^k) = J^k$	Bits I^k	RVLC code $\mathbf{RVLC}^k$
1	$(2 \times 2) = 4$	1	0
2	$(1 \times 2) = 2$	2	11
3	$(1 \times 1) = 1$	3	101
4	$(1 \times 1) = 1$	4	1001
5	$(1 \times 1) = 1$	5	10001

VQ tiles: **VQ^1**   **VQ^2**

$\{VQ_1^1, VQ_2^1, VQ_3^1, VQ_4^1\}$ $\{VQ_1^2, VQ_2^2\}$

VQ^3   **VQ^4**   **VQ^5**

Video block, comprising (8×8) pixels.

$\{VQ_1^3\}$ $\{VQ_1^4\}$ $\{VQ_1^5\}$

Figure 6.5: Example of a $K = 5$-entry VDVQ/RVLC codebook ©IEEE [194] Maunder *et al.* 2007.

6.3.2   VDVQ/RVLC Codebook

As stated in Section 6.2, each FD sub-frame $\mathbf{e}^m$ is approximated by the reconstructed FD sub-frame $\hat{\mathbf{e}}^m$ and represented by the bit-based transmission sub-frame $\mathbf{u}^m$ on the basis of joint VDVQ [290] and RVLC [276]. More specifically, the reconstructed FD sub-frame $\hat{\mathbf{e}}^m$ is formed as a tessellation of VQ tiles from the K-entry VDVQ/RVLC codebook, whilst the transmission sub-frame $\mathbf{u}^m$ is formed as the concatenation of the corresponding RVLC codewords from the VDVQ/RVLC codebook. The same K-entry VDVQ/RVLC codebook is employed for all VDVQ/RVLC operations of the proposed video codec, which comprises a set of VQ tiles addressed by their RVLC index representations. These are known to both the video encoder and the decoder. An example of a $K = 5$-entry VDVQ/RVLC codebook is provided in Figure 6.5. In the VQ tiles of this figure, dark pixels indicate negative FD values and light pixels represent positive FD values.

As exemplified in Figure 6.5, the K VQ tiles in the VDVQ/RVLC codebook may have a variety of dimensions, which must be multiples of the (8×8)-pixel VB dimensions. This allows the adequate representation both of small areas of high video activity and of large areas of low video activity. In this way, coding efficiency is maintained. The VDVQ/RVLC codebook entry with index $k \in [1 \ldots K]$ is associated with the VQ tile $\mathbf{VQ}^k$. This has the dimensions of $(J_x^k \times J_y^k) = J^k$ number of (8×8)-pixel VBs, as exemplified in Figure 6.5. However, a VQ tile's dimensions must not exceed the MB dimensions $(J_x^{\mathrm{MB}} \times J_y^{\mathrm{MB}}) = (2 \times 2)$ that have been employed, as defined in Section 6.3.1. Each VB $VQ_{j^k}^k$, of the set of J^k VBs that constitute the VQ tile $\mathbf{VQ}^k = \{VQ^k_{j^k}\}^{J^k}_{j^k=1}$, is allocated an index $j^k \in [1 \dots J^k]$. Note that these indices may be arbitrarily allocated. To emphasize this point, a random allocation of VB indices is employed in Figure 6.5, as exemplified by the VQ tiles $\mathbf{VQ}^1$ and $\mathbf{VQ}^2$.

Additionally, the VDVQ/RVLC codebook entry index k is represented by the RVLC code $\mathbf{RVLC}^k$, as exemplified in Figure 6.5. Each RVLC code $\mathbf{RVLC}^k = \{RVLC^k_{i^k}\}^{I^k}_{i^k=1}$ comprises I^k number of bits having values of $RVLC^k_{i^k} \in \{0, 1\}$, where $i^k \in [1 \dots I^k]$ is the bit index. The employment of RVLC codes instead of their more efficient, higher coding-rate alternatives, such as Huffman codes [155], is justified in Section 6.3.3.

The VDVQ/RVLC codebook should be designed by considering the statistical properties of the FD sub-frames. This can be achieved by employing the Linde–Buzo–Gray (LBG) algorithm [170] to design the VQ tile set and a Huffman-coding-based algorithm [276] to design the RVLC code set. Note that an improved performance could be expected if the VQ tile set and the RVLC code set were jointly designed in a manner similar to that of the COVQ scheme in [168, 169]. In this way, similar VQ tiles would be allocated similar RVLC codewords. However, this approach was rejected owing to the high complexity that would be required to compare the similarity of the variable-dimension VQ tiles and the variable-length codewords in various alignments and to perform the corresponding allocation.

Recall from Section 6.3.1 that all FD sub-frames exhibit similar statistical properties, but different statistical properties are exhibited by each of the $J/J^{\mathrm{MB}} = 3$ MBs that comprise the FD sub-frames. In an alternative approach, a separate VDVQ/RVLC codebook could be designed and employed to model the statistical properties of each of these MBs. This alternative approach would allow the achievement of a higher coding efficiency. However, it was discarded for the sake of simplicity.

6.3.3   VDVQ/RVLC-Induced Code Constraints

Let us now elaborate on how the employment of joint VDVQ and RVLC induces code constraints that govern the formation of legitimate bit-sequences within the transmission sub-frames. In Figure 6.6 an example of a $J = 12$-block reconstructed FD sub-frame $\hat{\mathbf{e}}^m$ is provided. Here, we employed the FD decomposition example of Figure 6.4 and the $K = 5$-entry VDVQ/RVLC codebook example of Figure 6.5. The corresponding $I = 17$-bit transmission sub-frame $\mathbf{u}^m$ is subject to the VDVQ/RVLC-induced code constraints outlined below.

As stated in Section 6.3.1, each reconstructed FD sub-frame $\hat{\mathbf{e}}^m = \{\hat{e}^m_j\}^J_{j=1}$ comprises $J = 12$ (8×8)-pixel VBs. These $J = 12$ VBs constitute $J/J^{\mathrm{MB}} = 3$ MBs, as shown in Figure 6.6. Each of these MBs comprises an appropriate tessellation of VQ tiles from the K-entry VDVQ/RVLC codebook. The tiles $\mathbf{VQ}^k = \{VQ^k_{j^k}\}^{J^k}_{j^k=1}$ may cover regions of $(J^k_x \times J^k_y) = J^k$ number of (8×8)-pixel VBs. Note that in Figure 6.6 each of the entries k^T, where $T \in [T_a \dots T_f]$, invoked from the VDVQ/RVLC codebook example of Figure 6.5, provides J^{k^T} number of (8×8)-pixel VBs for the reconstructed FD sub-frame $\hat{\mathbf{e}}^m$.

A specific constraint is imposed that restricts the positioning of the tile $\mathbf{VQ}^k$ in legitimate tessellations. Specifically, the J^k number of VBs in $\hat{\mathbf{e}}^m = \{\hat{e}^m_j\}^J_{j=1}$ that are represented by the tile $\mathbf{VQ}^k$ must have consecutive VB indices j. Note that this is true in all cases in Figure 6.6. This so-called *VQ tile positioning constraint* is necessary to allow the formation of the novel trellis structure, as described in Section 6.3.4. In order that the video degradation

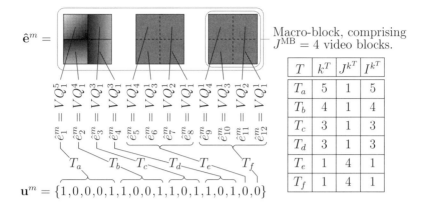

Figure 6.6: Example of a reconstructed FD sub-frame, comprising $J = 12$ VBs selected as exemplified in the FD decomposition of Figure 6.4, and the corresponding $I = 17$-bit transmission sub-frame. These comprise the entries k^T from the VDVQ/RVLC codebook example of Figure 6.5, where $T \in [T_a \ldots T_f]$ signifies the corresponding transition in the VDVQ/RVLC trellis structure example of Figure 6.7.

imposed by this constraint is minimized, the VQ tiles and the allocation of VB indices within the MBs should be jointly designed, as stated in Section 6.3.1. In view of this, the presence of the vertical VQ tile $\mathbf{VQ}^2$ in the VDVQ/RVLC codebook example of Figure 6.5 motivates the vertical allocation of consecutive VB indices within each MB, as seen in Figure 6.4.

The employment of the VQ tile $\mathbf{VQ}^k$ is expressed using J^k number of mappings of the form $\hat{e}_j^m = VQ_{j^k}^k$, as exemplified in Figure 6.6. With reference to the examples of Figures 6.4 and 6.5, we note that the resultant values of $j \in [1 \ldots J]$ and $j^k \in [1 \ldots J^k]$ in such mappings are dependent on the specific positioning of the VQ tile.

Each of the VQ tiles $\mathbf{VQ}^k$ that comprise the reconstructed FD sub-frame $\hat{\mathbf{e}}^m$ is associated with a RVLC code $\mathbf{RVLC}^k$, where k is the VDVQ/RVLC codebook entry index. These RVLC codes are concatenated in the order of the VB indices j that are associated with the employment of the corresponding VQ tiles in the reconstructed FD sub-frame $\hat{\mathbf{e}}^m = \{\hat{e}_j^m\}_{j=1}^J$. This is exemplified in Figure 6.6, in which each of the entries k^T invoked from the VDVQ/RVLC codebook example of Figure 6.5 is represented by I^{k^T} number of bits. Note that the superscript T, where $T \in [T_a \ldots T_f]$ in this case, denotes the corresponding transition in the associated VDVQ/RVLC trellis structure, as detailed in Section 6.3.4. In the proposed video codec each concatenation of RVLC codes is constrained to having a total length of $I = 17$ bits, since it constitutes the $I = 17$-bit transmission sub-frame $\mathbf{u}^m$.

It follows from the above discussions that each $I = 17$-bit transmission sub-frame must represent a legitimate tessellation of VQ tiles, comprising $J = 12$ VBs in total. These VDVQ/RVLC-induced code constraints represent redundancy within the bit-based transmission sub-frames. Note that the degree of this redundancy is dependent on the distance properties of the RVLC code set employed. As outlined in Section 6.6, the employment of

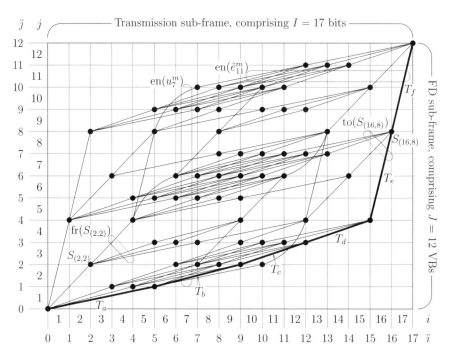

Figure 6.7: Example of a VDVQ/RVLC trellis structure for the FD decomposition example of Figure 6.4 and for the $K = 5$-entry VDVQ/RVLC codebook example of Figure 6.5. Trellis states occur between the consideration of specific bits and VBs; hence bit state indices $\ddot{\imath}$ and VB state indices $\ddot{\jmath}$ occur between the bit indices i and the VB indices j respectively. ©IEEE [194] Maunder *et al.* 2007.

RVLC codes rather than Huffman codes is justified, because their additional redundancy is desirable.

6.3.4   VDVQ/RVLC Trellis Structure

As described in Section 6.3.3, the formation of legitimate bit-sequences within each of the bit-based transmission sub-frames is governed by the VDVQ/RVLC-induced code constraints. During VDVQ/RVLC encoding and APP SISO decoding we consider only legitimate $I = 17$-bit transmission sub-frame permutations. Since satisfying this constraint is non-trivial, the employment of a trellis structure is proposed to describe the complete set of VDVQ/RVLC-induced code constraints. These VDVQ/RVLC-induced code constraints depend on the FD decomposition and the VDVQ/RVLC codebook employed, as described in Section 6.3.3. Hence, the design of the proposed VDVQ/RVLC trellis structure also depends on these aspects of the proposed video codec's operation. The VDVQ/RVLC trellis structure example of Figure 6.7 describes the complete set of VDVQ/RVLC-induced code constraints that correspond to the FD decomposition example of Figure 6.4 and to the $K = 5$-entry VDVQ/RVLC codebook example of Figure 6.5. In this example, each reconstructed FD sub-frame $\hat{\mathbf{e}}^m$ comprises $J = 12$ VBs and each transmission sub-frame $\mathbf{u}^m$ comprises $I = 17$ bits.

The proposed VDVQ/RVLC trellis structure comprises a set of transitions between trellis states, as exemplified in Figure 6.7. A novel block-based modification of the VLC symbol-level trellis structure described in [181] is employed. Whilst the bit index axis of [181] is retained in the proposed VDVQ/RVLC trellis structure, the VLC symbol index axis of [181] is replaced by a VB index axis. In contrast to [181], here transitions are permitted to skip a number of consecutive indices along this axis.

During a VDVQ/RVLC encoding or APP SISO decoding operation, there are a number of instances when it is possible to employ each VDVQ/RVLC codebook entry, as described below. The proposed VDVQ/RVLC trellis structure represents each of these legitimate possibilities with a transition, as exemplified in Figure 6.7. In this way, the proposed VDVQ/RVLC trellis structure describes the complete set of VDVQ/RVLC-induced code constraints. Each transition T represents a possible employment of the VDVQ/RVLC codebook entry having the index $k^T \in [1 \ldots K]$. This is associated with a unique combination of (a) the positioning of the VQ tile $\mathbf{VQ}^{k^T}$ within the reconstructed FD sub-frame $\hat{\mathbf{e}}^m$ and (b) the corresponding positioning of the associated RVLC code $\mathbf{RVLC}^{k^T}$ within the transmission sub-frame $\mathbf{u}^m$. Note that, for reasons to be discussed below, (a) is subject to the legitimate VQ tile positioning constraint that was stated in Section 6.3.3.

Each trellis state in the proposed VDVQ/RVLC trellis structure represents the progress made at a particular point during the VDVQ/RVLC encoding or APP SISO decoding operation. This point occurs immediately after the consideration of the first $\breve{j}$ number of VBs in the reconstructed FD sub-frame $\hat{\mathbf{e}}^m$ and the first $\breve{i}$ number of bits in the transmission sub-frame $\mathbf{u}^m$. Accordingly, trellis states are denoted $S_{(\breve{i},\breve{j})}$. Here, $\breve{j} \in [0 \ldots J]$ denotes a VB state index. These occur between the VB indices $j \in [1 \ldots J]$ that were introduced in Section 6.3.1. Likewise, bit state indices $\breve{i} \in [0 \ldots I]$ occur between the bit indices $i \in [1 \ldots I]$, as shown in Figure 6.7.

Each transition T represents the employment of the VDVQ/RVLC codebook entry with index $k^T \in [1 \ldots K]$ immediately after reaching a particular point during the VDVQ/RVLC encoding or APP SISO decoding operation. This point is identified by the state indices $\breve{i}^T \in [0 \ldots I]$ and $\breve{j}^T \in [0 \ldots J]$. Hence, the transition T emerges from the trellis state $S_{(\breve{i}^T,\breve{j}^T)}$. The VQ tile $\mathbf{VQ}^{k^T}$ provides J^{k^T} number of VBs for the reconstructed FD sub-frame $\hat{\mathbf{e}}^m$, as stated in Section 6.3.3. Additionally, the RVLC code $\mathbf{RVLC}^{k^T}$ provides I^{k^T} number of bits for the transmission sub-frame $\mathbf{u}^m$. Hence, the transition T merges into the trellis state $S_{(\breve{i}^T+J^{k^T},\breve{j}^T+I^{k^T})}$. Note that the VQ tile positioning constraint described in Section 6.3.3 is necessary to ensure that each transition is continuous with respect to the VB index axis. Also note that a particular employment of a VDVQ/RVLC codebook entry is only possible if the associated transition T contributes to a legitimate transition path between the trellis states $S_{(0,0)}$ and $S_{(I,J)}$. This condition is satisfied if at least one transition path exists between the trellis states $S_{(0,0)}$ and $S_{(\breve{i}^T,\breve{j}^T)}$ and between the trellis states $S_{(\breve{i}^T+I^{k^T},\breve{j}^T+J^{k^T})}$ and $S_{(I,J)}$.

Note that the reconstructed FD sub-frame and transmission sub-frame examples of Figure 6.6 correspond to the bold trellis path in Figure 6.7. Here, each transition $T \in [T_a \ldots T_f]$ corresponds to the similarly labeled employment of a VDVQ/RVLC codebook entry in Figure 6.6. With reference to Figure 6.6, observe that each transition $T \in [T_a \ldots T_f]$ encompasses I^{k^T} number of bit indices and J^{k^T} number of VB indices. Whilst the trellis path considered comprises six transitions, it should be noted that the trellis structure of

Figure 6.7 additionally contains trellis paths comprising seven transitions. In general, each transition path in the proposed VDVQ/RVLC trellis structure comprises a varying number of transitions.

Note furthermore that, owing to the diamond shape of the VDVQ/RVLC trellis structure, a single trellis considering $J \cdot M = 396$ VBs and $I \cdot M = 561$ bits would contain significantly more transitions than the combination of $M = 33$ VDVQ/RVLC trellis structures, each considering $J = 12$ VBs and $I = 17$ bits. This justifies the decomposition of the VDVQ/RVLC encoding and APP SISO decoding operations of an entire FD into $M = 33$ less complex trellis-based VDVQ/RVLC operations, as described in Section 6.2.2. This decomposition is associated with a reduced grade of VDVQ/RVLC encoding freedom and hence a reduced video reconstruction quality. However, this slight video degradation is insignificant compared with the resultant computational complexity reduction.

For the benefit of the following sections, we now introduce some trellis-transition set notation, which is exemplified in Figure 6.7. The set of all transitions that encompasses (en) the VB $j \in [1 \ldots J]$ of the reconstructed FD sub-frame $\hat{\mathbf{e}}^m$ is termed $\mathrm{en}(\hat{e}_j^m)$. Furthermore, $\mathrm{en}(\hat{e}_j^m = VQ_{j^k}^k)$ is the specific subset of $\mathrm{en}(\hat{e}_j^m)$ that maps the VB $j^k \in [1 \ldots J^k]$ of the VQ tile $\mathbf{VQ}^k$ onto $\hat{e}_j^m$, where $k \in [1 \ldots K]$. Note that some of these subsets may be empty. This is a consequence of the VQ tile positioning constraint described in Section 6.3.3. The set of all transitions that encompasses the bit $i \in [1 \ldots I]$ of the transmission sub-frame $\mathbf{u}^m$ is termed $\mathrm{en}(u_i^m)$. Additionally, $\mathrm{en}(u_i^m = b)$ is the specific subset of $\mathrm{en}(u_i^m)$ that maps the bit value $b \in \{0, 1\}$ onto u_i^m. Note that, for a particular transition T, we have $RVLC_{i-iT}^{k^T} = b$ if $T \in \mathrm{en}(u_i^m = b)$. Finally, the set of transitions emerging from (fr) the trellis state $S_{(\tilde{i},\tilde{j})}$ is denoted as $\mathrm{fr}(S_{(\tilde{i},\tilde{j})})$, whilst the set merging into (to) that trellis state is represented as $\mathrm{to}(S_{(\tilde{i},\tilde{j})})$.

6.4   VDVQ/RVLC Encoding

In the proposed VDVQ/RVLC-TCM scheme's transmitter of Figure 6.1, VDVQ/RVLC encoding is performed separately for each of the M FD sub-frames. Each VDVQ/RVLC encoder operates on the basis of the proposed VDVQ/RVLC trellis structure described in Section 6.3.4 and exemplified in Figure 6.7. Since it describes the complete set of VDVQ/RVLC-induced code constraints, the employment of the proposed VDVQ/RVLC trellis structure represents the consideration of every legitimate FD sub-frame encoding. This allows us to find the MMSE approximation of the FD sub-frame $\mathbf{e}^m$. The result is the optimal reconstructed FD sub-frame $\hat{\mathbf{e}}^m$ and the corresponding bit-based transmission sub-frame $\mathbf{u}^m$.

We quantize the video sequence in a novel manner, which is reminiscent of [291], considering the tessellation of potentially differently sized VQ tiles. The philosophy of Viterbi decoding [193] is employed, with a survivor path being selected between the trellis state $S_{(0,0)}$ and every other state $S_{(\tilde{i},\tilde{j})}$ in the trellis, as described in Chapter 3. This selection yields the $\tilde{i}$-bit encoding of the first $\tilde{j}$ number of VBs in the FD sub-frame $\mathbf{e}^m$ that introduces the lowest possible cumulative video distortion $D(S_{(\tilde{i},\tilde{j})})$, as detailed below. This is exemplified in Figure 6.8, where a survivor path is indicated in bold between the trellis state $S_{(0,0)}$ and every other state $S_{(\tilde{i},\tilde{j})}$ in the trellis.

As stated in Section 6.3.4, each transition T in the proposed VDVQ/RVLC trellis structure represents the employment of the VDVQ/RVLC codebook entry with index k^T during

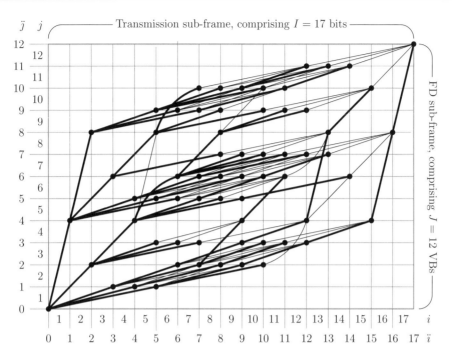

Figure 6.8: Example of MMSE VDVQ/RVLC encoding using the VDVQ/RVLC trellis structure example of Figure 6.7. The survivor path selected between the state $S_{(0,0)}$ and each other state $S_{(\tilde{i},\tilde{j})}$ is indicated using a bold line. Note that the survivor path selected between the trellis states $S_{(0,0)}$ and $S_{(I,J)}$ yields the FD encoding exemplified in Figure 6.6. ©IEEE [194] Maunder *et al.* 2007.

VDVQ/RVLC encoding. This corresponds to employing the VQ tile $\mathbf{VQ}^{k^T}$ to represent a total of J^{k^T} number of (8×8)-pixel VBs of the FD sub-frame $\mathbf{e}^m$. The distortion $d(T)$ associated with the transition T is the sum of the squared difference between $\mathbf{VQ}^{k^T}$ and the corresponding J^{k^T} number of VBs of $\mathbf{e}^m$.

The survivor path between the trellis state $S_{(0,0)}$ and a particular state $S_{(\tilde{i},\tilde{j})}$ is deemed to be that associated with the specific merging transition $T \in \text{to}(S_{(\tilde{i},\tilde{j})})$ having the minimum cumulative video distortion $D(T) = d(T) + D(S_{(\tilde{i}^T,\tilde{j}^T)})$, where $D(S_{(0,0)}) = 0$. Note that each trellis state in Figure 6.8 has exactly one merging transition. These are indicated in bold. Having determined the survivor path at the trellis state $S_{(I,J)}$, the MMSE VDVQ/RVLC encoding of the FD sub-frame $\mathbf{e}^m$ has been found. The reconstructed FD sub-frame $\hat{\mathbf{e}}^m$ is formed as the tessellation of the VQ tiles associated with the survivor path transitions. Additionally, the I-bit transmission sub-frame $\mathbf{u}^m$ is formed as the concatenation of the associated RVLC codes, as described in Section 6.3.3. Note in Figure 6.8 that the survivor path selected between the trellis states $S_{(0,0)}$ and $S_{(I,J)}$ yields the reconstructed FD sub-frame $\hat{\mathbf{e}}^m$ and transmission sub-frame $\mathbf{u}^m$ of Figure 6.6.

6.5   APP SISO VDVQ/RVLC Decoding

In the proposed VDVQ/RVLC-TCM scheme's receiver of Figure 6.2, APP SISO VDVQ/-RVLC decoding is performed separately for each of the M number of I-bit transmission sub-frames. Each of the M APP SISO VDVQ/RVLC decoders operates on the basis of the proposed VDVQ/RVLC trellis structure. The recovery of legitimate – although not necessarily error-free – video information is therefore guaranteed, since the proposed VDVQ/RVLC trellis structure describes the complete set of VDVQ/RVLC-induced code constraints. As stated in Section 6.3.3, these VDVQ/RVLC-induced code constraints impose redundancy within each transmission sub-frame. This redundancy may be exploited to assist during APP SISO VDVQ/RVLC decoding by invoking the BCJR [48] algorithm on the basis of the proposed VDVQ/RVLC trellis structure. Additionally, residual redundancy is exhibited by the transmission sub-frame $\mathbf{u}^m$ owing to the simplicity of the proposed video codec. This may be exploited by the BCJR algorithm with consideration of the statistical properties of FD sub-frames, as detailed below.

Each APP SISO VDVQ/RVLC decoder is provided with the *a priori* LLR sub-frame $L_a^o(\mathbf{u}^m)$ and generates the *a posteriori* LLR sub-frame $L_p^o(\mathbf{u}^m)$, as stated in Section 6.2.3. Additionally, each APP SISO VDVQ/RVLC decoder recovers the optimal MMSE-based reconstructed FD sub-frame estimate $\tilde{\mathbf{e}}^m$. This is achieved using soft APP-based reconstruction, as follows.

A novel block-based modification of the BCJR algorithm of [174] is invoked in the proposed VDVQ/RVLC trellis structure. This obtains an APP for each transition T, giving the probability that this specific transition was in the survivor path during the corresponding VDVQ/RVLC encoding operation, as described in Section 6.4. These APPs are calculated as

$$P_p(T) = \frac{1}{C_1} \cdot \alpha(S_{(i^T, j^T)}) \cdot \gamma(T) \cdot \beta(S_{(i^T + I^{k^T}, j^T + J^{k^T})}). \tag{6.1}$$

The specific value of the normalization factor C_1 in Equation (6.1) may be ignored in this application, since we are only concerned with the ratios of *a posteriori* transition probabilities, as we show below. The terms $\alpha(S_{(i^T, j^T)})$, $\gamma(T)$ and $\beta(S_{(i^T + I^{k^T}, j^T + J^{k^T})})$ consider the probability of the requisite trellis activity before, during and after the occurrence of the transition T, respectively.

Specifically, the term $\gamma(T)$ in (6.1) is calculated as

$$\gamma(T) = \frac{P(k^T)}{C_2(S_{(i^T, j^T)})} \cdot \prod_{i^k=1}^{I^{k^T}} P_a(u_{i^T+i^k}^m = RVLC_{i^k}^{k^T}), \tag{6.2}$$

where $P_a(u_i^m = b)$ is the *a priori* probability that the transmission sub-frame bit u_i^m has a value of $b \in \{0, 1\}$. This is obtained from the *a priori* LLR $L_a^o(u_i^m) = \ln(P_a(u_i^m = 0)/P_a(u_i^m = 1))$ [292]. Furthermore, $P(k)$ in (6.1) is the probability of occurrence for the VDVQ/RVLC codebook entry with index k. This is obtained based on knowledge of the statistical properties of the FD sub-frames. As stated in Section 6.3.1, different statistical properties are associated with each of the J/J^{MB} number of MBs constituting a FD sub-frame. Hence, different values of $P(k)$ are employed, depending on the specific MB of which T is a constituent. Finally, the normalization factor $C_2(S_{(\check{i},\check{j})}) = \sum_{T \in \text{fr}(S_{(\check{i},\check{j})})} P(k^T)$ was proposed in [174] for the sake of normalizing the VDVQ/RVLC codebook entries' probabilities of occurrence $P(k)$. This is necessary in the case where the VDVQ/RVLC-induced code constraints prevent the employment of one or more VDVQ/RVLC codebook entries from the state $S_{(\check{i},\check{j})}$. For example, this is the case for the state $S_{(2,2)}$ of Figure 6.7, which has just four emerging transitions, indicating that only four of the $K = 5$ VDVQ/RVLC codebook entries can be employed.

The BCJR algorithm's forward recursion emerging from the trellis state $S_{(0,0)}$ is employed to obtain the values of

$$\alpha(S_{(\check{i},\check{j})}) = \sum_{T \in \text{to}(S_{(\check{i},\check{j})})} \gamma(T) \cdot \alpha(S_{(\check{i}^T,\check{j}^T)}), \tag{6.3}$$

where $\alpha(S_{(0,0)}) = 1$. Similarly, a backward recursion from the trellis state $S_{(I,J)}$ is employed to obtain the values of

$$\beta(S_{(\check{i},\check{j})}) = \sum_{T \in \text{fr}(S_{(\check{i},\check{j})})} \gamma(T) \cdot \beta(S_{(\check{i}^T + I^k{}^T, \check{j}^T + J^k{}^T)}), \tag{6.4}$$

where $\beta(S_{(I,J)}) = 1$.

Having determined the *a posteriori* transition probabilities, the method of [174] is employed for obtaining bit-based soft outputs for each of the I number of bits in the transmission sub-frame $\mathbf{u}^m = \{u_i^m\}_{i=1}^I$. More specifically, an *a posteriori* LLR $L_p^o(u_i^m)$ is obtained for the particular bit u_i^m with consideration of the *a posteriori* probabilities of all trellis transitions that may influence its value. This set of transitions $\text{en}(u_i^m)$ includes all transitions that bisect a cross-section of the trellis structure that is perpendicular to the bit index axis at the particular index i, as exemplified for the bit u_7^m in the trellis structure of Figure 6.7. The associated *a posteriori* LLRs are calculated as

$$L_p^o(u_i^m) = \ln\left(\frac{\sum_{T \in \text{en}(u_i^m=0)} P_p(T)}{\sum_{T \in \text{en}(u_i^m=1)} P_p(T)}\right), \tag{6.5}$$

for all $i \in [1 \dots I]$.

The recovery of the reconstructed FD sub-frame estimate $\tilde{\mathbf{e}}^m$ is performed on an individual block-by-block basis. A soft APP-based output is obtained for each of the J number of (8×8)-pixel VBs in the reconstructed FD sub-frame $\hat{\mathbf{e}}^m = \{\hat{e}_j^m\}_{j=1}^J$. Again, a novel modification of the method of [174] is employed for obtaining these block-based soft outputs. In analogy to the generation of the previously mentioned bit-based soft outputs, the reconstruction of the particular VB $\hat{e}_j^m$ is performed with consideration of the *a posteriori* probabilities of all trellis transitions that may influence its value. This set of transitions $\text{en}(\hat{e}_j^m)$ includes all transitions that bisect a cross-section of the trellis structure that is perpendicular to the VB index axis at the particular index j, as exemplified for the VB $\hat{e}_{11}^m$ in the trellis structure of Figure 6.7. The APPs of the VB $\hat{e}_j^m$ being provided by a particular one of the J^k number of VBs in the VQ tile $\mathbf{VQ}^k = \{VQ_{j^k}^k\}_{j^k=1}^{J^k}$ are calculated as

$$P_p(\hat{e}_j^m = VQ_{j^k}^k) = \frac{\sum_{T\in\text{en}(\hat{e}_j^m = VQ_{j^k}^k)} P_p(T)}{\sum_{T\in\text{en}(\hat{e}_j^m)} P_p(T)}, \tag{6.6}$$

for all $j \in [1\ldots J]$, $k \in [1\ldots K]$ and $j^k \in [1\ldots J^k]$. Note that some of the sets $\text{en}(\hat{e}_j^m = VQ_{j^k}^k)$ may be empty, as described in Section 6.3.4. In this case, the corresponding APP is zero-valued.

Each of the J (8×8)-pixel VBs that constitute the reconstructed FD sub-frame estimate $\tilde{\mathbf{e}}^m = \{\tilde{e}_j^m\}_{j=1}^J$ is obtained using optimal MMSE estimation according to the weighted average

$$\tilde{e}_j^m = \sum_{k=1}^{K}\sum_{j^k=1}^{J^k} P_p(\hat{e}_j^m = VQ_{j^k}^k) \cdot VQ_{j^k}^k. \tag{6.7}$$

As described in Chapter 4, the logarithmic *A Posteriori* Probability (log-APP) algorithm [295][2] is employed to reduce the computational complexity of APP SISO VDVQ/RVLC decoding. Specifically, the above-mentioned calculations are performed in the logarithmic domain by employing the Jacobian approximation with an eight-entry table-lookup correction factor [296]. This reduces the number of multiplications required by the BCJR algorithm. Additionally, an appropriately modified reduced-complexity version of the BCJR algorithm [297] of Chapter 4 is employed to prune insignificant transitions from the proposed VDVQ/RVLC trellis structure, where the transition paths passing through the trellis state $S_{(\check{i},\check{j})}$ are pruned during the forward recursion if

$$\frac{\alpha(S_{(\check{i},\check{j})})}{\sum_{i'=0}^{I}\alpha(S_{(i',\check{j})})} < 0.001. \tag{6.8}$$

In the next section we consider the performance of the proposed VDVQ/RVLC-TCM scheme, where the above-mentioned complexity-reduction methods were observed to impose no significant performance degradation.

6.6   Simulation Results

In this section we assess the performance of the proposed VDVQ/RVLC-TCM scheme. We transmitted 250 video frames of the 'Lab' video-sequence [151]. This 10 frames/s grayscale head-and-shoulders (176×144)-pixel Quarter Common Intermediate Format (QCIF) video sequence exhibits a moderate level of video activity.

The proposed video codec was designed to achieve an attractive trade-off between the conflicting design requirements associated with bit rate, video reconstruction quality and computational complexity. The 396-block FDs were decomposed into $M = 33$ FD sub-frames, each comprising $J = 12$ (8×8)-pixel VBs. This was performed exactly as exemplified in Figure 6.4. However, in contrast to our simplified example of Figure 6.7 using

---

[2]The log-APP algorithm is also known as the logarithmic Maximum *A Posteriori* (log-MAP) algorithm.

$I = 17$ bits per sub-frame, each transmission sub-frame comprised a length of $I = 45$ bits in the scheme implemented. A $K = 512$-entry VDVQ/RVLC codebook was employed. This comprised the five VQ tiles shown in Figure 6.5 and an additional 507 single-VB VQ tiles. The corresponding RVLC codes were designed to have a free distance lower bound of $\bar{d}_{\text{free}} = 2$, since this supports iterative decoding convergence to an infinitesimally low probability of error [298], as described in Chapter 4. The lossless component of the proposed video codec's coding rate is defined here as $R_{\text{VDVQ/RVLC}} = E/I$, where I is the number of bits employed to represent each sub-frame and E is its expected sub-frame entropy, which may be estimated as

$$E = -\sum_{\forall T} P(T) \log_2 \left(\frac{P(k^T)}{C_2(S_{(i^T,j^T)})} \right), \tag{6.9}$$

where $-\log_2(P(k^T)/C_2(S_{(i^T,j^T)}))$ estimates the amount of information represented by the transition T in terms of bits, and $P(T)$ is the probability of occurrence for the transition T, which is may be estimated as

$$P(T) = P(S_{(i^T,j^T)}) \frac{P(k^T)}{C_2(S_{(i^T,j^T)})}, \tag{6.10}$$

where $P(S_{(i,j)})$ is the probability of occurrence for the trellis state $S_{(i,j)}$, which may be estimated as

$$P(S_{(i,j)}) = \sum_{T \in \text{to}(S_{(i,j)})} P(T). \tag{6.11}$$

For our scenario of $K = 512$, an approximate coding rate of $R_{\text{VDVQ/RVLC}} \approx 0.666$ is obtained.

Two VDVQ/RVLC-TCM schemes associated with different latencies were employed. The first scheme imposed a low latency equal to the duration of a single video frame, namely 0.1 s at 10 frames/s. This is suitable for real-time interactive video-telephony applications. In this scheme the length of each transmission frame $\mathbf{u}$ and that of the interleaver π equals $M \cdot I = 1485$ bits. The second VDVQ/RVLC-TCM scheme had a high latency of 50 video frames, i.e. 5 s at 10 frames/s. This is suitable for non-real-time video streaming and wireless-Internet download applications. Here, 50 transmission frames $\mathbf{u}$ are concatenated before interleaving, giving an interleaver length of $50 \cdot M \cdot I = 74\,250$ bits. Note that both schemes have a video-encoded bit rate of 14.85 kbit/s.

Both VDVQ/RVLC-TCM schemes employed the same TCM codec, having the Linear Feedback Shift Register (LFSR) schematic of Figure 6.9. As shown in Figure 6.9, the TCM encoder generates a set of four bits to represent each set of three input bits, giving a coding rate of $R_{\text{TCM}} = 3/4$. Three of the four output bits are systematic replications of the three input bits, whilst the fourth output bit is generated with the aid of the $L_{\text{TCM}} = 6$ modulo-2 memory elements. Note that the TCM codec is a recursive component having an infinite impulse response, since feedback is employed in the shift register of Figure 6.9. As a result, the TCM codec supports iterative decoding convergence to an infinitesimally low probability of error [299], as is the case for our VDVQ/RVLC codec, as described above. Hence, we may expect the proposed VDVQ/RVLC-TCM scheme to achieve iterative decoding convergence to an infinitesimally low probability of error, provided that the channel quality is sufficiently

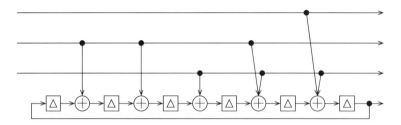

Figure 6.9: LFSR encoder schematic for the recursive systematic $R_{\mathrm{TCM}} = 3/4$-rate TCM codec employing a coding memory of $L_{\mathrm{TCM}} = 6$.

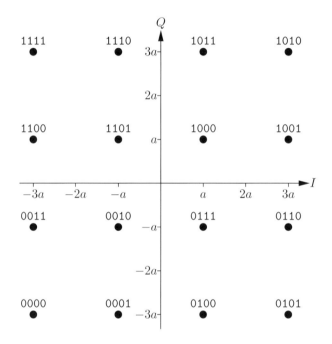

Figure 6.10: Constellation diagram for the $M_{\mathrm{TCM}} = 16$-level set partitioned QAM modulation scheme. For an average transmit energy of unity, $a = 1/\sqrt{10}$.

high to create an open EXIT chart tunnel and the iterative decoding trajectory approaches the inner and outer EXIT functions sufficiently closely, as discussed in Chapter 4. Furthermore, Figure 6.10 provides the constellation diagram for the $M_{\mathrm{TCM}} = 16$-ary set-partitioned [277] Quadrature Amplitude Modulation (QAM) modulation scheme of the TCM codec. This was employed together with In-phase Quadrature-phase (IQ)-interleaving [293] for transmission over an uncorrelated narrowband Rayleigh fading channel.

The effective throughput of the proposed VDVQ/RVLC-TCM scheme is given by $\eta = R_{\mathrm{VDVQ/RVLC}} \times R_{\mathrm{TCM}} \times \log_2(M_{\mathrm{TCM}}) \approx 2.00$ bits per channel use. Note that for $\eta = 2.00$ bits per channel use, the uncorrelated Rayleigh-fading channel's capacity for 16-ary

Quadrature Amplitude Modulation (16QAM) is reached for a Rayleigh fading channel SNR of 6.96 dB [300].

In the EXIT chart of Figure 6.11 we provide the EXIT functions [294] for APP SISO TCM decoding at various channel SNR values, as described in Chapter 4. These were generated using Gaussian distributed *a priori* LLRs and all mutual information measurements were made using the histogram-based approximation of the LLR Probability Density Functions (PDFs) [294], as described in Chapter 4. Note that, owing to its recursive nature, the APP SISO TCM decoder is capable of achieving unity extrinsic mutual information I_e^i for unity *a priori* mutual information I_a^i [299]. Additionally, Figure 6.11 provides the inverted APP SISO VDVQ/RVLC decoding EXIT function. Note that, according to the proof of [204], the area beneath the inverted APP SISO VDVQ/RVLC decoding EXIT function is equal to the lossless coding rate of the VDVQ/RVLC codec $R_{\mathrm{VDVQ/RVLC}}$. This area was found to be 0.666, which confirms the calculation provided above. Similarly to APP SISO TCM decoding, APP SISO VDVQ/RVLC decoding achieves unity extrinsic mutual information I_e^o for unity *a priori* mutual information I_a^o, owing to the employment of a RVLC code set having a free distance lower bound of $\bar{d}_{\mathrm{free}} = 2$ [298], as discussed in Chapter 3. Hence an open EXIT chart tunnel [267] can be created, if the channel SNR is sufficiently high. If the iterative decoding trajectory approaches the inner and outer EXIT functions sufficiently closely, iterative decoding convergence to unity mutual information will be achieved in this case. This is associated with an infinitesimally low probability of error, as described in Chapter 4. Since our objective is to mitigate all transmission errors in this way, it is for this reason that we do not employ any additional measures to counteract the propagation of distortion amongst consecutive differentially encoded video frames, as described in Section 6.2.1.

As shown in Figure 6.11, the proposed VDVQ/RVLC-TCM scheme exhibits an open EXIT chart tunnel for channel SNRs above a threshold of 8.25 dB, which is just 1.29 dB from the proposed VDVQ/RVLC-TCM scheme's SNR capacity bound of 6.96 dB. As described in Chapter 4, an open EXIT chart tunnel implies that iterative decoding convergence to an infinitesimally low video degradation may be achieved, again provided that the iterative decoding trajectory approaches the inner and outer EXIT functions sufficiently closely. Note that an open EXIT tunnel would be achieved for a lower channel SNR that is closer to the channel's SNR capacity bound of 6.96 dB, if the shape of the APP SISO VDVQ/RVLC decoder's inverted EXIT function was better matched to that of the APP SISO TCM decoder's EXIT function, as discussed in Chapter 4. This would naturally result in an EXIT chart tunnel that was narrow all along its length. Despite this, however, the APP SISO VDVQ/RVLC decoder's inverted EXIT function of Figure 6.11 does offer a fairly good match with the APP SISO TCM decoder's EXIT function for a channel SNR of 8.25 dB, resulting in an EXIT chart tunnel that is fairly narrow along its entire length.

Figure 6.11 additionally provides EXIT trajectories [294] that were averaged over all 250 source video frames for both the 1485- and the 74 250-bit interleaver VDVQ/RVLC-TCM schemes at a channel SNR of 9 dB. The 1485-bit interleaver trajectory can be seen to deviate from the EXIT functions and fails to converge to the desired unity mutual information. By contrast, the 74 250-bit interleaver trajectory offers an improved correlation with these EXIT functions and converges to unity mutual information. The improved performance of the 74 250-bit interleaver scheme is a benefit of its more uncorrelated *a priori* information [292], justified by the reasons noted in Section 6.2.3. Note however that an iterative decoding

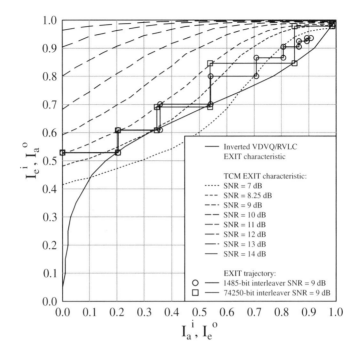

Figure 6.11: The proposed VDVQ/RVLC-TCM scheme's EXIT chart for the 'Lab' video sequence and a $3/4$-rate TCM scheme. ©IEEE [194] Maunder *et al.* 2007.

trajectory offering an even better match with the inner and outer EXIT functions may be expected for interleaver lengths higher than $74\,250$ bits.

In Figure 6.12, the video reconstruction quality of both the 1485- and the $74\,250$-bit interleaver VDVQ/RVLC-TCM schemes is assessed after a number of decoding iterations and for a range of channel SNR values. We employ the Peak Signal-to-Noise Ratio (PSNR) [151] as the objective video reconstruction quality metric. In this application, a PSNR of 29.5 dB is associated with an aesthetically pleasing video reconstruction quality. The proposed VDVQ/RVLC-TCM scheme exhibited substantial iteration gains, as shown in Figure 6.12. After eight decoding iterations a SNR iteration gain of 4.34 dB was achieved by the 1485-bit interleaver scheme at a PSNR of 29.5 dB. In the case of the $74\,250$-bit interleaver scheme, the corresponding iteration gain was 5.61 dB. The $74\,250$-bit interleaver scheme was seen to outperform the 1485-bit interleaver scheme, regardless of the number of decoding iterations. This is in agreement with the findings of the EXIT chart analysis of Figure 6.11, as discussed above. For the 1485-bit interleaver scheme, a PSNR higher than 29.5 dB was achieved after eight decoding iterations for channel SNRs greater than 10.00 dB, as shown in Figure 6.12. This is 3.04 dB from the proposed VDVQ/RVLC-TCM scheme's channel's SNR capacity bound of 6.96 dB. In the case of the $74\,250$-bit interleaver scheme, this is achieved for channel SNRs higher than 8.75 dB, which is just 1.79 dB from the channel's capacity bound. Note that if an even longer interleaver were employed, a PSNR in excess of 29.5 dB might be

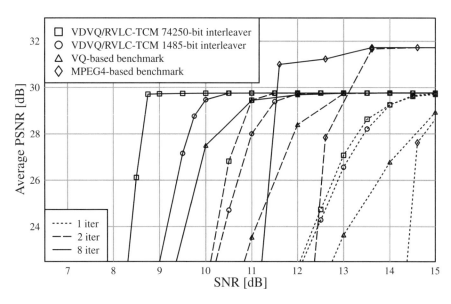

Figure 6.12: PSNR performance of the proposed VDVQ/RVLC-TCM scheme as well as of the VQ- and MPEG-4-based benchmarks for the 'Lab' video sequence, when communicating over an uncorrelated narrowband Rayleigh fading channel. ©IEEE [194] Maunder *et al.* 2007.

expected for channel SNRs that are even closer to the 8.25 dB threshold at which an open EXIT chart tunnel is achieved, as discussed above. Naturally, this would be associated with an even higher latency, however.

We now compare the performance of the proposed VDVQ/RVLC-TCM scheme with two benchmarks. Similarly to the proposed VDVQ/RVLC-TCM scheme, the scheme of [273] is also VQ-based and it is hence referred to as the *VQ-based benchmark* here. In contrast, the IQ-TCM-RVLC scheme of [274] is based upon the MPEG-4 video coding standard [158], and is referred to here as the *MPEG-4-based benchmark*. Similarly to the proposed VDVQ/RVLC-TCM scheme, the benchmarks employ both RVLC and TCM. However, they adopt Shannon's source and channel coding separation philosophy [133], with video decoding being performed independently of iterative channel decoding, as described in Section 6.1. Since the same input video sequence, latency, coding rate, bit rate and channel were employed for both of the benchmarks as well as for the 1485-bit interleaver VDVQ/RVLC-TCM scheme, their direct comparison is justified.

In Figure 6.12, the 1485-bit interleaver VDVQ/RVLC-TCM scheme can be seen to outperform the VQ-based benchmark, regardless of the number of decoding iterations. Although the MPEG-4-based benchmark achieves a slightly higher error-free video reconstruction quality than does the proposed VDVQ/RVLC-TCM scheme, it is outperformed by the scheme advocated at low values of channel SNR. At a PSNR of 29.5 dB, the proposed VDVQ/RVLC-TCM scheme offers a consistent improvement over both the VQ- and the MPEG-4-based benchmarks, amounting to about 1.1 dB and 1.6 dB respectively. This is a benefit of the iterative exploitation of the video codec's code constraints, as outlined in Section 6.5. Additionally, this approach guarantees the recovery of legitimate video information. Hence, the proposed VDVQ/RVLC-TCM scheme is never forced to discard any video information. In contrast, both the benchmarks of [273] and [274] drop video frames when the iterative decoding process does not recover legitimate video information, reducing the attainable video reconstruction quality.

The improved performance offered by the proposed VDVQ/RVLC-TCM scheme is achieved at the cost of an increased computational complexity. A simple metric of computational complexity considers the number of BCJR trellis transitions encountered during iterative decoding. Note that this ignores the complexity contribution imposed by trellis pruning and video reconstruction. Although an equal number of BCJR transitions are employed during APP SISO TCM decoding in the proposed VDVQ/RVLC-TCM scheme and in the benchmarks, the complexities of the respective VDVQ/RVLC and RVLC APP SISO decoding operations are different. More explicitly, in the proposed VDVQ/RVLC-TCM scheme the APP SISO VDVQ/RVLC decoding employs a channel-condition-dependent number of BCJR transitions, which is typically equal to the number of trellis transitions encountered during APP SISO TCM decoding. In contrast, the benchmarks' bit-level RVLC decoding trellises [301] have approximately a quarter of this number of trellis transitions, as discussed in Chapter 3. Hence, the proposed VDVQ/RVLC-TCM scheme has a computational complexity that may be deemed to be approximately 1.6 times that of the benchmarks.

6.7   Summary and Conclusions

In this chapter we have considered the application of joint source and channel coding for video transmission. More specifically, a novel VDVQ- and RVLC-based video codec was serially concatenated with TCM and was iteratively decoded, as described in Section 6.2. The proposed video codec imposed the code constraints of Section 6.3.3, specifying which particular VQ tiles and which RVLC codewords may be employed at which stages of the video coding process. Since the set of valid RVLC codewords varies throughout the video coding process, this chapter therefore exemplifies the application of IrVLC schemes in the context of joint source and channel coding.

In the proposed video codec, the complete set of VDVQ/RVLC-induced code constraints was described by the novel VDVQ/RVLC trellis structure of Section 6.3.4. Hence, the employment of the VDVQ/RVLC trellis structure allows the consideration of all legitimate transmission frame permutations. This fact was exploited in the transmitter of the VDVQ/RVLC-TCM scheme in order to perform novel MMSE VDVQ/RVLC encoding, as described in Section 6.4. Additionally, the employment of the VDVQ/RVLC trellis structure during VDVQ/RVLC decoding was shown in Section 6.5 to guarantee the recovery of legitimate – although not necessarily error-free – video information. This ensured that useful video information was never discarded, unlike in conventional video decoders, where a single transmission error may render an entire transmission frame invalid. A novel modification of the BCJR algorithm was employed during APP SISO VDVQ/RVLC decoding in order to facilitate the iterative exchange of soft information with the serially concatenated APP SISO TCM decoder and to perform the soft MMSE reconstruction of the video sequence. Since the VDVQ/RVLC trellis structure describes the complete set of VDVQ/RVLC-induced code constraints, all of the associated redundancy was beneficially exploited with the aid of the modified BCJR algorithm.

In Figure 6.11 the proposed VDVQ/RVLC-TCM scheme was shown to support iterative decoding convergence to an infinitesimally low video degradation for channel SNRs in excess of 8.25 dB, provided that the iterative decoding trajectory approaches the inner and outer EXIT functions sufficiently closely. Note that this is just 1.29 dB from the 16QAM-based Rayleigh fading channel's SNR capacity bound of 6.96 dB, which corresponds to an effective throughput of 2.00 bits per channel use. It was noted however that iterative decoding convergence to an infinitesimally low video degradation could be supported at channel SNRs that are closer to the SNR capacity bound, if the APP SISO VDVQ/RVLC decoder's inverted EXIT function shown in Figure 6.11 was better matched to the EXIT function of the APP SISO TCM decoder. Hence, we design IrVLCs with the aid of EXIT-chart matching in Chapters 7–9 using a procedure that was briefly introduced in Section 5.5.3 in the context of IRCCs.

Using an interleaver length of 74 250 bits, it was shown in Figure 6.12 that the 'Lab' video sequence encoded at a bit rate of 14.85 kb/s was reconstructed with a PSNR of at least 29.5 dB for channel SNRs in excess of 8.75 dB, which is 1.79 dB from the channel's capacity bound. It was also noted that if an even longer interleaver were employed, then the proposed scheme would support operation at channel SNRs that are even closer to the threshold at which an open EXIT chart tunnel can be obtained. This is because a longer interleaver would facilitate the mitigation of more correlation within the iteratively exchanged extrinsic information, resulting a better match between the iterative decoding trajectory and the inner and outer EXIT functions of Figure 6.11. Hence, the employment of longer interleavers is investigated in Chapters 7–9.

In Figure 6.12, the joint source and channel coding approach of the proposed VDVQ/RVLC-TCM scheme was shown to consistently outperform two powerful benchmarks, both employing the Shannonian source and channel coding separation philosophy. However, this performance improvement was found to accrue at the cost of a 1.6 times higher computational complexity. This was attributed to the APP SISO VDVQ/RVLC decoder's employment of the VDVQ/RVLC trellis structure of Section 6.3.4, which resembles a symbol-based VLC trellis, as described in Chapter 3. By contrast, the benchmarks employ bit-based VLC trellises in the corresponding role, having a lower complexity, as discussed in Chapter 3. In the light of this, we characterize and compare the complexity associated with employing bit- and symbol-based trellises for APP SISO VLC decoding in Chapter 7.

Chapter 7

Irregular Variable-Length Codes for EXIT-Chart Matching[1]

7.1 Introduction

As demonstrated in Section 6.6, a serially concatenated [275] transmission scheme is capable of achieving iterative decoding [278] convergence to an infinitesimally low probability of error at near-capacity Signal-to-Noise Ratios (SNRs), if the EXtrinsic Information Transfer (EXIT) functions of the inner and outer codecs are well matched. This motivated the design of Irregular Convolutional Code (IrCC) schemes in [149], as described in Section 5.5.

The inverted EXIT function of an outer IrCC channel codec can be specifically shaped in order to match the EXIT function of a serially concatenated inner codec. This is possible because IrCCs amalgamate a number of component Convolutional Codes (CC) [140] having different coding rates, each of which is employed to generate a specific fraction of the IrCC-encoded bit stream. As described in Section 5.5, the composite inverted IrCC EXIT function is given as a weighted average of the inverted EXIT functions of the individual component CCs, where each weight is given by the particular fraction of the IrCC-encoded bit stream that is generated by the corresponding component CC. Hence, it is the specific selection of these fractions that facilitates the shaping of the inverted composite IrCC EXIT function. Using the EXIT-chart matching algorithm of [149], the inverted IrCC EXIT chart may be matched to the EXIT function of the inner codec in this way. This facilitates the creation of an open EXIT-chart tunnel [267] at low channel SNRs that approach the channel's capacity bound.

However, the constituent bit-based CCs [140] of the IrCC codec of [149] are unable to exploit the unequal source symbol occurrence probabilities that are typically associated with audio, speech, image and video sources [151, 152]. Note that unequal source symbol occurrence probabilities were exemplified in Section 6.3.2. Since the exploitation of all available redundancy is required for near-capacity operation [133], the Huffman source encoder [155] of Chapter 2 must be employed to remove this source redundancy before IrCC encoding commences. However, the reconstruction of the Huffman-encoded bits with

---

[1]Part of this chapter is based on [202] © IEEE (2009).

Near-Capacity Variable-Length Coding   Lajos Hanzo, Robert G. Maunder, Jin Wang and Lie-Liang Yang
© 2011 John Wiley & Sons, Ltd

a particularly low Bit-Error Ratio (BER) is required in order that Huffman decoding [155] can achieve a low Symbol-Error Ratio (SER), owing to its high error sensitivity, which often leads to loss of synchronization.

This motivates the application of the Variable-Length Error-Correction (VLEC) code [179] and Reversible Variable-Length Code (RVLC) [276] classes of Variable-Length Codes (VLCs) as an alternative to the concatenated Huffman and CC coding of sequences of source symbols having values with unequal probabilities of occurrence. Unlike CCs, these joint source and channel coding VLC schemes are capable of exploiting unequal source symbol occurrence probabilities, as described in Chapter 3. More specifically, source symbols having indices of $k \in [1 \ldots K]$ and associated with unequal probabilities of occurrence $\{P(k)\}_{k=1}^{K}$ are mapped to binary codewords of varying lengths $\{I^k\}_{k=1}^{K}$ from a K-entry codebook **VLC** during VLC encoding. Typically, the more frequently a particular source symbol value occurs, the shorter its VLC codeword, resulting in a reduced average codeword length of

$$L(\mathbf{VLC}) = \sum_{k=1}^{K} P(k) \cdot I^k. \tag{7.1}$$

In order that each valid VLC codeword sequence may be uniquely decoded, a lower bound equal to the source entropy of

$$E = - \sum_{k=1}^{K} P(k) \cdot \log_2(P(k)) \tag{7.2}$$

is imposed upon the average codeword length $L(\mathbf{VLC})$. Any discrepancy between $L(\mathbf{VLC})$ and E is quantified by the coding rate of

$$R(\mathbf{VLC}) = \frac{E}{L(\mathbf{VLC})} \tag{7.3}$$

and may be attributed to the intentional introduction of redundancy into the VLEC or RVLC codewords. Naturally, this intentionally introduced redundancy imposes code constraints that limit the set of legitimate sequences of VLC-encoded bits. Like the code constraints of CCs [140], the VLC code constraints may be exploited in order to provide an error-correcting capability during VLC decoding [179]. Note that the lower the VLC coding rate, the higher the associated potential error correction capability, as described in Chapter 3. Furthermore, *unlike* in CC decoding, any redundancy owing to the unequal occurrence probabilities of the source symbol values may also be exploited during VLC decoding [179].

Depending on the coding rate $R(\mathbf{VLC})$ of the VLECs or RVLCs, the associated code constraints render their decoding substantially less sensitive to bit errors than Huffman decoding is, as described in Chapter 3. Hence, a coding gain of 1 dB at a SER of 10^{-5} has been observed by employing VLEC coding having a particular coding rate instead of a concatenated Huffman and Bose–Chaudhuri–Hocquenghem (BCH) [302, 303] coding scheme having the same coding rate [179].

Hence the application of EXIT-chart matching invoking Irregular Variable-Length Codes (IrVLCs) is motivated for the sake of near-capacity joint source and channel coding of source

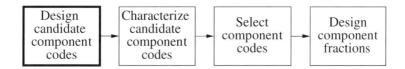

Figure 7.1: Conventional irregular coding design process. This chapter presents modifications to the aspects of this process that are indicated using a bold box.

symbol sequences having values exhibiting unequal occurrence probabilities. In this chapter we therefore employ a novel IrVLC scheme as our outer source codec, which we serially concatenate [275, 278] with an inner channel codec for the sake of exchanging extrinsic information. As shown in Figure 7.1, instead of the component CCs employed in IrCC schemes, the proposed IrVLC scheme employs component VLC codebooks. These have different coding rates and are used for encoding appropriately selected fractions of the input source symbol stream. In this way the resultant composite inverted EXIT function may be shaped in order to ensure that it does not cross the EXIT function of the inner channel codec.

Note that the proposed scheme has an Unequal Error Protection (UEP) capability [304], since different fractions of the input source symbol stream are protected by different VLC codebooks having different coding rates and, hence, different error-correction capabilities. In a manner similar to that of [189–191], for example, this UEP capability may be employed to appropriately protect audio-, speech-, image- and video-encoded bit sequences, which are typically generated using diverse encoding techniques and exhibit various error sensitivities. For example, video coding typically achieves compression by employing Motion Compensation (MC) [154] to exploit the characteristic inter-frame redundancy of video information and the Discrete Cosine Transform (DCT) [153] to exploit the intra-frame redundancy, as described in Section 6.1. As noted in [151], typically a higher degree of video reconstruction distortion results from transmission errors that affect the MC-generated motion vectors than from those inflicted on the DCT-encoded information. Hence, the proposed scheme's UEP capability may be employed to protect the MC-encoded information with a relatively strong error-correction capability, whilst employing a relatively weak error-correction code to protect the DCT-encoded information. This approach may hence be expected to yield a lower degree of video reconstruction distortion than would equal protection, as noted in [189–191], for example.

The rest of this chapter is outlined as follows. In Section 7.2 we propose iteratively decoded schemes, in which we opt for serially concatenating IrVLC with Trellis-Coded Modulation (TCM) [277]. Furthermore, Section 7.2 additionally introduces our benchmark schemes, where IrVLC is replaced by regular VLCs having the same coding rate. The design and EXIT-chart-aided characterization of these schemes is detailed in Section 7.3. In Section 7.4 we quantify the attainable performance improvements offered by the proposed IrVLC arrangements compared with that of the regular VLC benchmark schemes. Furthermore, in Section 7.4 we additionally consider a Huffman coding and IrCC-based benchmark. Section 7.4 also employs a novel method of quantifying the computational complexity required for the schemes considered in order to achieve different source sample reconstruction qualities at a range of Rayleigh fading channel SNRs. This method is employed to select our preferred scheme and to characterize the benefits of UEP. Finally, we offer our conclusions in Section 7.5.

7.2   Overview of Proposed Schemes

In this section we provide an overview of a number of serially concatenated [275] and iteratively decoded [278] joint source and channel coding schemes. Whilst the novel schemes introduced in this chapter may be tailored for operating in conjunction with any inner channel codec, we opt for employing TCM [277] in each of our considered schemes. This provides error protection without any bandwidth expansion or effective bit-rate reduction by accommodating the additional redundancy by transmitting more bits per channel symbol. The choice of TCM is further justified, since *A Posteriori* Probability (APP) TCM Soft-Input Soft-Output (SISO) decoding, similarly to APP SISO IrVLC decoding, operates on the basis of Add, Compare and Select (ACS) operations within a trellis structure. Hence, the APP SISO IrVLC and TCM decoders can share resources in systolic-array-based chips, facilitating a cost-effective implementation. Furthermore, we will show that TCM exhibits attractive EXIT characteristics in the proposed IrVLC context even without requiring Turbo Trellis-Coded Modulation (TTCM)- or Bit-Interleaved Coded Modulation (BICM)-style internal iterative decoding [296].

Our considered schemes differ in their choice of the outer source codec. Specifically, we consider a novel IrVLC codec and an equivalent regular VLC-based benchmark in this role. In both cases we employ both Symbol-Based (SB) [181] and Bit-Based (BB) [180] VLC decoding, resulting in a total of four different configurations. We refer to these four schemes as the SBIrVLC-, BBIrVLC-, SBVLC- and BBVLC-TCM arrangements, as appropriate. A schematic that is common to each of these four considered schemes is provided in Figure 7.2.

7.2.1   Joint Source and Channel Coding

The schemes considered are designed for facilitating the near-capacity detection of source samples received over an uncorrelated narrowband Rayleigh fading channel. We consider the case of independent identically distributed (i.i.d.) source samples, which may represent the prediction residual error that remains following the predictive coding of audio, speech, image or video information [151, 152], for example. Note that this was exemplified in the novel video codec of Chapter 6, in which frame differencing was employed, as depicted in Figure 6.1. A Gaussian source sample distribution is assumed here, since this has widespread applications owing to the wide applicability of the central limit theorem [305]. Additionally, a zero mean and unity source sample variance was assumed, resulting in the Probability Density Function (PDF) shown in Figure 7.3. Note however that, with the aid of suitable adaptation, the techniques proposed in this chapter may be just as readily applied to arbitrary source sample distributions.

In the block Q of the transmitter depicted in Figure 7.2, each real-valued source sample of the source sample frame e is quantized [164, 165] to one of the $K = 16$ quantization levels $\{\hat{e}^k\}_{k=1}^K$ provided in Figure 7.3. In each case, the selected quantization level is that which represents the source sample with the minimum squared error. Figure 7.3 provides decision boundaries, which are located halfway between each adjacent pair of quantization levels. Each pair of adjacent decision boundaries specifies the range of source sample values that

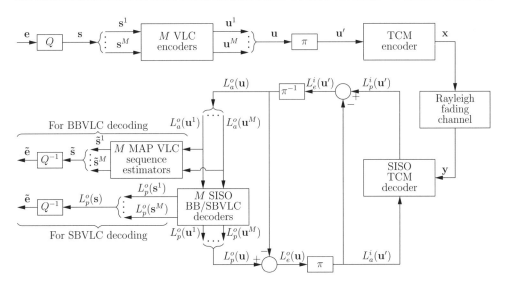

Figure 7.2: Schematic of the SBIrVLC-, BBIrVLC-, SBVLC- and BBVLC-TCM schemes. In the IrVLC schemes, the M VLC encoders, APP SISO decoders and MAP sequence estimators are each based upon one of N component VLC codebooks. By contrast, in the VLC benchmarks, all of the M VLC encoders, decoders and sequence estimators are based upon the same VLC codebook.©IEEE [196] Maunder *et al.* 2008.

are quantized to the quantization level at the center of gravity of this interval, resulting in the minimum squared error. Following quantization, each source sample in the source sample frame e is represented by a symbol in the source symbol frame s that represents the index of the selected quantization level $\hat{e}^k$ and has a value of $k \in [1 \dots K]$.

Owing to the lossy nature of quantization, distortion is imposed upon the reconstructed source sample frame ê that is obtained following inverse quantization in the block, as described in Section 3.2. Note that the set of quantization levels depicted in Figure 7.3 represents those of Lloyd–Max quantization [164, 165]. This employs the K-means algorithm [233] to search for the set of quantization levels that minimizes the expected quantization distortion. In the case of the quantization levels seen in Figure 7.3, the expected Signal-to-Quantization-Noise Ratio (SQNR) is about 20 dB. Note however that, again, with the aid of suitable adaptation, the techniques advocated in this chapter may be just as readily applied to arbitrary quantizers.

Also note that Lloyd–Max quantization results in a large variation in the occurrence probabilities of the resultant source symbol values. These occurrence probabilities are given by integrating the source PDF between each pair of adjacent decision boundaries, resulting in the values provided in Figure 7.3. These source symbol values' occurrence probabilities $\{P(k)\}_{k=1}^K$ are repeated in Table 7.1 and can be seen to vary by more than an order of magnitude. These probabilities correspond to the varying source symbol informations $\{-\log_2(P(k))\}_{k=1}^K$ provided in Table 7.1, motivating the application of VLC and giving a source entropy of $E = 3.77$ bits per source symbol, according to Equation (7.2).

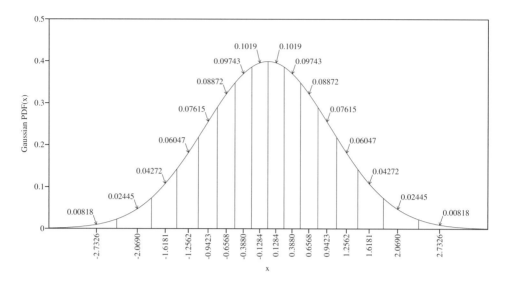

Figure 7.3: Gaussian PDF for unity mean and variance. The x-axis is labeled with the $K = 16$ Lloyd–Max quantization levels $\{\hat{e}^k\}_{k=1}^K$ as provided in [164]. The decision boundaries are employed to decompose the Gaussian PDF into $K = 16$ sections. The integral of the PDF between each pair of adjacent decision boundaries is provided.

In the transmitter of the proposed scheme, the Lloyd–Max quantized source symbol frame s is decomposed into $M = 300$ sub-frames $\{\mathbf{s}^m\}_{m=1}^M$, as shown in Figure 7.2. In the case of the SBIrVLC- and SBVLC-TCM schemes, this decomposition is necessary for the sake of limiting the computational complexity of VLC decoding, since the number of transitions in the symbol-based VLC trellis is inversely proportional to the number of sub-frames in this case [181], as described in Section 3.4.1. We opt for employing the same decomposition of the source symbol frames into sub-frames in the case of the BBIrVLC- and BBVLC-TCM schemes for the sake of ensuring that we make a fair comparison with the SBIrVLC- and SBVLC-TCM schemes. This is justified, since the decomposition considered benefits the performance of the BBIrVLC- and BBVLC-TCM schemes, as will be detailed below. Each source symbol sub-frame $\mathbf{s}^m$ comprises $J = 100$ source symbols. Hence, the total number of source symbols in a source symbol frame becomes $M \cdot J = 30\,000$. As described above, each Lloyd–Max quantized source symbol in the sub-frame $\mathbf{s}^m$ has a K-ary value $s_j^m \in [1 \ldots K]$, where we have $j \in [1 \ldots J]$.

As described in Section 7.1, we employ N component VLC codebooks to encode the source symbols, where we opted for $N = 15$ for the SBIrVLC and BBIrVLC schemes and $N = 1$ for the regular SBVLC and BBVLC schemes. Each Lloyd–Max quantized source symbol sub-frame $\mathbf{s}^m$ is VLC-encoded using a single component VLC codebook $\mathbf{VLC}^n$, where we have $n \in [1 \ldots N]$. In the case of the SBIrVLC and BBIrVLC schemes, the particular fraction C^n of the set of source symbol sub-frames that is encoded by the specific component VLC codebook $\mathbf{VLC}^n$ is fixed and will be derived in Section 7.3. The specific Lloyd–Max quantized source symbols having the value of $k \in [1 \ldots K]$ and encoded by the

Table 7.1: The probabilities of occurrence $P(k)$ and informations $-\log_2(P(k))$ of the $K = 16$ source symbol values $k \in [1 \ldots K]$ that result from the Lloyd–Max quantization of Gaussian distributed source samples (the corresponding source symbol entropy is $E = 3.77$ bits per source symbol, according to Equation (7.2))

k	$P(k)$	$-\log_2(P(k))$	I^k	$\mathbf{Huff}^{k\dagger}$
1	0.0082	6.93	7	0000000
2	0.0244	5.35	6	000001
3	0.0427	4.55	5	00010
4	0.0605	4.05	4	0010
5	0.0762	3.72	4	0100
6	0.0887	3.49	4	0110
7	0.0974	3.36	3	101
8	0.1019	3.29	3	110
9	0.1019	3.29	3	111
10	0.0974	3.36	3	100
11	0.0887	3.49	4	0111
12	0.0762	3.72	4	0101
13	0.0605	4.05	4	0011
14	0.0427	4.55	5	00011
15	0.0244	5.35	5	00001
16	0.0082	6.93	7	0000001

† Also provided is the composition of the $K = 16$ codewords in the corresponding Huffman codebook $\mathbf{Huff} = \{\mathbf{Huff}^k\}_{k=1}^{K}$ [155], having the codeword lengths $\{I^k\}_{k=1}^{K}$. According to Equation (7.1), the average Huffman codeword length is $L(\mathbf{Huff}) = 3.81$ bits per source symbol, which corresponds to a Huffman coding rate of $R_{\mathrm{Huff}} = 0.99$, according to Equation (7.3).

specific component VLC codebook $\mathbf{VLC}^n$ are represented by the codeword $\mathbf{VLC}^{n,k}$, which has a length of $I^{n,k}$ bits. The $J = 100$ VLC codewords that represent the $J = 100$ Lloyd–Max quantized source symbols in each source symbol sub-frame $\mathbf{s}^m$ are concatenated to provide the transmission sub-frame

$$\mathbf{u}^m = \{\mathbf{VLC}^{n,s_j^m}\}_{j=1}^{J}.$$

Owing to the variable lengths of the VLC codewords, each of the $M = 300$ transmission sub-frames typically consists of a different number of bits. In order to facilitate the VLC decoding of each transmission sub-frame $\mathbf{u}^m$, it is necessary to explicitly convey its length

$$I^m = \sum_{j=1}^{J} I^{n,s_j^m}$$

to the receiver. Furthermore, this highly error-sensitive side information must be reliably protected against transmission errors. This may be achieved using a low rate block code, for example. For the sake of avoiding obfuscation, this is not shown in Figure 7.2. Note that the choice of the specific number of sub-frames M in each frame constitutes a trade-off between the computational complexity of SBVLC decoding or the performance of BBVLC decoding and the amount of side information that must be conveyed. In Section 7.3 we comment on the amount of side information that is required for reliably conveying the specific number of bits in each transmission sub-frame to the decoder.

In the scheme's transmitter, the $M = 300$ number of transmission sub-frames $\{\mathbf{u}^m\}_{m=1}^{M}$ are concatenated. As shown in Figure 7.2, the resultant transmission frame $\mathbf{u}$ has a length of $\sum_{m=1}^{M} I^m$ bits.

In the proposed scheme, the VLC codec is protected by a serially concatenated TCM codec. Following VLC encoding, the bits of the transmission frame $\mathbf{u}$ are interleaved using the function π in order to provide the interleaved transmission frame $\mathbf{u}'$, which is TCM encoded in order to obtain the channel's input symbols $\mathbf{x}$, as shown in Figure 7.2. These are transmitted over an uncorrelated narrowband Rayleigh fading channel and are received as the channel's output symbols $\mathbf{y}$, as seen in Figure 7.2.

7.2.2 Iterative Decoding

In the receiver, APP SISO TCM and VLC decoding are performed iteratively, as shown in Figure 7.2. Both of these decoders invoke the Bahl–Cocke–Jelinek–Raviv (BCJR) algorithm [48] on the basis of their trellises. Symbol-based trellises are employed in the case of TCM [277], SBIrVLC and SBVLC [181] decoding, whilst BBIrVLC and BBVLC decoding rely on bit-based trellises [180]. All BCJR calculations are performed in the logarithmic probability domain and using an eight-entry lookup table for correcting the Jacobian approximation in the Log-MAP algorithm [296]. The proposed approach requires only Add, Compare and Select (ACS) computational operations during iterative decoding, which will be used as our complexity measure, since it is characteristic of the complexity/area/speed trade-offs in systolic-array-based chips.

As usual, extrinsic soft information, represented in the form of Logarithmic Likelihood Ratios (LLRs) [292], is iteratively exchanged between the TCM and VLC decoding stages for the sake of assisting each other's operation [275, 278], as described in Section 4.1. In Figure 7.2, $L(\cdot)$ denotes the LLRs of the bits concerned (or the log-APPs of the specific symbols as appropriate), where the superscript i indicates inner TCM decoding, while o corresponds to outer VLC decoding. Additionally, a subscript denotes the dedicated role of the LLRs (or log-APPs), with a, p and e indicating *a priori*, *a posteriori* and extrinsic information respectively.

During each decoding iteration, the inner TCM decoder is provided with *a priori* LLRs pertaining to the interleaved transmission frame $L_a^i(\mathbf{u}')$, as shown in Figure 7.2. These LLRs are obtained from the most recent operation of the outer VLC decoding stage, as will be highlighted below. In the case of the first decoding iteration, no previous VLC decoding has been performed and hence the *a priori* LLRs $L_a^i(\mathbf{u}')$ provided for TCM decoding are all zero-valued, corresponding to a probability of 0.5 for both '0' and '1'. Given the channel's output symbols $\mathbf{y}$ and the *a priori* LLRs $L_a^i(\mathbf{u}')$, the BCJR algorithm is employed for obtaining the *a posteriori* LLRs $L_p^i(\mathbf{u}')$, as shown in Figure 7.2.

During iterative decoding, it is necessary to prevent the reuse of already exploited information, since this would limit the attainable iteration gain [296], as described in Section 4.2. This is achieved following TCM decoding by the subtraction of $L_a^i(\mathbf{u}')$ from $L_p^i(\mathbf{u}')$, as shown in Figure 7.2. The resultant extrinsic LLRs $L_e^i(\mathbf{u}')$ are de-interleaved in the block π^{-1} and forwarded as *a priori* LLRs for VLC decoding. As described in Section 4.1, interleaving is employed in order to mitigate correlation within the *a priori* LLR frames. This is necessary since the BCJR algorithm assumes that all *a priori* LLRs that can influence any particular decoding decision are uncorrelated.

Just as $M = 300$ separate VLC encoding processes are employed in the proposed scheme's transmitter, $M = 300$ separate VLC decoding processes are employed in its receiver. In parallel to the composition of the bit-based transmission frame $\mathbf{u}$ from its $M = 300$ sub-frames, the *a priori* LLRs $L_a^o(\mathbf{u})$ are decomposed into $M = 300$ sub-frames, as shown in Figure 7.2. This is achieved with the aid of the explicit side information that conveys the number of bits I^m in each transmission sub-frame $\mathbf{u}^m$. Each of the $M = 300$ VLC decoding processes is provided with the *a priori* LLR sub-frame $L_a^o(\mathbf{u}^m)$ and in response it generates the *a posteriori* LLR sub-frame $L_p^o(\mathbf{u}^m)$, $m \in [1 \ldots M]$. These *a posteriori* LLR sub-frames are concatenated in order to provide the *a posteriori* LLR frame $L_p^o(\mathbf{u})$, as shown in Figure 7.2. Following the subtraction of the *a priori* LLRs $L_a^o(\mathbf{u})$, the resultant extrinsic LLRs $L_e^o(\mathbf{u})$ are interleaved and forwarded as *a priori* information to the next TCM decoding iteration.

In the case of SBIrVLC and SBVLC decoding, each of the $M = 300$ VLC decoding processes additionally provides log-APPs pertaining to the corresponding source symbol sub-frame $L_p^o(\mathbf{s}^m)$. This comprises a set of K number of log-APPs for each source symbol s_j^m in the sub-frame $\mathbf{s}^m$, where $j \in [1 \ldots J]$. Each of these log-APPs provides the logarithmic probability that the corresponding source symbol s_j^m has the particular value $k \in [1 \ldots K]$. In the receiver of Figure 7.2, the source symbols' log-APP sub-frames are concatenated to provide the source symbol log-APP frame $L_p^o(\mathbf{s})$. By inverse-quantizing this soft information in the block Q^{-1}, we may obtain a frame of Minimum Mean Square Error (MMSE) source sample estimates $\tilde{\mathbf{e}}$, which approximates the reconstructed source sample frame $\hat{\mathbf{e}}$ described in Section 7.2.1. More specifically, each source sample estimate is obtained by using the corresponding set of K source symbol value probabilities to find the weighted average of the K number of quantization levels $\{\hat{e}^k\}_{k=1}^K$.

Conversely, in the case of BBIrVLC and BBVLC decoding, no symbol-based *a posteriori* output is available. In this case, each source symbol sub-frame $\mathbf{s}^m$ is estimated from the corresponding *a priori* LLR sub-frame $L_a^o(\mathbf{u}^m)$. This may be achieved by employing Maximum A posteriori Probability (MAP) sequence estimation operating on a bit-based trellis structure, as shown in Figure 7.2. Unlike APP SISO SBIrVLC and SBVLC decoding, bit-based MAP sequence estimation cannot exploit the knowledge that each sub-frame $\mathbf{s}^m$ comprises $J = 100$ source symbols. For this reason, the resultant hard decision estimate $\tilde{\mathbf{s}}^m$ of each source symbol sub-frame $\mathbf{s}^m$ may or may not contain $J = 100$ source symbols.

In order that we may prevent the loss of synchronization that this would imply, source symbol estimates are removed from, or appended to the end of, each source symbol sub-frame estimate $\tilde{s}^m$ in order to ensure that they each comprise exactly $J = 100$ source symbol estimates. Note that it is the decomposition of the source symbol frame s into sub-frames that provides this opportunity to mitigate the loss of synchronization that is associated with bit-based MAP VLC sequence estimation. Hence the decomposition of the source symbol frame s into sub-frames benefits the performance of the BBIrVLC- and BBVLC-TCM schemes, as mentioned above.

Following MAP sequence estimation, the adjusted source symbol sub-frame estimates $\tilde{s}^m$ are concatenated for the sake of obtaining the source symbol frame estimate $\tilde{s}$. This may be inverse-quantized in order to obtain the source sample frame estimate $\tilde{e}$. Note that for the reconstruction of a source sample frame estimate $\tilde{e}$ from a given *a priori* LLR frame $L_a^o(\mathbf{u})$, a higher level of source distortion may be expected in the BBIrVLC- and BBVLC-TCM schemes than in the corresponding SBIrVLC- and SBVLC-TCM schemes. This is due to the BBIrVLC- and BBVLC-TCM schemes' reliance on hard decisions as opposed to the soft decisions of the SBIrVLC- and SBVLC-TCM schemes. However, this reduced performance substantially benefits us in terms of a reduced complexity, since the bit-based VLC decoding trellis employed during APP SISO BBIrVLC and BBVLC decoding and MAP sequence estimation contains significantly fewer transitions than the symbol-based VLC decoding trellis of APP SISO SBIrVLC and SBVLC decoding, as described in Section 3.4.

In the next section we detail the design of our IrVLC scheme and characterize each of the SBIrVLC-, BBIrVLC-, SBVLC- and BBVLC-TCM schemes with the aid of EXIT-chart analysis.

7.3   Parameter Design for the Proposed Schemes

7.3.1   Scheme Hypothesis and Parameters

As described in Section 7.1, the SBIrVLC and BBIrVLC schemes may be constructed by employing a number of component VLC codebooks having different coding rates, each of which encodes an appropriately chosen fraction of the input source symbols. We opted for using $N = 15$ component VLC codebooks $\{\mathbf{VLC}^n\}_{n=1}^{15}$ that were specifically designed for encoding $K = 16$-level Lloyd–Max quantized Gaussian i.i.d. source samples. As shown in Figure 7.1, these $N = 15$ component VLC codebooks were selected from a large number of candidates using a significant amount of 'trial-and-error'-based human interaction in order to provide a suite of 'similarly spaced' EXIT functions. More specifically, the $N = 15$ component VLC codebooks comprised 13 different Variable-Length Error-Correcting (VLEC) designs having various so-called minimum block, convergence and divergence distances as defined in Section 3.3, complemented by a Symmetric Reversible Variable-Length Coding (SRVLC) and an Asymmetric Reversible Variable-Length Coding (ARVLC) design. These codebooks were designed using Algorithms C and E of Section 3.3.

As described in Section 3.3, the free distance lower bound of a VLC codebook $\mathbf{VLC}^n$ can be calculated as

$$\bar{d}_{\text{free}}(\mathbf{VLC}^n) = \min(d_{b_{\min}}(\mathbf{VLC}^n), d_{d_{\min}}(\mathbf{VLC}^n) + d_{c_{\min}}(\mathbf{VLC}^n)),$$

where $d_{b_{\min}}(\mathbf{VLC}^n)$ is defined as the minimum block distance between any pair of equal-length codewords in the VLC codebook $\mathbf{VLC}^n$, whilst $d_{d_{\min}}(\mathbf{VLC}^n)$ and $d_{c_{\min}}(\mathbf{VLC}^n)$ are the minimum divergence and convergence distances between any pair of unequal-length codewords, respectively. In all codebooks, a free distance lower bound of $\bar{d}_{\text{free}}(\mathbf{VLC}^n) \geq 2$ was employed, since this supports iterative decoding convergence to an infinitesimally low probability of error [298], as described in Section 4.3. The resultant average VLC codeword lengths were found to range from 3.94 to 12.18 bits/symbol, according to Equation (7.1). When compared with the source symbol entropy of $E = 3.77$ bits per source symbol, these correspond to coding rates spanning the range of 0.31 to 0.96, according to Equation (7.3). The properties and composition of the $N = 15$ component VLC codebooks $\{\mathbf{VLC}^n\}_{n=1}^{15}$ are summarized in Table 7.2.

As detailed below, our SBIrVLC and BBIrVLC schemes were designed under the constraint that they have an overall coding rate of $R_{\text{IrVLC}} = 0.52$. This value was chosen because it is the coding rate of the VLC codebook $\mathbf{VLC}^{10}$, which we employ in our SBVLC and BBVLC benchmarks using $N = 1$ codebook. This coding rate results in an average inter-leaver length of $M \cdot J \cdot E / R_{\text{IrVLC}} = 217\,500$ bits for all the schemes considered. Note that this interleaver length is nearly three times longer than any of those considered in Chapter 6.

Each of the schemes considered employs the same TCM codec, having the Linear Feedback Shift Register (LFSR) schematic of Figure 6.9. As shown in Figure 6.9, the TCM encoder generates a set of four bits to represent each set of three input bits, giving a coding rate of $R_{\text{TCM}} = 3/4$. Three of the four output bits are systematic replications of the three input bits, whilst the fourth output bit is generated with the aid of the $L_{\text{TCM}} = 6$ modulo-2 memory elements. Note that the TCM codec is a recursive component having an infinite impulse response, since feedback is employed in the shift register of Figure 6.9. As a result, the TCM codec supports iterative decoding convergence to an infinitesimally low probability of error [299], as is the case for our component VLC codebooks, as described above. Hence, we may expect the proposed scheme to achieve iterative decoding convergence to an infinitesimally low probability of error, provided that the channel quality is sufficiently high to create an open EXIT-chart tunnel and the iterative decoding trajectory approaches the inner and outer codecs' EXIT functions sufficiently closely, as discussed in Section 4.3. Furthermore, Figure 6.10 provides the constellation diagram for the $M_{\text{TCM}} = 16$-ary set-partitioned [277] QAM scheme of the TCM codec. This was employed together with In-phase Quadrature-phase (IQ)-interleaving [293] for transmission over an uncorrelated narrowband Rayleigh fading channel.

Ignoring the modest bit-rate contribution of conveying the side information, the effective throughput of the schemes considered is $\eta = R_{\text{IrVLC}} \cdot R_{\text{TCM}} \cdot \log_2(M_{\text{TCM}}) = 1.56$ bits per channel use. This implies that iterative decoding convergence to an infinitesimally low probability of error cannot be achieved when channel capacities of less than 1.56 bits per channel use [133] are attained at low E_c/N_0 values, where E_c is the transmit energy per Rayleigh fading channel use and N_0 is the average noise energy. Note that the uncorrelated narrowband Rayleigh fading channel's capacity for 16QAM is less than 1.56 bits per channel use for E_b/N_0 values below 2.6 dB [300], where $E_b = E_c/\eta$ is the transmit energy per bit of source entropy. Given this point on the corresponding channel capacity versus E_b/N_0 function, we will be able to quantify how closely the proposed schemes may approach this ultimate limit.

Table 7.2: Properties and composition of the 15 component VLC codebooks $\{\mathbf{VLC}^n\}_{n=1}^{15}$

$\mathbf{VLC}^n$	Properties[†]	Composition[‡]
$\mathbf{VLC}^1$	(VLEC,2,1,1,2,0.96)	6,6,5,5,4,4,3,3,3,3,4,4,5,5,6,6, 857E1FD3074A55133A
$\mathbf{VLC}^2$	(ARVLC,2,1,1,2,0.91)	6,6,5,5,4,4,4,3,3,4,4,4,5,5,6,6, 1EB624C9A1D58F6E4A1
$\mathbf{VLC}^3$	(SRVLC,2,1,1,2,0.86)	7,6,6,5,5,4,4,3,3,4,4,5,5,6,6,7, 7D9C248FCAAC0EDBC641
$\mathbf{VLC}^4$	(VLEC,3,1,1,2,0.81)	8,7,7,6,6,5,4,2,3,3,4,5,6,7,7,8, 81F6F9E86322ACEDE0E77E
$\mathbf{VLC}^5$	(VLEC,4,1,1,2,0.75)	8,8,7,6,6,5,4,2,3,4,5,6,7,7,8,8, 36EF61EB5BA44D179F5D7E81
$\mathbf{VLC}^6$	(VLEC,2,2,1,2,0.70)	8,7,7,6,6,6,4,4,4,4,6,6,6,7,7,8, E6C99FCADB9035628FF0E2EA
$\mathbf{VLC}^7$	(VLEC,3,2,1,3,0.64)	8,8,7,7,6,6,6,4,5,5,6,6,7,7,7,9, 7FDE5CD3E65403625A267AAD7C
$\mathbf{VLC}^8$	(VLEC,3,2,2,3,0.60)	9,8,8,7,7,6,6,4,6,6,6,6,7,8,8,9, 696F594FCBA5A03159B3F8B35583
$\mathbf{VLC}^9$	(VLEC,5,2,2,4,0.57)	10,10,9,8,8,7,6,4,5,5,6,7,8,9,9,10, 126307A57CE367501B2AAC9A69CF9ED
$\mathbf{VLC}^{10}$	(VLEC,4,3,2,4,0.52)	11,10,9,8,8,7,7,6,6,6,7,7,8,9,9,11, 1673E8F0CB2DAAA401F9CC68CD55E37BF
$\mathbf{VLC}^{11}$	(VLEC,4,3,3,4,0.47)	11,11,10,9,9,8,7,6,6,7,8,8,9,10,10,12, 11FA38AB9536B72B800F4D67B3355A655663
$\mathbf{VLC}^{12}$	(VLEC,7,3,3,6,0.43)	12,12,11,10,10,9,8,6,7,7,8,9,11,11,12,13, 2F696B8EC5D38F93A5007715A363233BBA2B899
$\mathbf{VLC}^{13}$	(VLEC,5,4,3,5,0.39)	13,12,11,10,10,9,9,8,9,9,9,10,10,11,11,14, 17455A1FFED72B7CC9380079C479A5F32C9555A A4D
$\mathbf{VLC}^{14}$	(VLEC,9,4,4,8,0.35)	15,14,14,12,12,11,10,8,9,9,10,11,13,13,14,15, 18DA499F59CAB71C9B55C9C003DE1361552D2CD 7ACFB4D3B
$\mathbf{VLC}^{15}$	(VLEC,8,5,5,8,0.31)	16,15,15,13,13,12,12,10,10,11,12,12,14,14,15,16, 31D97570AE9A5A9C6A59664D4003FE87CE53537 C671CE53464F3A

[†] The properties of each component VLC codebook $\mathbf{VLC}^n$ are provided using the format (Type, $d_{b_{\min}}(\mathbf{VLC}^n)$, $d_{d_{\min}}(\mathbf{VLC}^n)$, $d_{c_{\min}}(\mathbf{VLC}^n)$, $\bar{d}_{\text{free}}(\mathbf{VLC}^n)$, $R(\mathbf{VLC}^n)$).

[‡] The composition of each component VLC codebook $\mathbf{VLC}^n$ is specified by providing the $K = 16$ codeword lengths $\{I^{n,k}\}_{k=1}^K$, together with the hexadecimal representation of the ordered concatenation of the $K = 16$ VLC codewords in the codebook.

Recall from Section 7.2 that it is necessary to convey the length of each transmission sub-frame $\mathbf{u}^m$ to the receiver in order to facilitate its VLC decoding. The amount of side information required may be determined by considering the range of transmission sub-frame lengths that can result from VLC encoding using each of the $N = 15$ component codebooks. When all $J = 100$ source symbols in a particular source symbol sub-frame $\mathbf{s}^m$ are represented by the codeword from the component VLC codebook $\mathbf{VLC}^n$ having the maximal length $\max_{k\in[1\ldots K]} I^{n,k}$, a maximal transmission sub-frame length of

$$I^n_{\max} = J \cdot \max_{k\in[1\ldots K]} I^{n,k}$$

results. Similarly, a minimal transmission sub-frame length of

$$I^n_{\min} = J \cdot \min_{k\in[1\ldots K]} I^{n,k}$$

results, when all source symbols are represented by the minimal-length VLC codeword. A transmission sub-frame $\mathbf{u}^m$ encoded using the component VLC codebook $\mathbf{VLC}^n$ will therefore have one of $(I^n_{\max} - I^n_{\min} + 1)$ number of lengths in the range $I^m \in [I^n_{\min} \ldots I^n_{\max}]$. Hence, the length of the transmission sub-frame I^m can be represented using a fixed-length codeword comprising $\lceil \log_2(I^n_{\max} - I^n_{\min} + 1) \rceil$ number of bits. When considering the VLC codeword lengths provided in Table 7.2, it was found for all schemes that a single 10-bit fixed-length codeword of side information is sufficient for conveying the length of each of the $M = 300$ transmission sub-frames $\mathbf{u}^m$ in each transmission frame $\mathbf{u}$. As suggested in Section 7.2, this error-sensitive side information may be protected by a low-rate block code in order to ensure its reliable transmission. Using a $R_{\rm rep} = 1/3$-rate repetition code results in a total of $10 \cdot M/R_{\rm rep} = 9\,000$ bits of side information per frame, which represents an average of just 4% of the transmitted information, when appended to the transmission frame $\mathbf{u}$, which has an average length of $M \cdot J \cdot E/R_{\rm IrVLC} = 217\,500$ bits for all of the schemes considered.

7.3.2   EXIT-Chart Analysis and Optimization

We now consider the EXIT characteristics of the various components of our various schemes. In all cases, EXIT functions were generated using uncorrelated Gaussian distributed *a priori* LLRs and all mutual information measurements were made using the histogram-based approximation of the LLR PDFs [294].

In Figures 7.4 and 7.5, we provide the EXIT functions $I^i_e(I^i_a, E_b/N_0)$ of the TCM scheme for a number of E_b/N_0 values above the channel capacity bound of 2.6 dB. Note that, owing to its recursive nature, the APP SISO TCM decoder can be seen to achieve unity extrinsic mutual information I^i_e for unity *a priori* mutual information I^i_a [299]. Additionally, the inverted EXIT functions $I^{o,n}_a(I^o_e)$ plotted for the $N = 15$ component VLC codebooks, together with their coding rates $R(\mathbf{VLC}^n)$, are given in Figure 7.4 for symbol-based APP SISO VLC decoding and in Figure 7.5 for bit-based APP SISO VLC decoding. Similarly to APP SISO TCM decoding, APP SISO VLC decoding achieves unity extrinsic mutual information I^o_e for unity *a priori* mutual information I^o_a in all cases, owing to the employment of codebooks having a free distance lower bound of $d_{\rm free} \geq 2$ [298], as discussed in Section 4.3. Note that the EXIT functions obtained for symbol- and bit-based APP SISO VLC decoding are slightly different. This is because, unlike the bit-based APP SISO VLC decoder, the symbol-based APP SISO VLC decoder is capable of exploiting the knowledge that there are $J = 100$ source symbols in each source symbol sub-frame $\mathbf{s}^m$, as described in Section 3.4.

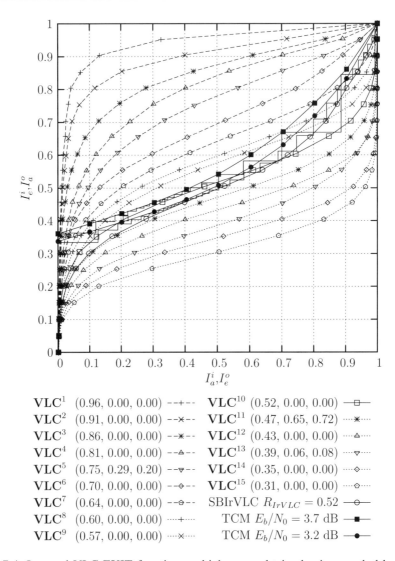

VLC^1	$(0.96, 0.00, 0.00)$ --+--	**VLC**^{10}	$(0.52, 0.00, 0.00)$ --□--	
VLC^2	$(0.91, 0.00, 0.00)$ --×--	**VLC**^{11}	$(0.47, 0.65, 0.72)$ ····*····	
VLC^3	$(0.86, 0.00, 0.00)$ --*--	**VLC**^{12}	$(0.43, 0.00, 0.00)$ ····△····	
VLC^4	$(0.81, 0.00, 0.00)$ --△--	**VLC**^{13}	$(0.39, 0.06, 0.08)$ ····▽····	
VLC^5	$(0.75, 0.29, 0.20)$ --▽--	**VLC**^{14}	$(0.35, 0.00, 0.00)$ ····◇····	
VLC^6	$(0.70, 0.00, 0.00)$ --◇--	**VLC**^{15}	$(0.31, 0.00, 0.00)$ ····○····	
VLC^7	$(0.64, 0.00, 0.00)$ --○--	SBIrVLC $R_{IrVLC} = 0.52$ --○--		
VLC^8	$(0.60, 0.00, 0.00)$ ····+····	TCM $E_b/N_0 = 3.7$ dB --■--		
VLC^9	$(0.57, 0.00, 0.00)$ ····×····	TCM $E_b/N_0 = 3.2$ dB --●--		

Figure 7.4: Inverted VLC EXIT functions, which were obtained using symbol-based APP SISO VLC decoding. The inverted EXIT function is provided for the corresponding SBIrVLC arrangement, together with TCM EXIT functions for a number of E_b/N_0 values. Decoding trajectories are provided for the SBIrVLC-TCM scheme at a channel E_b/N_0 value of 3.2 dB, as well as for the SBVLC-TCM scheme at a channel E_b/N_0 value of 3.7 dB. Inverted VLC EXIT functions are labeled using the format **VLC**^{n} $(R(\mathbf{VLC}^n), C_{SB}^n, \alpha_{SB}^n)$. ©IEEE [196] Maunder *et al.* 2008.

The inverted EXIT function of an IrVLC scheme $I_a^o(I_e^o)$ can be obtained as the appropriately weighted superposition of the $N = 15$ component VLC codebooks' EXIT

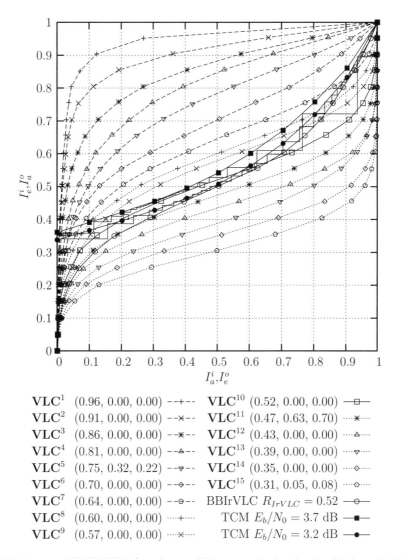

Figure 7.5: Inverted VLC EXIT functions, which were obtained using bit-based APP SISO VLC decoding. The inverted EXIT function is provided for the corresponding BBIrVLC arrangement, together with TCM EXIT functions for a number of E_b/N_0 values. Decoding trajectories are provided for the BBIrVLC-TCM scheme at a channel E_b/N_0 value of 3.2 dB, as well as for the BBVLC-TCM scheme at a channel E_b/N_0 value of 3.7 dB. Inverted VLC EXIT functions are labeled using the format $\mathbf{VLC}^n$ $(R(\mathbf{VLC}^n), C_{BB}^n, \alpha_{BB}^n)$.

functions,

$$I_a^o(I_e^o) = \sum_{n=1}^{N} \alpha^n I_a^{o,n}(I_e^o), \tag{7.4}$$

where α^n is the fraction of the transmission frame $\mathbf{u}$ that is generated by the specific component codebook $\mathbf{VLC}^n$. Note that since all of the $N = 15$ component VLC codebooks' EXIT functions achieve unity extrinsic mutual information I_e^o for unity *a priori* mutual information I_a^o, the same is true for the composite IrVLC EXIT function. Also note that the values of α^n are subject to the constraints

$$\sum_{n=1}^{N} \alpha^n = 1, \quad \alpha^n \geq 0 \,\forall\, n \in [1 \ldots N]. \tag{7.5}$$

The specific fraction of source symbol sub-frames $\mathbf{s}^m$ that should be encoded by the specific component codebook $\mathbf{VLC}^n$ in order that it generates a fraction α^n of the transmission frame $\mathbf{u}$ is given by

$$C^n = \alpha^n \cdot R(\mathbf{VLC}^n)/R_{\text{IrVLC}}, \tag{7.6}$$

where $R_{\text{IrVLC}} = 0.52$ is the desired overall coding rate. Again, the specific values of C^n are subject to the constraints

$$\sum_{n=1}^{N} C^n = \sum_{n=1}^{N} \alpha^n \cdot R(\mathbf{VLC}^n)/R_{\text{IrVLC}} = 1, \quad C^n \geq 0 \,\forall\, n \in [1 \ldots N]. \tag{7.7}$$

As described in Section 4.3, an open EXIT-chart tunnel [267] can be achieved at sufficiently high channel E_b/N_0 values, since both the VLC and the TCM APP SISO decoders support iterative decoding convergence to unity mutual information. Hence, beneficial values of $\{C^n\}_{n=1}^{N}$ may be chosen by ensuring that there is an open EXIT-chart tunnel between the inverted IrVLC EXIT function and the EXIT function of TCM at an E_b/N_0 value that is close to the channel capacity bound. This may be achieved using the iterative EXIT-chart matching process of [149] to adjust the values of $\{C^n\}_{n=1}^{N}$ under the constraints of Equations (7.5) and (7.7) for the sake of minimizing the error function

$$\{C^n\}_{n=1}^{N} = \operatorname*{argmin}_{\{C^n\}_{n=1}^{N}} \left(\int_0^1 e(I)^2 \, dI \right), \tag{7.8}$$

where

$$e(I) = I_e^i(I, E_b/N_0) - I_a^o(I) \tag{7.9}$$

is the difference between the inverted IrVLC EXIT function and the EXIT function of TCM at a particular target E_b/N_0 value. Note that in order to ensure that the design results in an open EXIT tunnel, we must impose the additional constraint of

$$e(I) > 0 \,\forall\, I \in [0, 1]. \tag{7.10}$$

Open EXIT tunnels were found to be achievable for both the SBIrVLC- and the BBIrVLC-TCM schemes at a threshold E_b/N_0 value of 3.1 dB, which is just 0.5 dB from the channel capacity bound of 2.6 dB. The inverted SBIrVLC EXIT function is shown in Figure 7.4, which is slightly different from the BBIrVLC EXIT function shown in Figure 7.5, owing to the slight differences in the EXIT functions obtained for bit- and symbol-based APP

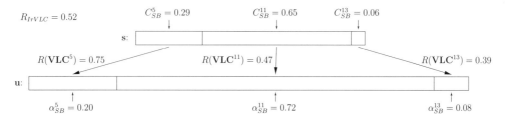

Figure 7.6: Illustration depicting the corresponding fractions of the source symbol frame **s** and the transmission frame **u** that are encoded using the three component VLC codebooks $\mathbf{VLC}^5$, $\mathbf{VLC}^{11}$ and $\mathbf{VLC}^{13}$ in the SBIrVLC-TCM scheme.

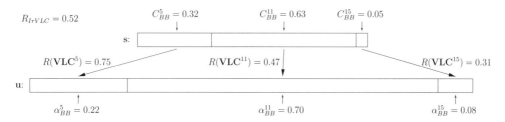

Figure 7.7: Illustration depicting the corresponding fractions of the source symbol frame **s** and the transmission frame **u** that are encoded using the three component VLC codebooks $\mathbf{VLC}^5$, $\mathbf{VLC}^{11}$ and $\mathbf{VLC}^{15}$ in the BBIrVLC-TCM scheme.

SISO decoding, as described above. The corresponding values of C^n and α^n are provided for the SBIrVLC- and the BBIrVLC-TCM schemes in Figures 7.4 and 7.5 respectively, and illustrated in Figures 7.6 and 7.7 respectively. Note that in both the SBIrVLC- and BBIrVLC-TCM schemes there are just three activated component VLC codebooks, which have corresponding values of C^n and α^n that are higher than zero.

The source symbol frame **s** and the transmission frame **u** are depicted in Figures 7.6 and 7.7. Note that, in both cases, the horizontal bar representing the source symbol frame **s** is $R_{\mathrm{IrVLC}} = 0.52$ times as long as that representing the transmission frame **u**, since an overall coding rate of $R_{\mathrm{IrVLC}} = 0.52$ is employed. Each bar is decomposed into three sections, representing the three activated component VLCs, namely $\mathbf{VLC}^5$, $\mathbf{VLC}^{11}$ and $\mathbf{VLC}^{13}$ in the case of the SBIrVLC-TCM scheme and $\mathbf{VLC}^5$, $\mathbf{VLC}^{11}$ and $\mathbf{VLC}^{15}$ in the case of the BBIrVLC-TCM scheme. The length of each section corresponds to the fraction C^n of the source symbol frame **s** or the fraction α^n of the transmission frame **u** that is coded using the associated component VLC codebook.

In the case of the SBVLC- and BBVLC-TCM benchmarks, an open EXIT-chart tunnel between the inverted EXIT function of their only component VLC codebook $\mathbf{VLC}^{10}$ and the TCM EXIT function was only found to be achieved for E_b/N_0 values above a threshold value of 3.6 dB. This E_b/N_0 value is 1.0 dB from the channel capacity bound of 2.6 dB, a discrepancy that is twice that of the SBIrVLC- and BBIrVLC-TCM schemes' 0.5 dB value. We can therefore expect our SBIrVLC- and BBIrVLC-TCM schemes to be capable

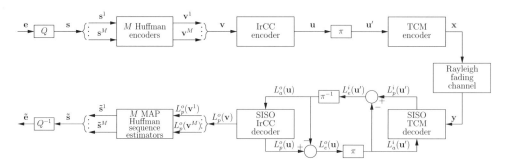

Figure 7.8: Schematic of the Huffman-IrCC-TCM scheme. All of the M Huffman encoders and MAP sequence estimators are based upon the same Huffman coding codebook.

of operating significantly closer to the channel's E_b/N_0 capacity bound in comparison with our benchmarks, achieving a gain of about 0.5 dB.

7.4 Simulation Results

In this section we discuss our findings when communicating over an uncorrelated narrowband Rayleigh fading channel having a range of E_b/N_0 values above the channel capacity bound of 2.6 dB. In all simulations we considered the transmission of a single source sample frame **e**, since this comprises a sufficiently large number of samples, namely $M \cdot J = 30\,000$.

7.4.1 IrCC-Based Benchmark

In addition to the proposed SBIrVLC-, BBIrVLC-, SBVLC- and BBVLC-TCM schemes, in this section we also consider the operation of an additional benchmark that we call the Huffman-IrCC-TCM scheme, as depicted in the schematic of Figure 7.8. In contrast to the SBIrVLC-, BBIrVLC-, SBVLC- and BBVLC-TCM schemes of Figure 7.2, in the Huffman-IrCC-TCM scheme the transmission frame **u** is generated by both Huffman and concatenated IrCC encoding the source symbol frame **s**, rather than by invoking VLC encoding.

In the Huffman-IrCC-TCM scheme, Huffman coding is employed on a sub-frame by sub-frame basis, as described in Section 7.2. Table 7.1 provides the composition of the $K = 16$ codewords in the Huffman codebook $\mathbf{Huff} = \{\mathbf{Huff}^k\}_{k=1}^{K}$, having the codeword lengths of $\{I^k\}_{k=1}^{K}$. Compared with the source symbol entropy of $E = 3.77$ bits per source symbol, the average Huffman codeword length is $L(\mathbf{Huff}) = 3.81$ bits per source symbol and the coding rate is $R_{\text{Huff}} = 0.99$, according to Equations (7.1) and (7.3) respectively.

As shown in Figure 7.8, the frame of Huffman-encoded bits **v** is protected by the $N = 17$-component IrCC scheme of [306], which employs a coding memory of $L_{\mathbf{IrCC}} = 4$. The inverted EXIT functions of the $N = 17$ component CC codes are provided in Figure 7.9. The EXIT-chart matching algorithm of [149] was employed to design the IrCC scheme. This was tailored to have an overall coding rate of $R_{\text{IrCC}} = 0.525$ so that the combined Huffman coding and IrCC coding rate $R_{\text{Huff}} \cdot R_{\text{IrCC}} = 0.52$ equals that of the outer codecs in the SBIrVLC-, BBIrVLC-, SBVLC- and BBVLC-TCM schemes. As with the SBIrVLC and BBIrVLC designs detailed in Section 7.3, an open EXIT-chart tunnel was found to be

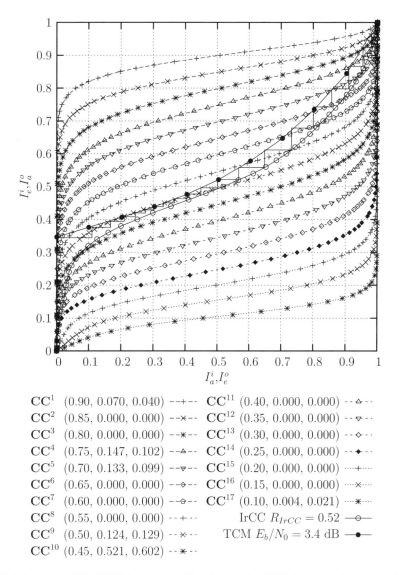

Figure 7.9: Inverted CC EXIT functions. The inverted EXIT function is provided for the corresponding IrCC arrangement, together with the TCM EXIT function corresponding to an E_b/N_0 value of 3.4 dB. A decoding trajectory is provided for the Huffman-IrCC-TCM scheme at a channel E_b/N_0 value of 3.4 dB. Inverted CC EXIT functions are labeled using the format $\mathbf{CC}^n$ ($R(\mathbf{CC}^n)$, C^n, α^n).

achievable between the inverted IrCC EXIT function and the TCM EXIT function at an E_b/N_0 value of 3.1 dB, resulting in the inverted IrCC EXIT function of Figure 7.9.

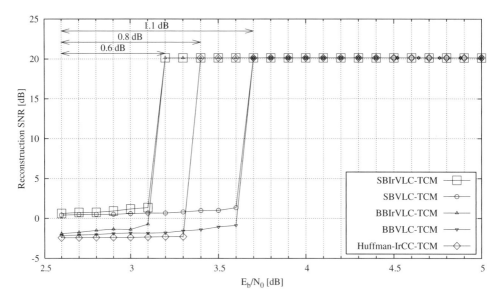

Figure 7.10: Reconstruction SNR versus E_b/N_0 for a Gaussian source using $K = 16$-level Lloyd–Max quantization for the SBIrVLC-, BBIrVLC-, SBVLC- and BBVLC-TCM schemes, as well as for the Huffman-IrCC-TCM scheme, communicating over an uncorrelated narrowband Rayleigh fading channel following iterative decoding convergence. ©IEEE [196] Maunder *et al.* 2008.

In the Huffman-IrCC-TCM receiver, iterative APP SISO IrCC and TCM decoding proceeds as described in Section 7.2. Note that in addition to the *a posteriori* LLR frame $L_p^o(\mathbf{u})$ pertaining to the transmission frame $\mathbf{u}$, the APP SISO IrCC decoder can additionally provide the *a posteriori* LLR frame $L_p^o(\mathbf{v})$ pertaining to the frame of Huffman-encoded bits $\mathbf{v}$. It is on the basis of this that bit-based MAP Huffman sequence estimation may be invoked on a sub-frame-by-sub-frame basis in order to obtain the source symbol frame estimate $\tilde{\mathbf{s}}$, as shown in Figure 7.8.

7.4.2 Iterative Decoding Convergence Performance

For each of our schemes and for each value of E_b/N_0 investigated, we consider the reconstructed source sample frame $\tilde{\mathbf{e}}$ and evaluate the SNR associated with the ratio of the source signal's energy and the reconstruction error energy that may be achieved following iterative decoding convergence. This relationship is plotted for each of the SBIrVLC-, BBIrVLC-, SBVLC- and BBVLC-TCM schemes, as well as for the Huffman-IrCC-TCM scheme, in Figure 7.10.

As shown in Figure 7.10, the source sample reconstruction SNR attained following the achievement of iterative decoding convergence increases as the channel's E_b/N_0 value increases, for all schemes considered. This may be explained by considering the associated EXIT-chart tunnels, which gradually open and become wider as the E_b/N_0 value is increased from the channel's capacity bound, allowing the iterative decoding trajectory to progress further, as explained in Section 4.3. Note that an open EXIT-chart tunnel implies that iterative decoding convergence to an infinitesimally low probability of error can be achieved, provided that the iterative decoding trajectory approaches the inner and outer codecs' EXIT functions sufficiently closely, as described in Section 4.3. However, it can be seen in Figure 7.10 that high source sample reconstruction SNRs were not achieved at the threshold E_b/N_0 values, for which open EXIT-chart tunnels may be created. This is because our 217 500-bit interleaver is unable to entirely eradicate the correlation within the *a priori* LLR frames $L_a^o(\mathbf{u})$ and $L_a^i(\mathbf{u}')$, which the BCJR algorithm assumes to be uncorrelated [48]. As a result, the iterative decoding trajectory does not perfectly match with the inner and outer codecs' EXIT functions and the EXIT-chart tunnel must be further widened before the iterative decoding trajectory can reach the top right-hand corner of the EXIT chart, which is associated with an infinitesimally low probability of error, as described in Section 4.3.

For sufficiently high E_b/N_0 values, the iterative decoding trajectory of all considered schemes was found to approach the top right-hand corner of the EXIT chart, yielding source sample reconstruction SNRs of 20 dB. As described in Section 7.2.1, this represents the infinitesimally low probability of error scenario, where quantization noise provides the only significant degradation. As shown in Figure 7.10, source sample reconstruction SNRs of 20 dB may be achieved by the SBIrVLC- and BBIrVLC-TCM schemes at E_b/N_0 values above 3.2 dB, which is just 0.1 dB from the corresponding threshold E_b/N_0 value of 3.1 dB, as described in Section 7.3.2. In the case of the SBVLC- and BBVLC-TCM schemes, Figure 7.10 also shows a 0.1 dB discrepancy between the threshold E_b/N_0 value of 3.6 dB and the lowest E_b/N_0 value, for which a source sample reconstruction SNR of 20 dB may be achieved, namely 3.7 dB. By contrast, Figure 7.10 shows a 0.3 dB discrepancy between the threshold E_b/N_0 value of 3.1 dB and the lowest E_b/N_0 value for which the Huffman-IrCC-TCM scheme may achieve a source sample reconstruction SNR of 20 dB, namely 3.4 dB.

For each of our schemes, the iterative decoding trajectory that reaches the $(1, 1)$ point of the EXIT chart at the lowest channel E_b/N_0 value considered is provided in Figure 7.4, 7.5 or 7.9, as appropriate. Note that the iterative decoding trajectories of the SBIrVLC-, BBIrVLC-, SBVLC- and BBVLC-TCM schemes approach the corresponding inner and outer EXIT functions fairly closely, facilitating iterative decoding convergence to the $(1, 1)$ point of the EXIT chart at a channel E_b/N_0 value that is just 0.1 dB above the threshold value. This is in contrast to the iterative decoding trajectories of Figure 6.11, which did not exhibit a close match with the inner and outer EXIT functions, requiring an E_b/N_0 value that is 0.5 dB above the threshold value in order that the $(1, 1)$ point of the EXIT chart may be reached. The improved matching of the SBIrVLC-, BBIrVLC-, SBVLC- and BBVLC-TCM schemes' iterative decoding trajectories is a benefit of employing an interleaver that is nearly three times longer than any of those employed in Chapter 6, facilitating the improved mitigation of correlation within the iteratively exchanged extrinsic information. However, the iterative decoding trajectory of the Huffman-IrCC-TCM scheme does not approach the inner and outer EXIT functions as closely as those of the SBIrVLC-, BBIrVLC-, SBVLC- and BBVLC-TCM schemes. As a result, the channel's E_b/N_0 value must be increased by 0.3 dB beyond the threshold E_b/N_0 value before the EXIT-chart tunnel becomes sufficiently wide for the iterative decoding trajectory to reach the $(1, 1)$ point of the EXIT chart. This may be attributed to the APP SISO IrCC decoder's relatively high sensitivity to any residual correlation within

Figure 7.11: Dependencies between sets of $1/R_{\mathrm{IrCC}} = 1.92$ IrCC-encoded bits, for a coding memory of $L_{\mathrm{IrCC}} = 4$.

the *a priori* LLR frame $L_a^o(\mathbf{u})$ that was insufficiently mitigated by the 217 500-bit interleaver, as will be detailed in Section 7.4.3.

7.4.3   Interleaver Length and Latency

As described in Section 7.2.2, interleaving is employed before the *a priori* LLR frame $L_a^o(\mathbf{u})$ is forwarded to the outer APP SISO decoder of each of the schemes considered. This is necessary, since the BCJR algorithm employed by the APP SISO decoders assumes that all *a priori* LLRs that can influence any particular decoding decision are uncorrelated, as described in Section 4.3. However, despite the employment of a long average interleaver length of 217 500 bits, APP SISO IrCC decoding applied to the Huffman-IrCC-TCM scheme is still sensitive to the residual correlation within the *a priori* LLR frame $L_a^o(\mathbf{u})$. As a result, the Huffman-IrCC-TCM scheme suffers from a gradually eroding iterative decoding performance, when the EXIT-chart tunnel is narrow, as explained above. Let us now consider the relatively high sensitivity of APP SISO IrCC decoding to the residual correlation within the *a priori* LLR frame $L_a^o(\mathbf{u})$ in greater detail.

In the IrCC encoder [306] of the Huffman-IrCC-TCM scheme, which employs a coding memory of $L_{\mathrm{IrCC}} = 4$, each bit of the Huffman-encoded frame $\mathbf{v}$ is encoded in conjunction with the preceding $L_{\mathrm{IrCC}} = 4$ bits, in order to generate an average of $1/R_{\mathrm{IrCC}} = 1.92$ bits for the transmission frame $\mathbf{u}$ [140]. Hence, each set of 1.92 bits in the transmission frame $\mathbf{u}$ is directly influenced by the values of the preceding $L_{\mathrm{IrCC}} = 4$ sets of 1.92 bits, which are each in turn directly influenced by their preceding $L_{\mathrm{IrCC}} = 4$ sets of 1.92 bits and so on, providing indirect influences. Similarly, each set of 1.92 bits in the transmission frame $\mathbf{u}$ has a direct influence upon the values of the following $L_{\mathrm{IrCC}} = 4$ sets of 1.92 bits, each of which in turn has a direct influence upon their following $L_{\mathrm{IrCC}} = 4$ sets of 1.92 bits and so on, providing further indirect influences. These dependencies between the sets of $1/R_{\mathrm{IrCC}} = 1.92$ bits are illustrated in Figure 7.11.

The aforementioned influences amongst the bits in the transmission frame $\mathbf{u}$ are exploited during APP SISO IrCC decoding, by employing the BCJR algorithm in order to consider the *a priori* LLRs in the frame $L_a^o(\mathbf{u})$ that pertain to both the preceding and the following bits of $\mathbf{u}$. However, the BCJR algorithm assumes that all *a priori* LLRs in the frame $L_a^o(\mathbf{u})$ that can influence a particular decoding decision are uncorrelated, as described in Section 4.3. Since all bits in the transmission frame $\mathbf{u}$ are either directly or indirectly influenced by each other, we could argue that APP SISO IrCC decoding is sensitive to *all* correlation within the *a priori* LLR frame $L_a^o(\mathbf{u})$. However, each set of 1.92 bits in the transmission frame $\mathbf{u}$ is only directly influenced by the values of the preceding $L_{\mathrm{IrCC}} = 4$ sets of 1.92 bits and only has direct influence upon the values of the following $L_{\mathrm{IrCC}} = 4$ sets of 1.92 bits in the Huffman-IrCC-TCM scheme. Hence, we can say that APP SISO IrCC decoding is only directly sensitive to correlation within the sets of $1/R_{\mathrm{IrCC}} \times L_{\mathrm{IrCC}} + 1/R_{\mathrm{IrCC}} + 1/R_{\mathrm{IrCC}} \times L_{\mathrm{IrCC}} = 17.28$

consecutive *a priori* LLRs. We may therefore conclude that the sensitivity of APP SISO IrCC decoding to correlation within the *a priori* LLR frame $L_a^o(\mathbf{u})$ is dependent on both the IrCC coding memory L_{IrCC} and the coding rate R_{IrCC}. Note that this implies that a shorter interleaver and latency may be afforded, provided that a higher IrCC coding rate and/or a lower memory was employed.

By contrast, during VLC encoding in the context of the SBIrVLC-, BBIrVLC-, SBVLC- and BBVLC-TCM schemes, the source symbols of the source symbol frame $\mathbf{s}$ are encoded in isolation using VLC codewords having an average length of $E/R_{\mathrm{IrVLC}} = 7.25$ bits, which are concatenated to provide the transmission frame $\mathbf{u}$. During APP SISO VLC decoding using the BCJR algorithm, all *a priori* LLRs in the frame $L_a^o(\mathbf{u})$ are considered for the sake of investigating the lengths of the VLC codewords. Despite this, however, we could argue that only the *a priori* LLRs in the frame $L_a^o(\mathbf{u})$ that pertain to a particular VLC codeword have a direct influence upon its APP SISO decoding. We can therefore say that APP SISO VLC decoding in the SBIrVLC-, BBIrVLC-, SBVLC- and BBVLC-TCM schemes is only particularly sensitive to correlation within the sets of 7.25 consecutive *a priori* LLRs. Additionally, we may conclude that the sensitivity of APP SISO VLC decoding to correlation within the *a priori* LLR frame $L_a^o(\mathbf{u})$ is dependent only on the VLC coding rate. Again, this implies that a shorter interleaver and latency might be afforded, if a higher VLC coding rate were employed.

Whilst APP SISO VLC decoding applied in the context of the SBIrVLC-, BBIrVLC-, SBVLC- and BBVLC-TCM schemes is only particularly sensitive to correlation within sets of 7.25 consecutive *a priori* LLRs in the frame $L_a^o(\mathbf{u})$, APP SISO IrCC decoding in the Huffman-IrCC-TCM scheme is particularly sensitive to correlation within sets of 17.28 consecutive *a priori* LLRs, which are about 2.4 times longer. This therefore explains the observation that the Huffman-IrCC-TCM scheme would require a longer interleaver and latency to achieve iterative decoding convergence to an infinitesimally low probability of error for channel E_b/N_0 values between 3.2 dB and 3.4 dB.

7.4.4 Performance During Iterative Decoding

The achievement of iterative decoding convergence requires the completion of a sufficiently high number of decoding iterations. Clearly, each decoding iteration undertaken is associated with a particular computational complexity, the sum of which represents the total computational complexity of the iterative decoding process. Hence, the completion of a sufficiently high number of decoding iterations in order to achieve iterative decoding convergence may be associated with a high computational complexity. In order to quantify how this computational complexity scales as iterative decoding proceeds, we recorded the total number of ACS operations performed per source sample during APP SISO decoding and MAP sequence estimation.

Furthermore, the performance of the considered schemes was also assessed *during* the iterative decoding process, not only after its completion once convergence had been achieved. This was done by evaluating the source sample reconstruction SNR following the completion of *each* decoding iteration. The total computational complexity associated with this SNR was calculated as the sum of the computational complexities associated with all decoding iterations completed so far during the iterative decoding process. Clearly, as more and more decoding iterations are completed, the resultant source sample reconstruction SNR can be expected to increase until iterative decoding convergence is achieved. However, the associated total computational complexity will also increase as more and more decoding iterations are completed. Hence this approach allows the characterization of the trade-off between reconstruction quality and computational complexity.

For each considered Rayleigh channel E_b/N_0 value, a set of source sample reconstruction SNRs and their corresponding computational complexities was obtained, as described above. Note that the size of these sets was equal to the number of decoding iterations required to achieve iterative decoding convergence at the particular E_b/N_0 value. It would therefore be possible to display the source sample reconstruction SNR versus both the E_b/N_0 and the computational complexity in a three-dimensional surface plot, for each of the SBIrVLC-, BBIrVLC-, SBVLC- and BBVLC-TCM schemes. For clarity, however, these surfaces are projected in the direction of the source sample reconstruction SNR axis into two dimensions in the novel plot of Figure 7.12. We employ contours of constant source sample reconstruction SNR, namely 15 dB and 20 dB, to parameterize the relationship between the Rayleigh fading channel's E_b/N_0 value and the associated computational complexity. Note that the plot of Figure 7.10 may be thought of as a cross-section through the surfaces represented by Figure 7.12, perpendicular to the computational complexity axis at $1 \cdot 10^7$ ACS operations per source sample. Note that this particular value of computational complexity is sufficiently high to achieve iterative decoding convergence at all values of E_b/N_0, in each of the considered schemes.

Note that the SBIrVLC and SBVLC decoders have a computational complexity per source sample that depends on the number of symbols in each source symbol sub-frame $\mathbf{s}^m$, namely J. This is because the number of transitions in their symbol-based trellises is proportional to J^2 [181], as described in Section 3.4. Hence the results provided in Figure 7.12 for the SBIrVLC- and SBVLC-TCM schemes are specific to the $J = 100$ scenario. By contrast, the TCM, BBIrVLC, BBVLC and IrCC decoders have a computational complexity per source sample that is independent of the number of symbols in each source symbol sub-frame $\mathbf{s}^m$, namely J. This is because the number of transitions in their trellises is proportional to J [277, 296,301], as described in Section 3.4. Hence the results for the BBIrVLC- and BBVLC-TCM schemes, as well as for the Huffman-IrCC-TCM scheme, provided in Figure 7.12 are *not* specific for the $J = 100$ case.

As shown in Figure 7.12, source sample reconstruction SNRs of up to 20 dB can be achieved within 0.6 dB of the channel's E_b/N_0 capacity bound of 2.6 dB for the SBIrVLC- and BBIrVLC-TCM schemes, within 1.1 dB for the SBVLC- and BBVLC-TCM schemes and within 0.8 dB for the Huffman-IrCC-TCM scheme. Note that these findings agree with those of the EXIT-chart analysis and the asymptotic performance analysis.

7.4.5 Complexity Analysis

We now comment on the computational complexities of the considered schemes and select our preferred arrangement.

In all considered schemes and at all values of E_b/N_0, a source sample reconstruction SNR of 15 dB can be achieved at a lower computational complexity than a SNR of 20 dB can, as shown in Figure 7.12. This is because a reduced number of decoding iterations is required for achieving the extrinsic mutual information value associated with a lower reconstruction quality, as stated above. However, for all considered schemes operating at high values of

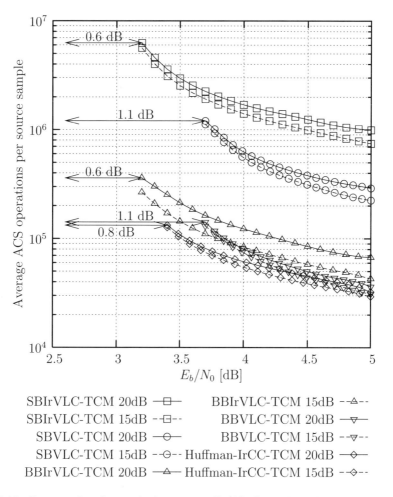

Figure 7.12: Computational complexity versus E_b/N_0 for a Gaussian source using $K = 16$-level Lloyd–Max quantization for the SBIrVLC-, BBIrVLC-, SBVLC- and BBVLC-TCM schemes, as well as for the Huffman-IrCC-TCM scheme, communicating over an uncorrelated narrowband Rayleigh fading channel, parameterized with the source sample reconstruction SNR. ©IEEE [196] Maunder *et al.* 2008.

E_b/N_0, this significant 5 dB reduction in source sample reconstruction SNR facilitates only a relatively modest reduction of the associated computational complexity, which was between 9% in the case of the Huffman-IrCC-TCM scheme and 36% for the BBIrVLC-TCM scheme. Hence we may conclude that the continuation of iterative decoding until near-perfect convergence is achieved can be justified at all values of E_b/N_0.

Additionally, it may be seen that a given source sample reconstruction SNR may be achieved at a reduced computational complexity for all considered schemes as the E_b/N_0 value increases. This may be explained by the widening of the EXIT-chart tunnel as the

E_b/N_0 value increases. As a result, fewer decoding iterations are required in order to reach the extrinsic mutual information that is associated with the specific source sample reconstruction SNR considered.

In each of the considered schemes it was found that VLC and CC decoding is associated with a higher contribution to the total computational complexity than TCM decoding. Indeed, in the case of the SBIrVLC- and SBVLC-TCM schemes, it was found that VLC decoding accounts for about 97% of the numbers of ACS operations per source sample, having a complexity of about 32.3 times higher than that of TCM decoding. In contrast, in the BBIrVLC- and BBVLC-TCM schemes, VLC decoding accounts for only 70% of the operations, having a complexity of about 2.3 times that of TCM decoding. Similarly, CC decoding accounts for only 60% of the ACS operations in the Huffman-IrCC-TCM scheme, having a complexity of about 1.4 times that of TCM decoding.

The high complexity of the SBIrVLC and SBVLC decoders may be attributed to the specific structure of their trellises, which contain significantly more transitions than those of the BBIrVLC, BBVLC and IrCC decoders [181], as described in Section 3.4. As a result, the SBIrVLC- and SBVLC-TCM schemes have a complexity that is about an order of magnitude higher than that of the BBIrVLC- and BBVLC-TCM schemes, as well as the Huffman-IrCC-TCM scheme, as shown in Figure 7.12. In the light of this, the employment of the SBIrVLC- and SBVLC-TCM schemes cannot be readily justified.

Observe in Figure 7.12 that at high E_b/N_0 values the SBIrVLC- and BBIrVLC-TCM schemes have a higher computational complexity than do the corresponding SBVLC- or BBVLC-TCM schemes. This is due to the influence of their low-rate component VLC codebooks. These codebooks comprise codewords with many different lengths, which introduce many transitions when represented in a trellis structure, as described in Section 3.4. The observed computational complexity discrepancy is particularly high in the case of the schemes that employ the symbol-based VLC trellis, owing to its particular nature. For this reason, the SBIrVLC-TCM scheme has a computational complexity that is 240% higher than that of the SBVLC-TCM scheme.

In contrast, we note that at high values of E_b/N_0 the BBIrVLC-TCM scheme has only about a 60% higher computational complexity than that of the BBVLC-TCM scheme. Similarly, the BBIrVLC-TCM scheme has only twice the computational complexity of the Huffman-IrCC-TCM scheme. Coupled with the BBIrVLC-TCM scheme's ability to operate within 0.6 dB of the Rayleigh fading channel's E_b/N_0 capacity bound, we are able to identify this as our preferred arrangement.

7.4.6 Unequal Error-Protection Performance

Let us now examine the UEP performance of our preferred BBIrVLC-TCM scheme. As described in Section 7.1, the UEP capability of IrVLC is manifested because different fractions of the source symbol frame **s** are encoded with different component VLC codebooks having a variety of coding rates and, hence, error-correction capabilities. More specifically, the lower the coding rate of a component VLC codebook, the higher the associated potential error-correction capability, as described in Section 3.3.

As argued above, the composite source sample reconstruction SNR was evaluated following the completion of each decoding iteration during our simulations. The total computational complexity associated with this SNR was calculated as the sum of the computational complexities associated with all decoding iterations completed so far during the iterative decoding process. These computational complexities were plotted against E_b/N_0 and parameterized by the source sample reconstruction SNR in Figure 7.12. Note that the composite BBIrVLC-TCM ACS-complexity verses E_b/N_0 plots are repeated in Figure 7.13. In addition to recording the composite source sample reconstruction SNR after each decoding iteration, we also recorded the reconstruction SNRs associated with the fractions of the source sample frame e that were protected by each of the three activated component VLC codebooks $\mathbf{VLC}^5$, $\mathbf{VLC}^{11}$ and $\mathbf{VLC}^{15}$. For each case, the associated computational complexities are plotted against E_b/N_0 and parameterized by the source sample reconstruction SNR in Figure 7.13.

As shown in Figure 7.13, the lower the coding rate $R(\mathbf{VLC}^n)$ of the component VLC codebook $\mathbf{VLC}^n$ that is employed to protect a fraction C_{BB}^n of the source sample frame e, the lower the computational complexity that is required to reconstruct it with a particular reconstruction SNR at a particular E_b/N_0 value. Indeed, at high E_b/N_0 values the complexity associated with reconstructing the fraction of the source sample frame e that is protected by the $R(\mathbf{VLC}^5) = 0.75$ coding rate component VLC codebook $\mathbf{VLC}^5$ is about twice as high as that associated with the $R(\mathbf{VLC}^{11}) = 0.47$ coding rate component VLC codebook $\mathbf{VLC}^{11}$. This is, in turn, about 1.5 times as high as that associated with the component VLC codebook $\mathbf{VLC}^{15}$, having a coding rate of $R(\mathbf{VLC}^{11}) = 0.31$. In the scenario where only a limited iterative decoding computational complexity can be afforded at the receiver, the fractions of the source sample frame e that are protected by the different component VLC codebooks would be reconstructed with SNRs that are commensurate with the associated coding rates, demonstrating the UEP capability of the BBIrVLC-TCM scheme.

As described in Section 7.1, each of the activated component VLC codebooks $\mathbf{VLC}^5$, $\mathbf{VLC}^{11}$ and $\mathbf{VLC}^{15}$ in the BBIrVLC-TCM scheme is employed to protect a different fraction of the source sample frame e. More specifically, the component VLC codebooks $\mathbf{VLC}^5$, $\mathbf{VLC}^{11}$ and $\mathbf{VLC}^{15}$ each protect a fraction $C_{BB}^5 = 0.32$, $C_{BB}^{11} = 0.63$ and $C_{BB}^{15} = 0.05$ of the source sample frame e, respectively. Note that the composite computational complexity versus E_b/N_0 plots depend on each of the component plots. Furthermore, we may expect the composite plots to be dominated by the component plots associated with the largest fraction of the source sample frame. Specifically, these are the component plots associated with the component VLC codebook $\mathbf{VLC}^{11}$, which is employed to protect a fraction $C_{BB}^{11} = 0.63$ of the source sample frame e. However, Figure 7.13 shows that the composite plots are actually dominated by the component plots associated with the component VLC codebook $\mathbf{VLC}^5$, which is employed to protect only a fraction $C_{BB}^5 = 0.32$ of the source sample frame e. This may be explained as follows.

A composite error-free reconstruction SNR of 20 dB can only be achieved if error-free reconstruction is attained for all three of the fractions of the source sample frame e that are protected by the three activated component VLC codebooks $\mathbf{VLC}^5$, $\mathbf{VLC}^{11}$ and $\mathbf{VLC}^{15}$. Hence, the composite computational complexity versus E_b/N_0 plot that is parameterized by an error-free reconstruction SNR of 20 dB is dominated by that associated with the specific component VLC codebook having the weakest error-correction capability, namely $\mathbf{VLC}^5$, as shown in Figure 7.13 and observed above. Note that the component VLC codebook $\mathbf{VLC}^5$ has the weakest error-correction capability of the three activated codebooks, since it has the highest coding rate of $R(\mathbf{VLC}^5) = 0.75$, as shown in Figure 7.13. This effect may also

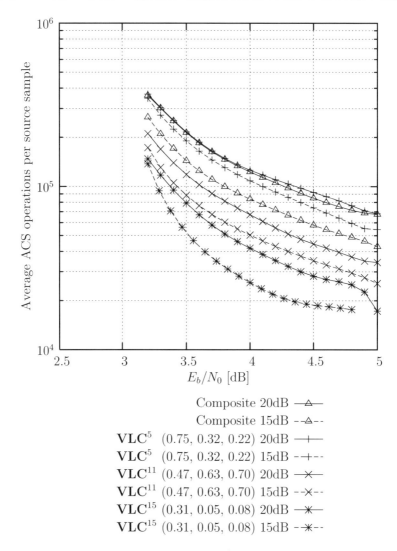

Figure 7.13: Computational complexity versus E_b/N_0 for a Gaussian source using $K = 16$-level Lloyd–Max quantization for the BBIrVLC-TCM scheme, communicating over an uncorrelated narrowband Rayleigh fading channel, parameterized with the source sample reconstruction SNR. Separate plots are provided for the quantized source samples that are VLC encoded using each of the component VLC codebooks $\mathbf{VLC}^5$, $\mathbf{VLC}^{11}$ and $\mathbf{VLC}^{15}$, together with the composite BBIrVLC-TCM plots of Figure 7.12. Components are labeled using the format $\mathbf{VLC}^n$ $(R(\mathbf{VLC}^n), C_{BB}^n, \alpha_{BB}^n)$.

explain the domination of the composite plot that is parameterized by a reconstruction SNR of 15 dB corresponding with that associated with the component VLC codebook $\mathbf{VLC}^5$, despite a relatively low fraction of $C^5 = 0.32$ being protected by this codebook.

7.5 Summary and Conclusions

In this chapter we have investigated the application of IrVLCs for EXIT-chart matching. This was prompted by the observation that the serially concatenated video transmission scheme of Chapter 6 could have facilitated operation at channel E_b/N_0 values that are closer to the capacity bound, if the EXIT functions of its inner and outer codecs had been better matched. More specifically, this would have facilitated the creation of an open EXIT-chart tunnel at near-capacity E_b/N_0 values, implying that iterative decoding convergence to an infinitesimally low probability of error may be achieved, if the iterative decoding trajectory approaches the inner and outer codecs' EXIT functions sufficiently closely.

In analogy with IrCCs, the novel IrVLC scheme of this chapter employs a number of component VLC codebooks having different coding rates for the sake of generating particular fractions of the transmission frame, as described in Section 7.1. We demonstrated that this provides a UEP capability, which may be employed to appropriately protect the various components of audio-, speech-, image- and video-coded information, which typically have different error sensitivities. Furthermore, we showed in Figures 7.4 and 7.5 that the composite inverted IrVLC EXIT function is given by a weighted average of the inverted EXIT functions of the individual component VLC codebooks, where each weight is given by the specific fraction of the transmission frame that is generated by the corresponding component. Finally, we demonstrated that this inverted IrVLC EXIT function may be shaped to match the EXIT function of a serially concatenated TCM codec using the EXIT-chart matching algorithm of [149].

It was noted that an IrVLC scheme's component VLC codebooks should have a suite of widely varying inverted EXIT functions in order that accurate EXIT-chart matching can be performed. Hence, a significant amount of 'trial-and-error'-based human interaction was required in order to select our component VLC codebooks. In Chapter 8 we therefore propose and characterize an efficient technique for designing high-quality suites of component VLC codebooks that does not require 'trial-and-error'-based human interaction. In addition to this, Chapter 8 investigates the relationship between the suite of component VLC codebooks and the resultant IrVLC EXIT-chart matching accuracy. Furthermore, the application of IrVLCs for EXIT-chart matching is further explored in Chapter 9, where the EXIT functions of IrVLCs and of novel Irregular Unity-Rate Codes (IrURCs) are jointly matched to each other, facilitating the creation of an open EXIT-chart tunnel at channel E_b/N_0 values that are even closer to the channel's capacity bound.

During the EXIT-chart matching investigations of this chapter, an open EXIT-chart tunnel was created in Figures 7.4 and 7.5 for the IrVLC-TCM schemes considered and in Figure 7.9 for the Huffman-IrCC-TCM benchmark for channel E_b/N_0 values above a threshold of 3.1 dB. This is just 0.5 dB from the Rayleigh fading channel's E_b/N_0 capacity bound of 2.6 dB, which corresponds to our schemes' effective throughput of 1.56 bits per channel use. By contrast, an open EXIT-chart tunnel was only facilitated in Figures 7.4 and 7.5 for the conventional regular VLC-TCM benchmarks for the increased channel E_b/N_0 values in excess of a threshold of 3.6 dB, which is 1.0 dB from the channel's capacity bound, corresponding to twice the discrepancy of the IrVLC-TCM schemes. Note that the above-mentioned discrepancy of the VLC-TCM benchmarks is similar to the 1.29 dB discrepancy of the VDVQ/RVLC-TCM scheme of Chapter 6, which also does not employ irregular coding techniques.

The iterative decoding performance and computational complexity of the considered schemes was investigated in a novel context using plots of the computational complexity required to achieve particular source sample reconstruction SNRs as a function of the channel's E_b/N_0 value in Figure 7.12. Recall that we observed that the iteratively decoded video transmission scheme of Chapter 6 would have been capable of achieving iterative decoding convergence to an infinitesimally low probability of error at channel E_b/N_0 values that are closer to the threshold at which an open EXIT-chart tunnel can be achieved, if a longer interleaver had been employed. This prompted the consideration in this chapter of an interleaver having a length of $217\,500$ bits, which is nearly three times longer than any of those considered in Chapter 6. Indeed, it was found that the IrVLC- and VLC-TCM schemes were capable of achieving a high-quality source sample reconstruction within 0.1 dB of the threshold channel E_b/N_0 values, which were the lowest values at which an open EXIT-chart tunnel was achievable.

However, in the case of the Huffman-IrCC-TCM scheme, high-quality source sample reconstruction was only achievable for channel E_b/N_0 values above 3.4 dB, which is 0.3 dB above the threshold at which an open EXIT-chart tunnel may be achieved. This was explained in Section 7.4.3 by the relatively high sensitivity of the APP SISO IrCC decoder to any residual correlation within the *a priori* LLRs that was insufficiently mitigated by the $217\,500$-bit interleaver. This resulted in a poor match between the iterative decoding trajectory and the inverted IrCC EXIT function. More specifically, we concluded that an APP SISO IrCC decoder's sensitivity to this correlation depends on both its coding rate and, in particular, its coding memory, which had the relatively high value of $L_{\mathrm{IrCC}} = 4$ in the IrCC scheme considered. We additionally concluded that an APP SISO VLC decoder's sensitivity to the aforementioned correlation depends only on its coding rate, and that shorter interleavers and latencies could be afforded if a higher coding rate was employed. Note that the effect of the VLC coding rate upon its sensitivity to correlation within the *a priori* LLR's frame is investigated in Chapter 8, whilst the effect of the interleaver length upon iterative decoding is investigated in greater detail in Chapter 9.

Recall that the outer APP SISO video decoder of the iteratively decoded video transmission scheme of Chapter 6 operated on the basis of the trellis outlined in Section 6.3.4, which is reminiscent of the symbol-based VLC trellis of [181]. This was the reason why this video transmission scheme was associated with a higher computational complexity than the benchmarks, which employed the bit-based VLC trellis of [180] as the basis of their outer APP SISO decoders. Hence, in this chapter we characterized the computational complexity associated with employing both symbol- and bit-based trellises as the basis of APP SISO VLC decoding. In both cases we concluded that the computational complexity associated with continuing iterative decoding until convergence is achieved is justified, owing to the significantly improved reconstruction quality that results. As predicted in Chapter 6, the schemes that employed bit-based trellises for APP SISO VLC decoding were found to achieve iterative decoding convergence with a significantly lower computational complexity than the schemes employing symbol-based trellises in Figure 7.12.

Owing to its reduced iterative decoding computational complexity, the IrVLC-TCM scheme employing the bit-based VLC trellis as the basis of APP SISO VLC decoding was identified as our preferred arrangement in Section 7.4.5. Additionally, for this reason, only bit-based trellises are employed as the basis of APP SISO VLC decoding in Chapters 8 and 9. Note that in this chapter the source symbol frame was decomposed into $M = 300$ sub-frames in order that the computational complexity associated with the symbol-based VLC trellis could be limited. However, explicit side information was required in order to convey the length of each of the corresponding transmission sub-frames to the receiver, resulting in a trade-off between the computational complexity associated with the symbol-based VLC trellis and the amount of side information required. Indeed, in all IrVLC-TCM parameterizations considered, the required side information was found to account for 4% of the total information conveyed in Section 7.3. In this chapter the source symbol frame was also decomposed into $M = 300$ sub-frames when the bit-based VLC trellis was employed, in order that a fair comparison could be obtained. However, since Chapters 8 and 9 only consider the employment of the bit-based VLC trellis rather than the symbol-based VLC trellis, a significant reduction in the amount of required side information is achieved there by employing a single source symbol sub-frame per activated component VLC codebook.

Genetic Algorithm-Aided Design of Irregular Variable-Length Coding Components[1]

8.1 Introduction

In Chapter 7 we proposed the employment of Irregular Variable-Length Codes (IrVLCs) for the near-capacity transmission of source symbols having values that occur with unequal probabilities. In this scenario, an outer joint source- and channel-coding-based IrVLC codec is serially concatenated [275] with an inner channel codec and they are iteratively decoded [278]. In the transmitter, the IrVLC encoder employs N component Variable-Length Coding codebooks $\{\mathbf{VLEC}^n\}_{n=1}^{N}$ of the Variable-Length Error-Correcting (VLEC) [179] class. These are employed to generate particular fractions $\{\alpha^n\}_{n=1}^{N}$ of the transmission frame, as described in Section 5.5. In the receiver, the iterative exchange of extrinsic information [278] between the inner and outer *A Posteriori* Probability (APP) Soft-Input Soft-Output (SISO) decoders may be characterized using an EXtrinsic Information Transfer (EXIT) chart [294], as described in Section 4.2. The composite inverted IrVLC EXIT function $I_a^o(I_e^o)$ may be obtained as a weighted average of the various inverted EXIT functions $\{I_a^{o,n}(I_e^o)\}_{n=1}^{N}$ corresponding to the N component VLEC codebooks $\{\mathbf{VLEC}^n\}_{n=1}^{N}$ according to

$$I_a^o(I_e^o) = \sum_{n=1}^{N} \alpha^n I_a^{o,n}(I_e^o), \tag{8.1}$$

where the weights are provided by the fractions $\{\alpha^n\}_{n=1}^{N}$. The EXIT-chart matching algorithm of [149] may be employed to select the appropriate values for these fractions $\{\alpha^n\}_{n=1}^{N}$ in order to shape the inverted IrVLC EXIT function so that it does not intersect the inner codec's EXIT function, as described in Section 5.5. In this way, an open EXIT-chart tunnel [267] may be formed at near-capacity channel Signal-to-Noise Ratios (SNRs), which

[1]Part of this chapter is based on [200] © IEEE (2009).

implies that iterative decoding convergence to an infinitesimally low probability of error can be achieved, provided that the iterative decoding trajectory approaches the inner and outer codecs' EXIT functions sufficiently closely, as described in Section 4.3.

In Chapter 7 we speculated that the distance from the channel's SNR capacity bound at which an open EXIT-chart tunnel can be achieved by EXIT-chart matching depends on the degree of variety exhibited in terms of the shapes of the inverted component EXIT functions $\{I_a^{o,n}(I_e^o)\}_{n=1}^N$. That issue is therefore investigated in this chapter. Furthermore, this chapter demonstrates that the inverted EXIT function $I_a^{o,n}(I_e^o)$ that corresponds to a particular component VLEC codebook $\mathbf{VLEC}^n$ depends on both its coding rate $R(\mathbf{VLEC}^n)$ and its error-correction capability.

It was shown in [204] that the area beneath the inverted EXIT function $I_a^o(I_e^o)$ of a particular VLEC codebook $\mathbf{VLEC}$ is approximately equal to its coding rate $R(\mathbf{VLEC})$, as described in Section 5.5. This coding rate depends on the probabilities of occurrence of the various source symbol values and on the length of the corresponding VLEC codewords, as described in Section 3.3. More specifically, the K-ary source symbols have values of $k \in [1, K]$ that occur with the probabilities $\{P(k)\}_{k=1}^K$, resulting in the entropy of

$$E = -\sum_{k=1}^{K} P(k) \cdot \log_2[P(k)]. \tag{8.2}$$

During VLEC encoding using a K-entry VLEC codebook $\mathbf{VLEC}$, the source symbols are mapped to binary codewords $\{\mathbf{VLEC}^k\}_{k=1}^K$ having lengths of $\{I^k\}_{k=1}^K$. In order to ensure that each valid VLEC codeword sequence may be uniquely decoded, a lower bound equal to the source symbol entropy E is imposed upon the average VLEC codeword length of

$$L(\mathbf{VLEC}) = \sum_{k=1}^{K} P(k) \cdot I^k, \tag{8.3}$$

as described in Section 3.3. Any discrepancy between $L(\mathbf{VLEC})$ and E is quantified by the coding rate of

$$R(\mathbf{VLEC}) = \frac{E}{L(\mathbf{VLEC})} \tag{8.4}$$

and may be attributed to the intentional introduction of redundancy into the VLEC codewords. Naturally, this redundancy imposes code constraints that limit the set of legitimate sequences of VLEC-encoded bits. These VLEC code constraints may be exploited in order to provide an error-correcting capability during VLEC decoding.

The error-correction capability of a VLEC codebook $\mathbf{VLEC}$ is typically characterized by its free distance $d_{\text{free}}(\mathbf{VLEC})$, which is equal to the minimum number of differing bits in any pair of equal-length legitimate VLEC-encoded bit sequences [179], as described in Section 3.3. This is because transmission errors that transform the transmitted sequence of VLEC-encoded bits into any other legitimate sequence of VLEC-encoded bits cannot be detected during VLEC decoding [179]. Hence, the free distance characterizes the probability of occurrence of the most likely undetectable transmission error scenario, namely that requiring the lowest number of corrupted bits.

However, the free distance of a VLEC codebook $d_{\text{free}}(\textbf{VLEC})$ is typically difficult to determine. Fortunately, Buttigieg and Farrell formulated a simple free distance lower bound $\bar{d}_{\text{free}}(\textbf{VLEC})$ in [179],

$$d_{\text{free}}(\textbf{VLEC}) \geq \bar{d}_{\text{free}}(\textbf{VLEC}) = \min(d_{b_{\min}}(\textbf{VLEC}), d_{d_{\min}}(\textbf{VLEC}) + d_{c_{\min}}(\textbf{VLEC})), \tag{8.5}$$

where $d_{b_{\min}}(\textbf{VLEC})$ was defined as the minimum block-distance between any pair of equal-length codewords in the VLEC codebook $\textbf{VLEC}$, whilst $d_{d_{\min}}(\textbf{VLEC})$ and $d_{c_{\min}}(\textbf{VLEC})$ were termed the minimum divergence and convergence distances between any pair of unequal-length codewords, respectively, in [179]. As a benefit of its ease of calculation, typically the free distance lower bound $\bar{d}_{\text{free}}(\textbf{VLEC})$ of (8.5) is employed instead of the free distance $d_{\text{free}}(\textbf{VLEC})$ to characterize the error-correction capability of a VLEC codebook $\textbf{VLEC}$.

Note that VLEC codebooks having different error-correction capabilities may have the same Integer-Valued Free Distance (IV-FD) lower bound $\bar{d}_{\text{free}}(\textbf{VLEC})$. Hence these error-correction capabilities cannot be compared by using the IV-FD lower bound $\bar{d}_{\text{free}}(\textbf{VLEC})$ as a metric, owing to its non-unique integer value. For this reason, the IV-FD lower bound $\bar{d}_{\text{free}}(\textbf{VLEC})$ is unsuitable for employment as the objective function or fitness function of a Genetic Algorithm (GA) [307] designed to generate VLEC codebooks having strong error-correcting capabilities. This observation motivates the introduction of a novel Real-Valued Free Distance Metric (RV-FDM) $D(\textbf{VLEC})$ in this chapter in order to characterize the error-correction capability of a VLEC codebook $\textbf{VLEC}$. Since this RV-FDM is defined within the real-valued domain, it facilitates the unambiguous comparison of diverse VLEC codebooks' error-correction capabilities, even if they happen to have the same IV-FD lower bound.

As argued above, the shape of the inverted EXIT function $I_a^o(I_e^o)$ corresponding to a particular VLEC codebook $\textbf{VLEC}$ depends on its error-correction capability. In this chapter we show that the number of points of inflection in the inverted EXIT function $I_a^o(I_e^o)$ depends on the RV-FDM $D(\textbf{VLEC})$. This complements the proof of [298], which shows that the inverted EXIT function $I_a^o(I_e^o)$ will reach the $[E(\textbf{VLEC}), E(\textbf{VLEC})]$ point of the EXIT chart, if the VLEC codebook $\textbf{VLEC}$ has an IV-FD $d_{\text{free}}(\textbf{VLEC}) \geq 2$. Here $E(\textbf{VLEC})$ is defined as the entropy of the VLEC-encoded bits, which is typically assumed to be unity and is given by

$$E(\textbf{VLEC}) = -\sum_{b=0}^{1} \frac{L_b(\textbf{VLEC})}{L(\textbf{VLEC})} \cdot \log_2 \frac{L_b(\textbf{VLEC})}{L(\textbf{VLEC})}, \tag{8.6}$$

where

$$L_b(\textbf{VLEC}) = \sum_{k=1}^{K} P(k) \cdot I_b^k \tag{8.7}$$

and I_b^k is the number of bits in the VLEC codeword $\textbf{VLEC}^k$ that assume a value of $b \in \{0, 1\}$.

A Heuristic Algorithm (HA) was proposed for the construction of VLEC codes in [205], which was later refined in [206] and Section 3.3 of this book. These algorithms attempt to maximize the coding rate $R(\textbf{VLEC})$ of a VLEC codebook $\textbf{VLEC}$ satisfying certain minimum integer-valued block, divergence and convergence distances of $d_{b_{\min}}(\textbf{VLEC})$,

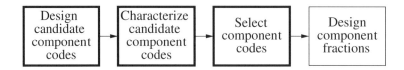

Figure 8.1: Conventional irregular coding design process. This chapter presents modifications to the aspects of this process that are indicated using a bold box.

$d_{d_{\min}}(\textbf{VLEC})$ and $d_{c_{\min}}(\textbf{VLEC})$, respectively. However, the nature of these heuristic algorithms does not facilitate the direct control or prediction of the resultant VLEC codebook's coding rate $R(\textbf{VLEC})$ or RV-FDM $D(\textbf{VLEC})$. Hence these HAs have a limited control over the inverted EXIT function's shape. Therefore, their employment to design a suite of IrVLC component VLEC codebooks having a wide range of inverted EXIT function shapes typically requires a significant amount of 'trial-and-error'-based human interaction. Note that this was the case in Section 7.3, where Algorithm 3.3 of Section 3.3 was employed to design IrVLC component codebooks.

In contrast with the HAs of [205, 206] and Section 3.3, in this chapter we propose a novel GA [307], which is capable of generating VLEC codebooks having specific coding rates and RV-FDMs, associated with desirable bit entropies and decoding complexities. This algorithm is therefore capable of generating a wide range of inverted VLEC EXIT function shapes. For this reason, our proposed GA is better suited for the design of IrVLC component VLEC codebooks than are the HAs of [205, 206] and Section 3.3. In other words, the GA-based codebook design is capable of providing a higher design flexibility, and hence facilitating the appropriate tailoring of the VLEC codebook characteristics without requiring 'trial-and-error'-based human interaction to select a suite of component codebooks from a large number of candidates, as shown in Figure 8.1. This readily facilitates the design of IrVLCs having inverted EXIT functions that closely match the serially concatenated inner codec's EXIT function, allowing operation at channel SNRs that approach the channel's capacity bound.

The employment of the proposed GA for the design of component IrVLC codebooks will also be exemplified in this chapter. This will be achieved in the context of a scheme facilitating the near-capacity transmission of source symbols having values that occur with unequal probabilities over an uncorrelated narrowband Rayleigh fading channel. Similarly to Chapter 7's IrVLC-TCM scheme, the proposed scheme will employ the serial concatenation [275] and iterative decoding [278] of an outer IrVLC codec with an inner channel codec. However, in contrast to the IrVLC-TCM scheme of Chapter 7, our so-called IrVLC-URC scheme will employ a Unity-Rate Code (URC) as the inner channel codec, instead of Trellis-Coded Modulation (TCM). This is because the maximum effective throughput is limited in any transmission scheme that employs an inner codec having a coding rate of less than unity, such as TCM. For example, in the case of Chapter 7's IrVLC-TCM scheme, which employs a $R_{\text{TCM}} = 3/4$-rate TCM codec and $M_{\text{TCM}} = 16$-ary Quadrature Amplitude Modulation (16QAM), the maximum effective throughput is $R_{\text{TCM}} \cdot \log_2(M_{\text{TCM}}) = 3$ bits per channel use. Hence, an effective throughput loss will occur for high channel SNRs, where the capacity of the 16QAM-based channel will exceed the maximum effective throughput of 3 bits per channel use attained by the $R_{\text{TCM}} = 3/4$-rate TCM scheme and will approach $\log_2(M_{\text{TCM}}) = 4$ bits per channel use.

In Section 7.4.5 we observed that the computational complexity of APP SISO VLEC decoding and Maximum *A posteriori* Probability (MAP) VLEC sequence estimation is lower when these are performed on the basis of a bit-based trellis [180] rather than a symbol-based trellis [181]. Therefore in the IrVLC-URC scheme of this chapter, APP SISO VLEC decoding and MAP VLEC sequence estimation are performed on the basis of the bit-based trellis. Note that the computational complexities of APP SISO VLEC decoding and MAP VLEC sequence estimation are monotonically increasing functions of the number of bit-based trellis transitions per VLEC-encoded bit $T(\textbf{VLEC})$, as described in Section 3.4. Also note that the selection of the bit-based trellis allows us to employ just one transmission sub-frame per component VLEC codebook, as described in Section 7.5. This is in contrast to the IrVLC-TCM scheme of Chapter 7, where a number of transmission sub-frames were employed for each component VLEC codebook. As described in Section 7.2.1, the length of each transmission sub-frame must be explicitly conveyed to the receiver as side information in order to facilitate VLEC decoding. Hence, the approach of this chapter facilitates a significant reduction of the side information overhead compared with that of the IrVLC-TCM scheme of Chapter 7.

Additionally, Section 7.4.3 demonstrated that APP SISO VLEC decoding is sensitive to correlation within the *a priori* Logarithmic Likelihood Ratio (LLR) frame. This can result in a mismatch between the simulation-based iterative decoding trajectory and the inverted VLEC EXIT function. We speculated that the APP SISO VLEC decoder's sensitivity to this correlation is dependent on the coding rate of the corresponding VLEC codebook. This relationship is investigated in this chapter by comparing the performance of a number of different parameterizations of the IrVLC-URC scheme, employing different IrVLC coding rates.

The rest of this chapter is organized as follows. In Section 8.2 we detail our novel RV-FDM, which allows the comparison of the error-correction capability of two VLEC codebooks having equal IV-FD lower bounds. We also show in Section 8.2 that the shape of the inverted EXIT function $I_a^o(I_e^o)$ of a VLEC codebook $\textbf{VLEC}$ depends on its RV-FDM $D(\textbf{VLEC})$. Our novel GA used for designing VLEC codebooks having both specific coding rates and RV-FDMs, as well as desirable bit entropies and bit-based trellis complexities, is detailed in Section 8.3. In Section 8.4 we introduce the proposed IrVLC-URC scheme. Our parameterizations of this scheme are developed in Section 8.5. Here, we investigate the suitability of both the proposed GA and Algorithm 3.3 of Section 3.3 for designing the suites of IrVLC component VLEC codebooks required. In Section 8.6 both the achievable Bit-Error Ratio (BER) performance and the computational complexity imposed by our IrVLC-URC parameterizations are characterized and compared. Finally, we offer our conclusions in Section 8.7.

8.2   The Free Distance Metric

In this section we introduce our RV-FDM $D(\textbf{VLEC})$, which allows the comparison of the error-correction capability of two VLEC codebooks having equal IV-FD lower bounds $\bar{d}_{\text{free}}(\textbf{VLEC})$.

As described in Section 8.1, the IV-FD lower bound $\bar{d}_{\text{free}}(\mathbf{VLEC})$ of a VLEC codebook $\mathbf{VLEC}$ depends on the specific constituent codeword pairs that have the minimal block, divergence and convergence distances, $d_{b_{\min}}(\mathbf{VLEC})$, $d_{d_{\min}}(\mathbf{VLEC})$ and $d_{c_{\min}}(\mathbf{VLEC})$, respectively. We refer to these codeword pairs as 'similar' pairs since they are responsible for the most likely undetectable decoding errors, as described in Section 8.1. The remaining so-called 'dissimilar' VLEC codeword pairs do not contribute to the IV-FD lower bound $\bar{d}_{\text{free}}(\mathbf{VLEC})$, since they have block, divergence and convergence distances that are higher than the minima $d_{b_{\min}}(\mathbf{VLEC})$, $d_{d_{\min}}(\mathbf{VLEC})$ and $d_{c_{\min}}(\mathbf{VLEC})$, respectively. Hence, we assume that the error-correction capability of a VLEC codebook may be adequately characterized by modifying the IV-FD lower bound $\bar{d}_{\text{free}}(\mathbf{VLEC})$ to yield the RV-FDM according to

$$D(\mathbf{VLEC}) = \bar{d}_{\text{free}}(\mathbf{VLEC}) + F(\mathbf{VLEC}), \tag{8.8}$$

where $F(\mathbf{VLEC}) \in [0, 1)$ quantifies the average similarity of the various VLEC codeword pairs, as we shall detail below. Note that $\lfloor D(\mathbf{VLEC}) \rfloor = \bar{d}_{\text{free}}(\mathbf{VLEC})$, since we have $F(\mathbf{VLEC}) \in [0, 1)$. Hence, the value of a VLEC codebook's RV-FDM $D(\mathbf{VLEC})$ is dominated by its IV-FD lower bound, which was appropriately adjusted in Equation (8.8).

The average similarity of the various VLEC codeword pairs of a particular K-entry VLEC codebook $\mathbf{VLEC} = \{\mathbf{VLEC}^k\}_{k=1}^K$ may be quantified by

$$F(\mathbf{VLEC}) = \frac{\sum_{k^1=1}^{K-1} \sum_{k^2=k^1+1}^{K} P(k^1) \cdot P(k^2) \cdot F(\mathbf{VLEC}^{k^1}, \mathbf{VLEC}^{k^2})}{\sum_{k^1=1}^{K-1} \sum_{k^2=k^1+1}^{K} P(k^1) \cdot P(k^2)}, \tag{8.9}$$

where $F(\mathbf{VLEC}^{k^1}, \mathbf{VLEC}^{k^2})$ quantifies the similarity of the specific pair of VLEC codewords $\mathbf{VLEC}^{k^1}$ and $\mathbf{VLEC}^{k^2}$, as will be detailed below. Note that the calculation of $F(\mathbf{VLEC})$ in Equation (8.9) takes into account the probabilities of occurrence $P(k^1)$ and $P(k^2)$ of the specific VLEC codewords $\mathbf{VLEC}^{k^1}$ and $\mathbf{VLEC}^{k^2}$, respectively. These must be considered, because the error correction capability of the VLEC codebook is related to the occurrence probabilities of those specific VLEC codewords that contribute to the IV-FD lower bound $\bar{d}_{\text{free}}(\mathbf{VLEC})$. For example, a stronger error-correcting capability can be expected if only pairs of infrequently occurring VLEC codewords are similar. Note that in order to ensure that $F(\mathbf{VLEC}) \in [0, 1)$, VLEC codewords are not paired with themselves and a normalization factor of $\sum_{k^1=1}^{K-1} \sum_{k^2=k^1+1}^{K} P(k^1) \cdot P(k^2) < 1$ is employed in the calculation of $F(\mathbf{VLEC})$ in Equation (8.9).

As described above, the similarity of a specific pair of VLEC codewords $\mathbf{VLEC}^{k^1}$ and $\mathbf{VLEC}^{k^2}$ depends on the degree to which they contribute to the IV-FD lower bound. This may be determined according to Table 8.1, in which I^k is the length of the VLEC codeword $\mathbf{VLEC}^k$ and $d_b(\mathbf{VLEC}^{k^1}, \mathbf{VLEC}^{k^2})$ is the block distance between a pair of equal-length VLEC codewords $\mathbf{VLEC}^{k^1}$ and $\mathbf{VLEC}^{k^2}$. Additionally, $d_d(\mathbf{VLEC}^{k^1}, \mathbf{VLEC}^{k^2})$ and $d_c(\mathbf{VLEC}^{k^1}, \mathbf{VLEC}^{k^2})$ are the divergence and convergence distances between a specific pair of unequal-length codewords $\mathbf{VLEC}^{k^1}$ and $\mathbf{VLEC}^{k^2}$, respectively [179].

We now consider the effect that the RV-FDM $D(\mathbf{VLEC})$ of a VLEC codebook $\mathbf{VLEC}$ has on the corresponding inverted EXIT function shape, which may be employed to

Table 8.1: The similarity of a specific pair of VLEC codewords $\mathbf{VLEC}^{k^1}$ and $\mathbf{VLEC}^{k^2}$ depending on the degree to which they contribute to the IV-FD lower bound. ©IEEE [200] Maunder and Hanzo, 2008.

Condition	$F(\mathbf{VLEC}^{k^1}, \mathbf{VLEC}^{k^2})$
if $I^{k^1} = I^{k^2}$ and $d_{b_{\min}}(\mathbf{VLEC}) \leq$ $d_{d_{\min}}(\mathbf{VLEC}) + d_{c_{\min}}(\mathbf{VLEC})$ and $d_b(\mathbf{VLEC}^{k^1}, \mathbf{VLEC}^{k^2}) = d_{b_{\min}}(\mathbf{VLEC})$	0
if $I^{k^1} \neq I^{k^2}$ and $d_{d_{\min}}(\mathbf{VLEC}) + d_{c_{\min}}(\mathbf{VLEC}) \leq$ $d_{b_{\min}}(\mathbf{VLEC})$ and $d_d(\mathbf{VLEC}^{k^1}, \mathbf{VLEC}^{k^2}) = d_{d_{\min}}(\mathbf{VLEC})$ and $d_c(\mathbf{VLEC}^{k^1}, \mathbf{VLEC}^{k^2}) = d_{c_{\min}}(\mathbf{VLEC})$	0
if $I^{k^1} \neq I^{k^2}$ and $d_{d_{\min}}(\mathbf{VLEC}) + d_{c_{\min}}(\mathbf{VLEC}) \leq$ $d_{b_{\min}}(\mathbf{VLEC})$ and $\left(d_d(\mathbf{VLEC}^{k^1}, \mathbf{VLEC}^{k^2}) = d_{d_{\min}}(\mathbf{VLEC}) \right.$ xor $\left. d_c(\mathbf{VLEC}^{k^1}, \mathbf{VLEC}^{k^2}) = d_{c_{\min}}(\mathbf{VLEC}) \right)$	0.5
otherwise	1

characterize the APP SISO decoding performance. This is investigated with the aid of the rate-0.5 $K = 4$-ary VLEC codebooks $\mathbf{VLEC}^a$, $\mathbf{VLEC}^b$ and $\mathbf{VLEC}^c$ of Table 8.2, which have different RV-FDMs of $D(\mathbf{VLEC}^a) = 2.143$, $D(\mathbf{VLEC}^b) = 2.300$ and $D(\mathbf{VLEC}^c) = 2.943$, respectively. Furthermore, Figure 8.2 provides four examples of the VLEC-encoded bit sequence estimation process, which will also aid our discussions. Each example of Figure 8.2 depicts a sequence of VLEC-encoded bits, which is formed as a concatenation of VLEC codewords from either the relatively high Free Distance Metric (FDM) codebook $\mathbf{VLEC}^c$ or the relatively low FDM codebook $\mathbf{VLEC}^a$ of Table 8.2. In each case, Figure 8.2 provides a set of *a priori* conjectures about the VLEC-encoded bit values, containing five errors. One of the *a priori* conjectures in each example is associated with a particularly high confidence, which is either justified if the conjecture is correct, or unjustified if it is incorrect. The set of *a posteriori* bit estimates of each example is obtained as the particular concatenation of VLEC codewords that matches that *a priori* conjecture at a high confidence, as well as matching the maximum number of the other *a priori* conjectures. These hard-decision-based bit sequence estimation processes operate on the same basis as APP SISO decoding and these simple examples are therefore relevant to the context of the inverted VLEC EXIT functions.

During APP SISO decoding, the RV-FDM $D(\mathbf{VLEC})$ of the corresponding VLEC codebook $\mathbf{VLEC}$ affects the degree to which the resultant extrinsic LLRs are dominated

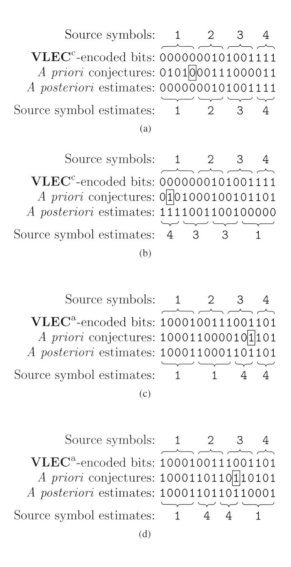

Source symbols: 1   2   3   4

VLEC^c-encoded bits: 0000000101001111
A priori conjectures: 0101000111000011
A posteriori estimates: 0000000101001111

Source symbol estimates: 1   2   3   4

(a)

Source symbols: 1   2   3   4

VLEC^c-encoded bits: 0000000101001111
A priori conjectures: 0101000100101101
A posteriori estimates: 1111001100100000

Source symbol estimates: 4   3   3   1

(b)

Source symbols: 1   2   3   4

VLEC^a-encoded bits: 1000100111001101
A priori conjectures: 1000110000101101
A posteriori estimates: 1000110001101101

Source symbol estimates: 1   1   4   4

(c)

Source symbols: 1   2   3   4

VLEC^a-encoded bits: 1000100111001101
A priori conjectures: 1000110110110101
A posteriori estimates: 1000110110110001

Source symbol estimates: 1   4   4   1

(d)

Figure 8.2: Examples of obtaining *a posteriori* estimates of VLEC-encoded bits from sets of *a priori* conjectures containing five errors. In (a) and (b), the VLEC codebook **VLEC**^{c} having a relatively high RV-FDM of $D(\mathbf{VLEC}^c) = 2.943$ is employed. By contrast, the VLEC codebook **VLEC**^{a} having a relatively low RV-FDM of $D(\mathbf{VLEC}^a) = 2.143$ is employed in (c) and (d). In all cases, the boxed *a priori* conjecture is associated with a particularly high confidence. This confidence is justified in the case of (a) and (c), since the associated conjecture is correct. By contrast, the conjecture is incorrect in (b) and (d) and so the associated high confidence is not justified.

Table 8.2: Composition of three rate-0.5 $K = 4$-ary VLEC codebooks $\mathbf{VLEC}^a$, $\mathbf{VLEC}^b$ and $\mathbf{VLEC}^c$, which have RV-FDMs of $D(\mathbf{VLEC}^a) = 2.143$, $D(\mathbf{VLEC}^b) = 2.300$ and $D(\mathbf{VLEC}^c) = 2.943$, respectively

k	$P(k)$	$\mathbf{VLEC}^{a,k}$	$\mathbf{VLEC}^{b,k}$	$\mathbf{VLEC}^{c,k}$
1	0.1	10001	00001	00000
2	0.2	0011	0110	0010
3	0.3	1001	1010	1001
4	0.4	101	110	111

by the specific *a priori* LLRs having a relatively high magnitude. These relatively high-magnitude *a priori* LLRs represent relatively high confidences (be they justified or not) in the values of the corresponding bits. During APP SISO decoding, relatively high levels of confidence will typically be associated with those sequences of VLEC codewords that have bit values that agree with the specific polarities of the relatively high-magnitude *a priori* LLRs.

In the case of VLEC codebooks having a high RV-FDM, such as $\mathbf{VLEC}^c$ of Table 8.2, typically there will only be a relatively low number of VLEC codeword sequences that are associated with a relatively high confidence. This is because typically a low number of identical bit values are shared by any pair of high RV-FDM VLEC codeword sequences. Hence, the relatively few high-confidence VLEC codeword sequences will typically unanimously agree and thus associate a high confidence with the value of a relatively high number of the VLEC-encoded bits. Therefore in the case of VLEC codebooks having a high RV-FDM, the extrinsic LLRs are dominated by the specific *a priori* LLRs having a relatively high magnitude. This is exemplified in Figures 8.2a and 8.2b, where the high-confidence *a priori* bit value conjecture causes five and six, respectively, of the other conjectures to be overthrown, when the *a posteriori* estimates are obtained.

By contrast, the extrinsic LLRs are not dominated by the *a priori* LLRs having a high magnitude in the case of VLEC codebooks having a low RV-FDM, such as $\mathbf{VLEC}^a$ of Table 8.2. This is because a relatively high number of identical bit values are shared by the pairs of low RV-FDM VLEC codeword sequences. Hence, a relatively high number of VLEC codeword sequences will be associated or credited with a high confidence during APP SISO decoding and these will typically unanimously agree on the value of a relatively low number of the VLEC-encoded bits. Therefore, in the case of VLEC codebooks having a low RV-FDM, each extrinsic LLR is dominated only by the corresponding *a priori* LLR. This is exemplified in Figures 8.2c and 8.2d, where the high-confidence *a priori* bit value conjecture causes only one of the other conjectures to be overthrown when the *a posteriori* estimates are obtained.

When the *a priori* LLRs have a low mutual information I_a, a high proportion of them will have 'unjustly' high magnitudes, representing unjustified confidence in the values of the corresponding VLEC-encoded bits. During the APP SISO decoding of VLEC codes having a high RV-FDM, these *a priori* LLRs having unjustly high magnitudes will influence the resultant extrinsic LLRs, and, as a result, they will have a lower mutual information I_e than

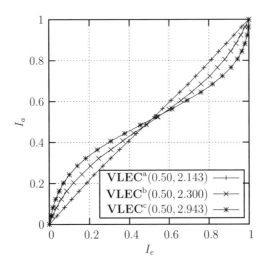

Figure 8.3: Inverted EXIT functions, characterizing the APP SISO decoding performance of the VLEC codebooks **VLEC**[a], **VLEC**[b] and **VLEC**[c]. The functions are labeled using the format **VLEC** $[R(\mathbf{VLEC}),\ D(\mathbf{VLEC})]$. ©IEEE [200] Maunder and Hanzo, 2008.

that of the *a priori* LLRs. This is exemplified in Figure 8.2b, where the *a priori* bit value conjecture having an unjustly high confidence induces six more errors within the *a posteriori* estimates.

By contrast, a low proportion of the *a priori* LLRs will have unjustly high magnitudes, when they have a high mutual information I_a and therefore extrinsic LLRs having a higher mutual information I_e will result. This is exemplified in Figure 8.2a, where the *a priori* bit value conjecture having a justifiable high confidence succeeds in correcting all five errors within the other conjectures, when the *a posteriori* estimates are obtained.

Owing to the two above-mentioned effects, an 'S'-shaped inverted EXIT function having two points of inflection may be expected for a VLEC codebook having a high RV-FDM. This is illustrated in Figure 8.3 for the VLEC codebook **VLEC**[c] of Table 8.2.

By contrast, the APP SISO decoding of a VLEC code having a lower RV-FDM is not so dominated by the *a priori* LLRs having high magnitudes (be they justified or not), and hence an extrinsic mutual information I_e that is more commensurate with the *a priori* mutual information I_a is yielded. Consequently, the inverted EXIT function will no longer have such a pronounced 'S' shape, as shown in Figure 8.3 for the VLEC codebooks **VLEC**[a] and **VLEC**[b] of Table 8.2. Note that since the VLEC codebook **VLEC**[b] has a higher RV-FDM than **VLEC**[a] does, its inverted EXIT function has a more pronounced 'S' shape.

8.3 Overview of the Proposed Genetic Algorithm

In this section we assume a basic familiarity with GAs [307] and we introduce a novel GA that facilitates the design of VLEC codebooks having arbitrary coding rates and RV-FDMs, as well as maintaining near-unity bit entropies and bit-based trellis representations comprising only a low number of trellis transitions, and hence yielding a low decoding complexity. Figure 8.4 provides a flowchart for this GA, which operates on the basis of a list of VLEC codebooks **L**, initially containing only the so-called parent VLEC codebook **VLEC**^{0}. As the GA proceeds, the list **L** is populated with mutations of its constituent VLEC codebooks. In this way, the GA attempts to find VLEC codebooks **VLEC** having coding rates $R(\textbf{VLEC})$ and RV-FDMs $D(\textbf{VLEC})$ that are either increased or decreased with respect to those of the parent VLEC codebook, tending towards the specified limits R^{lim} and D^{lim} respectively. Additionally, they aim for near-unity bit entropies $E(\textbf{VLEC})$ and bit-based trellis representations comprising a low number of trellis transitions $T(\textbf{VLEC})$. The benefits of a mutated VLEC codebook **VLEC** are assessed using a composite quality metric $M(\textbf{VLEC})$, which considers the codebook's coding rate $R(\textbf{VLEC})$, RV-FDM $D(\textbf{VLEC})$, bit entropy $E(\textbf{VLEC})$ and bit-based trellis complexity $T(\textbf{VLEC})$. Naturally, a VLEC codebook having a coding rate $R(\textbf{VLEC})$ and a RV-FDM $D(\textbf{VLEC})$ close to their specified limits R^{lim} and D^{lim} respectively, as well as having a near-unity bit entropy $E(\textbf{VLEC})$ and a low bit-based trellis complexity $T(\textbf{VLEC})$, is regarded as having a high merit and has a high-valued quality metric $M(\textbf{VLEC})$.

The proposed heuristic quality metric of a VLEC codebook **VLEC** is defined here as

$$M(\textbf{VLEC}) = \begin{array}{l} \alpha^D \beta^D \frac{D(\textbf{VLEC})-D^{\text{best}}}{D^{\text{best}}} \quad + \quad \alpha^R \beta^R \frac{R(\textbf{VLEC})-R^{\text{best}}}{R^{\text{best}}} \quad + \\ \alpha^E \frac{E(\textbf{VLEC})-E^{\text{best}}}{E^{\text{best}}} \quad + \quad \alpha^T \frac{T^{\text{best}}-T(\textbf{VLEC})}{T^{\text{best}}} \end{array}, \quad (8.10)$$

where α^D, α^R, α^E, α^T, β^D and β^R are described in Table 8.3. The values of D^{best}, R^{best}, E^{best} and T^{best} are obtained by taking into consideration the RV-FDMs, coding rates, bit entropies and bit-based trellis complexities of all VLEC codebooks that have been incorporated into the list **L** so far. More specifically, D^{best} and R^{best} are the RV-FDM and coding rate that are closest to the specified limits D^{lim} and R^{lim} respectively. Additionally, E^{best} and T^{best} are the highest bit entropy and lowest bit-based trellis complexity respectively. These values are employed to normalize the RV-FDM, coding rate, bit entropy and bit-based trellis complexity of the VLEC codebook **VLEC** and are constantly updated during the operation of the proposed GA. It is necessary to employ these constantly updated values, since it is often difficult to predict in advance the values of the best RV-FDM, coding rate, bit entropy and bit-based trellis complexity that will be found during the operation of the proposed GA.

At each stage of the proposed GA's operation shown in Figure 8.4, the list **L** is searched to find the worst constituent VLEC codebook **VLEC**^{worst}, namely that having the lowest quality metric

$$\textbf{VLEC}^{\text{worst}} = \underset{\textbf{VLEC}\in\textbf{L}}{\text{argmin}} \ M(\textbf{VLEC}). \quad (8.11)$$

Next, P mutations of this VLEC codebook **VLEC**^{worst} are generated, as detailed below. As shown in Figure 8.4, a mutation **VLEC**$'$ is inserted into the list **L** if it

- has a higher quality metric than the VLEC codebook **VLEC**^{worst}, or more specifically if $M(\textbf{VLEC}') > M(\textbf{VLEC}^{\text{worst}})$;
- has a RV-FDM $D(\textbf{VLEC}')$ that does not exceed the specified limit D^{lim}, or more specifically if we have $\beta^D D(\textbf{VLEC}') < \beta^D D^{\text{lim}}$;

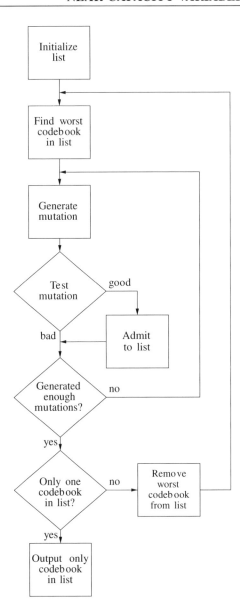

Figure 8.4: Flowchart of the proposed genetic algorithm.

- has a coding rate $R(\mathbf{VLEC}')$ that does not exceed the specified limit R^{lim}, or more specifically if $\beta^R R(\mathbf{VLEC}') < \beta^R R^{\mathrm{lim}}$;

- is different with respect to all VLEC codebooks in the list $\mathbf{L}$, i.e. we have $\mathbf{VLEC}' \notin \mathbf{L}$.

Finally, the VLEC codebook $\mathbf{VLEC}^{\mathrm{worst}}$ is removed from the list $\mathbf{L}$, unless it is the only codebook in the list, in which case the GA's operation is concluded and $\mathbf{VLEC}^{\mathrm{worst}}$ is output as the designed VLEC codebook, as shown in Figure 8.4.

The number of mutations that are generated at each stage of the proposed GA's operation, P, is constantly varied in an attempt to maintain a constant list length $L(\mathbf{L})$ equal to a specified target L^{tar}. This approach prevents the list $\mathbf{L}$ and hence the complexity of the proposed GA from growing without bound, whilst avoiding directly limiting the number of mutations of any VLEC codebook that may be admitted to the list $\mathbf{L}$. Following each stage of the proposed GA, the number of mutations to be generated in the next stage is calculated as $\max[1, \min(P^{\mathrm{next}}, P^{\mathrm{max}})]$, where P^{max} is a parameter that limits the complexity of the next stage, and

$$P^{\mathrm{next}} = P(1 + 0.05([L^{\mathrm{tar}} - L(\mathbf{L})] - [L(\mathbf{L}) - L^{\mathrm{prev}}])). \tag{8.12}$$

Here, $L(\mathbf{L}) - L^{\mathrm{prev}}$ is the change in list length that resulted from the current stage of the proposed GA, whilst $L^{\mathrm{tar}} - L(\mathbf{L})$ is the list length change that is desired during the next stage. Note that the constant value of 0.05 in Equation (8.12) was experimentally found to successfully maintain a wide range of desired list lengths L^{tar}.

During the mutation of a particular VLEC codebook $\mathbf{VLEC}$, the proposed GA employs four different types of operation, as follows.

- A randomly selected bit value of either zero or one is **inserted** into a randomly selected position in a randomly selected codeword of the VLEC codebook $\mathbf{VLEC}$.
- The value of a randomly selected bit in a randomly selected codeword of the VLEC codebook $\mathbf{VLEC}$ is **toggled** from zero to one or from one to zero, as appropriate.
- A randomly selected bit in a randomly selected codeword of the VLEC codebook $\mathbf{VLEC}$ is **removed**.
- Two randomly selected codewords in the VLEC codebook $\mathbf{VLEC}$ are **swapped**.

During the mutation of a particular VLEC codebook $\mathbf{VLEC}$, Q operations are performed on the VLEC codebook $\mathbf{VLEC}$, where the type of each operation is randomly selected from the above list. The number Q of GA operations performed on the VLEC codebook $\mathbf{VLEC}$ is randomly selected, where the probability that Q takes a particular value $q \geq 1$ is given as $P(Q = q) = 2^{-q}$. In this way, invoking a high number of mutations of the VLEC codebook $\mathbf{VLEC}$ is possible but infrequent, since this is less likely to result in improved quality metrics. However, this measure allows the GA to generate diverse codebooks in the interest of exploring the entire search space sufficiently densely. The mutation of the example VLEC codebook $\mathbf{VLEC}^{\mathrm{b}}$ from Table 8.2 is exemplified in Figure 8.5.

The proposed GA is detailed in Algorithm 8.1 and its parameters are described in Table 8.3.

8.4 Overview of Proposed Scheme

In this section we provide an overview of a transmission scheme that facilitates the joint source and channel coding of a sequence of source symbols having values with unequal probabilities of occurrence for near-capacity transmission over an uncorrelated narrowband Rayleigh fading channel. As demonstrated in Chapter 7, this application motivates the employment of IrVLCs, which are used here in order to exemplify the application of the

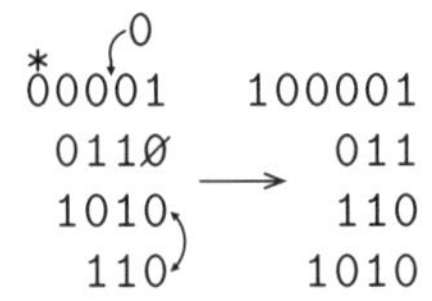

$$
\begin{array}{ll}
\overset{*}{\underset{}{0}}0\overset{\curvearrowleft 0}{0}01 & 100001 \\
011\cancel{0} & 011 \\
1010\rlap{\raise-1pt{\Large\diagdown}} & 110 \\
110\rlap{\raise3pt{\Large\diagup}} & 1010
\end{array}
$$

Figure 8.5: Example of the VLEC codebook mutation. A total of four operations are performed during the mutation: in the first codeword, the bit marked by the asterisk is toggled and a zero-valued bit is inserted; a zero-valued bit is removed from the second codeword, while the third and fourth codewords are swapped.

Table 8.3: Parameters that control the operation of the proposed GA

VLEC^0	Is the parent VLEC codebook.
$\alpha^D \geq 0$, $\alpha^R \geq 0$, $\alpha^E \geq 0$ and $\alpha^T \geq 0$	Specify the relative importance of optimizing the RV-FDM, coding rate, bit entropy and bit-based trellis complexity in Equation (8.10), respectively.
$\beta^D \in \{-1, +1\}$ and $\beta^R \in \{-1, +1\}$	Specify, in Equation (8.10), whether it is desirable to design VLEC codebooks that have RV-FDMs and coding rates, respectively, that are higher than or lower than those of the parent VLEC codebook. A value of $+1$ specifies that it is desirable to increase the RV-FDM or coding rate, whilst a value of -1 indicates that a reduction is desirable.
D^{lim} and R^{lim}	Specify the maximum desirable RV-FDM when $\beta^D = +1$ and the maximum desirable coding rate when $\beta^R = +1$, respectively. By contrast, they specify the minimum desirable RV-FDM when $\beta^D = -1$ and the minimum desirable coding rate when $\beta^R = -1$.
L^{tar}	Specifies the target length of the list **L**. Note that improved solutions can be expected for longer target list lengths, although this is associated with a greater GA computational complexity.
P^{max}	Specifies the maximum number of mutations of any VLEC codebook that may be considered at each stage of the proposed GA. Note that improved choices can be expected for higher maximum mutation limits, although this is associated with a higher GA computational complexity.

Algorithm 8.1 Genetic algorithm for the design of VLEC codebooks ©IEEE [200] Maunder and Hanzo, 2008.

1: Initialize the list $\mathbf{L}$ with the parent VLEC codebook $\mathbf{VLEC}^0$, in order that it has a length of $L(\mathbf{L}) = 1$.

2: $\hat{D}^{\text{best}} \leftarrow D(\mathbf{VLEC}^0)$, $\quad \hat{R}^{\text{best}} \leftarrow R(\mathbf{VLEC}^0)$, $\quad \hat{E}^{\text{best}} \leftarrow E(\mathbf{VLEC}^0)$, $\quad \hat{T}^{\text{best}} \leftarrow T(\mathbf{VLEC}^0)$

3: $P \leftarrow P^{\text{max}}$

4: **loop**

5: $\quad D^{\text{best}} \leftarrow \hat{D}^{\text{best}}, R^{\text{best}} \leftarrow \hat{R}^{\text{best}}, E^{\text{best}} \leftarrow \hat{E}^{\text{best}}, T^{\text{best}} \leftarrow \hat{T}^{\text{best}}$

6: $\quad L^{\text{prev}} \leftarrow L(\mathbf{L})$

7: $\quad \mathbf{VLEC}^{\text{worst}} \leftarrow \text{argmin}_{\mathbf{VLEC} \in \mathbf{L}} M(\mathbf{VLEC})$

8: $\quad$ **for** $p \leftarrow 1$ to P **do**

9: $\qquad$ Generate a random mutation $\mathbf{VLEC}'$ of $\mathbf{VLEC}^{\text{worst}}$.

10: $\qquad$ **if** $(M(\mathbf{VLEC}') > M(\mathbf{VLEC}^{\text{worst}})$ **and** $\beta^D D(\mathbf{VLEC}') < \beta^D D^{\text{lim}}$ **and** $\beta^R R(\mathbf{VLEC}') < \beta^R R^{\text{lim}}$ **and** $\mathbf{VLEC}' \notin \mathbf{L})$ **then**

11: $\qquad\quad$ Insert $\mathbf{VLEC}'$ into the list $\mathbf{L}$, in order that its length $L(\mathbf{L})$ is incremented.

12: $\qquad\quad$ **if** $\beta^D D(\mathbf{VLEC}') > \beta^D \hat{D}^{\text{best}}$ **then** $\hat{D}^{\text{best}} \leftarrow D(\mathbf{VLEC}')$

13: $\qquad\quad$ **if** $\beta^R R(\mathbf{VLEC}') > \beta^R \hat{R}^{\text{best}}$ **then** $\hat{R}^{\text{best}} \leftarrow R(\mathbf{VLEC}')$

14: $\qquad\quad$ **if** $E(\mathbf{VLEC}') > \hat{E}^{\text{best}}$ **then** $\hat{E}^{\text{best}} \leftarrow E(\mathbf{VLEC}')$

15: $\qquad\quad$ **if** $T(\mathbf{VLEC}') < \hat{T}^{\text{best}}$ **then** $\hat{T}^{\text{best}} \leftarrow T(\mathbf{VLEC}')$

16: $\qquad$ **end if**

17: $\quad$ **end for**

18: $\quad$ **if** $L(\mathbf{L}) = 1$ **then**

19: $\qquad$ Return the single remaining VLEC codebook in the list $\mathbf{VLEC}^{\text{worst}}$.

20: $\quad$ **else**

21: $\qquad$ Remove $\mathbf{VLEC}^{\text{worst}}$ from the list $\mathbf{L}$, in order that its length $L(\mathbf{L})$ is decremented.

22: $\quad$ **end if**

23: $\quad P \leftarrow \max(1, \min(P^{\text{next}}, P^{\text{max}}))$

24: **end loop**

GA proposed in Section 8.3. In the proposed scheme, an IrVLC outer codec is serially concatenated [275] and iteratively decoded [278] with an inner codec, as in the BBIrVLC-TCM scheme of Section 7.3. However, in contrast to the BBIrVLC-TCM scheme of Section 7.3, our proposed scheme does not employ a TCM inner codec, since this limits the maximum effective throughput and hence fails to approach the achievable capacity, as described in Section 8.1. Instead, we employ a URC-based inner codec [308], which does not suffer from an effective throughput loss [204]. For the sake of simplicity, Binary Phase-Shift Keying (BPSK) modulation is employed, facilitating an unambiguous mapping of the URC-encoded bits to the modulation constellation points [300]. A schematic of the proposed IrVLC-URC scheme is provided in Figure 8.6.

Similarly to the BBIrVLC-TCM scheme detailed in Section 7.2, APP SISO IrVLC decoding is achieved by applying the Bahl–Cocke–Jelinek–Raviv (BCJR) algorithm [48]

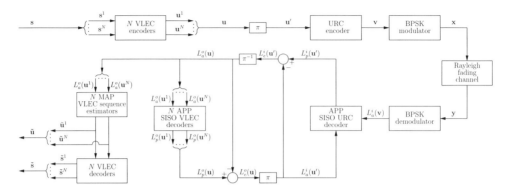

Figure 8.6: Schematic of IrVLC-URC scheme. The N number of VLEC encoders, APP SISO decoders and MAP sequence estimators are each based upon a different component VLEC codebook. ©IEEE [200] Maunder and Hanzo, 2008.

to bit-based trellises [180]. Note that bit-based trellises are chosen instead of symbol-based trellises [181], since they exhibit a lower computational complexity, as observed in Figure 7.12 of Section 7.4.4. A trellis [141] is also employed as the basis of APP SISO URC decoding using the BCJR algorithm. Hence, the APP SISO IrVLC and TCM decoders can share resources in systolic-array-based chips, facilitating a cost-effective implementation. Note that the BCJR algorithm requires only Add, Compare and Select (ACS) operations when decoding operations are carried out in the logarithmic domain, while using a lookup table to correct the Jacobian approximation in the Log-MAP algorithm [296], as described in Section 3.4.

Recall that the BBIrVLC-TCM scheme of Section 7.3 employed a high number of transmission sub-frames. In order that these sub-frames could be decoded, it was necessary to explicitly convey their lengths to the receiver as side information. In the proposed IrVLC-URC scheme, we shall attain a significant reduction in the amount of side information that must be conveyed by employing only a single transmission sub-frame per component VLEC codebook.

8.4.1 Joint Source and Channel Coding

As described above, the proposed IrVLC-URC scheme facilitates the near-capacity trans-mission of source symbols having values with unequal probabilities of occurrence over an uncorrelated narrowband Rayleigh fading channel. In the transmitter of the proposed IrVLC-URC scheme portrayed in Figure 8.6, the source symbol frame $\mathbf{s}$ comprises J^{sum} number of source symbols and is decomposed into N number of sub-frames $\{\mathbf{s}^n\}_{n=1}^N$. Each source symbol sub-frame $\mathbf{s}^n = \{s_{j^n}^n\}_{j^n=1}^{J^n}$ comprises J^n number of source symbols, where we have

$$\sum_{n=1}^N J^n = J^{\text{sum}}. \tag{8.13}$$

Note that the fractions $\{C^n\}_{n=1}^N$, where $C^n = J^n/J^{\mathrm{sum}}$, may be specifically chosen in order to shape the inverted IrVLC EXIT function, as seen in Figures 7.4 and 7.5 of Section 7.3.2.

Each source symbol sub-frame $\mathbf{s}^n$ is VLEC-encoded using a corresponding component VLEC codebook $\mathbf{VLEC}^n$, having a particular coding rate $R(\mathbf{VLEC}^n)$ and RV-FDM $D(\mathbf{VLEC}^n)$. The specific source symbols having the value of $k \in [1 \ldots K]$ and encoded by the specific component VLEC codebook $\mathbf{VLEC}^n$ are represented by the codeword $\mathbf{VLEC}^{n,k}$, which has a length of $I^{n,k}$ bits. The J^n number of VLEC codewords that represent the J^n number of source symbols in each particular source symbol sub-frame $\mathbf{s}^n$ are concatenated to provide the transmission sub-frame $\mathbf{u}^n = \{\mathbf{VLEC}^{n,s_{j^n}^n}\}_{j^n=1}^{J^n}$, having a length of

$$I^n = \sum_{j^n=1}^{J^n} I^{n,s_{j^n}^n} \tag{8.14}$$

bits. Note that these transmission sub-frame lengths will typically vary from frame to frame, owing to the variable lengths of the VLEC codewords.

In order to facilitate the VLEC decoding of each transmission sub-frame $\mathbf{u}^n$, it is necessary to explicitly convey its length I^n to the receiver. Furthermore, this highly error-sensitive side information must be reliably protected against transmission errors. This may be achieved using a low-rate block code, for example. For the sake of avoiding obfuscation, this is not shown in Figure 8.6. In Section 8.5.5 we comment on the amount of side information that is required for reliably conveying the specific number of bits in each transmission sub-frame to the decoder. As described above, however, we may expect this to be significantly less than the amount of side information required in the BBIrVLC-TCM scheme of Section 7.2.

In the scheme's transmitter, the N transmission sub-frames $\{\mathbf{u}^n\}_{n=1}^N$ are concatenated, as shown in Figure 8.6. The resultant transmission frame $\mathbf{u}$ has a length of $I^{\mathrm{sum}} = \sum_{n=1}^N I^n$ bits. Note that each component VLEC codebook $\mathbf{VLEC}^n$ generates a fraction $\alpha^n = I^n/I^{\mathrm{sum}}$ of the transmission frame $\mathbf{u}$.

Following IrVLC encoding, the bits of the transmission frame $\mathbf{u}$ are interleaved in the block π of Figure 8.6 in order to provide the interleaved transmission frame $\mathbf{u}'$. This is URC encoded in order to obtain the URC-encoded frame $\mathbf{v}$, and BPSK modulation is employed to generate the channel's input symbols $\mathbf{x}$, as shown in Figure 8.6. These symbols are transmitted over an uncorrelated narrowband Rayleigh fading channel and are received as the channel's output symbols $\mathbf{y}$, as seen in Figure 8.6.

8.4.2 Iterative Decoding

In the receiver of Figure 8.6, APP SISO URC and VLEC decoding are performed iteratively. As usual, extrinsic soft information, represented in the form of LLRs [292], is iteratively exchanged between the URC and VLEC decoding stages for the sake of assisting each other's operation [275, 278], as described in Section 4.3. In Figure 8.6, $L(\cdot)$ denotes the LLRs of the bits concerned, where the superscript i indicates inner URC decoding, while o corresponds to outer VLEC decoding. Additionally, a subscript denotes the dedicated role of the LLRs, with a, p and e indicating *a priori*, *a posteriori* and extrinsic information respectively.

During each decoding iteration, the inner URC decoder is provided with *a priori* LLRs pertaining to the interleaved transmission frame $L_a^i(\mathbf{u}')$, as shown in Figure 8.6. These LLRs are gleaned from the most recent iteration of the outer VLEC decoding stage's operation, as highlighted below. In the case of the first decoding iteration, no previous VLEC decoding has been performed and hence the *a priori* LLRs $L_a^i(\mathbf{u}')$ provided for URC decoding are all zero valued, corresponding to a probability of 0.5 for both '0' and '1.' Given the *a priori* LLRs $L_a^i(\mathbf{v})$ provided by the soft BPSK demodulator and the *a priori* LLRs $L_a^i(\mathbf{u}')$, the BCJR algorithm is employed for obtaining the *a posteriori* LLRs $L_p^i(\mathbf{u}')$, as shown in Figure 8.6.

During iterative decoding it is necessary to prevent the reuse of already exploited information, since this would limit the attainable iteration gain [296], as described in Section 4.2. This is achieved following URC decoding by the subtraction of $L_a^i(\mathbf{u}')$ from $L_p^i(\mathbf{u}')$, as observed in Figure 8.6. The resultant extrinsic LLRs $L_e^i(\mathbf{u}')$ are de-interleaved in the block π^{-1} and forwarded as *a priori* LLRs to the VLEC decoder. As described in Section 4.3, interleaving is employed in order to mitigate correlation within the *a priori* LLR frames. This is necessary since the BCJR algorithm assumes that all *a priori* LLRs that can influence any particular decoding decision are uncorrelated.

Since N separate VLEC encoding processes are employed in the proposed scheme's transmitter, N separate VLEC decoding processes are employed in its receiver. In parallel with the composition of the bit-based transmission frame $\mathbf{u}$ from its N sub-frames, the *a priori* LLRs $L_a^o(\mathbf{u})$ are decomposed into N sub-frames, as seen in Figure 8.6. This is achieved with the aid of the explicit side information that conveys the number of bits I^n in each transmission sub-frame $\mathbf{u}^n$. Each of the N VLEC decoding processes is provided with the *a priori* LLR sub-frame $L_a^o(\mathbf{u}^n)$ and in response it generates the *a posteriori* LLR sub-frame $L_p^o(\mathbf{u}^n)$, $n \in [1 \dots N]$. These *a posteriori* LLR sub-frames are concatenated in order to provide the *a posteriori* LLR frame $L_p^o(\mathbf{u})$, as portrayed in Figure 8.6. Following the subtraction of the *a priori* LLRs $L_a^o(\mathbf{u})$, the resultant extrinsic LLRs $L_e^o(\mathbf{u})$ are interleaved and forwarded as *a priori* information to the next URC decoding iteration.

Following the completion of iterative decoding, each transmission sub-frame $\mathbf{u}^n$ is estimated from the corresponding *a priori* LLR sub-frame $L_a^o(\mathbf{u}^n)$. This may be achieved by employing MAP sequence estimation operating on a bit-based trellis structure, as shown in Figure 8.6. The resultant transmission sub-frame estimate $\tilde{\mathbf{u}}^n$ is VLEC decoded using the corresponding component VLEC codebook $\mathbf{VLEC}^n$ in order to provide the source symbol sub-frame estimate $\tilde{\mathbf{s}}^n$, as seen in Figure 8.6. Note that this approach cannot exploit the knowledge that each sub-frame $\mathbf{s}^n$ comprises a particular number J^n of source symbols. For this reason, the source symbol sub-frame estimate $\tilde{\mathbf{s}}^n$ of each source symbol sub-frame $\mathbf{s}^n$ may or may not contain J^n source symbols. In order that we may prevent the loss of synchronization that this would imply, dummy source symbol estimates are removed from, or appended to the end of, each source symbol sub-frame estimate $\tilde{\mathbf{s}}^n$ in order to ensure that they each comprise the correct number of source symbol estimates. Finally, the source symbol frame estimate $\tilde{\mathbf{s}}$ is obtained by concatenating the N source symbol sub-frame estimates, as observed in Figure 8.6.

8.5   Parameter Design for the Proposed Scheme

In this section we demonstrate the employment of both Algorithm 3.3 of Section 3.3 and the GA of Section 8.3 to design suites of IrVLC component VLEC codebooks. The APP SISO decoding and MAP sequence estimation performances associated with these codebooks are characterized, and we investigate the suitability of the two suites of component VLEC codebooks for use in the IrVLC-URC scheme of Section 8.4. Finally, we detail a number of parameterizations of the IrVLC-URC scheme of Section 8.4, which employ various IrVLC coding rates.

8.5.1 Design of IrVLC Component VLEC Codebook Suites

As described above, both Algorithm 3.3 of Section 3.3 and the GA of Section 8.3 were employed to design suites of IrVLC component VLEC codebooks. All component VLEC codebooks were designed for the encoding of $K = 16$-ary source symbol values, having the specific probabilities of occurrence that result from the Lloyd–Max quantization [164, 165] of independent Gaussian distributed continuous-valued source samples, as exemplified in Section 7.2.1. These occurrence probabilities of $\{P(k)\}_{k=1}^{K} = \{0.008, 0.024, 0.043, 0.060,$ $0.076, 0.089, 0.097, 0.102, 0.102, 0.097, 0.089, 0.076, 0.060, 0.043, 0.024, 0.008\}$ can be seen to vary by more than an order of magnitude. Since these probabilities correspond to varying numbers of $\{-\log_2(P(k))\}_{k=1}^{K} = \{6.93, 5.35, 4.55, 4.05, 3.72, 3.49, 3.36, 3.29,$ $3.29, 3.36, 3.49, 3.72, 4.05, 4.55, 5.35, 6.93\}$ bits of information per symbol, the application of VLCs is motivated, since the source entropy is $E = 3.77$ bits of information per symbol, according to Equation (8.2). Furthermore, all component VLEC codebooks were designed to have RV-FDMs, and hence IV-FD lower bounds, of at least two. This ensures that the resultant IrVLC schemes will have inverted EXIT functions that reach the top right-hand corner of the EXIT chart and will therefore support iterative decoding convergence to an infinitesimally low probability of error, as described in Section 4.3.

As argued in Section 8.1, employing Algorithm 3.3 of Section 3.3 to design a suite of component VLEC codebooks having a wide range of inverted EXIT function shapes typically requires a significant amount of 'trial-and-error'-based human interaction. As we demonstrate in this section, this is avoided when employing the GA of Section 8.3. Hence, in order to ensure a fair comparison, the trial and error was avoided when designing a suite of component VLEC codebooks for use in IrVLCs using Algorithm 3.3 of Section 3.3. Instead, 11 component VLEC codebooks $\{\mathbf{VLEC}^n\}_{n=1}^{11}$ having IV-FD lower bounds given by $\bar{d}_{\text{free}}(\mathbf{VLEC}^n) = n + 1 \, \forall n \in [1 \ldots 11]$ were generated. This was achieved in accordance with Equation (8.5) by specifying a target minimum block distance of

$$d_{b_{\min}}(\mathbf{VLEC}^n) = \bar{d}_{\text{free}}(\mathbf{VLEC}^n), \qquad (8.15)$$

a target minimum divergence distance of

$$d_{d_{\min}}(\mathbf{VLEC}^n) = \lceil \bar{d}_{\text{free}}(\mathbf{VLEC}^n)/2 \rceil \qquad (8.16)$$

and a target minimum convergence distance of

$$d_{c_{\min}}(\mathbf{VLEC}^n) = \lfloor \bar{d}_{\text{free}}(\mathbf{VLEC}^n)/2 \rfloor. \qquad (8.17)$$

This approach is not labor-intensive, but results in a suite of component VLEC codebooks having only a **limited** variety of inverted EXIT function shapes, as will be shown in Section 8.5.2. The coding rate $R(\mathbf{VLEC}^n)$, RV-FDM $D(\mathbf{VLEC}^n)$ and the composition of each resultant component VLEC codebook $\mathbf{VLEC}^n$ is provided in Table 8.4. Note that, in all cases, the VLEC-encoded bit entropy $E(\mathbf{VLEC}^n)$ was found to be only a negligible distance away from unity.

Table 8.4: Properties and composition of the 11 component VLEC codebooks $\{\mathbf{VLEC}^n\}_{n=1}^{11}$ designed using Algorithm 3.3 of Section 3.3

$\mathbf{VLEC}^n$	Properties[†]	Composition[‡]
$\mathbf{VLEC}^1$	(0.956, 2.395)	6,6,5,5,4,4,3,3,3,4,4,5,5,6,6, E97E2BD3314A4F1321
$\mathbf{VLEC}^2$	(0.637, 3.786)	9,8,7,7,6,6,6,5,4,5,6,6,7,7,7,8, 3EBBB53335228C1853E597BAFB
$\mathbf{VLEC}^3$	(0.603, 4.354)	10,10,9,8,7,6,5,4,4,5,6,7,8,8,9,10, 9AA5CA652B36B3E065564E5554E65
$\mathbf{VLEC}^4$	(0.468, 5.775)	12,11,10,9,9,8,8,6,5,7,8,9,9,10,11,12, 8DC2F2F996FDB346A3815B72DA394EEFC4FF
$\mathbf{VLEC}^5$	(0.446, 6.388)	12,12,11,10,10,9,7,6,6,7,8,9,10,11,11,12, 2D917354DA4CDC9A71FC01CACCB694ED6233A9
$\mathbf{VLEC}^6$	(0.376, 7.815)	14,13,12,12,11,10,9,8,7,9,10,11,11,12,13,13, 56D59BBBD526C778ACB683C00D4E69D35594EBA D8663
$\mathbf{VLEC}^7$	(0.371, 8.331)	13,13,12,12,11,10,9,8,8,9,10,11,12,12,13,13, C959B2B1AA8D5A9B65E1FF000F29A6B1664F198 E9395
$\mathbf{VLEC}^8$	(0.311, 9.523)	17,16,15,14,13,12,11,10,8,11,12,13,14,15,16,16, 16CEAAE7BB6CB5D2BB5DF1AE68F800CEB4E735D B29F51E7AD4D5EA
$\mathbf{VLEC}^9$	(0.308, 10.359)	17,16,15,14,13,12,11,10,9,11,12,13,14,15,16,17, 1E7E727198F26E57CE7A6D5E41F002C6BCB4D72 ED5AD732EE3ABA7
$\mathbf{VLEC}^{10}$	(0.267, 11.512)	19,18,17,16,15,14,13,12,11,12,14,15,16,17,18,19, 3560BB748EDE0D768C5B9AEAB1CE01F8007C1E5 636CAAAD6E73653E1CD66F4
$\mathbf{VLEC}^{11}$	(0.265, 12.404)	19,18,17,16,15,14,13,12,11,13,14,15,16,17,18,19, 35EB5B3BC1D9DAEB73994EB37CF183F00038E2D 9276B2FD1B6736BDA6AE3E8

[†] The properties of each component VLEC codebook $\mathbf{VLEC}^n$ are provided using the format $(R(\mathbf{VLEC}^n), D(\mathbf{VLEC}^n))$.

[‡] The composition of each component VLEC codebook $\mathbf{VLEC}^n$ is specified by providing the $K = 16$ codeword lengths $\{I^{n,k}\}_{k=1}^K$, together with the hexadecimal representation of the ordered concatenation of the $K = 16$ VLEC codewords $\{\mathbf{VLEC}^{n,k}\}_{k=1}^K$ in the codebook.

Similarly, the GA of Section 8.3 was employed to generate a suite of 11 component VLEC codebooks $\{\mathbf{VLEC}^n\}_{n=12}^{22}$. However, this suite comprises two categories of component VLEC codebook, namely $\{\mathbf{VLEC}^n\}_{n=12}^{18}$ and $\{\mathbf{VLEC}^n\}_{n=19}^{22}$. We demonstrate in Section 8.5.2 that these two categories provide some component VLEC codebooks having RV-FDMs that correspond to 'S'-shaped inverted EXIT functions having two points of inflection and some that have only one point of inflection, as described in Section 8.2. More specifically, the GA attempted to maximize the RV-FDM $D(\mathbf{VLEC})$ in order to provide 'S'-shaped

Table 8.5: Parameters employed by the proposed GA during the generation of the 11 component VLEC codebooks $\{\mathbf{VLEC}^n\}_{n=12}^{22}$ (the variables are defined in Table 8.3)

VLEC	$\mathbf{VLEC}^0$	β^D	β^R	D^{lim}	R^{lim}
$\mathbf{VLEC}^{12}$	$\mathbf{VLEC}^1$	$+1$	-1	∞	0.95
$\mathbf{VLEC}^{13}$	$\mathbf{VLEC}^1$	$+1$	-1	∞	0.85
$\mathbf{VLEC}^{14}$	$\mathbf{VLEC}^1$	$+1$	-1	∞	0.75
$\mathbf{VLEC}^{15}$	$\mathbf{VLEC}^1$	$+1$	-1	∞	0.65
$\mathbf{VLEC}^{16}$	$\mathbf{VLEC}^3$	$+1$	-1	∞	0.50
$\mathbf{VLEC}^{17}$	$\mathbf{VLEC}^7$	$+1$	-1	∞	0.35
$\mathbf{VLEC}^{18}$	$\mathbf{VLEC}^{11}$	$+1$	-1	∞	0.25
$\mathbf{VLEC}^{19}$	$\mathbf{VLEC}^1$	-1	-1	2	0.65
$\mathbf{VLEC}^{20}$	$\mathbf{VLEC}^1$	-1	-1	2	0.50
$\mathbf{VLEC}^{21}$	$\mathbf{VLEC}^1$	-1	-1	2	0.35
$\mathbf{VLEC}^{22}$	$\mathbf{VLEC}^1$	-1	-1	2	0.25

inverted EXIT functions having two points of inflection during the generation of the first seven component VLEC codebooks $\{\mathbf{VLEC}^n\}_{n=12}^{18}$, having coding rates $R(\mathbf{VLEC})$ in the range of $[0.25, 0.95]$. In contrast, during the generation of the remaining four component VLEC codebooks $\{\mathbf{VLEC}^n\}_{n=19}^{22}$, which have coding rates $R(\mathbf{VLEC})$ in the range of $[0.25, 0.65]$, RV-FDMs $D(\mathbf{VLEC})$ as close to, but no less than, two were sought in order to provide inverted EXIT functions that have only one point of inflection. Naturally, maximal VLEC-encoded bit entropies and minimal bit-based trellis complexities were sought in all cases. The GA parameters employed to design the two categories of component VLEC codebooks $\{\mathbf{VLEC}^n\}_{n=12}^{18}$ and $\{\mathbf{VLEC}^n\}_{n=19}^{22}$ are detailed in Table 8.5. Note that the specified coding rates were chosen in order to provide a suite of component VLEC codebooks having a wide variety of inverted EXIT function shapes.

During all iterations of the GA, a target list length of $L^{\text{tar}} = 1000$ was employed. Furthermore, in order to limit the GA's complexity, the value of N^{max} was set only just high enough to ensure that the list length reached $L^{\text{tar}} = 1000$.

Finally, the generation of component VLEC codebooks using the GA of Section 8.3 was divided into two phases.

During the first phase of the GA's operation, each parent VLEC codebook $\mathbf{VLEC}^0$ was provided by one of the codebooks designed using Algorithm 3.3 of Section 3.3 $\{\mathbf{VLEC}^n\}_{n=1}^{11}$, as shown in Table 8.5. More specifically, for each of the seven component VLEC codebooks $\{\mathbf{VLEC}^n\}_{n=12}^{18}$, the heuristically designed VLEC codebook from $\{\mathbf{VLEC}^n\}_{n=1}^{11}$ having the coding rate $R(\mathbf{VLEC})$ that is closest to, but not lower than, the target coding rate R^{lim} was employed as the parent codebook $\mathbf{VLEC}^0$. By contrast, the high-rate VLEC codebook $\mathbf{VLEC}^1$ was employed as the parent codebook during the generation of each of the four component VLEC codebooks $\{\mathbf{VLEC}^n\}_{n=19}^{22}$.

Note that each first phase of the GA's operation was employed to provide a VLEC codebook $\mathbf{VLEC}$ having an optimized RV-FDM $D(\mathbf{VLEC})$ and a coding rate $R(\mathbf{VLEC})$

Table 8.6: Properties and composition of the 11 component VLEC codebooks $\{\mathbf{VLEC}^n\}_{n=12}^{22}$ designed using the GA of Section 8.3

$\mathbf{VLEC}^n$	Properties[†]	Composition[‡]
$\mathbf{VLEC}^{12}$	(0.950, 2.412)	6,6,5,5,3,4,3,3,3,4,4,4,5,5,6,6,
		317E29FEA2490F1EA1
$\mathbf{VLEC}^{13}$	(0.850, 2.617)	6,7,5,5,4,4,3,3,5,4,4,5,5,6,7,8,
		140DE2BCB304A54F6D818
$\mathbf{VLEC}^{14}$	(0.750, 2.793)	7,8,7,6,6,4,5,4,5,4,5,3,6,5,8,7,
		1761E3F330FB2A494F8AD1C
$\mathbf{VLEC}^{15}$	(0.650, 2.981)	7,7,8,7,7,5,6,4,6,4,6,5,6,7,8,7,
		3540C30FF578C04E753C1862D
$\mathbf{VLEC}^{16}$	(0.501, 4.807)	13,11,9,11,7,8,6,5,5,6,8,7,9,11,12,13,
		13192CCD6E0F556B3F80CA55438997333A69
$\mathbf{VLEC}^{17}$	(0.350, 8.475)	15,14,16,14,12,10,9,8,8,9,10,11,13,13,15,16,
		1522D8CA63D49CD714DB2F0FF8007A4D358B39A
		6353669C2B
$\mathbf{VLEC}^{18}$	(0.251, 12.449)	25,18,17,16,15,14,13,12,11,13,14,15,28,17,18,19,
		1670497B3BC1D9DAEB73994EB37CF183F00038E
		2D9276B2CE0FB5B6736BDA6AE3E8
$\mathbf{VLEC}^{19}$	(0.653, 2.269)	14,11,11,7,9,5,3,3,3,3,5,7,5,9,16,12,
		1558553AA17EA7350C850E542AB42AA
$\mathbf{VLEC}^{20}$	(0.501, 2.133)	11,17,8,8,9,5,6,2,2,3,12,18,5,15,11,14,
		0DA1B6DCD0DCD843181B61B6D8736D86DCDB7
$\mathbf{VLEC}^{21}$	(0.350, 2.155)	13,19,7,17,10,10,7,3,3,3,13,37,6,16,16,24,
		36D1B6D9346DB7368D932C536C9B6D804764936
		D936DA36DB01
$\mathbf{VLEC}^{22}$	(0.252, 2.117)	8,14,8,9,12,6,5,2,3,2,56,62,5,15,11,11,
		06E6DB9A1B0DB0C22236DB34343BEB6C36DB343
		43BEB6DB0E6DB0DB9B4

[†] The properties of each component VLEC codebook $\mathbf{VLEC}^n$ are provided using the format $(R(\mathbf{VLEC}^n), D(\mathbf{VLEC}^n))$.

[‡] The composition of each component VLEC codebook $\mathbf{VLEC}^n$ is specified by providing the $K = 16$ codeword lengths $\{I^{n,k}\}_{k=1}^K$, together with the hexadecimal representation of the ordered concatenation of the $K = 16$ VLEC codewords $\{\mathbf{VLEC}^{n,k}\}_{k=1}^K$ in the codebook.

that is reduced from that of the parent code $\mathbf{VLEC}^0$ towards the target coding rate R^{lim}. This was achieved using the parameter values of $\alpha^D = 1$ and $\alpha^R = \alpha^E = \alpha^T = 0$. Note that whilst this specific choice of the parameters appears to optimize only the RV-FDM $D(\mathbf{VLEC})$, this naturally results in a reduced coding rate $R(\mathbf{VLEC})$, as desired.

During the second phase of the GA's operation, the result of the first phase was employed as the parent code $\mathbf{VLEC}^0$. Furthermore, the parameter values of $\alpha^D = 6$, $\alpha^R = 2$, $\alpha^E = 3$ and $\alpha^T = 1$ were employed in order to jointly optimize the resultant RV-FDM $D(\mathbf{VLEC})$, the coding rate $R(\mathbf{VLEC})$, the VLEC-encoded bit entropy $E(\mathbf{VLEC})$ and the bit-based trellis complexity $T(\mathbf{VLEC})$. The coding rates, the RV-FDMs and the composition of the 11 resultant component VLEC codebooks are provided in Table 8.6. Note that in all cases the VLEC-encoded bit entropy $E(\mathbf{VLEC}^n)$ was found to be only a negligible distance away from unity. Rather than providing the bit-based trellis complexity $T(\mathbf{VLEC}^n)$ of each component VLEC codebook $\mathbf{VLEC}^n$, we consider the computational complexity associated with APP SISO decoding and MAP sequence estimation in Section 8.5.2.

8.5.2 Characterization of Component VLEC Codebooks

In this section we characterize the component VLEC codebooks of the two suites that were introduced in Section 8.5.1, which were designed using Algorithm 3.3 of Section 3.3 and the GA of Section 8.3. More specifically, we consider both the APP SISO decoding and the MAP sequence estimation performances that are associated with these component VLEC codebooks.

The APP SISO decoding and MAP sequence estimation performances of the component VLEC codebooks $\{\mathbf{VLEC}^n\}_{n=1}^{22}$ were investigated by simulating the VLEC encoding and decoding of 100 frames of $10\,000$ randomly generated source symbols. As described in Section 8.5.1, these source symbols have $K = 16$-ary values with the specific probabilities of occurrence that result from the Lloyd–Max quantization of independent Gaussian distributed source samples. The corresponding frames of VLEC-encoded bits were employed to generate uncorrelated Gaussian distributed *a priori* LLR frames having a range of mutual informations $I_a \in [0, E(\mathbf{VLEC})]$, where $E(\mathbf{VLEC})$ is the VLEC-encoded bit entropy of the VLEC codebook $\mathbf{VLEC}$. Note that the VLEC-encoded bit entropy $E(\mathbf{VLEC})$ was found to be only at a negligible distance from unity in the case of all the component VLEC codebooks considered, as described in Section 8.5.1. During APP SISO decoding and MAP sequence estimation, all calculations were performed in the logarithmic probability domain, while using an eight-entry lookup table for correcting the Jacobian approximation in the Log-MAP algorithm [296]. This approach requires only ACS computational operations, which will be used as our computational complexity measure, since this is characteristic of the complexity/area/speed trade-offs in systolic-array-based chips.

The APP SISO decoding performances of the component VLEC codebooks designed using Algorithm 3.3 of Section 3.3 and the GA of Section 8.3 are characterized by the inverted EXIT functions of Figures 8.7 and 8.8 respectively. Here, all mutual information measurements were made using the histogram-based approximation of the LLR Probability Density Functions (PDFs) [294]. For each considered VLEC codebook $\mathbf{VLEC}$, the coding rate $R(\mathbf{VLEC})$ and RV-FDM $D(\mathbf{VLEC})$ are provided in Figures 8.7 and 8.8. Furthermore, the computational complexity of APP SISO decoding is characterized by the average number $O^{\mathrm{APP}}(\mathbf{VLEC})$ of ACS operations performed per source symbol.

As shown in Figure 8.7, the suite of component VLEC codebooks $\{\mathbf{VLEC}^n\}_{n=1}^{11}$ generated using Algorithm 3.3 of Section 3.3 provides only a limited variety of 'S'-shaped inverted EXIT functions having as many as two points of inflection, as predicted in Section 8.5.1. These 'S'-shaped functions are generated by the HA when seeking a minimal coding rate for a particular IV-FD lower bound, which results in a relatively high RV-FDM for that specific coding rate, as described in Section 8.2. By contrast, the GA of Section 8.3 provided a suite of component VLEC codebooks $\{\mathbf{VLEC}^n\}_{n=12}^{22}$ having the wide variety of inverted EXIT function shapes shown in Figure 8.8, some having an 'S' shape with as many as two points of inflection, while others have no more than one point of inflection, as predicted in Section 8.5.1.

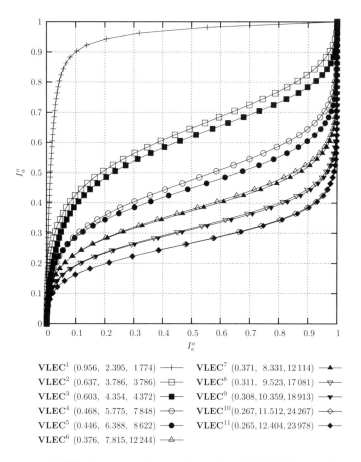

VLEC[1] $(0.956, 2.395, 1\,774)$ —+— VLEC[7] $(0.371, 8.331, 12\,114)$ —▲—
VLEC[2] $(0.637, 3.786, 3\,786)$ —⊟— VLEC[8] $(0.311, 9.523, 17\,081)$ —▽—
VLEC[3] $(0.603, 4.354, 4\,372)$ —■— VLEC[9] $(0.308, 10.359, 18\,913)$ —▼—
VLEC[4] $(0.468, 5.775, 7\,848)$ —○— VLEC[10] $(0.267, 11.512, 24\,267)$ —◇—
VLEC[5] $(0.446, 6.388, 8\,622)$ —●— VLEC[11] $(0.265, 12.404, 23\,978)$ —◆—
VLEC[6] $(0.376, 7.815, 12\,244)$ —△—

Figure 8.7: Inverted EXIT functions characterizing the APP SISO decoding performance of the component VLEC codebooks that were designed using Algorithm 3.3 of Section 3.3. Functions are labeled using the format **VLEC** $(R(\mathbf{VLEC}), D(\mathbf{VLEC}), O^{\mathrm{APP}}(\mathbf{VLEC}))$. ©IEEE [200] Maunder and Hanzo, 2008.

Note that for all component VLEC codebooks having relatively high RV-FDMs $\{\mathbf{VLEC}^n\}_{n=1}^{18}$, the RV-FDM $D(\mathbf{VLEC})$ can be seen to increase as the coding rate $R(\mathbf{VLEC})$ decreases. By contrast, the RV-FDM $D(\mathbf{VLEC})$ can be seen to decrease as the coding rate $R(\mathbf{VLEC})$ decreases in the case of the component VLEC codebooks $\{\mathbf{VLEC}^n\}_{n=19}^{22}$ designed using the GA of Section 8.3 to have minimized RV-FDMs. These results can be explained by the increased degree of design freedom that is facilitated by having reduced coding rates $R(\mathbf{VLEC})$. Finally, note that all inverted EXIT functions of Figures 8.7 and 8.8 reach the $(1, 1)$ point of the EXIT chart, since all component VLEC codebooks $\{\mathbf{VLEC}^n\}_{n=1}^{22}$ have RV-FDMs $D(\mathbf{VLEC})$ and hence IV-FD lower bounds $\bar{d}_{\mathrm{free}}(\mathbf{VLEC})$ of at least two [298], as described in Section 4.3.

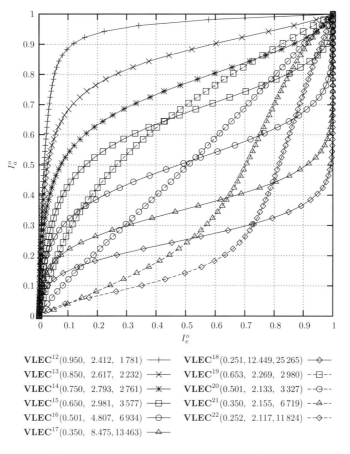

VLEC^{12}(0.950, 2.412, 1 781) —+—	VLEC^{18}(0.251, 12.449, 25 265) —◇—
VLEC^{13}(0.850, 2.617, 2 232) —×—	VLEC^{19}(0.653, 2.269, 2 980) --⊟--
VLEC^{14}(0.750, 2.793, 2 761) —✳—	VLEC^{20}(0.501, 2.133, 3 327) --⊖--
VLEC^{15}(0.650, 2.981, 3 577) —⊟—	VLEC^{21}(0.350, 2.155, 6 719) --△--
VLEC^{16}(0.501, 4.807, 6 934) —⊖—	VLEC^{22}(0.252, 2.117, 11 824) --◇--
VLEC^{17}(0.350, 8.475, 13 463) —△—	

Figure 8.8: Inverted EXIT functions characterizing the APP SISO decoding performance of the component VLEC codebooks that were designed using the GA of Section 8.3. Functions are labeled using the format **VLEC** $(R(\textbf{VLEC}),\ D(\textbf{VLEC}),\ O^{\text{APP}}(\textbf{VLEC}))$.

Let us now consider the MAP sequence estimation performance that is associated with each component VLEC codebook designed by Algorithm 3.3 of Section 3.3 and the GA of Section 8.3. This may be achieved by evaluating the BER between the transmission sub-frame $\mathbf{u}^n$ and the transmission sub-frame estimate $\tilde{\mathbf{u}}^n$ that is obtained following the consideration of the *a priori* LLR sub-frame $L^o_a(\mathbf{u}^n)$ by the corresponding VLEC MAP sequence estimator of Figure 8.6. Note that we opted for employing the BER associated with the transmission sub-frame estimate $\tilde{\mathbf{u}}^n$ as our performance metric, rather than the Symbol-Error Ratio (SER) associated with the source symbol sub-frame estimate $\tilde{\mathbf{s}}^n$. This is because, owing to the nature of VLEC codes, synchronization can be occasionally lost during the error-infested reconstruction of some source symbol frame estimates, resulting in a particularly poor SER [179]. Hence, the transmission of a prohibitively high number of source symbol frames would have to be simulated in order that the SER could be

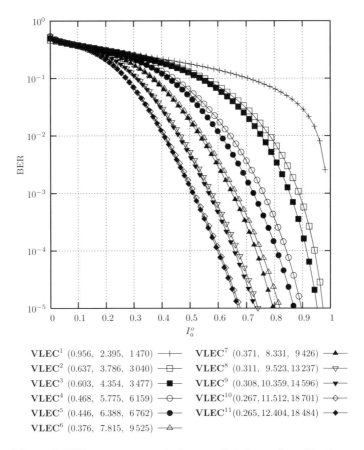

VLEC^1 (0.956, 2.395, 1 470) $-\!\!+\!\!-$ VLEC^7 (0.371, 8.331, 9 426) $-\!\!\blacktriangle\!\!-$

VLEC^2 (0.637, 3.786, 3 040) $-\!\!\boxminus\!\!-$ VLEC^8 (0.311, 9.523, 13 237) $-\!\!\triangledown\!\!-$

VLEC^3 (0.603, 4.354, 3 477) $-\!\!\blacksquare\!\!-$ VLEC^9 (0.308, 10.359, 14 596) $-\!\!\blacktriangledown\!\!-$

VLEC^4 (0.468, 5.775, 6 159) $-\!\!\ominus\!\!-$ VLEC^{10}(0.267, 11.512, 18 701) $-\!\!\lozenge\!\!-$

VLEC^5 (0.446, 6.388, 6 762) $-\!\!\bullet\!\!-$ VLEC^{11}(0.265, 12.404, 18 484) $-\!\!\blacklozenge\!\!-$

VLEC^6 (0.376, 7.815, 9 525) $-\!\!\triangle\!\!-$

Figure 8.9: Plots of BER versus *a priori* mutual information I_a^o characterizing the MAP sequence estimation performance of the component VLEC codebooks that were designed using Algorithm 3.3 of Section 3.3. Plots are labeled using the format **VLEC** $(R(\mathbf{VLEC}), \ D(\mathbf{VLEC}), \ O^{\mathrm{MAP}}(\mathbf{VLEC}))$.

averaged and employed as a reliable performance metric. Plots of the BER attained during our simulations versus the mutual information I_a^o associated with the *a priori* LLR sub-frame $L_a^o(\mathbf{u}^n)$ are provided in Figures 8.9 and 8.10 for the component VLEC codebooks designed by Algorithm 3.3 of Section 3.3 and the GA of Section 8.3 respectively. For each specific VLEC codebook **VLEC**, the coding rate $R(\mathbf{VLEC})$ and RV-FDM $D(\mathbf{VLEC})$ are also provided in Figures 8.9 and 8.10. Furthermore, the computational complexity of MAP sequence estimation is characterized by the average number $O^{\mathrm{MAP}}(\mathbf{VLEC})$ of ACS operations performed per source symbol. As may be expected, the MAP sequence estimator's computational complexities recorded in Figures 8.9 and 8.10 are lower than those of APP SISO decoding, as recorded in Figures 8.7 and 8.8.

Observe in Figures 8.9 and 8.10 that for all component VLEC codebooks having relatively high RV-FDMs $\{\mathbf{VLEC}^n\}_{n=1}^{18}$, the BER performance can be seen to improve as the coding

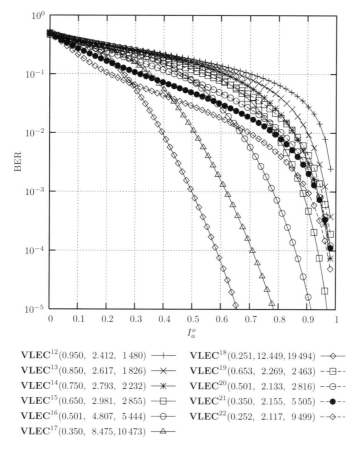

Figure 8.10: Plots of BER versus *a priori* mutual information I_a^o characterizing the MAP sequence estimation performance of the component VLEC codebooks that were designed using the GA of Section 8.3. Plots are labeled using the format **VLEC** ($R(\textbf{VLEC})$, $D(\textbf{VLEC})$, $O^{\text{MAP}}(\textbf{VLEC})$).

rate $R(\textbf{VLEC})$ decreases, as may be expected. Note however that a poor BER performance can be seen in Figure 8.10 for the minimized RV-FDM component VLEC codebooks $\{\textbf{VLEC}^n\}_{n=19}^{22}$ generated using the GA of Section 8.3. As a result, an IrVLC scheme comprising the component VLEC codebooks $\{\textbf{VLEC}^n\}_{n=12}^{22}$ can be expected to have a worse BER performance than that designed using the components $\{\textbf{VLEC}^n\}_{n=1}^{11}$. However, we concluded in Chapter 7 that the computational complexity associated with continuing iterative decoding until we achieve perfect convergence at the $(1, 1)$ point of the EXIT chart is justified, since an infinitesimally low BER is achieved. Hence the poor BER performance of the minimized RV-FDM component VLEC codebooks $\{\textbf{VLEC}^n\}_{n=19}^{22}$ becomes irrelevant, since it is circumvented.

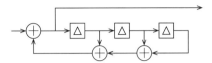

Figure 8.11: LFSR encoder schematic for the recursive URC codec employing a coding memory of $L_{\mathrm{URC}} = 3$.

Let us now comment on the APP SISO decoding and MAP sequence estimation computational complexities that are associated with the component VLEC codebooks of Section 8.5.1, as recorded in Figures 8.7–8.10. Observe that for all the specific component VLEC codebooks $\{\mathbf{VLEC}^n\}_{n=1}^{18}$ having relatively high RV-FDMs, the computational complexity of both the APP SISO decoding and the MAP sequence estimation can be seen to increase as the coding rate is reduced. This is due to the higher codeword lengths and hence the increased number of bit-based trellis transitions that are required at reduced coding rates. Whilst the computational complexities associated with the component VLEC codebooks $\{\mathbf{VLEC}^n\}_{n=19}^{22}$ having relatively low RV-FDMs also increase as the corresponding coding rates are reduced, these are lower than those of the codebooks $\{\mathbf{VLEC}^n\}_{n=1}^{18}$ having relatively high RV-FDMs. Owing to the presence of these low-complexity components within the suite of component VLEC codebooks $\{\mathbf{VLEC}^n\}_{n=12}^{22}$ generated using the GA of Section 8.3, a corresponding IrVLC scheme can be expected to have a lower computational complexity than one designed using the components $\{\mathbf{VLEC}^n\}_{n=1}^{11}$ generated using Algorithm 3.3 of Section 3.3.

8.5.3 Suitability of IrVLC Component Codebook Suites

Let us now consider the suitability of the component VLEC codeword suites of Section 8.5.1 designed using Algorithm 3.3 of Section 3.3 and the GA of Section 8.3 for near-capacity IrVLC coding in the proposed IrVLC-URC scheme of Section 8.4. In this scheme, an outer IrVLC codec having a parameterized composite coding rate of R_{IrVLC} is serially concatenated with an inner URC codec, which has a coding rate of $R_{\mathrm{URC}} = 1$ by definition. Here, $M_{\mathrm{BPSK}} = 2$-ary Phase-Shift Keying (PSK), or BPSK, modulation is employed for transmission over an uncorrelated narrowband Rayleigh fading channel, as described in Section 8.4. Note that the effective throughput $\eta = R_{\mathrm{IrVLC}} \cdot R_{\mathrm{URC}} \cdot \log_2(M_{\mathrm{BPSK}})$ of the IrVLC-URC scheme is equal to the composite IrVLC coding rate R_{IrVLC}, since $R_{\mathrm{URC}} = \log_2(M_{\mathrm{BPSK}}) = 1$. We opted for employing the Linear Feedback Shift Register (LFSR) structure shown in Figure 8.11 for our URC encoder. This employs a coding memory of $L_{\mathrm{URC}} = 3$ and recursive feedback, as shown in Figure 8.11. Note that a recursive URC codec was chosen, since these have EXIT functions that reach the top right-hand corner of the EXIT chart [299], as do all of our component VLEC codebooks. Hence we can expect to achieve an open EXIT-chart tunnel for the proposed IrVLC-URC scheme, if the channel quality is sufficiently high. As discussed in Section 4.3, this implies that iterative decoding convergence to an infinitesimally low probability of error is facilitated, provided that the iterative decoding trajectory approaches the inner and outer EXIT functions sufficiently closely.

A range of Rayleigh fading channel SNRs $E_c/N_0 \in [-3, 16]$ dB were considered, and in each case appropriate IrVLC schemes were designed using either the suite of component VLEC codebooks $\{\mathbf{VLEC}^n\}_{n=1}^{11}$ of Section 8.5.1 that were designed using Algorithm 3.3

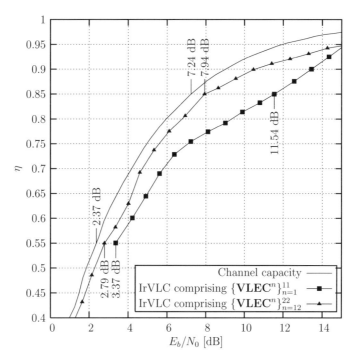

Figure 8.12: The maximum effective throughputs η of IrVLC schemes based on the component VLEC codebook suites $\{\mathbf{VLEC}^n\}_{n=1}^{11}$ and $\{\mathbf{VLEC}^n\}_{n=12}^{22}$ that yield an open EXIT-chart tunnel when serially concatenated with a URC that protects transmission over a BPSK-modulated Rayleigh fading channel having a particular E_b/N_0 value.

of Section 3.3 or the set of $\{\mathbf{VLEC}^n\}_{n=12}^{22}$ designed using the GA of Section 8.3. For each IrVLC scheme, the specific fraction α^n of the transmission frame $\mathbf{u}$ that is generated by each component VLEC codebook $\mathbf{VLEC}^n$ was designed using the EXIT-chart matching algorithm of [149] in order to shape the composite inverted IrVLC EXIT function, as described in Section 8.1. More specifically, the algorithm of [149] ensures that the inverted IrVLC EXIT function does not cross the URC EXIT function corresponding to the particular Rayleigh fading channel SNR E_c/N_0. In each case, we recorded the maximum effective throughput η, which ensured that the corresponding inverted EXIT function did not cross the URC's EXIT function, resulting in an open EXIT-chart tunnel.

The recorded effective throughputs η are plotted in Figure 8.12 together with the BPSK-modulated Rayleigh fading channel's capacity, which dictates the highest possible effective throughput η of the IrVLC-URC scheme that will permit iterative decoding convergence to an infinitesimally low probability of error [133]. Rather than employing the Rayleigh fading channel SNR E_c/N_0, Figure 8.12 plots the recorded effective throughputs against E_b/N_0, where the transmit energy per bit of source entropy is given as $E_b = E_c/\eta$, with E_c specifying the transmit energy per Rayleigh fading channel use and N_0 denoting the average noise energy.

As shown in Figure 8.12, the suite of component VLEC codebooks $\{\mathbf{VLEC}^n\}_{n=12}^{22}$ that was designed using the GA of Section 8.3 tended to be more suitable for use in IrVLCs than was the set $\{\mathbf{VLEC}^n\}_{n=1}^{11}$ designed using Algorithm 3.3 of Section 3.3. In both cases, open EXIT-chart tunnels were created at threshold channel E_b/N_0 values that were closer to the capacity bound when lower effective throughputs η were employed. The IrVLC scheme comprising the component VLEC codebooks $\{\mathbf{VLEC}^n\}_{n=12}^{22}$ is capable of maintaining open EXIT-chart tunnels for E_b/N_0 values within 1 dB of the BPSK-modulated Rayleigh fading channel's capacity bound for effective throughputs in the range of $\eta \in [0.4, 0.86]$ bits per channel use. By contrast, the IrVLC scheme comprising the component VLEC codebooks $\{\mathbf{VLEC}^n\}_{n=1}^{11}$ only becomes capable of maintaining open EXIT-chart tunnels within a substantially higher margin of 4.4 dB away from the channel's E_b/N_0 capacity bound for effective throughputs of up to 0.86 bits per channel use. Furthermore, this arrangement is unable to maintain open EXIT-chart tunnels for effective throughputs below 0.55 bits per channel use, as shown in Figure 8.12.

The relatively good performance of the IrVLC scheme comprising the component VLEC codebooks $\{\mathbf{VLEC}^n\}_{n=12}^{22}$ may be explained by the wide variety of the corresponding inverted component EXIT functions, some having 'S' shapes with as many as two points of inflection, whilst others have no more than one point of inflection, as shown in Figure 8.8. This facilitates accurate EXIT-chart matching to a wide range of inner decoder EXIT function shapes. By contrast, the IrVLC scheme comprising the component VLEC codebooks $\{\mathbf{VLEC}^n\}_{n=1}^{11}$ is associated with the more limited variety of inverted component EXIT function shapes shown in Figure 8.7.

8.5.4 Parameterizations of the Proposed Scheme

In this section we detail four parameterizations of the IrVLC-URC scheme introduced in Section 8.4. The IrVLC scheme of two of these parameterizations is based upon the suite of component VLEC codebooks $\{\mathbf{VLEC}^n\}_{n=1}^{11}$ that was designed using Algorithm 3.3 of Section 3.3, as detailed in Section 8.5.1. The other two parameterizations employ the suite of component VLEC codebooks $\{\mathbf{VLEC}^n\}_{n=12}^{22}$ that was designed using the GA of Section 8.3. In each pair of parameterizations, one employs an IrVLC coding rate of $R_{\mathrm{IrVLC}} = 0.85$, whilst the other employs $R_{IrVLC} = 0.55$. This will therefore illustrate the effect that the IrVLC coding rate has upon the associated APP SISO decoder's sensitivity to correlation within the *a priori* LLR frame $L_a^o(\mathbf{u})$, as described in Section 8.1. We refer to our four parameterizations as the $(\{\mathbf{VLEC}^n\}_{n=1}^{11}, R_{\mathrm{IrVLC}} = 0.85)$, $(\{\mathbf{VLEC}^n\}_{n=12}^{22}, R_{\mathrm{IrVLC}} = 0.85)$, $(\{\mathbf{VLEC}^n\}_{n=1}^{11}, R_{\mathrm{IrVLC}} = 0.55)$ and $(\{\mathbf{VLEC}^n\}_{n=12}^{22}, R_{\mathrm{IrVLC}} = 0.55)$ arrangements accordingly.

For each parameterization, the EXIT-chart matching algorithm of [149] was employed to shape the inverted IrVLC EXIT function under the constraint of achieving the above-mentioned IrVLC coding rates R_{IrVLC}. In each case, the lowest Rayleigh fading channel E_b/N_0 value for which an open EXIT-chart tunnel may be created was recorded. In Figure 8.12, these threshold E_b/N_0 values are employed to plot the parameterizations' effective throughputs η, which are equal to the IrVLC coding rates R_{IrVLC}, as described in Section 8.5.3.

As shown in Figure 8.12, the EXIT-chart matching algorithm of [149] was capable of achieving an open EXIT-chart tunnel for the $(\{\mathbf{VLEC}^n\}_{n=1}^{11}, R_{\mathrm{IrVLC}} = 0.85)$ and

($\{\mathbf{VLEC}^n\}_{n=12}^{22}$, $R_{\mathrm{IrVLC}} = 0.85$) arrangements at threshold channel E_b/N_0 values of 11.54 dB and 7.94 dB respectively. As described in Section 4.3, an open EXIT-chart tunnel implies that iterative decoding at an infinitesimally low probability of error can be achieved, if the iterative decoding trajectory approaches the inner and outer codecs' EXIT functions sufficiently closely. The threshold channel E_b/N_0 value of the ($\{\mathbf{VLEC}^n\}_{n=1}^{11}$, $R_{\mathrm{IrVLC}} = 0.85$) arrangement is 4.3 dB away from the capacity bound of 7.24 dB, as shown in Figure 8.12. As argued in Section 8.5.3, this relatively high discrepancy is a consequence of having a poor variety of inverted component EXIT functions in the suite $\{\mathbf{VLEC}^n\}_{n=1}^{11}$. By contrast, the threshold channel E_b/N_0 value of the ($\{\mathbf{VLEC}^n\}_{n=12}^{22}$, $R_{\mathrm{IrVLC}} = 0.85$) arrangement is only 0.7 dB from the channel's capacity bound, owing to the wide variety of inverted component EXIT functions in the suite $\{\mathbf{VLEC}^n\}_{n=12}^{22}$. With reference to Figure 8.12, note that the parameterizations employing an IrVLC coding rate of $R_{\mathrm{IrVLC}} = 0.85$ demonstrate the potential gain that is offered by employing the suite of component VLEC codebooks $\{\mathbf{VLEC}^n\}_{n=12}^{22}$ that was designed using the GA of Section 8.3 instead of the suite $\{\mathbf{VLEC}^n\}_{n=1}^{11}$ that was generated using Algorithm 3.3 of Section 3.3.

Similarly, an open EXIT-chart tunnel was created for the ($\{\mathbf{VLEC}^n\}_{n=1}^{11}$, $R_{\mathrm{IrVLC}} = 0.55$) and ($\{\mathbf{VLEC}^n\}_{n=12}^{22}$, $R_{\mathrm{IrVLC}} = 0.55$) arrangements at threshold channel E_b/N_0 values of 3.37 dB and 2.79 dB respectively, as shown in Figure 8.12. Whilst the threshold E_b/N_0 value of the ($\{\mathbf{VLEC}^n\}_{n=1}^{11}$, $R_{\mathrm{IrVLC}} = 0.55$) arrangement is 1 dB from the channel's capacity bound of 2.37 dB, that of the ($\{\mathbf{VLEC}^n\}_{n=12}^{22}$, $R_{\mathrm{IrVLC}} = 0.55$) arrangement is just 0.42 dB away, as shown in Figure 8.12. With reference to Figure 8.12, note that the nearest-capacity threshold E_b/N_0 values result when employing an IrVLC coding rate of $R_{\mathrm{IrVLC}} = 0.55$ both for the suite of component VLEC codebooks $\{\mathbf{VLEC}^n\}_{n=12}^{22}$ that was designed using the GA of Section 8.3 and for the suite $\{\mathbf{VLEC}^n\}_{n=1}^{11}$ that was generated using Algorithm 3.3 of Section 3.3.

Note that the 0.5 dB discrepancy between the threshold channel E_b/N_0 values and the capacity bounds of the IrVLC-TCM schemes of Section 7.3.2 are similar to those of the ($\{\mathbf{VLEC}^n\}_{n=12}^{22}$, $R_{\mathrm{IrVLC}} = 0.85$) and ($\{\mathbf{VLEC}^n\}_{n=12}^{22}$, $R_{\mathrm{IrVLC}} = 0.55$) arrangements, which are 0.7 dB and 0.42 dB respectively. This may be expected, because, like these parameterizations, the IrVLC-TCM schemes of Section 7.3.2 employed a suite of component VLC codebooks having a wide variety of inverted EXIT function shapes.

As described in Section 8.5.3, the EXIT-chart matching algorithm of [149] shaped the inverted IrVLC EXIT functions of our four IrVLC-URC scheme parameterizations by appropriately selecting the fraction C^n of the source symbol frame $\mathbf{s}$ and the fraction α^n of the transmission frame $\mathbf{u}$ that is encoded by each component VLEC codebook $\mathbf{VLEC}^n$. These fractions are summarized for the ($\{\mathbf{VLEC}^n\}_{n=1}^{11}$, $R_{\mathrm{IrVLC}} = 0.85$) and the ($\{\mathbf{VLEC}^n\}_{n=1}^{11}$, $R_{\mathrm{IrVLC}} = 0.55$) arrangements in Table 8.7 and for the ($\{\mathbf{VLEC}^n\}_{n=12}^{22}$, $R_{\mathrm{IrVLC}} = 0.85$) and the ($\{\mathbf{VLEC}^n\}_{n=12}^{22}$, $R_{\mathrm{IrVLC}} = 0.55$) arrangements in Table 8.8.

Let us now discuss the properties of each parameterization's activated component VLEC codebooks. These are employed to encode a nonzero fraction C^n of the source symbol frame $\mathbf{s}$, and, hence, a nonzero fraction α^n of the transmission frame $\mathbf{u}$. Observe in Tables 8.7 and 8.8 that in the ($\{\mathbf{VLEC}^n\}_{n=1}^{11}$, $R_{\mathrm{IrVLC}} = 0.85$), ($\{\mathbf{VLEC}^n\}_{n=12}^{22}$, $R_{\mathrm{IrVLC}} = 0.85$) and ($\{\mathbf{VLEC}^n\}_{n=1}^{11}$, $R_{\mathrm{IrVLC}} = 0.55$) arrangements, each activated component VLEC codebook $\mathbf{VLEC}^n$ has a RV-FDM $D(\mathbf{VLEC}^n)$ that may be considered to be high for its coding rate $R(\mathbf{VLEC}^n)$, as described in Section 8.5.1. By contrast, in the ($\{\mathbf{VLEC}^n\}_{n=12}^{22}$, $R_{\mathrm{IrVLC}} =$

Table 8.7: Properties of the 11 component VLEC codebooks $\{\mathbf{VLEC}^n\}_{n=1}^{11}$ designed using Algorithm 3.3 of Section 3.3, together with the fractions $\{C^n\}_{n=1}^{11}$ of the source symbol frame **s** and the fractions $\{\alpha^n\}_{n=1}^{11}$ of the transmission frame **u** that they encode in both the $(\{\mathbf{VLEC}^n\}_{n=1}^{11}, R_{\mathrm{IrVLC}} = 0.85)$ and the $(\{\mathbf{VLEC}^n\}_{n=1}^{11}, R_{\mathrm{IrVLC}} = 0.55)$ arrangements of the IrVLC-URC scheme

		Fractions[‡]	
$\mathbf{VLEC}^n$	Properties[†]	$R_{\mathrm{IrVLC}} = 0.85$	$R_{\mathrm{IrVLC}} = 0.55$
$\mathbf{VLEC}^1$	(0.956, 2.395)	(**0.766**, **0.681**)	(**0.109**, **0.063**)
$\mathbf{VLEC}^2$	(0.637, 3.786)	(**0.227**, **0.303**)	(**0.744**, **0.643**)
$\mathbf{VLEC}^3$	(0.603, 4.354)	(0.001, 0.002)	(0.003, 0.003)
$\mathbf{VLEC}^4$	(0.468, 5.775)	(0.001, 0.002)	(0.002, 0.003)
$\mathbf{VLEC}^5$	(0.446, 6.388)	(0.001, 0.002)	(0.002, 0.003)
$\mathbf{VLEC}^6$	(0.376, 7.815)	(0.001, 0.002)	(0.002, 0.003)
$\mathbf{VLEC}^7$	(0.371, 8.331)	(0.001, 0.002)	(0.002, 0.003)
$\mathbf{VLEC}^8$	(0.311, 9.523)	(0.001, 0.002)	(0.002, 0.003)
$\mathbf{VLEC}^9$	(0.308, 10.359)	(0.001, 0.002)	(0.001, 0.003)
$\mathbf{VLEC}^{10}$	(0.267, 11.512)	(0.001, 0.002)	(0.001, 0.003)
$\mathbf{VLEC}^{11}$	(0.265, 12.404)	(0.001, 0.002)	(**0.130**, **0.270**)

[†] The properties of each component VLEC codebook $\mathbf{VLEC}^n$ are provided using the format $(R(\mathbf{VLEC}^n), D(\mathbf{VLEC}^n))$.

[‡] The fractions are specified using the format (C^n, α^n).

0.55) arrangement, the particular activated component VLEC codebooks $\mathbf{VLEC}^{20}$ and $\mathbf{VLEC}^{22}$ have RV-FDMs that may be considered to be low for their coding rates, as shown in Table 8.8. The consequences of this are discussed below.

The inverted IrVLC EXIT functions of the four IrVLC-URC parameterizations are provided in Figure 8.13, together with the URC EXIT functions corresponding to the threshold channel E_b/N_0 values. The inverted IrVLC EXIT functions were generated by simulating the APP SISO decoding of 100 frames of 10 000 randomly generated source symbols. Meanwhile, the URC EXIT functions were generated by simulating the APP SISO decoding of 100 frames of 10 000 randomly generated bits. Here, Gaussian distributed *a priori* LLR values were employed and all mutual information measurements were made using the histogram-based approximation of the LLR PDFs [294]. During IrVLC and URC APP SISO decoding, all calculations were performed in the logarithmic probability domain, while using an eight-entry lookup table for correcting the Jacobian approximation in the Log-MAP algorithm [296]. This approach requires only ACS operations, which will be used as our computational complexity measure, since this is characteristic of the complexity/area/speed trade-offs in systolic-array-based chips. Figure 8.13 therefore also provides the average number $O_{\mathrm{IrVLC}}^{\mathrm{APP}}$ of ACS operations performed per source symbol during APP SISO IrVLC decoding.

Table 8.8: Properties of the 11 component VLEC codebooks $\{\mathbf{VLEC}^n\}_{n=12}^{22}$ designed using the GA of Section 8.3, together with the fractions $\{C^n\}_{n=12}^{22}$ of the source symbol frame s and the fractions $\{\alpha^n\}_{n=12}^{22}$ of the transmission frame u that they encode in both the $(\{\mathbf{VLEC}^n\}_{n=12}^{22}, R_{\mathrm{IrVLC}} = 0.85)$ and the $(\{\mathbf{VLEC}^n\}_{n=12}^{22}, R_{\mathrm{IrVLC}} = 0.55)$ arrangements of the IrVLC-URC scheme

		Fractions[‡]	
$\mathbf{VLEC}^n$	Properties[†]	$R_{\mathrm{IrVLC}} = 0.85$	$R_{\mathrm{IrVLC}} = 0.55$
$\mathbf{VLEC}^{12}$	$(0.950,\ 2.412)$	$(\mathbf{0.467},\ \mathbf{0.418})$	$(0.050,\ 0.029)$
$\mathbf{VLEC}^{13}$	$(0.850,\ 2.617)$	$(\mathbf{0.161},\ \mathbf{0.161})$	$(\mathbf{0.188},\ \mathbf{0.122})$
$\mathbf{VLEC}^{14}$	$(0.750,\ 2.793)$	$(\mathbf{0.372},\ \mathbf{0.421})$	$(0.075,\ 0.055)$
$\mathbf{VLEC}^{15}$	$(0.650,\ 2.981)$	$(0.000,\ 0.000)$	$(0.010,\ 0.008)$
$\mathbf{VLEC}^{16}$	$(0.501,\ 4.807)$	$(0.000,\ 0.000)$	$(0.070,\ 0.076)$
$\mathbf{VLEC}^{17}$	$(0.350,\ 8.475)$	$(0.000,\ 0.000)$	$(0.034,\ 0.053)$
$\mathbf{VLEC}^{18}$	$(0.251, 12.449)$	$(0.000,\ 0.000)$	$(0.000,\ 0.000)$
$\mathbf{VLEC}^{19}$	$(0.653,\ 2.269)$	$(0.000,\ 0.000)$	$(0.000,\ 0.000)$
$\mathbf{VLEC}^{20}$	$(0.501,\ 2.133)$	$(0.000,\ 0.000)$	$(\mathbf{0.549},\ \mathbf{0.603})$
$\mathbf{VLEC}^{21}$	$(0.350,\ 2.155)$	$(0.000,\ 0.000)$	$(0.000,\ 0.000)$
$\mathbf{VLEC}^{22}$	$(0.252,\ 2.117)$	$(0.000,\ 0.000)$	$(0.024,\ 0.053)$

[†] The properties of each component VLEC codebook $\mathbf{VLEC}^n$ are provided using the format $(R(\mathbf{VLEC}^n), D(\mathbf{VLEC}^n))$.
[‡] The fractions are specified using the format (C^n, α^n).

The observed discrepancies between the threshold E_b/N_0 values at which open EXIT-chart tunnels emerge and the channel capacity bounds can be explained by considering the EXIT charts seen in Figure 8.13. More specifically, these discrepancies are proportional to the area [204] between the inverted IrVLC EXIT functions and the URC EXIT functions that correspond to the associated threshold E_b/N_0 values, as described in Section 5.5. As shown in Figure 8.13, the width of the EXIT-chart tunnel varies along its length for the parameterizations employing the suite of component VLEC codebooks $\{\mathbf{VLEC}^n\}_{n=1}^{11}$ designed using Algorithm 3.3 of Section 3.3. In contrast, improved EXIT-chart matches are achieved for the parameterizations employing the suite of component VLEC codebooks $\{\mathbf{VLEC}^n\}_{n=12}^{22}$ designed using the GA of Section 8.3, owing to its wide variety of component EXIT functions, as described in Section 8.5.2. For these parameterizations, the EXIT-chart tunnel is narrow in all regions, except when the mutual information I_a^i of the inner APP SISO URC decoder's *a priori* LLR frame $L_a^i(\mathbf{u'})$ is low. This is because the inner APP SISO URC decoder generates extrinsic LLRs $L_e^i(\mathbf{u'})$ having a nonzero mutual information I_e^i even when the mutual information I_a^i of its *a priori* LLR frame $L_a^i(\mathbf{u'})$ is zero, as shown in Figure 8.13. By contrast, the inverted EXIT function of the APP SISO IrVLC decoder emerges from a point close to the $(0, 0)$ corner of the EXIT chart, as described in Section 4.3. Note that we could therefore expect to achieve improved EXIT chart matches if the EXIT

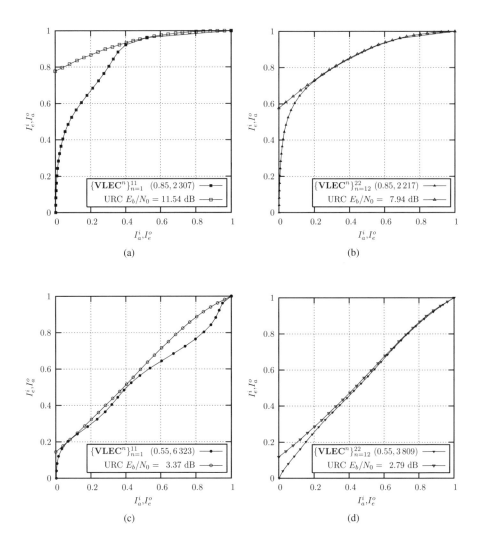

Figure 8.13: EXIT charts for the parameterizations of the IrVLC-URC scheme, characterizing the IrVLC and URC APP SISO decoding performance: (a) ($\{\mathbf{VLEC}^n\}_{n=1}^{11}$, $R_{\mathrm{IrVLC}} = 0.85$) arrangement; (b) ($\{\mathbf{VLEC}^n\}_{n=12}^{22}$, $R_{\mathrm{IrVLC}} = 0.85$) arrangement; (c) ($\{\mathbf{VLEC}^n\}_{n=1}^{11}$, $R_{\mathrm{IrVLC}} = 0.55$) arrangement; (d) ($\{\mathbf{VLEC}^n\}_{n=12}^{22}$, $R_{\mathrm{IrVLC}} = 0.55$) arrangement. The inverted IrVLC EXIT functions are labeled using the format $\{\mathbf{VLEC}^n\}$ (R_{IrVLC}, $O_{\mathrm{IrVLC}}^{\mathrm{APP}}$). The URC EXIT functions are provided for the threshold channel E_b/N_0 values, as specified in the URC EXIT function labels.

function of the APP SISO URC decoder could be shaped to start closer to the $(0,0)$ point of the EXIT chart. This would reduce the area between the inverted IrVLC EXIT function and the URC EXIT function, and hence would result in a reduced discrepancy between the threshold E_b/N_0 value and the channel capacity bound.

Note that the $(\{\mathbf{VLEC}^n\}_{n=12}^{22}, R_{\mathrm{IrVLC}} = 0.55)$ arrangement's narrow EXIT-chart tunnel of Figure 8.13d is facilitated by the activation of the particular component VLEC codebooks $\mathbf{VLEC}^{20}$ and $\mathbf{VLEC}^{22}$, as shown in Table 8.8. This is because these component VLEC codebooks have low RV-FDMs and are therefore associated with inverted EXIT functions having no more than a single point of inflection, as shown in Figure 8.8 and described in Section 8.2. This is in contrast to the $(\{\mathbf{VLEC}^n\}_{n=1}^{11}, R_{\mathrm{IrVLC}} = 0.55)$ arrangement, which activates component VLEC codebooks that are associated with inverted EXIT functions having up to two points of inflection, as shown in Table 8.7. As a result, the composite inverted IrVLC EXIT function of Figure 8.13c has many points of inflection and the width of the EXIT-chart tunnel varies along its length, as described above.

Note that the areas beneath the URC EXIT functions provided in Figure 8.13 were found to be equal to the corresponding channel capacities, which specify the maximum effective throughput for which iterative decoding convergence to an infinitesimally low probability of error is theoretically possible. This is a consequence of the area property of EXIT charts [204] and shows that there is no capacity loss in the IrVLC-URC scheme, as described in Section 5.5. Note that this is in contrast to the schemes of Chapter 7, which employed a $R_{\mathrm{TCM}} = 3/4$-rate TCM inner codec. In this case, the maximum achievable effective throughput was 3 bits per $M_{\mathrm{TCM}} = 16$-ary QAM channel use, whilst the capacity approached 4 bits per channel use for high channel E_b/N_0 values, as described in Section 8.1.

Let us now characterize the performance of the IrVLC MAP sequence estimators of each of our four IrVLC-URC parameterizations. This may be achieved by evaluating the BER between the transmission frame $\mathbf{u}$ and the transmission frame estimate $\tilde{\mathbf{u}}$ that is obtained following the consideration of the *a priori* LLR frame $L_a^o(\mathbf{u})$ by the IrVLC MAP sequence estimator of Figure 8.6. Note that we opted for employing the BER associated with the transmission frame estimate $\tilde{\mathbf{u}}$ as our performance metric, rather than the SER associated with the source symbol frame estimate $\tilde{\mathbf{s}}$ for the reasons discussed in Section 8.5.2. Plots of the BER versus the mutual information I_a^o associated with the *a priori* LLR frame $L_a^o(\mathbf{u})$ are provided for each of our four IrVLC-URC parameterizations in Figure 8.14. These were obtained by simulating the MAP sequence estimation of 100 frames of 10 000 randomly generated source symbols, using Gaussian distributed *a priori* LLR values [294]. During MAP sequence estimation, all calculations were performed in the logarithmic probability domain, while using an eight-entry lookup table for correcting the Jacobian approximation in the Log-MAP algorithm [296]. Again, this approach requires only ACS computational operations, which will be used as our computational complexity measure, since this is characteristic of the complexity/area/speed trade-offs in systolic-array-based chips. Figure 8.14 therefore also provides the average number $O_{\mathrm{IrVLC}}^{\mathrm{MAP}}$ of ACS operations performed per source symbol during IrVLC MAP sequence estimation. As may be expected, the MAP sequence estimation computational complexities recorded in Figure 8.14 are lower than those of APP SISO decoding, as recorded in Figure 8.13.

By comparing plots of Figure 8.14 with those of Figures 8.9 and 8.10, we may observe that the MAP sequence estimation performance of each parameterization is dominated by that associated with a particular one of its activated component VLEC codebooks. In all four cases, the dominant component VLEC codebook is employed to encode a large fraction C^n

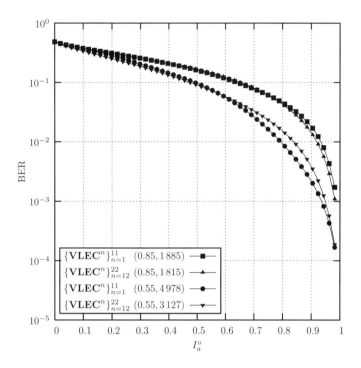

Figure 8.14: Plots of BER versus *a priori* mutual information I_a^o, characterizing the IrVLC MAP sequence estimation performance in the four IrVLC-URC parameterizations. Plots are labeled using the format $\{\mathbf{VLEC}^n\}$ (R_{IrVLC}, $O_{\text{IrVLC}}^{\text{MAP}}$).

of the source symbol frame **s** and has a relatively low RV-FDM, resulting in a poor BER performance, as described in Section 8.5.2. This is because the number of bit errors produced by these codebooks eclipses or dominates those of all other activated component VLEC codebooks. For the ($\{\mathbf{VLEC}^n\}_{n=1}^{11}$, $R_{\text{IrVLC}} = 0.85$), ($\{\mathbf{VLEC}^n\}_{n=12}^{22}$, $R_{\text{IrVLC}} = 0.85$), ($\{\mathbf{VLEC}^n\}_{n=1}^{11}$, $R_{\text{IrVLC}} = 0.55$) and ($\{\mathbf{VLEC}^n\}_{n=12}^{22}$, $R_{\text{IrVLC}} = 0.55$) arrangements, the dominant component VLEC codebooks can be identified as $\mathbf{VLEC}^1$, $\mathbf{VLEC}^{12}$, $\mathbf{VLEC}^2$ and $\mathbf{VLEC}^{20}$, respectively.

As may be expected, the IrVLC-URC parameterizations employing a coding rate of $R_{\text{IrVLC}} = 0.55$ offered a lower BER at all values of *a priori* mutual information I_a^o than those employing a coding rate of $R_{\text{IrVLC}} = 0.85$, as shown in Figure 8.14. Whilst the two parameterizations employing a coding rate of $R_{\text{IrVLC}} = 0.85$ exhibit similar BER performances, the ($\{\mathbf{VLEC}^n\}_{n=1}^{11}$, $R_{\text{IrVLC}} = 0.55$) arrangement offers a better BER performance than the ($\{\mathbf{VLEC}^n\}_{n=12}^{22}$, $R_{\text{IrVLC}} = 0.55$) arrangement. This may be attributed to the activation of the component VLEC codebooks that were designed by the GA of Section 8.3 to have minimal RV-FDMs in the ($\{\mathbf{VLEC}^n\}_{n=12}^{22}$, $R_{\text{IrVLC}} = 0.55$) arrangement, as predicted in Section 8.5.2. Note however that we concluded in Chapter 7 that the computational complexity associated with continuing iterative decoding until perfect convergence at the $(1, 1)$ point of the EXIT chart is achieved may be justified owing to the corresponding infinitesimally low BER achieved. Hence, the poor BER performance of the $(\{\mathbf{VLEC}^n\}_{n=12}^{22}, R_{\mathrm{IrVLC}} = 0.55)$ arrangement becomes irrelevant.

Let us now comment on both the APP SISO IrVLC decoding and the MAP IrVLC sequence estimation computational complexities that are associated with our four IrVLC-URC parameterizations, as recorded in Figures 8.13 and 8.14. Observe that the IrVLC-URC parameterizations employing a coding rate of $R_{\mathrm{IrVLC}} = 0.55$ are associated with higher computational complexities than those employing a coding rate of $R_{\mathrm{IrVLC}} = 0.85$. This may be explained by the high computational complexities that are associated with the low-rate component VLEC codebooks that are activated in the context of the $R_{\mathrm{IrVLC}} = 0.55$ parameterizations, as detailed in Section 8.5.2. Whilst the two parameterizations employing a coding rate of $R_{\mathrm{IrVLC}} = 0.85$ exhibit similar computational complexities, the $(\{\mathbf{VLEC}^n\}_{n=12}^{22}, R_{\mathrm{IrVLC}} = 0.55)$ arrangement incurs lower computational complexities than the $(\{\mathbf{VLEC}^n\}_{n=1}^{11}, R_{\mathrm{IrVLC}} = 0.55)$ parameterization. As predicted in Section 8.5.2, this may be attributed to the activation of the low-complexity component VLEC codebooks that were designed by the GA of Section 8.3 to have minimal RV-FDMs in the $(\{\mathbf{VLEC}^n\}_{n=12}^{22}, R_{\mathrm{IrVLC}} = 0.55)$ arrangement.

Note that the average number of ACS operations performed per LLR during APP SISO URC decoding was found to be approximately equal to 200 in all parameterizations. This corresponds to an average of $O_{\mathrm{URC}}^{\mathrm{APP}} = 200 \cdot E/R_{\mathrm{IrVLC}} = 887$ ACS operations per source symbol in the case of the $R_{\mathrm{IrVLC}} = 0.85$ parameterizations and $O_{\mathrm{URC}}^{\mathrm{APP}} = 1370$ ACS operations per source symbol for the $R_{\mathrm{IrVLC}} = 0.55$ parameterizations. By comparing these complexities with those provided in Figures 8.13 and 8.14 for APP SISO IrVLC decoding and MAP IrVLC sequence estimation, we may expect these latter processes to incur between 70% and 80% of the total iterative decoding complexity.

8.5.5 Interleaver Length

We opted for employing an average interleaver length of $I^{\mathrm{sum}} \approx 100\,000$ bits in each of our four IrVLC-URC parameterizations. Since this is less than half as long as the interleaver employed in Chapter 7, we can expect it to mitigate the correlation within the a priori LLR frames $L_a^o(\mathbf{u})$ and $L_a^i(\mathbf{u}')$ to a lesser degree, as described in Section 4.3. Hence, this choice of interleaver length will allow us to characterize the effect that the IrVLC coding rate $R_{\mathbf{IrVLC}}$ has upon the APP SISO IrVLC decoder's sensitivity to the aforementioned correlation, as discussed in Section 7.4.3.

In order to achieve an average interleaver length of $I^{\mathrm{sum}} \approx 100\,000$ bits, we must employ source sample frames s comprising the following number of source symbols:

$$J^{\mathrm{sum}} = I^{\mathrm{sum}} \cdot \frac{R_{\mathrm{IrVLC}}}{E}, \tag{8.18}$$

where $E = 3.77$ bits/symbol is the entropy of the source symbols. Hence, our IrVLC-URC parameterizations having coding rates of $R_{\mathbf{IrVLC}} = 0.85$ and $R_{\mathbf{IrVLC}} = 0.55$ employ source symbol frames comprising $J^{\mathrm{sum}} = 22\,556$ and $J^{\mathrm{sum}} = 14\,595$ symbols, respectively.

As described in Section 8.4.1, each source symbol frame s is decomposed into $N = 11$ sub-frames in our IrVLC-URC parameterizations. Each source symbol sub-frame $\mathbf{s}^n$ comprises a specific number $J^n \approx C^n \cdot J^{\mathrm{sum}}$ of source symbols, according to the fractions C^n specified in Tables 8.7 and 8.8. These fractions are also provided together with the

Table 8.9: Properties of the 11 component VLEC codebooks $\{\mathbf{VLEC}^n\}_{n=1}^{11}$ designed using Algorithm 3.3 of Section 3.3, together with the sub-frame parameters in both the $(\{\mathbf{VLEC}^n\}_{n=1}^{11}, R_{\mathrm{IrVLC}} = 0.85)$ and the $(\{\mathbf{VLEC}^n\}_{n=1}^{11}, R_{\mathrm{IrVLC}} = 0.55)$ arrangements of the IrVLC-URC scheme

		Sub-frame parameters[‡]	
$\mathbf{VLEC}^n$	Properties[†]	$R_{\mathrm{IrVLC}} = 0.85$	$R_{\mathrm{IrVLC}} = 0.55$
$\mathbf{VLEC}^1$	(3, 6)	(0.766, 17 282, 16)	(0.109, 1 591, 13)
$\mathbf{VLEC}^2$	(4, 9)	(0.227, 5 114, 15)	(0.744, 10 864, 16)
$\mathbf{VLEC}^3$	(4, 10)	(0.001, 27, 8)	(0.003, 46, 9)
$\mathbf{VLEC}^4$	(5, 12)	(0.001, 22, 8)	(0.002, 36, 8)
$\mathbf{VLEC}^5$	(6, 12)	(0.001, 21, 7)	(0.002, 34, 8)
$\mathbf{VLEC}^6$	(7, 14)	(0.001, 18, 7)	(0.002, 29, 8)
$\mathbf{VLEC}^7$	(8, 13)	(0.001, 17, 7)	(0.002, 29, 8)
$\mathbf{VLEC}^8$	(8, 17)	(0.001, 15, 8)	(0.002, 24, 8)
$\mathbf{VLEC}^9$	(9, 17)	(0.001, 14, 7)	(0.001, 24, 8)
$\mathbf{VLEC}^{10}$	(11, 19)	(0.001, 12, 7)	(0.001, 21, 8)
$\mathbf{VLEC}^{11}$	(11, 19)	(0.001, 14, 7)	(0.130, 1 897, 14)
	Sums:	(1.000, 22 556, 97)	(1.000, 14 595, 108)

[†] The properties of each component VLEC codebook $\mathbf{VLEC}^n$ are provided using the format $(\min_{k \in [1...K]} I^{n,k}, \max_{k \in [1...K]} I^{n,k})$.

[‡] The sub-frame parameters are specified using the format $(C^n, J^n, \lceil \log_2(I^n_{\max} - I^n_{\min} + 1) \rceil)$.

corresponding values of J^n in Tables 8.9 and 8.10 for the parameterizations employing the suite of component VLEC codebooks $\{\mathbf{VLEC}^n\}_{n=1}^{11}$ designed using Algorithm 3.3 of Section 3.3 and the suite $\{\mathbf{VLEC}^n\}_{n=12}^{22}$ designed using the GA of Section 8.3, respectively. As described in Section 8.4.1, the corresponding component VLEC codebook $\mathbf{VLEC}^n$ is employed to encode each source symbol sub-frame $\mathbf{s}^n$, yielding a transmission sub-frame $\mathbf{u}^n$ comprising I^n number of bits. Since the lengths of the transmission sub-frames typically vary from frame to frame, it is necessary to explicitly convey them as side information in order to facilitate their decoding in the receiver, as described in Section 8.4.1.

The amount of side information required may be determined by considering the range of transmission sub-frame lengths that can result from VLEC encoding using each of the component codebooks. When all J^n source symbols in a particular source symbol sub-frame $\mathbf{s}^n$ are represented by the particular codeword from the component VLC codebook $\mathbf{VLEC}^n$ having the maximal length $\max_{k \in [1...K]} I^{n,k}$, a maximal transmission sub-frame length of

$$I^n_{\max} = J^n \cdot \max_{k \in [1...K]} I^{n,k} \tag{8.19}$$

Table 8.10: Properties of the 11 component VLEC codebooks $\{\mathbf{VLEC}^n\}_{n=12}^{22}$ designed using the GA of Section 8.3, together with the sub-frame parameters in both the $(\{\mathbf{VLEC}^n\}_{n=12}^{22}, R_{\mathrm{IrVLC}} = 0.85)$ and the $(\{\mathbf{VLEC}^n\}_{n=12}^{22}, R_{\mathrm{IrVLC}} = 0.55)$ arrangements of the IrVLC-URC scheme

		Sub-frame parameters[‡]	
$\mathbf{VLEC}^n$	Properties[†]	$R_{\mathrm{IrVLC}} = 0.85$	$R_{\mathrm{IrVLC}} = 0.55$
$\mathbf{VLEC}^{12}$	(3, 6)	(0.467, 10 540, 15)	(0.050, 724, 12)
$\mathbf{VLEC}^{13}$	(3, 8)	(0.161, 3 635, 15)	(0.188, 2 747, 14)
$\mathbf{VLEC}^{14}$	(3, 8)	(0.372, 8 381, 16)	(0.075, 1 096, 13)
$\mathbf{VLEC}^{15}$	(4, 8)	(0.000, 0, 0)	(0.010, 146, 10)
$\mathbf{VLEC}^{16}$	(5, 13)	(0.000, 0, 0)	(0.070, 1 015, 13)
$\mathbf{VLEC}^{17}$	(8, 16)	(0.000, 0, 0)	(0.034, 497, 12)
$\mathbf{VLEC}^{18}$	(11, 28)	(0.000, 0, 0)	(0.000, 2, 6)
$\mathbf{VLEC}^{19}$	(3, 16)	(0.000, 0, 0)	(0.000, 0, 0)
$\mathbf{VLEC}^{20}$	(2, 18)	(0.000, 0, 0)	(0.549, 8 013, 17)
$\mathbf{VLEC}^{21}$	(3, 37)	(0.000, 0, 0)	(0.000, 1, 6)
$\mathbf{VLEC}^{22}$	(2, 62)	(0.000, 0, 0)	(0.024, 354, 15)
	Sums:	(1.000, 22 556, 46)	(1.000, 14 595, 118)

[†] The properties of each component VLEC codebook $\mathbf{VLEC}^n$ are provided using the format $(\min_{k \in [1...K]} I^{n,k}, \max_{k \in [1...K]} I^{n,k})$.

[‡] The sub-frame parameters are specified using the format $(C^n, J^n, \lceil \log_2(I_{\max}^n - I_{\min}^n + 1) \rceil)$.

results. Similarly, a minimal transmission sub-frame length of

$$I_{\min}^n = J^n \cdot \min_{k \in [1...K]} I^{n,k} \tag{8.20}$$

results, when all source symbols are represented by the minimal length VLC codeword. A transmission sub-frame $\mathbf{u}^n$ encoded using the component VLC codebook $\mathbf{VLEC}^n$ will therefore have one of $(I_{\max}^n - I_{\min}^n + 1)$ number of lengths in the range $I^m \in [I_{\min}^n \ldots I_{\max}^n]$. Hence, the length of the transmission sub-frame I^n can be represented using a fixed-length codeword comprising $\lceil \log_2(I_{\max}^n - I_{\min}^n + 1) \rceil$ number of bits. The values of $\min_{k \in [1...K]} I^{n,k}$, $\max_{k \in [1...K]} I^{n,k}$ and $\lceil \log_2(I_{\max}^n - I_{\min}^n + 1) \rceil$ are provided in Tables 8.9 and 8.10 for the suite of component VLEC codebooks $\{\mathbf{VLEC}^n\}_{n=1}^{11}$ designed using Algorithm 3.3 of Section 3.3 and the suite $\{\mathbf{VLEC}^n\}_{n=12}^{22}$ designed using the GA of Section 8.3, respectively.

As shown in Tables 8.9 and 8.10, the $(\{\mathbf{VLEC}^n\}_{n=12}^{22}, R_{\mathrm{IrVLC}} = 0.55)$ arrangement requires the most side information, employing a total of 118 bits per frame to convey the lengths of the $N = 11$ transmission sub-frames. As suggested in Section 8.4.1, this error-sensitive side information may be protected by a low-rate block code in order to ensure its reliable transmission. Using a $R_{\text{rep}} = 1/3$-rate repetition code results in a total of $118/R_{\text{rep}} = 354$ bits of side information per frame, which represents an average of just 0.35% of the transmitted information, when appended to the transmission frame $\mathbf{u}$, which has an average length of $I^{\text{sum}} \approx 100\,000$ bits. Note that this is significantly less side information than that required by the schemes of Chapter 7, where 4% of the transmitted information was constituted by side information. As described in Section 8.1, this is a benefit of the IrVLC-URC scheme's employment of only a single transmission sub-frame per component VLEC codebook.

8.6 Simulation Results

In this section we characterize and compare the achievable performance and the computational complexity imposed by our four IrVLC-URC parameterizations of Section 8.5.4. The transmission of 200 frames was simulated for a range of E_b/N_0 values above the capacity bounds when communicating over uncorrelated narrowband Rayleigh fading channels. In each case, iterative decoding was continued until convergence was achieved and no further BER improvements could be attained. After each decoding iteration, we recorded the BER between the transmission frame $\mathbf{u}$ and the transmission frame estimate $\tilde{\mathbf{u}}$ that is obtained following the consideration of the *a priori* LLR frame $L_a^o(\mathbf{u})$ by the IrVLC MAP sequence estimator of Figure 8.6. Again, we opted for employing the BER associated with the transmission frame estimate $\tilde{\mathbf{u}}$ as our performance metric, rather than the SER associated with the source symbol frame estimate $\tilde{\mathbf{s}}$ for the reasons discussed in Section 8.5.2. For each BER value recorded, we also recorded the total number of ACS operations performed during IrVLC MAP sequence estimation and during the iterative IrVLC and URC APP SISO decoding process so far. The corresponding plots of the BER attained following the achievement of iterative decoding convergence versus E_b/N_0 are provided for the IrVLC-URC parameterizations having coding rates of $R_{\text{IrVLC}} = 0.85$ and $R_{\text{IrVLC}} = 0.55$ in Figures 8.15 and 8.16 respectively. Meanwhile, Figures 8.17 and 8.18 provide the corresponding plots of the average number of ACS operations required per source symbol in order to achieve a BER of 10^{-5} versus E_b/N_0 for the IrVLC-URC parameterizations employing coding rates of $R_{\text{IrVLC}} = 0.85$ and $R_{\text{IrVLC}} = 0.55$ respectively.

As shown in Figures 8.15 and 8.16, the BER attained following the achievement of iterative decoding convergence reduces as the channel's E_b/N_0 value increases, for all parameterizations considered. This may be explained by considering the EXIT-chart tunnel, which gradually opens and becomes wider as the E_b/N_0 value is increased from the channel's capacity bound, allowing the iterative decoding trajectory to progress further, as explained in Section 4.3. Note that an open EXIT-chart tunnel implies that iterative decoding convergence to an infinitesimally low BER can be achieved, provided that the iterative decoding trajectory approaches the inner and outer codecs' EXIT functions sufficiently closely, as described in Section 4.3. However, it can be seen in Figures 8.15 and 8.16 that the low predicted BERs were not actually achieved at the threshold E_b/N_0 values, where the EXIT-chart tunnels open. This is because our relatively short $100\,000$-bit interleaver is unable to mitigate all of the correlation within the *a priori* LLR frames $L_a^o(\mathbf{u})$ and $L_a^i(\mathbf{u}')$, which the BCJR algorithm assumes to be uncorrelated [48]. As a result, the iterative decoding trajectory does not match perfectly with the inner and outer codecs' EXIT functions and the tunnel must be further

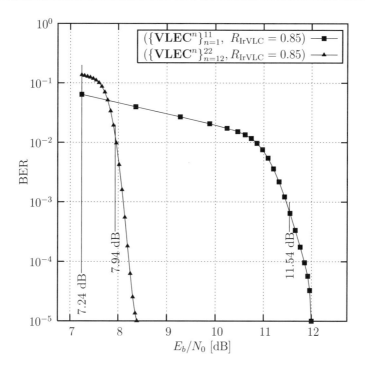

Figure 8.15: Plots of the BER attained following the achievement of iterative decoding convergence versus E_b/N_0 for the IrVLC-URC parameterizations employing a coding rate of $R_{\text{IrVLC}} = 0.85$.

widened before the trajectory can reach the top right-hand corner of the EXIT chart, which is associated with an infinitesimally low BER, as described in Section 4.3.

In the case of the ($\{\mathbf{VLEC}^n\}_{n=1}^{11}$, $R_{\text{IrVLC}} = 0.85$) arrangement, a BER of less than 10^{-5} can only be achieved for channel E_b/N_0 values in excess of 12 dB, as shown in Figure 8.15. This value is 4.76 dB away from the E_b/N_0 capacity bound of 7.24 dB and 0.46 dB away from the E_b/N_0 threshold of 11.54 dB, at which the EXIT-chart tunnel becomes open. Meanwhile, a BER of less than 10^{-5} can be achieved for channel E_b/N_0 values in excess of 8.39 dB in the case of the ($\{\mathbf{VLEC}^n\}_{n=12}^{22}$, $R_{\text{IrVLC}} = 0.85$) arrangement, which is a significantly lower discrepancy of 1.15 dB from the channel's capacity bound. Furthermore, a BER of less than 10^{-5} can be achieved within 0.45 dB of this parameterization's threshold E_b/N_0 value of 7.94 dB, as shown in Figure 8.15.

Similarly, a BER of less than 10^{-5} can only be achieved by the ($\{\mathbf{VLEC}^n\}_{n=1}^{11}$, $R_{\text{IrVLC}} = 0.55$) arrangement for channel E_b/N_0 values in excess of 3.95 dB, as shown in Figure 8.16. This value is 1.58 dB from the E_b/N_0 capacity bound of 2.37 dB and 0.58 dB from the E_b/N_0 threshold of 3.37 dB. Meanwhile, the ($\{\mathbf{VLEC}^n\}_{n=12}^{22}$, $R_{\text{IrVLC}} = 0.55$) arrangement achieves a BER below 10^{-5} for channel E_b/N_0 values in excess of 8.78 dB, which constitutes a slightly lower discrepancy of 1.41 dB from the channel's capacity

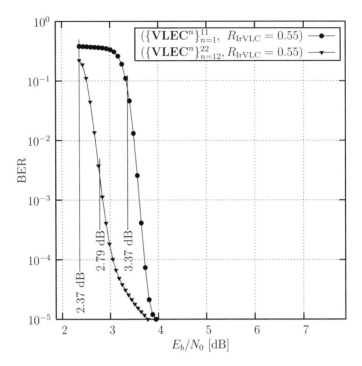

Figure 8.16: Plots of the BER attained following the achievement of iterative decoding convergence versus E_b/N_0 for the IrVLC-URC parameterizations employing a coding rate of $R_{\mathrm{IrVLC}} = 0.55$.

bound. Furthermore, a BER of less than 10^{-5} can be achieved within 0.99 dB of this parameterization's threshold E_b/N_0 value of 7.94 dB, as shown in Figure 8.16.

Note that the discrepancy between the E_b/N_0 value at which a BER of less than 10^{-5} can be achieved and that at which the EXIT-chart tunnel becomes open is higher for the parameterizations employing a coding rate of $R_{\mathrm{IrVLC}} = 0.55$ than for those employing a coding rate of $R_{\mathrm{IrVLC}} = 0.85$. This implies that the parameterizations employing a coding rate of $R_{\mathrm{IrVLC}} = 0.55$ are more sensitive to correlation within the *a priori* LLR frames $L_a^o(\mathbf{u})$ and $L_a^i(\mathbf{u}')$, resulting in a poor match between the iterative decoding trajectory and the inner and outer EXIT functions, as described above. These findings agree with the suggestion of Section 7.4.3 that the sensitivity of an APP SISO VLC decoder to correlation within the LLR frame $L_a^o(\mathbf{u})$ increases as the corresponding coding rate is decreased.

Let us now consider the iterative decoding complexities of our four parameterizations. Note that between 70% and 80% of this complexity was found to be incurred by APP SISO IrVLC decoding and MAP IrVLC sequence estimation, with the remainder being incurred by APP SISO URC decoding, as predicted in Section 8.5.4. We can therefore conclude that efforts to optimize the complexity associated with the IrVLC scheme are more likely to yield a significant reduction in the overall complexity than those directed at the URC scheme.

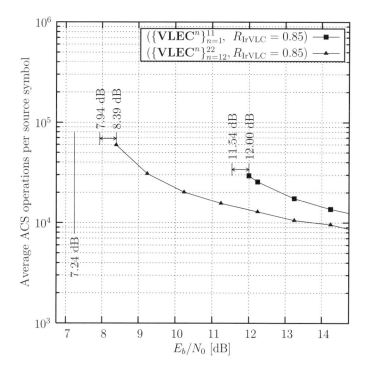

Figure 8.17: Plots of the average number of ACS operations required per source symbol in order to achieve a BER of 10^{-5} versus E_b/N_0 for the IrVLC-URC parameterizations employing a coding rate of $R_{IrVLC} = 0.85$.

We therefore investigate methods of reducing the complexity associated with IrVLCs in Chapter 9.

As shown in Figures 8.17 and 8.18, the iterative decoding computational complexity required to achieve a BER of 10^{-5} reduces as the channel's E_b/N_0 value is increased. This is because the EXIT-chart tunnel widens as the E_b/N_0 value increases, requiring fewer decoding iterations to reach the top right-hand corner of the EXIT chart. By comparing Figures 8.17 and 8.18, it can be seen that the computational complexity required to achieve a BER of 10^{-5} at an E_b/N_0 value that is a particular distance from the associated E_b/N_0 capacity bound is higher for the parameterizations employing a coding rate of $R_{IrVLC} = 0.55$ compared with those employing a coding rate of $R_{IrVLC} = 0.85$. For example, the complexity associated with the $R_{IrVLC} = 0.55$ parameterizations at an E_b/N_0 value that is 7 dB above the associated channel's capacity bound of 7.24 dB is about 75% higher than that associated with the corresponding $R_{IrVLC} = 0.85$ parameterizations at an E_b/N_0 value that is 7 dB above the associated capacity bound of 2.37 dB. This may be explained by the $R_{IrVLC} = 0.55$ parameterizations' activation of low-rate component VLEC codebooks that are associated with high computational complexities, as shown in Tables 8.7 and 8.8 as well as Figures 8.7–8.10.

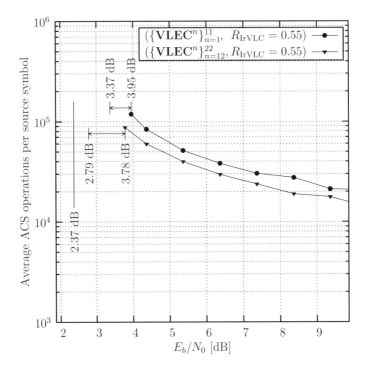

Figure 8.18: Plots of the average number of ACS operations required per source symbol in order to achieve a BER of 10^{-5} versus E_b/N_0 for the IrVLC-URC parameterizations employing a coding rate of $R_{\mathrm{IrVLC}} = 0.55$.

As portrayed in Figure 8.18, the $(\{\mathbf{VLEC}^n\}_{n=1}^{11}, R_{\mathrm{IrVLC}} = 0.55)$ arrangement is associated with a 35% higher complexity than that of the $(\{\mathbf{VLEC}^n\}_{n=12}^{22}, R_{\mathrm{IrVLC}} = 0.55)$ arrangement, when considering the high channel E_b/N_0 value range. As predicted in Section 8.5.2, this may be attributed to the $(\{\mathbf{VLEC}^n\}_{n=12}^{22}, R_{\mathrm{IrVLC}} = 0.55)$ arrangement's activation of component VLEC codebooks that were designed to have a low RV-FDM by the GA of Section 8.3, which are associated with low computational complexities, as shown in Table 8.8 as well as Figures 8.8 and 8.10.

Similarly, the $(\{\mathbf{VLEC}^n\}_{n=1}^{11}, R_{\mathrm{IrVLC}} = 0.85)$ arrangement is associated with a 38% higher complexity at high channel E_b/N_0 values than that of the corresponding $(\{\mathbf{VLEC}^n\}_{n=12}^{22}, R_{\mathrm{IrVLC}} = 0.85)$ arrangement, as shown in Figure 8.17. This may be attributed to the activation of low-rate component VLEC codebooks that are associated with a high computational complexity in the $(\{\mathbf{VLEC}^n\}_{n=1}^{11}, R_{\mathrm{IrVLC}} = 0.85)$ arrangement, as shown in Table 8.7 as well as Figures 8.7 and 8.9. We may therefore identify the $(\{\mathbf{VLEC}^n\}_{n=12}^{22}, R_{\mathrm{IrVLC}} = 0.85)$ arrangement as our preferred parameterization owing to its relatively low computational complexity and the near-capacity operation that it facilitates.

8.7 Summary and Conclusions

In this chapter we have proposed the novel RV-FDM of Equation (8.8) as an alternative to the IV-FD lower bound of Equation (8.5) for the characterization of the error-correction capability that is associated with VLEC codebooks. Unlike the IV-FD lower bound, the RV-FDM exists within the real domain, allowing the comparison of the error-correction capability of two VLEC codebooks having equal IV-FD lower bounds. Furthermore in Figure 8.3 of Section 8.2 we showed that a VLEC codebook's RV-FDM affects the number of inflection points appearing in the corresponding inverted EXIT function. This complements the property [204] that the area below an inverted VLEC EXIT function equals the corresponding coding rate and the property that a RV-FDM of at least two yields an inverted VLEC EXIT function that reaches the top right-hand corner of the EXIT chart. Hence, we have shown in Figure 8.3 that the shape of an inverted VLEC EXIT function depends on both the corresponding coding rate and the RV-FDM.

This motivated our novel GA design of Section 8.3, which facilitates the design of VLEC codebooks having arbitrary coding rates as well as RV-FDMs and hence inverted EXIT function shapes, while seeking desirable bit entropies and bit-based trellis complexities. This is in contrast to Algorithm 3.3 of Section 3.3, which was employed to design component VLC codebooks in Chapter 7 and is incapable of designing codebooks having arbitrary coding rates and RV-FDMs without imposing a significant level of 'trial-and-error'-based human interaction. Recall that in Chapter 7 we speculated that our ability to match the inverted EXIT function of an IrVLC to the EXIT function of a serially concatenated inner codec depends on how much variety is exhibited within the shape of the inverted EXIT functions of the component VLC codebooks. For this reason, our GA may be considered to be more suitable for the design of IrVLC component codebooks than is Algorithm 3.3 of Section 3.3.

This was investigated in Section 8.5 by employing both our novel GA and Algorithm 3.3 of Section 3.3 to design suites of $N = 11$ component VLC codebooks. The suite of component VLC codebooks designed in Section 8.5.1 by our novel GA had a wide variety of inverted EXIT functions. This was achieved by seeking component VLC codebooks having a wide variety of coding rates and RV-FDMs. In some cases, high RV-FDMs were sought, resulting in inverted EXIT functions having up to two points of inflection, whilst low RV-FDMs were sought for the remaining component VLC codebooks, which were associated with inverted EXIT functions having no more than one point of inflection. In order to facilitate a fair comparison, trial and error was not employed when Algorithm 3.3 of Section 3.3 was used to design component VLC codebooks. Instead, each component VLC codebook was designed to have a different IV-FD lower bound. As a result, however, the suite of component VLC codebooks had only a limited variety of coding rates and RV-FDMs, resulting in the limited variety of inverted EXIT functions shown in Figure 8.7.

The suitability of the component VLC codebook suites for use in IrVLCs was investigated by employing the algorithm of [149] to match the corresponding inverted IrVLC EXIT functions to a selection of inner codec EXIT functions. These were provided by the serially concatenated URC codec of Figure 8.6, employing BPSK modulation for transmission over an uncorrelated narrowband Rayleigh fading channel having a range of E_b/N_0 values. As predicted by the EXIT chart's area property of [204], the areas beneath the URC EXIT functions were found to be equal to the corresponding channel capacities, showing that the amalgamation of URCs and BPSK modulation does not suffer from the effective throughput loss of Chapter 7's TCM inner codec.

The suite of component VLEC codebooks designed by our novel GA in Section 8.5.1 was found to be more suitable for use in IrVLCs than that designed using Algorithm 3.3 of Section 3.3. More specifically, an open IrVLC-URC EXIT-chart tunnel was found to be achievable in Figure 8.12 at channel E_b/N_0 values within 1 dB of the Rayleigh fading channel's capacity bound for a wide range of effective throughputs, when employing the suite of component VLEC codebooks generated using our GA. By contrast, open EXIT-chart tunnels could only be achieved when employing the suite designed by Algorithm 3.3 of Section 3.3 for a limited range of effective throughputs and within a significantly higher margin of 4.4 dB from the E_b/N_0 capacity bound, as seen in Figure 8.12. This confirms the speculation that our ability to perform EXIT-chart matching depends on how much variety is exhibited within the suite of inverted EXIT functions.

However, regardless of the component VLEC codebook suite employed, the high starting point of the URC's EXIT function along the EXIT chart's I_e^i axis in Figure 8.13 was found to limit the EXIT-chart matching accuracy. This is because the inverted IrVLC EXIT function always emerges from the $(0,0)$ point of the EXIT chart. Hence, in Chapter 9 we investigate the employment of Irregular Unity-Rate Codes (IrURCs), having composite EXIT functions that may be shaped so that they start from a point along the EXIT chart's I_e^i axis that is closer to the $(0,0)$ point. In this way, the IrVLC and IrURC EXIT functions will be jointly matched to each other, creating an EXIT-chart tunnel that is narrow along its entire length and facilitating operation at E_b/N_0 values that are even closer to the channel's capacity bound.

It was found in Figure 8.12 that lower IrVLC coding rates facilitated improved EXIT-chart matching, yielding open EXIT-chart tunnels for E_b/N_0 values in excess of thresholds that are closer to the channel's capacity bound. For our IrVLC-URC parameterizations outlined in Table 8.8 and employing the suite of component VLEC codebooks designed using our novel GA, an open EXIT-chart tunnel was achieved in Figure 8.12 within 0.7 dB of the E_b/N_0 capacity bound when employing a coding rate of $R_{\mathrm{IrVLC}} = 0.85$. An even lower discrepancy of 0.42 dB was seen in Figure 8.12 for a coding rate of $R_{\mathrm{IrVLC}} = 0.55$. Note that the effective EXIT-chart matching of the latter case was attributed to its activation of component VLEC codebooks having low RV-FDMs, which were shown in Figure 8.8 to have inverted EXIT functions exhibiting no more than one point of inflection. In the case of parameterizations employing the suite of component VLEC codebooks outlined in Table 8.7 and designed using Algorithm 3.3 of Section 3.3, an open EXIT-chart tunnel was created in Figure 8.12 within 4.3 dB of the E_b/N_0 capacity bound for a coding rate of $R_{\mathrm{IrVLC}} = 0.85$ and within 1 dB for a coding rate of $R_{\mathrm{IrVLC}} = 0.55$.

However, we observed a discrepancy in Figures 8.15–8.18 between the E_b/N_0 value at which a BER of 10^{-5} could be achieved and that at which an open EXIT-chart tunnel was achieved. This may be explained by the APP SISO IrVLC decoder's sensitivity to residual correlation within the *a priori* LLR frame that was not entirely eliminated by the interleaver having a length of 100 000 bits. As a result, the match between the iterative decoding trajectory and the inner and outer EXIT functions was poor, requiring the widening of the EXIT-chart tunnel in order for the trajectory to reach the top right-hand corner of the EXIT chart. As speculated in Section 7.4.3, the APP SISO IrVLC decoder's sensitivity to this extrinsic information correlation was found to increase as the IrVLC coding rate was reduced. Indeed, a discrepancy of about 0.45 dB was observed in Figures 8.15 and 8.17 for our IrVLC-URC parameterizations employing a coding rate of $R_{\mathrm{IrVLC}} = 0.85$, which compared favorably with the discrepancies of 0.58 dB and 0.99 dB observed in Figures 8.16 and 8.18 for the $R_{\mathrm{IrVLC}} = 0.55$ parameterizations.

Whilst the most accurate EXIT-chart matching was achieved in Figure 8.13d by the $R_{\mathrm{IrVLC}} = 0.55$ IrVLC-URC parameterization employing the suite of component VLEC codebooks designed using our novel GA of Section 8.3, it did not offer the nearest-capacity operation. More specifically, this parameterization could only achieve a BER of less than 10^{-5} for E_b/N_0 values in excess of a limit within 1.41 dB of the channel's E_b/N_0 capacity bound, owing to its sensitivity to correlation within the *a priori* LLR frame, as described above. By contrast, the equivalent parameterization having a coding rate of $R_{\mathrm{IrVLC}} = 0.85$ suffered the least sensitivity to this correlation and therefore achieved a BER of less than 10^{-5} for E_b/N_0 values in excess of a limit that was within a smaller discrepancy of 1.15 dB from the channel's capacity bound.

Furthermore, the iterative decoding computational complexity required to achieve a BER of 10^{-5} at a particular distance from the channel's E_b/N_0 capacity bound was found to decrease as the coding rate was increased. Indeed, observe in Figure 8.17 for the $R_{\mathrm{IrVLC}} = 0.55$ parameterizations, that the complexity associated with an E_b/N_0 value that is 7 dB above the associated channel's capacity bound was found to be about 75% higher than that of the $R_{\mathrm{IrVLC}} = 0.85$ parameterizations. Note that the relatively low complexity of the $R_{\mathrm{IrVLC}} = 0.55$ parameterization employing the suite of component VLEC codebooks designed using our novel GA was attributed to its activation of component VLEC codebooks having low RV-FDMs, which were shown to be associated with low complexities in Figures 8.8 and 8.10. Owing to its near-capacity operation and low computational complexity, the $R_{\mathrm{IrVLC}} = 0.85$ parameterization employing the suite of component VLEC codebooks designed using our novel GA and summarized in Table 8.8 was identified as our preferred arrangement.

In each IrVLC-URC parameterization considered, the IrVLC's MAP sequence estimator's BER performance was found to be dominated by that of a single activated component VLEC codebook. In each case, this particular component VLEC codebook encoded a significant fraction of the source symbols and had a low RV-FDM, which was found to be associated with a relatively poor BER performance in Figures 8.9 and 8.10. As a result, the particular component VLEC codebook was responsible for a high number of bit errors, which eclipsed or dominated those of any other activated codebook. However, we concluded that we may justify the computational complexity required to continue iterative decoding until convergence to an infinitesimally low BER is achieved.

Finally, we demonstrated in Section 8.5.5 that the amount of side information required to facilitate the IrVLC decoding can be reduced to less than 0.35% of the transmitted information, by employing only a single sub-frame per component VLEC codebook. This compares favorably with the 4% of the transmitted information that was required by the IrVLC-TCM schemes of Chapter 7, owing to their employment of multiple transmission sub-frames per component codebook.

Chapter 9

Joint EXIT-Chart Matching of Irregular Variable-Length Coding and Irregular Unity-Rate Coding[1]

9.1 Introduction

In Chapters 7 and 8 we characterized the serial concatenation [275] of an irregular outer codec with a regular inner codec, as proposed in [149]. Whilst Chapters 7 and 8 employed different regular inner codecs, they both employed Irregular Variable-Length Coding (IrVLC) as their irregular outer codec. This scheme comprises N component Variable-Length Error-Correction (VLEC) codebooks [179] $\{\mathbf{VLEC}^n\}_{n=1}^N$ having a variety of coding rates $\{R(\mathbf{VLEC}^n)\}_{n=1}^N$, as described in Section 5.5. At the transmitter, each component VLEC codebook $\mathbf{VLEC}^n$ is invoked to generate a specific fraction of the transmission frame, given by the component's weight α^n, where we have

$$\alpha^n \in [0, 1] \ \forall \ n \in [1 \ldots N], \tag{9.1}$$

$$\sum_{n=1}^N \alpha^n = 1. \tag{9.2}$$

At the receiver, an inner *A Posteriori* Probability (APP) Soft-Input Soft-Output (SISO) decoder [292] iteratively exchanges extrinsic information with the APP SISO IrVLC decoder [278], as described in Section 4.1. As usual, the EXtrinsic Information Transfer (EXIT) function [294] $I_e^i(I_a)$ of the inner APP SISO decoder may be employed to characterize its operation. Similarly, the operation of the APP SISO IrVLC decoder may be characterized by its inverted EXIT function $I_a^o(I_e)$. As described in Section 5.5, this may be obtained [149] as a weighted average of the inverted EXIT functions $\{I_a^{o,n}(I_e)\}_{n=1}^N$

[1]Part of this chapter is based on [202] © IEEE (2009).

corresponding to the N component VLEC codebooks, according to

$$I_a^o(I_e) = \sum_{n=1}^{N} \alpha^n \cdot I_a^{o,n}(I_e).$$ (9.3)

As demonstrated in Chapters 7 and 8, the EXIT-chart matching algorithm of [149] is capable of shaping the inverted IrVLC EXIT function $I_a^o(I_e)$ by specifically selecting the fractions $\{\alpha^n\}_{n=1}^{N}$, subject to the constraints of Equations (9.1) and (9.2), and to

$$\sum_{n=1}^{N} \alpha^n \cdot R(\mathbf{VLEC}^n) = R_{\mathrm{IrVLC}},$$ (9.4)

where R_{IrVLC} is a specified target IrVLC coding rate. In this way, the inverted IrVLC EXIT function $I_a^o(I_e)$ may be shaped so that it does not intersect with an inner EXIT function that corresponds to a near-capacity channel Signal-to-Noise Ratio (SNR). Note that the resultant open EXIT-chart tunnel [267] implies that iterative decoding convergence to an infinitesimally low probability of error can be achieved, provided that the iterative decoding trajectory approaches the inner and outer EXIT functions sufficiently closely, as described in Section 4.3.

However, it was found in Figure 8.13 of Section 8.5.4 that the accuracy of EXIT-chart matching using the algorithm of [149] is limited by the typically high starting point of the EXIT functions $I_e^i(I_a)$ of the inner codes along the EXIT chart's I_e^i axis. This is because the inverted IrVLC EXIT function $I_a^o(I_e)$ is constrained to emerge from the $(0,0)$ point of the EXIT chart, as described in Section 4.3. The resultant non-negligible area between the inverted IrVLC EXIT function $I_a^o(I_e)$ and the inner code's EXIT function $I_e^i(I_a)$ limits the scheme's ability to approach the channel's capacity, owing to the area property of EXIT charts [204], as described in Section 5.5.

This observation motivates this chapter's consideration of an irregular inner codec, having an EXIT function $I_e^i(I_a)$ that may be shaped to emerge from a lower point along the EXIT chart's I_e^i axis. A higher degree of design freedom may be afforded by serially concatenating irregular inner and outer codecs in this way. Hence, in this chapter, we propose a novel joint EXIT-chart matching algorithm, which iteratively applies the algorithm of [149] in order to match the EXIT functions of two serially concatenated irregular codes to each other. This maximizes the effective throughput η by creating a narrow EXIT-chart tunnel, implying that near-capacity operation is facilitated, as described above.

In this chapter, we characterize the serial concatenation of an IrVLC with an inner codec based upon Unity-Rate Codes (URCs) [308], similarly to Chapter 8. However, in contrast with the IrVLC-URC scheme of Chapter 8, this chapter's so-called IrVLC-IrURC scheme employs an Irregular Unity-Rate Code (IrURC) instead of a regular URC. In analogy with IrVLCs, this IrURC scheme comprises R different component URC codes $\{\mathbf{URC}^r\}_{r=1}^{R}$, which are employed to encode particular fractions $\{\alpha^r\}_{r=1}^{R}$ of the transmission frame. Note that since all component URCs have unity-valued coding rates [308], the fractions of the IrURC-encoded frame generated by the components are equal to the fractions $\{\alpha^r\}_{r=1}^{R}$ of the transmission frame that they encode. Subject to constraints that are analogous to

Equations (9.1) and (9.2), the fractions $\{\alpha^r\}_{r=1}^R$ may be chosen in order to shape the IrURC EXIT function $I_e^i(I_a)$ as a weighted average of the component URC EXIT functions $\{I_e^{i,r}(I_a)\}_{r=1}^R$, according to

$$I_e^i(I_a) = \sum_{r=1}^R \alpha^r \cdot I_e^{i,r}(I_a). \tag{9.5}$$

Similarly to the URC scheme of Chapter 8, the employment of an IrURC in this chapter avoids the effective throughput loss of the $R_{\text{TCM}} = 3/4$-rate Trellis-Coded Modulation (TCM) scheme that was employed in Chapter 7. However, in contrast to the 1 bit per channel use Binary Phase-Shift Keying (BPSK) modulation scheme of Chapter 8, this chapter employs $M_{\text{QAM}} = 16$-ary Quadrature Amplitude Modulation (16QAM) in order to benefit from the higher throughput of $\log_2(M_{\text{QAM}}) = 4$ bits per channel employed in Chapter 7.

Note that while the EXIT-chart matching algorithm of [149] is capable of shaping the inverted EXIT functions of IrVLCs, which employ component VLEC codebooks having a diverse variety of coding rates $\{R(\mathbf{VLEC}^n)\}_{n=1}^N$, it must be modified to facilitate the design of IrURCs, in which all component codes have the *same* unity-valued coding rate. This motivates the novel modification of the EXIT-chart matching algorithm of [149] in this chapter, removing the constraint on the fractions $\{\alpha^r\}_{r=1}^R$ that would otherwise be analogous to Equation (9.4).

In Figure 8.12 of Section 8.5.3, we showed that reduced IrVLC coding rates are associated with improved EXIT-chart matching, yielding open EXIT-chart tunnels at channel SNRs that are closer to the channel's capacity bounds. This motivates the choice of a relatively low IrVLC coding rate of $R_{\text{IrVLC}} = 0.53$ in this chapter. However, in Section 8.6 reduced IrVLC coding rates were associated with increased sensitivity to correlation within frames of *a priori* Logarithmic Likelihood Ratios (LLRs) [292] during APP SISO IrVLC decoding. This degrades the degree to which the iterative decoding trajectory [294] approaches the inner and outer EXIT functions, limiting the near-capacity operation, as described in Section 4.3. In this chapter the sensitivity of our APP SISO IrVLC decoder to correlation within frames of *a priori* LLRs allows us to characterize the degree to which the interleaver's length affects its ability to mitigate this correlation. This is achieved by considering a number of different interleaver lengths. Furthermore, Figures 8.17 and 8.18 of Section 8.6 demonstrated that re-duced IrVLC coding rates are associated with increased iterative decoding complexities. This complexity is mitigated in this chapter by employing an additional novel modification of the EXIT-chart matching algorithm of [149]. More specifically, this modification simultaneously facilitates EXIT-chart matching and the quest for a reduced APP SISO decoding complexity. This is achieved by considering the APP SISO decoding complexity that is associated with each component code, when selecting the specific fraction of the transmission frame that it should encode. Note that Section 8.6 observed that the iterative decoding complexity of the IrVLC-URC scheme of Chapter 8 was dominated by that of the IrVLC decoder. Hence, the modification of the EXIT-chart matching algorithm of [149] focusses on limiting the APP SISO IrVLC decoding complexity in this chapter, ignoring that associated with the inner codec.

As shown in Figure 9.1, this chapter's modifications of the EXIT-chart matching algorithm of [149] represent alterations to the method of selecting the component fractions that is employed during conventional irregular codec design.

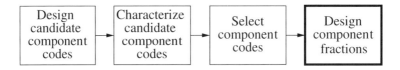

Figure 9.1: Conventional irregular coding design process. This chapter presents modifications to the aspects of this process that are indicated using a bold box.

This chapter is organized as follows. In Section 9.2 we detail our modifications of the EXIT-chart matching algorithm of [149]. Our novel joint EXIT-chart matching algorithm is introduced in Section 9.3. The application of this algorithm is exemplified in the context of our novel IrVLC-IrURC scheme, which is introduced in Section 9.4. We detail the choice of parameters for our proposed transmission scheme in Section 9.5. Here, we additionally introduce benchmark parameterizations, where either one or both of the IrVLC-IrURC scheme's irregular codecs is replaced by its regular equivalent. The achievable Bit-Error Ratio (BER) performance and computational complexity of our parameterizations are characterized and analyzed in Section 9.6. Finally, we offer our conclusions in Section 9.7.

9.2 Modifications of the EXIT-Chart Matching Algorithm

In this section we detail a number of modifications of the EXIT-chart matching algorithm of [149]. We assume a familiarity with Section VII and Table 1 of [149] and we adopt the notation of this reference for the remainder of this section. In the case of an outer IrVLC codec, the notation L is employed to represent the number N of component VLEC codebooks employed, while R_{IR} represents the composite IrVLC coding rate R_{IrVLC}. Furthermore, the notation α_i is employed to represent the fraction α^n of the transmission frame that is encoded using the particular component VLEC codebook $\mathbf{VLEC}^n$, while R_i is employed to represent the codebook's coding rate $R(\mathbf{VLEC}^n)$, where $i = n \in [1 \ldots N]$. In the case of an inner IrURC codec, the number R of component URC codes is represented using the notation L, while the fraction α^r of the transmission frame that is encoded using the particular component URC code $\mathbf{URC}^r$ is represented by α_i, where $i = r \in [1 \ldots R]$. Note that, since all component URC codes have the same unity-valued coding rate, the composite IrURC coding rate will also be equal to unity: $R_{IR} = R_1 = R_2 = \cdots = R_L = 1$. We therefore commence by detailing a modification to the EXIT-chart matching algorithm of [149] that eliminates the requirement for a target composite coding rate R_{IR} to be specified and removes the constraint that would otherwise be analogous to Equation (9.4), as described in Section 9.1.

In the EXIT-chart matching algorithm of [149], adherence to the constraints of Equations (9.2) and (9.4) is achieved by maintaining the relationship

$$\mathbf{C}\alpha = \mathbf{d}, \tag{9.6}$$

where

$$C = \begin{bmatrix} 1 & 1 & \cdots & 1 \\ R_1 & R_2 & \cdots & R_L \end{bmatrix}, \quad \alpha = \begin{bmatrix} \alpha_1 & \alpha_2 & \cdots & \alpha_L \end{bmatrix}^{\mathrm{T}}, \quad d = \begin{bmatrix} 1 \\ R_{\mathrm{IR}} \end{bmatrix}.$$

Here, the first and second rows of C and d maintain constraint (9.2) and constraint (9.4) respectively. Hence, constraint (9.4) may be removed by replacing C with a vector $c = \begin{bmatrix} 1 & 1 & \cdots & 1 \end{bmatrix}$ in which all L number of elements are equal to unity and by replacing d with a unity-valued scalar $d = 1$ throughout Section VII and Table 1 of [149]. Note that these modifications require that the matrix T of Section VII and Table 1 of [149] becomes a $L \times (L - 1)$ matrix, whose columns are a basis of the null space of c, of dimension $L - 1$.

The algorithm of [149] performs EXIT-chart matching by employing the steepest descent approach [309] to minimize a cost function $J(\alpha) = \|e\|_2$, where e is a vector of distances between the irregular outer code's inverted EXIT function and the serially concatenated inner code's EXIT function. As described in Section 9.1, the algorithm of [149] may be adapted to jointly perform EXIT-chart matching and simultaneously seek a reduced computational complexity for the resultant irregular coding scheme's APP SISO decoder. This may be achieved by specifying a vector o having a length L, comprising the average number of Add, Compare and Select (ACS) operations performed per LLR [292] during the APP SISO decoding of each component code using the Bahl–Cocke–Jelinek–Raviv (BCJR) algorithm [48] in the logarithmic probability domain and an eight-entry lookup table for correcting the Jacobian approximation [296]. With reference to Section VII of [149], the cost function may be adapted to consider the APP SISO decoder's computational complexity according to

$$J(\alpha) = \|e\|_2 + y^2 \|o^{\mathrm{T}}\alpha\|_2, \tag{9.7}$$

where y is a positive scalar value, which is commensurate with the relative importance of seeking a low computational complexity for the APP SISO decoder, or has a zero value if this complexity should not be considered during EXIT-chart matching.

During the steepest descent approach of [149], the gradient e_b of the cost function $J(\alpha)$ is evaluated with respect to a vector β that is dependent on the vector α listing the constituent codes' weights. This gradient e_b is employed to iteratively update the vector β, and hence the weight vector α, in order to converge towards a minimal cost function $J(\alpha)$. The EXIT-chart matching algorithm may be adapted to additionally seek a reduced APP SISO decoding computational complexity by employing the cost function of Equation (9.7) during the derivation of e_b and β. With reference to Section VII and Table 1 of [149], the revised e_b and β are specified by

$$e_b = (T^{\mathrm{T}}A^{\mathrm{T}}AT + y^2 T^{\mathrm{T}}oo^{\mathrm{T}}T)\beta - T^{\mathrm{T}}A^{\mathrm{T}}(b - A\alpha_c) + y^2 T^{\mathrm{T}}oo^{\mathrm{T}}\alpha_c \tag{9.8}$$

and

$$\beta = \beta - e_b^{\mathrm{T}}e_b / (e_b^{\mathrm{T}}(T^{\mathrm{T}}A^{\mathrm{T}}AT + y^2 T^{\mathrm{T}}oo^{\mathrm{T}}T)e_b) \cdot e_b \tag{9.9}$$

respectively.

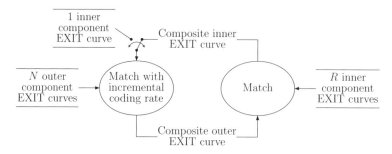

Figure 9.2: Data-flow diagram of the joint EXIT-chart matching algorithm. ©IEEE [202] Maunder and Hanzo, 2009.

9.3   Joint EXIT-Chart Matching

In this section we detail a novel joint EXIT-chart matching algorithm. As described in Section 9.1, this allows the joint EXIT-chart matching of a pair of serially concatenated irregular codes that both support iterative decoding convergence to an infinitesimally low probability of error, in order to obtain the maximal effective throughput that maintains an open EXIT-chart tunnel. The data-flow diagram of the algorithm is provided in Figure 9.2. The algorithm of [149] is iteratively applied in order to match a pair of composite EXIT functions that are formed as superpositions of N outer component EXIT functions and R inner component EXIT functions respectively. However, at the commencement of the algorithm no composite EXIT functions are available. For this reason, the algorithm begins with the matching of the composite outer EXIT function to a single one of the R inner component EXIT functions, as shown in Figure 9.2. The specific choice of the inner component EXIT function that fulfills this role is therefore a parameter of the joint EXIT-chart matching algorithm. Following this, the algorithm alternates between the matching of the composite inner EXIT function to the composite outer EXIT function and vice versa.

The proposed joint EXIT-chart matching algorithm assumes that all inner component codes have the same coding rate, as is the case if they are all URCs, for example. The algorithm must therefore employ the appropriate modification to the EXIT-chart matching algorithm of [149] that was introduced in Section 9.2. Note that the designed irregular inner code will have an overall coding rate that equals those of the component codes in every iteration of the joint EXIT-chart matching algorithm.

By contrast, it is assumed that the outer component codes have a variety of coding rates. Hence, a target composite coding rate must be specified each time, when the algorithm of [149] is invoked for matching the composite outer EXIT function to the inner. For this reason, the joint EXIT-chart matching algorithm is parameterized by a series of incremental target outer coding rates. As the overall coding rate increases with each iteration, the open EXIT tunnel becomes narrower owing to the area property of EXIT charts [204]. However, the alternate matching of the composite inner and outer EXIT functions ensures that they will tend to converge to near-parallel shapes. In this way, an open EXIT tunnel may be maintained even for near-capacity operation.

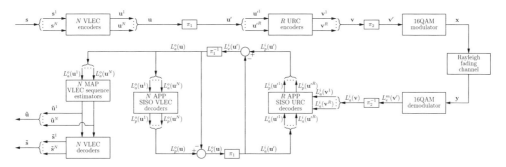

Figure 9.3: Schematic of the IrVLC-IrURC scheme. ©IEEE [202] Maunder and Hanzo, 2009.

The iterative joint EXIT-matching algorithm continues, until the designed composite inner and outer EXIT functions intersect. At this point, the algorithm outputs the inner and outer component fractions found to yield the maximal coding rate, whilst maintaining an open EXIT tunnel.

9.4 Overview of the Transmission Scheme Considered

In this section we consider a transmission scheme that facilitates the joint source and channel coding of a sequence of source symbols having values with unequal probabilities of occurrence for near-capacity transmission over an uncorrelated narrowband Rayleigh fading channel. This application motivates the employment of an IrVLC, as discussed in Section 7.1. This outer codec is serially concatenated and iteratively decoded by exchanging extrinsic information with a novel IrURC used as the inner codec. Owing to its unity coding rate, the IrURC code avoids the effective throughput loss of the $R_{\text{TCM}} = 3/4$-rate TCM codec employed in Chapter 7, as discussed in Section 8.1. Since this combined IrVLC-IrURC scheme employs two irregular codecs, it allows us to exemplify the joint EXIT-chart matching algorithm proposed in Section 9.3.

In the IrVLC-IrURC scheme, the outer IrVLC arrangement employs N component VLEC codebooks $\{\mathbf{VLEC}^n\}_{n=1}^N$. These codebooks have different coding rates of $\{R(\mathbf{VLEC}^n)\}_{n=1}^N$ and/or Real-Valued Free Distance Metrics (RV-FDMs) of $\{D(\mathbf{VLEC}^n)\}_{n=1}^N$, as described in Section 8.2. Furthermore, each component VLEC codebook is invoked for generating a particular fraction of the transmission frame, as described in Section 9.1. Similarly, the inner IrURC arrangement employs R component URC codes $\{\mathbf{URC}^r\}_{r=1}^R$, having different generator and/or feedback polynomials [308]. The schematic of the IrVLC-IrURC scheme is provided in Figure 9.3, which is reminiscent of the schematic of Figure 8.6 using a regular inner code.

9.4.1 Joint Source and Channel Coding

We employ $K = 16$-ary source symbols having values obeying the probabilities of occurrence that result from the Lloyd–Max quantization [164, 165] of independent Gaussian distributed continuous-valued source samples, as exemplified in Section 7.2.1. These occurrence probabilities of $\{P(k)\}_{k=1}^K = \{0.008, 0.024, 0.043, 0.060, 0.076, 0.089, 0.097, 0.102, 0.102,$

$0.097, 0.089, 0.076, 0.060, 0.043, 0.024, 0.008\}$ can be seen to vary by more than an order of magnitude. Since these probabilities correspond to varying numbers of $\{-\log_2(P(k))\}_{k=1}^K = \{6.93, 5.35, 4.55, 4.05, 3.72, 3.49, 3.36, 3.29, 3.29, 3.36, 3.49, 3.72, 4.05, 4.55, 5.35, 6.93\}$ bits of information per symbol, the application of VLCs is motivated, since the source entropy is $E = 3.77$ bits of information per symbol, according to Equation (8.2).

In the transmitter shown in Figure 9.3, the source symbol frame $\mathbf{s}$ is decomposed into N sub-frames $\{\mathbf{s}^n\}_{n=1}^N$, where each sub-frame $\mathbf{s}^n = \{s_{j^n}^n\}_{j^n=1}^{J^n}$ comprises a potentially different number J^n of K-ary source symbols $s_{j^n}^n \in [1 \ldots K]$. Hence, the total number of source symbols in the source symbol frame $\mathbf{s}$ is given by

$$J^{\text{sum}} = \sum_{n=1}^N J^n. \tag{9.10}$$

Each of the N source symbol sub-frames $\{\mathbf{s}^n\}_{n=1}^N$ is VLEC-encoded using the corresponding VLEC codebook from the set $\{\mathbf{VLEC}^n\}_{n=1}^N$, where each VLEC codebook $\mathbf{VLEC}^n = \{\mathbf{VLEC}^{n,k}\}_{k=1}^K$ comprises K VLEC codewords. During the VLEC encoding of the source symbol sub-frame $\mathbf{s}^n = \{s_{j^n}^n\}_{j^n=1}^{J^n}$, each constituent source symbol $s_{j^n}^n$ having a particular value $k \in [1 \ldots K]$ is represented by the corresponding VLEC codeword $\mathbf{VLEC}^{n,k}$, having a length of $I^{n,k}$ bits. Following this, the resultant set of J^n VLEC codewords $\{\mathbf{VLEC}^{n,s_{j^n}^n}\}_{j^n=1}^{J^n}$ is concatenated to obtain the transmission sub-frame $\mathbf{u}^n$, which comprises a number of bits given by

$$I^n = \sum_{j^n=1}^{J^n} I^{n,s_{j^n}^n}. \tag{9.11}$$

Owing to the variable lengths of the VLEC codewords, the transmission sub-frame lengths $\{I^n\}_{n=1}^N$ will typically vary from frame to frame. In order to facilitate the VLEC decoding of the transmission sub-frames $\{\mathbf{u}^n\}_{n=1}^N$, it is necessary to explicitly convey their lengths $\{I^n\}_{n=1}^N$ to the receiver. Furthermore, this highly error-sensitive side information must be reliably protected against transmission errors, using a powerful low-rate block code, for example. For the sake of avoiding obfuscation, this is not shown in Figure 9.3. In Section 8.5.5 we comment on the amount of side information that is required for reliably conveying the specific number of bits in each transmission sub-frame to the decoder. As described in Section 8.5.5, we may expect the required amount of side information to be negligible compared with the transmission frame length I^{sum}, since, as in the IrVLC-URC scheme of Section 8.4, we employ only a single transmission sub-frame per component VLEC codebook.

In the scheme's transmitter, the N transmission sub-frames $\{\mathbf{u}^n\}_{n=1}^N$ are concatenated. As shown in Figure 9.3, the resultant transmission frame $\mathbf{u}$ comprises a number of bits given by

$$I^{\text{sum}} = \sum_{n=1}^N I^n, \tag{9.12}$$

where I^{sum} will typically vary from frame to frame. Note that the average fractions $\alpha^n = I^n/I^{\mathrm{sum}}$ of the transmission frame $\mathbf{u}$ that are generated by each VLEC component code $\mathbf{VLEC}^n$ may be chosen in order to shape the inverted EXIT function of the IrVLC code, as detailed in Section 9.5. The corresponding fractions $C^n = J^n/J^{\mathrm{sum}}$ of the source symbol frame $\mathbf{s}$ that should be encoded by each VLEC component code $\mathbf{VLEC}^n$ may be obtained as

$$C^n = \alpha^n \cdot R(\mathbf{VLEC}^n)/R_{\mathrm{IrVLC}}, \qquad (9.13)$$

where R_{IrVLC} is the composite IrVLC coding rate.

In the transmitter shown in Figure 9.3, the transmission frame $\mathbf{u}$ is interleaved using a random I^{sum}-bit interleaver π_1 in order to obtain the interleaved transmission frame $\mathbf{u}'$. This is decomposed into R sub-frames $\{\mathbf{u}'^r\}_{r=1}^R$, where each sub-frame $\mathbf{u}'^r = \{u'^r_{ir}\}_{ir=1}^{I^r}$ comprises I^r bits. Each of the R interleaved transmission sub-frames $\{\mathbf{u}'^r\}_{r=1}^R$ is URC-encoded using the corresponding URC from the set $\{\mathbf{URC}^r\}_{r=1}^R$. Note that the fraction $\alpha^r = I^r/I^{\mathrm{sum}}$ of the interleaved transmission frame $\mathbf{u}'$ that is encoded by each component URC code $\mathbf{URC}^r$ may be chosen in order to shape the EXIT function of the IrURC codec, as detailed in Section 9.5. Since all URCs have a coding rate of unity, the resultant IrURC-encoded sub-frames $\{\mathbf{v}^r\}_{r=1}^R$ have the same lengths as the corresponding interleaved transmission sub-frames $\{\mathbf{u}'^r\}_{r=1}^R$. As shown in Figure 9.3, the IrURC-encoded sub-frames $\{\mathbf{v}^r\}_{r=1}^R$ are concatenated to provide the IrURC-encoded frame $\mathbf{v}$, having a length of I^{sum} bits.

As shown in Figure 9.3, the IrURC-encoded frame $\mathbf{v}$ is interleaved using a random I^{sum}-bit interleaver π_2 and modulated using Gray-coded [310] $M_{\mathrm{QAM}} = 16$QAM [300], as shown in the constellation diagram of Figure 9.4. As described in Section 9.1, we employ 16QAM in order to benefit from the throughput of $\log_2(M_{\mathrm{QAM}}) = 4$ bits per symbol, which is higher than the BPSK throughput of 1 bit per symbol employed in the IrVLC-URC scheme of Section 8.4. Note that Gray-coded 16QAM is employed since it minimizes the cost of omitting the iterative exchange of extrinsic information between the demodulator and the APP SISO URC decoder [311], as detailed in Section 9.5.2. This facilitates an implementational and computational complexity saving in the receiver of Figure 9.3, which does not perform iterative demodulation and APP SISO IrURC decoding. Following modulation, the resultant channel input symbols $\mathbf{x}$ are transmitted over an uncorrelated narrowband Rayleigh fading channel and are received as the channel output symbols $\mathbf{y}$, as seen in Figure 9.3.

9.4.2 Iterative Decoding

Although no extrinsic information is iteratively exchanged between the demodulator and the APP SISO IrURC decoder of Figure 9.3, the APP SISO IrURC and IrVLC decoding operations *are* performed iteratively. Both of these decoders invoke the BCJR algorithm [48] applied to bit-based trellises [141, 180]. All BCJR calculations are performed in the logarithmic probability domain and using an eight-entry lookup table for correcting the Jacobian approximation [296]. Note that this approach requires only the use of ACS operations. Extrinsic soft information, represented in the form of LLRs [292], is iteratively exchanged between the IrURC and IrVLC decoding stages of Figure 9.3 for the sake of assisting each other's operation. In Figure 9.3, $L(\cdot)$ denotes the LLRs of the bits concerned, where the superscript m denotes demodulation, i indicates inner IrURC decoding and

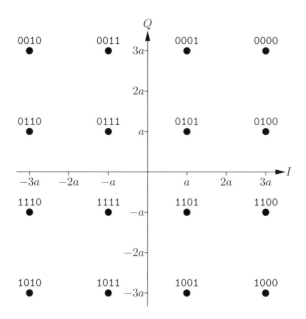

Figure 9.4: Constellation diagram for the $M_{\mathrm{QAM}} = 16$-level Gray-coded QAM modulation scheme. For an average transmit energy of unity, $a = 1/\sqrt{10}$.

o corresponds to outer IrVLC decoding. Additionally, the specific subscript denotes the dedicated role of the LLRs, with a, p and e indicating *a priori*, *a posteriori* and extrinsic information, respectively. Similarly, a tilde over the notation represents a reconstructed estimate of the bits or symbols concerned.

Following the demodulation of the channel output symbols $\mathbf{y}$ of Figure 9.3, the resultant LLR frame $L_e^m(\mathbf{v})$ is deinterleaved π_2^{-1} and provided to the APP SISO IrURC decoder as the *a priori* LLR frame $L_a^i(\mathbf{v})$. Just as R separate URC encoding processes are employed in the IrVLC-IrURC scheme's transmitter, R separate APP SISO URC decoding processes are employed in its receiver. In parallel with the composition of the IrURC-encoded frame $\mathbf{v}$ from R sub-frames, the corresponding *a priori* LLR frame $L_a^i(\mathbf{v})$ is decomposed into R number of sub-frames following deinterleaving π_2^{-1}, as shown in Figure 9.3. Similarly, the LLR frame $L_a^i(\mathbf{u}')$ representing the interleaved transmission frame is decomposed into R sub-frames. Each of the R APP SISO URC decoding processes is provided with the corresponding *a priori* LLR sub-frames $L_a^i(\mathbf{v}^r)$ and $L_a^i(\mathbf{u}'^r)$ and in response it generates the *a posteriori* LLR sub-frame $L_p^i(\mathbf{u}'^r)$. These *a posteriori* LLR sub-frames are concatenated in order to provide the *a posteriori* LLR frame $L_p^i(\mathbf{u}')$, as shown in Figure 9.3.

Similarly, N separate APP SISO VLEC decoding processes are employed in the IrVLC-IrURC receiver. In parallel with the composition of the transmission frame $\mathbf{u}$ from N sub-frames, the corresponding *a priori* LLR frame $L_a^o(\mathbf{u})$ is decomposed into N sub-frames, as shown in Figure 9.3. This is achieved with the aid of the explicit side information that conveys the number of bits I^n in each transmission sub-frame $\mathbf{u}^n$. Each of the N VLEC decoding processes is provided with the *a priori* LLR sub-frame $L_a^o(\mathbf{u}^n)$ and in response it

Table 9.1: Considered parameterizations of the IrVLC-IrURC scheme

Parameterization	N	R	IrVLC y	IrURC y	Threshold E_b/N_0 for $\eta = 2.12$
IrVLC-IrURC-high	30	10	0	0	4.61 dB
IrVLC-IrURC-low	30	10	10^{-5}	0	4.61 dB
IrVLC-URC-high	30	1	0	N/A	5.11 dB
VLC-URC	1	1	N/A	N/A	5.97 dB

generates the *a posteriori* LLR sub-frame $L_p^o(\mathbf{u}^n)$. These *a posteriori* LLR sub-frames are concatenated in order to provide the *a posteriori* LLR frame $L_p^o(\mathbf{u})$, as shown in Figure 9.3.

During the final decoding iteration, N bit-based Maximum *A posteriori* Probability (MAP) VLEC sequence estimation processes are invoked instead of APP SISO VLEC decoding, as shown in Figure 9.3. In this case, each transmission sub-frame $\mathbf{u}^n$ is estimated from the corresponding *a priori* LLR sub-frame $L_a^o(\mathbf{u}^n)$. The resultant transmission sub-frame estimates $\tilde{\mathbf{u}}^n$ are VLEC decoded to provide the source symbol sub-frame estimates $\tilde{\mathbf{s}}^n$. Note that the bit-based MAP sequence estimator is unable to exploit the knowledge that the particular source symbol sub-frame $\mathbf{s}^n$ should comprise J^n source symbols. In order that we may prevent the loss of synchronization that would result if the source symbol sub-frame estimate $\tilde{\mathbf{s}}^n$ did not comprise J^n source symbol estimates, an appropriate number of dummy symbols are removed from, or appended to, the end of $\tilde{\mathbf{s}}^n$, accordingly. Finally, the set of adjusted source symbol sub-frame estimates $\{\tilde{\mathbf{s}}^n\}_{n=1}^N$ may be concatenated to provide the source symbol frame estimate $\tilde{\mathbf{s}}$, as shown in Figure 9.3.

9.5 System Parameter Design

Throughout this section we introduce a number of different parameterizations of the IrVLC-IrURC scheme of Section 9.4. We refer to these as the IrVLC-IrURC-high, the IrVLC-IrURC-low and the IrVLC-URC-high parameterizations. In Section 9.5.4 we additionally introduce a fourth parameterization, which we refer to as the VLC-URC arrangement. Our four parameterizations are summarized in Table 9.1.

9.5.1 Component VLEC Codebooks

As described in Section 9.4, the IrVLC-IrURC scheme employs N component VLEC codebooks and R component URC codes. However, observe in Table 9.1 that the values of N and R vary amongst our parameterizations considered. In the IrVLC-IrURC-high, IrVLC-IrURC-low and IrVLC-URC-high parameterizations, the IrVLC arrangement employs $N = 30$ component VLEC codebooks $\{\mathbf{VLEC}^n\}_{n=1}^{30}$ for generating particular fractions $\{\alpha^n\}_{n=1}^{30}$ of the transmission frame $\mathbf{u}$, as described in Section 9.4. These $N = 30$ component VLEC codebooks $\{\mathbf{VLEC}^n\}_{n=1}^{30}$ were designed using the Genetic Algorithm (GA) of Section 8.3 to have a range of coding rates $\{R(\mathbf{VLEC}^n)\}_{n=1}^{30}$ and RV-FDMs $\{D(\mathbf{VLEC}^n)\}_{n=1}^{30}$. More specifically, maximal RV-FDMs were sought for coding rates in the range of $[0.25, 0.95]$ during the generation of the first 15 of the $N = 30$ component VLEC codebooks $\{\mathbf{VLEC}^n\}_{n=1}^{15}$, as described in Section 8.5.1. By contrast, RV-FDMs that are as close as possible to, but no less than, two were sought during the generation of the remaining 15 component VLEC codebooks $\{\mathbf{VLEC}^n\}_{n=16}^{30}$, which also have coding rates in the range of $[0.25, 0.95]$. Note that since all $N = 30$ component VLEC codebooks employ a RV-FDM of at least two, the corresponding inverted EXIT functions will reach the top right-hand corner of the EXIT chart [298], as detailed below. The coding rate $R(\mathbf{RVLC}^n)$, RV-FDM $D(\mathbf{RVLC}^n)$ and the composition of the $N = 30$ component VLEC codebooks $\{\mathbf{VLEC}^n\}_{n=1}^{30}$ are provided in Tables 9.2 and 9.3.

The inverted EXIT functions of the $N = 30$ component VLEC codebooks in the suite $\{\mathbf{VLEC}^n\}_{n=1}^{30}$ are provided in Figure 9.5. These were obtained by simulating the VLEC encoding and APP SISO decoding of 100 frames of 10 000 randomly generated source symbols. As described in Section 9.4, these source symbols have $K = 16$-ary values with the probabilities of occurrence that result from the Lloyd–Max quantization [164, 165] of independent Gaussian distributed source samples. The corresponding frames of VLEC-encoded bits were employed to generate uncorrelated Gaussian distributed *a priori* LLR frames having a range of mutual informations $I_a^o \in [0, E(\mathbf{VLEC}^n)]$, where $E(\mathbf{VLEC}^n)$ is the VLEC-encoded bit entropy of the considered component VLEC codebook $\mathbf{VLEC}^n$. Note that the VLEC-encoded bit entropy $E(\mathbf{VLEC}^n)$ was found to be only at a negligible distance from unity in the case of all considered component VLEC codebooks $\{\mathbf{VLEC}^n\}_{n=1}^{30}$. Finally, all mutual information measurements were carried out using the histogram-based approximation of the LLR PDFs [294]. For each considered component VLEC codebook $\mathbf{VLEC}^n$, the coding rate $R(\mathbf{VLEC}^n)$ and RV-FDM $D(\mathbf{VLEC}^n)$ are provided in Figure 9.5. Furthermore, the computational complexity of APP SISO decoding is characterized by the average number $O^{\mathrm{APP}}(\mathbf{VLEC}^n)$ of ACS operations performed per source symbol.

Observe in Figure 9.5 that the inverted EXIT functions of all $N = 30$ component VLEC codebooks $\{\mathbf{VLEC}^n\}_{n=1}^{30}$ reach the top right-hand corner of the EXIT chart, as described above. Note that 'S'-shaped inverted EXIT functions having up to two points of inflection are obtained in Figure 9.5 for the component VLEC codebooks $\{\mathbf{VLEC}^n\}_{n=1}^{15}$, which have high RV-FDMs, as explained in Section 8.2. By contrast, the low RV-FDM VLEC codebooks $\{\mathbf{VLEC}^n\}_{n=16}^{30}$ are associated with EXIT functions having no more than a single point of inflection. Since the suite of $N = 30$ VLEC codebooks $\{\mathbf{VLEC}^n\}_{n=1}^{30}$ is associated with a wide variety of inverted EXIT functions, it is particularly suitable for use in IrVLCs, as concluded in Chapter 8. Also note that the high RV-FDM VLEC codebooks $\{\mathbf{VLEC}^n\}_{n=1}^{15}$ are associated with higher APP SISO decoder complexities than are the low RV-FDM VLEC codebooks $\{\mathbf{VLEC}^n\}_{n=16}^{30}$, as was also observed in Section 8.5.1.

9.5.2 Component URC Codes

In the same way that the IrVLC-IrURC scheme employs N component VLEC codebooks, it also employs R component URC codes, as described in Section 9.4. However, the value of R varies amongst our considered parameterizations, as described above. More specifically, the IrVLC-IrURC-high and the IrVLC-IrURC-low parameterizations of Table 9.1 employ $R = 10$ component URC codes $\{\mathbf{URC}^r\}_{r=1}^{10}$. By contrast, the benchmark IrVLC-URC-high parameterization of Table 9.1 employs only the component URC code $\mathbf{URC}^1$ having the lowest coding memory of $L_{\mathrm{URC}} = 1$, and hence imposes the lowest APP SISO decoding complexity [141]. Since the IrVLC-URC-high parameterization employs just $R = 1$ component URC code, it can be thought of as employing a regular URC instead of an IrURC.

Table 9.2: Properties and composition of the 15 component VLEC codebooks $\{\mathbf{VLEC}^n\}_{n=1}^{15}$ that were designed to have high RV-FDMs

$\mathbf{VLEC}^n$	Properties†	Composition‡
$\mathbf{VLEC}^1$	(0.950, 2.412)	6,6,5,5,3,4,3,3,3,4,4,4,5,5,6,6, 317E29FEA2490F1EA1
$\mathbf{VLEC}^2$	(0.900, 2.518)	6,6,5,4,3,5,3,4,4,3,5,4,5,5,6,6, 3A5F8580E52ABCF1321
$\mathbf{VLEC}^3$	(0.850, 2.617)	6,7,5,5,4,4,3,3,5,4,4,5,5,6,7,8, 140DE2BCB304A54F6D818
$\mathbf{VLEC}^4$	(0.800, 2.710)	6,7,6,6,5,3,4,3,4,4,5,5,6,7,7,8, 23B98BDB09CA5D5B60BE3C
$\mathbf{VLEC}^5$	(0.750, 2.793)	7,8,7,6,6,4,5,4,5,4,5,3,6,5,8,7, 1761E3F330FB2A494F8AD1C
$\mathbf{VLEC}^6$	(0.701, 2.922)	7,7,7,6,6,6,5,4,4,5,6,6,5,5,6,7, E45F40FBC3DB413C43A3C12
$\mathbf{VLEC}^7$	(0.650, 2.981)	7,7,8,7,7,5,6,4,6,4,6,5,6,7,8,7, 3540C30FF578C04E753C1862D
$\mathbf{VLEC}^8$	(0.601, 4.364)	13,10,9,8,7,6,5,4,4,5,6,7,8,8,9,10, 49AA7CA652B36B3E065564E5554E43
$\mathbf{VLEC}^9$	(0.550, 4.628)	11,11,10,8,7,5,5,4,7,6,6,7,9,11,9,12, 1E73F274A5646CFC1295595C0C1211F4
$\mathbf{VLEC}^{10}$	(0.501, 4.807)	13,11,9,11,7,8,6,5,5,6,8,7,9,11,12,13, 13192CCD6E0F556B3F80CA55438997333A69
$\mathbf{VLEC}^{11}$	(0.450, 5.869)	14,10,10,9,9,8,8,6,7,7,8,9,11,10,11,12, 1FAEDF0F996FDB34623C056DCB66384ED78FF4
$\mathbf{VLEC}^{12}$	(0.400, 6.573)	15,15,11,10,13,10,7,6,6,9,8,9,12,14,13,18, 3AD43A94E6964B2A9B38FE0471599674D2B5581 C7587
$\mathbf{VLEC}^{13}$	(0.350, 8.475)	15,14,16,14,12,10,9,8,8,9,10,11,13,13,15,16, 1522D8CA63D49CD714DB2F0FF8007A4D358B39A 6353669C2B
$\mathbf{VLEC}^{14}$	(0.301, 10.384)	17,16,15,14,13,12,11,10,9,11,12,13,14,15,28,17, 367E727298F26E57CE7A6D5E41F002C6BCB4D72 ED5AD7393AAAE3ABA7
$\mathbf{VLEC}^{15}$	(0.251, 12.449)	25,18,17,16,15,14,13,12,11,13,14,15,28,17,18,19, 1670497B3BC1D9DAEB73994EB37CF183F00038E 2D9276B2CE0FB5B6736BDA6AE3E8

† The properties of each component VLEC codebook $\mathbf{VLEC}^n$ are provided using the format $(R(\mathbf{VLEC}^n), D(\mathbf{VLEC}^n))$.

‡ The composition of each component VLEC codebook $\mathbf{VLEC}^n$ is specified by providing the $K = 16$ codeword lengths $\{I^{n,k}\}_{k=1}^K$, together with the hexadecimal representation of the ordered concatenation of the $K = 16$ VLEC codewords $\{\mathbf{VLEC}^{n,k}\}_{k=1}^K$ in the codebook.

Note that since the IrVLC-IrURC-high and the IrVLC-IrURC-low parameterizations employ two irregular codecs, they can benefit from the joint matching of their EXIT

Table 9.3: Properties and composition of the 15 component VLEC codebooks $\{\mathbf{VLEC}^n\}_{n=16}^{30}$ that were designed to have low RV-FDMs

$\mathbf{VLEC}^n$	Properties[†]	Composition[‡]
$\mathbf{VLEC}^{16}$	(0.952, 2.394)	8,6,5,5,4,4,3,3,3,3,4,4,5,5,6,6, 0DD7E2A52878BCF1321
$\mathbf{VLEC}^{17}$	(0.901, 2.373)	6,7,5,6,4,4,3,3,3,3,4,5,6,4,8,7, 0C59CEBD331494BCAD1A
$\mathbf{VLEC}^{18}$	(0.851, 2.357)	9,4,5,8,5,4,3,3,3,3,5,4,6,6,7,10, 0B7FE2D1D2B1944BF4B16C
$\mathbf{VLEC}^{19}$	(0.800, 2.312)	7,8,8,5,5,6,3,3,3,3,5,5,6,6,8,7, E869D9CF9998A4394A1617
$\mathbf{VLEC}^{20}$	(0.750, 2.308)	7,8,5,7,6,6,3,3,3,3,6,6,9,6,4,7, 0A96B8B17533152456FD7A7
$\mathbf{VLEC}^{21}$	(0.700, 2.275)	16,11,7,9,5,5,3,3,3,3,5,7,5,9,13,15, 1555154E857FF350C850E542AA1554
$\mathbf{VLEC}^{22}$	(0.653, 2.269)	14,11,11,7,9,5,3,3,3,3,5,7,5,9,16,12, 1558553AA17EA7350C850E542AB42AA
$\mathbf{VLEC}^{23}$	(0.605, 2.257)	12,14,10,11,6,6,3,3,3,3,6,7,15,6,8,10, 1DD1DDE761DCF0C74628E1DDC3CE8EF
$\mathbf{VLEC}^{24}$	(0.553, 2.225)	10,14,8,12,5,6,3,3,3,3,9,15,5,13,11,16, 368DB73736C21998A6C36D8736D1B736DA
$\mathbf{VLEC}^{25}$	(0.501, 2.133)	11,17,8,8,9,5,6,2,2,3,12,18,5,15,11,14, 0DA1B6DCD0DCD843181B61B6D8736D86DCDB7
$\mathbf{VLEC}^{26}$	(0.451, 2.133)	11,14,6,15,9,5,2,5,3,2,12,21,8,18,14,17, 0DB9B6861B6C3612384DB0DB6D86E6DB61B6E6DB7
$\mathbf{VLEC}^{27}$	(0.407, 2.113)	8,17,9,12,21,5,6,2,3,2,12,24,5,15,14,18, 06E6DB7361B79B6DB0862236C36DB6C39B6C36DCDB6C
$\mathbf{VLEC}^{28}$	(0.350, 2.155)	13,19,7,17,10,10,7,3,3,3,13,37,6,16,16,24, 36D1B6D9346DB7368D932C536C9B6D804764936D936DA36DB01
$\mathbf{VLEC}^{29}$	(0.300, 2.123)	8,19,8,12,9,6,5,2,2,3,42,45,5,15,11,17, 06E6DB50D0DB0D861181B6D593DC586DB564F716C39B6C36E6DB4
$\mathbf{VLEC}^{30}$	(0.252, 2.117)	8,14,8,9,12,6,5,2,3,2,56,62,5,15,11,11, 06E6DB9A1B0DB0C22236DB34343BEB6C36DB34343BEB6DB0E6DB0DB9B4

[†] The properties of each component VLEC codebook $\mathbf{VLEC}^n$ are provided using the format $(R(\mathbf{VLEC}^n), D(\mathbf{VLEC}^n))$.

[‡] The composition of each component VLEC codebook $\mathbf{VLEC}^n$ is specified by providing the $K = 16$ codeword lengths $\{I^{n,k}\}_{k=1}^{K}$, together with the hexadecimal representation of the ordered concatenation of the $K = 16$ VLEC codewords $\{\mathbf{VLEC}^{n,k}\}_{k=1}^{K}$ in the codebook.

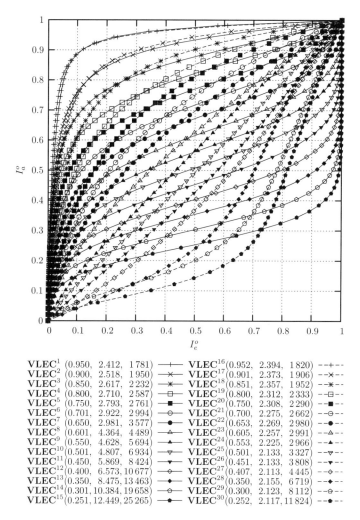

Figure 9.5: Inverted EXIT functions for the $N = 30$ component VLEC codebooks $\{\mathbf{VLEC}^n\}_{n=1}^{30}$. Curves are labeled using the format $\mathbf{VLEC}^n$ ($R(\mathbf{VLEC}^n)$, $D(\mathbf{VLEC}^n)$, $O^{\mathrm{APP}}(\mathbf{VLEC}^n)$). ©IEEE [202] Maunder and Hanzo, 2009.

functions, as described in Section 9.3. By contrast, the only irregular codec employed by the IrVLC-URC-high benchmark parameterization of Table 9.1 is an IrVLC, permitting only the matching of its inverted EXIT function to the fixed EXIT function of the URC codec, as demonstrated for the similar IrVLC-URC scheme of Section 8.5.3.

The Linear Feedback Shift Register (LFSR) designs of Figure 9.6 were selected for the $R = 10$ component URC codes $\{\mathbf{URC}^r\}_{r=1}^{10}$ from the set of all possible designs containing a number of memory elements in the set $L_{\mathrm{URC}} \in \{1, 2, 3\}$. The selected codes were chosen in order to yield the sufficiently diverse variety of EXIT function shapes that are provided

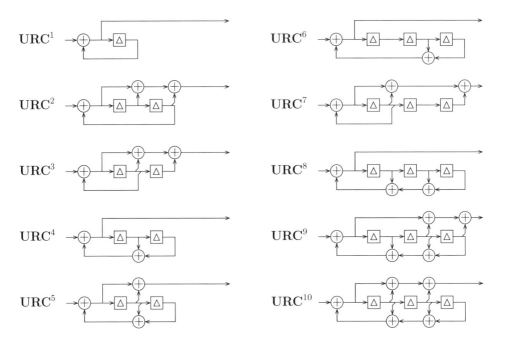

Figure 9.6: LFSR encoder schematics for the component URC codes $\{\mathbf{URC}^r\}_{r=1}^{10}$.

in Figure 9.7, together with the corresponding generator and feedback polynomials. These EXIT functions were obtained by simulating the URC encoding and APP SISO decoding of 100 frames of 10 000 randomly generated bits in the context of transmission over a $M_{\mathrm{QAM}} = 16\mathrm{QAM}$-modulated uncorrelated narrowband Rayleigh fading channel having a SNR of $E_c/N_0 = 8$ dB. The randomly generated bits were employed to generate uncorrelated Gaussian distributed *a priori* LLR frames having a range of mutual informations $I_a \in [0, 1]$ and all mutual information measurements were carried out using the histogram-based approximation of the LLR PDFs [294].

Note that some URC component EXIT functions of Figure 9.7 emerge from the $(0, 0)$ point of the EXIT chart. By contrast, the component URCs $\mathbf{URC}^1$, $\mathbf{URC}^4$, $\mathbf{URC}^6$ and $\mathbf{URC}^8$ offer some nonzero extrinsic mutual information I_e^i even in the absence of any *a priori* mutual information I_a^i. This may be attributed to the presence of only a single nonzero coefficient within their generator polynomials [312], as shown in Figures 9.6 and 9.7. Furthermore, since each of the $R = 10$ component URC codes $\{\mathbf{URC}^r\}_{r=1}^{10}$ employs recursive feedback in Figure 9.6, the EXIT functions of Figure 9.7 reach the top right-hand corner of the EXIT chart, as described in [299]. Since the inverted EXIT functions of our component VLEC codebooks also reach the top right-hand corner of the EXIT chart, we can expect to achieve open EXIT-chart tunnels [267] if the channel quality is sufficiently high. As discussed in Section 4.3, this implies that iterative decoding convergence to an infinitesimally low probability of error is facilitated, provided that the iterative decoding trajectory approaches the inner and outer EXIT functions sufficiently closely.

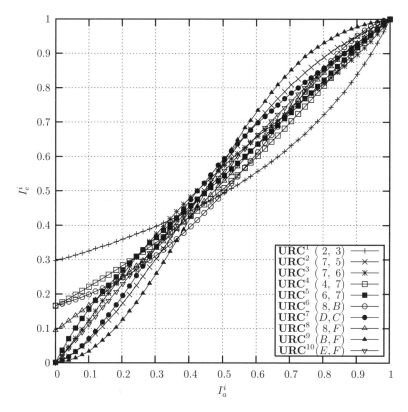

Figure 9.7: EXIT functions for the $R = 10$ component URC codes $\{\mathbf{URC}^r\}_{r=1}^{10}$ for a 16QAM-modulated Rayleigh fading channel SNR of $E_c/N_0 = 8$ dB. Curves are labeled using the format $\mathbf{URC}$ $(g(\mathbf{URC}), f(\mathbf{URC}))$, where $g(\mathbf{URC})$ and $f(\mathbf{URC})$ are the hexadecimal generator and feedback polynomials of the URC code, respectively [296]. ©IEEE [202] Maunder and Hanzo, 2009.

Note that the EXIT chart area property [204] suggests that, when multiplied by $\log_2(M_{\mathrm{QAM}}) = 4$, the area beneath the component URC EXIT functions should equal the channel's capacity, since the component URC codes have unity-valued coding rates. However, the average area beneath the component URC EXIT functions of Figure 9.7 was found to equal 0.538. Upon multiplying this value by $\log_2(M_{\mathrm{QAM}}) = 4$, the resultant value of 2.154 is found to be slightly lower than the 16QAM-modulated Rayleigh fading channel's capacity at the corresponding SNR of $E_c/N_0 = 8$ dB, which is 2.209 bits per channel symbol [300]. This discrepancy may be explained by the lack of an iterative exchange of extrinsic information between the demodulator and the URC-based inner APP SISO decoder, as shown in Figure 9.3. To illustrate this point, the EXIT function corresponding to the Gray-coded $M_{\mathrm{QAM}} = 16$QAM demodulator is provided in Figure 9.8 for a Rayleigh fading channel SNR of $E_c/N_0 = 8$ dB. Since iterative demodulation is not employed, the LLR frame $L_e^m(\mathbf{v})$ of Figure 9.3 will only have the mutual information I_e^m associated with a zero-valued *a priori*

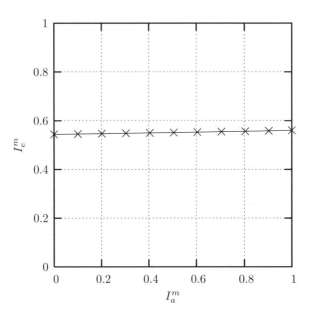

Figure 9.8: EXIT functions for the Gray-coded demodulator for a Rayleigh fading channel SNR of $E_c/N_0 = 8$ dB. ©IEEE [202] Maunder and Hanzo, 2009.

mutual information I_a^m and will be unable to reach that associated with a higher value of I_a^m. In the $E_c/N_0 = 8$ dB scenario shown in Figure 9.8, the LLR frame $L_e^m(\mathbf{v})$ will have a mutual information of $I_e^m = 0.544$ and will be unable to reach the maximal value of 0.560. Note that, upon multiplying the area beneath the demodulator's EXIT function by $\log_2(M_{\mathrm{QAM}}) = 4$, the channel capacity of 2.209 bits per symbol *is* obtained. Also note that these findings are contrary to those of Section 8.5.4, where the area beneath the URC EXIT functions was found to be equal to the corresponding channel capacities. However, in this case BPSK modulation was employed, for which there is no benefit in performing iterative demodulation [300].

9.5.3 EXIT-Chart Matching

The IrVLC arrangement of the benchmark IrVLC-URC-high parameterization of Table 9.1 was designed using the EXIT-chart matching algorithm of [149]. By contrast, the joint EXIT-chart matching algorithm of Section 9.3 was employed for the design of the IrVLC and IrURC arrangements of the IrVLC-IrURC-high and the IrVLC-IrURC-low parameterizations. In the case of the IrVLC-IrURC-low parameterization, the method of Section 9.2 was used for jointly performing EXIT-chart matching and for reducing the corresponding APP SISO decoder's computational complexity. The low complexity that we sought for this arrangement explains why we refer to it as the IrVLC-IrURC-low parameterization. Recall that Section 8.6 observed that the computational complexity of IrVLCs typically dominates that of iterative decoding. For this reason we opted for employing the method of Section 9.2 to reduce the computational complexity associated with only the IrVLC and not the IrURC scheme. Hence, a weighting coefficient of $y = 10^{-5}$ was employed when designing the IrVLC scheme, while a value of $y = 0$ was used for designing the IrURC scheme, as shown in Table 9.1. Note that the method of Section 9.2 invoked for jointly performing EXIT-chart matching and for reducing the corresponding APP SISO decoding complexity was not employed in the design of the therefore relatively high-complexity IrVLC-IrURC-high parameterization, as shown in Table 9.1. The comparison of the IrVLC-IrURC-high and IrVLC-IrURC-low parameterizations will therefore facilitate the characterization of the complexity reduction method of Section 9.2. Note that this modification of the EXIT-chart matching algorithm of [149] was found to offer no significant APP SISO decoder complexity reduction when regular URCs were employed. It is for this reason that we do not consider an 'IrVLC-URC-low' parameterization.

For the IrVLC-IrURC-high, IrVLC-IrURC-low and IrVLC-URC-high parameterizations, EXIT-chart matching was performed for a number of 16QAM Rayleigh fading channel SNRs in the range of $E_c/N_0 \in [3.5, 20.5]$ dB. In each case we sought the highest IrVLC coding rate R_{IrVLC} for which an open EXIT-chart tunnel could be achieved. Since the IrURC and URC arrangements have unity coding rates, and because $M_{\mathrm{QAM}} = 16$QAM is employed in all parameterizations, the effective throughput of each parameterization is given by $\eta = R_{\mathrm{IrVLC}} \cdot \log_2(M_{\mathrm{QAM}})$ bits per channel use, where R_{IrVLC} is the IrVLC coding rate. Additionally, note that the channel SNR values of E_c/N_0 may be converted to E_b/N_0 values according to $E_b/N_0 = E_c/N_0 \cdot 1/\eta$, where N_0 is the noise power spectral density, and E_c and E_b are the transmit energy per channel use and per bit of source information respectively. The maximum effective throughput η for which open EXIT-chart tunnels were created for a range of 16QAM Rayleigh fading channel E_b/N_0 values are plotted in Figure 9.9 for the IrVLC-IrURC-high, the IrVLC-IrURC-low and the IrVLC-URC-high parameterizations.

Note that Figure 9.9 also provides plots of the channel's capacity [300] and the channel's *attainable* capacity versus E_b/N_0, as described above. Here, the channel's attainable capacity is obtained for each particular E_b/N_0 value by multiplying the average area beneath the corresponding EXIT functions of the $R = 10$ component URC codes by $\log_2(M_{\mathrm{QAM}}) = 4$, as described above. Note that the channel's attainable capacity represents an upper bound to the maximum effective throughput η for which an open EXIT-chart tunnel can be achieved. This is because a parameterization's effective throughput may be approximated by multiplying the area beneath the inverted IrVLC EXIT function by $\log_2(M_{\mathrm{QAM}}) = 4$ [204]. Since this area must be lower than that beneath the inner EXIT function in order for an open EXIT chart to be facilitated, iterative decoding convergence to an infinitesimally low probability of error is prevented when the effective throughput η is higher than the channel's attainable capacity. The discrepancy between the channel's capacity and its *attainable* capacity therefore imposes an effective throughput loss. Note that we propose a method for mitigating this effective throughput loss during our discussion of future work in Section 12.12.

In Figure 9.9, the IrVLC-IrURC-high and IrVLC-IrURC-low parameterizations can be seen to consistently maintain open EXIT-chart tunnels at higher effective throughputs than the IrVLC-URC-high parameterization does, across the entire range of E_b/N_0 values considered. This may be explained by the higher degree of design freedom that is facilitated by the joint EXIT-chart matching algorithm that was employed to design the IrVLC-IrURC-high and IrVLC-IrURC-low parameterizations of Table 9.1. Whilst both the IrVLC-IrURC-high and the IrVLC-IrURC-low parameterizations are capable of creating open EXIT-chart tunnels for E_b/N_0 values above a threshold that is within 0.8 dB of the channel's E_b/N_0 capacity bound

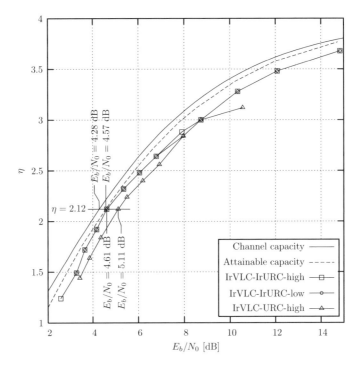

Figure 9.9: The maximum effective throughput for which open EXIT-chart tunnels could be achieved versus 16QAM-modulated Rayleigh fading channel E_b/N_0 for the IrVLC-IrURC-high, IrVLC-IrURC-low and IrVLC-URC-high parameterizations of Table 9.1. The channel's capacity and attainable capacity plots are additionally provided [300]. ©IEEE [202] Maunder and Hanzo, 2009.

for effective throughputs in the range [1.24,2.68], the IrVLC-URC-high parameterization of Table 9.1 can only achieve this above a less desirable threshold that is within about 1.2 dB of the channel's E_b/N_0 capacity bound. Furthermore, the IrVLC-IrURC-high and IrVLC-IrURC-low parameterizations of Table 9.1 can achieve this latter feat for more desirable effective throughputs of up to 3.06 bits per channel use.

9.5.4   Parameterizations of the Proposed Scheme

For the remainder of this chapter, we consider only the IrVLC-IrURC-high, IrVLC-IrURC-low and IrVLC-URC-high arrangements obtained in Section 9.5.3 that have an effective throughput of $\eta = 2.12$ bits per channel use. This effective throughput corresponds to an IrVLC coding rate of $R_{\mathrm{IrVLC}} = 0.53$, to an E_b/N_0 capacity bound of 4.28 dB [300] and an attainable E_b/N_0 capacity bound of 4.57 dB, as shown in Figure 9.9.

As described in Section 9.5.3, the EXIT-chart matching algorithms of [149] and Section 9.3 were employed to shape the inverted IrVLC EXIT functions of our arrangements by appropriately selecting the fraction C^n of the source symbol frame s and the fraction α^n of the transmission frame u that is encoded by each component VLEC codebook $\mathbf{VLEC}^n$.

These fractions are summarized for the IrVLC-IrURC-high, IrVLC-IrURC-low and IrVLC-URC-high arrangements in Table 9.4.

Let us now discuss the properties of each arrangement's activated component VLEC codebooks. These are employed to encode a nonzero fraction C^n of the source symbol frame s and hence, a nonzero fraction α^n of the transmission frame $\mathbf{u}$. Observe in Table 9.4 that while the IrVLC-URC-high arrangement activates only component VLEC codebooks from the set $\{\mathbf{VLEC}^n\}_{n=1}^{15}$ designed by the GA of Section 8.3 to have high RV-FDMs, the IrVLC-IrURC-high and IrVLC-IrURC-low arrangements additionally activate the low RV-FDM components $\{\mathbf{VLEC}^n\}_{n=16}^{30}$. Note that this is particularly true in the case of the IrVLC-IrURC-low arrangement, which encodes nearly 99% of the source symbols using component VLEC codebooks from the set $\{\mathbf{VLEC}^n\}_{n=16}^{30}$ designed by the GA of Section 8.3 to have low RV-FDMs. In the case of the IrVLC-IrURC-high arrangement, the corresponding percentage is just over 38%.

Furthermore, the EXIT-chart matching algorithm of Section 9.3 was employed to shape the inverted IrURC EXIT functions of our IrVLC-IrURC-high and IrVLC-IrURC-low arrangements by appropriately selecting the fraction α^n of the interleaved transmission frame $\mathbf{u}'$ that is encoded by each component URC code $\mathbf{URC}^r$. These fractions are summarized in Table 9.5.

Observe in Table 9.5 that in the IrVLC-IrURC-high arrangement nearly 93% of the IrURC-encoded bits are generated by the component URC codes $\{\mathbf{URC}^r\}_{r=2}^5$ employing $L_{\mathrm{URC}} = 2$ memory elements in their LFSRs, as shown in Figure 9.6. The remaining 7% of the IrURC-encoded bits are generated by the component URC codes $\{\mathbf{URC}^r\}_{r=6}^{10}$ employing $L_{\mathrm{URC}} = 3$ memory elements in their LFSRs. In contrast, in the IrVLC-IrURC-low arrangement, 78% and 22% of the IrURC-encoded bits are generated by the component URC codes employing $L_{\mathrm{URC}} = 2$ and $L_{\mathrm{URC}} = 3$ memory elements in their LFSRs, respectively. Since the APP SISO URC decoding complexity is proportional to $2^{L_{\mathrm{URC}}}$ [141], a lower APP SISO IrURC decoding complexity can be expected for the IrVLC-IrURC-high arrangement than the IrVLC-IrURC-low arrangement. Furthermore, the lowest inner APP SISO decoding complexity can be expected for the IrVLC-URC-high arrangement, in which all IrURC-encoded bits are generated by the component URC code $\mathbf{URC}^1$ employing just $L_{\mathrm{URC}} = 1$ memory element, as described in Section 9.5.2.

In this section we additionally introduce the VLC-URC benchmark of Table 9.1, which employs only $N = 1$ component VLEC codebook and $R = 1$ component URC codebook. This benchmark therefore represents the serial concatenation and iterative decoding of regular VLC and URC codecs. Like the IrVLC-URC-high parameterization, our so-called VLC-URC arrangement employs only the component URC code $\mathbf{URC}^1$ from Figure 9.6, which has the lowest coding memory of $L_{\mathrm{URC}} = 1$ and, hence, the lowest APP SISO decoding complexity [141]. The VLC-URC arrangement employs a VLEC codebook $\mathbf{VLEC}^{31}$ that was specially designed using the GA of Section 8.3 to have a coding rate $R(\mathbf{VLEC}^{31}) = 0.53$ equal to those of the above-mentioned IrVLC-IrURC-high, IrVLC-IrURC-low and IrVLC-URC-high arrangements. Furthermore, this codebook was designed to have a high RV-FDM of $D(\mathbf{VLEC}^{31}) = 4.665$ since the corresponding 'S'-shaped inverted EXIT function was found to offer a relatively good match with the EXIT function associated with the component URC code $\mathbf{URC}^1$, as will be shown in Figure 9.10. The composition of the VLEC codebook $\mathbf{VLEC}^{31}$ is provided in Table 9.6.

Table 9.4: Properties of the 30 component VLEC codebooks $\{\mathbf{VLEC}^n\}_{n=1}^{30}$, together with the fractions $\{C^n\}_{n=1}^{30}$ of the source symbol frame $\mathbf{s}$ and the fractions $\{\alpha^n\}_{n=1}^{30}$ of the transmission frame $\mathbf{u}$ that they encode in the $\eta = 2.12$ bits per channel symbol IrVLC-IrURC-high, IrVLC-IrURC-low and IrVLC-URC-high arrangements

		Fractions[‡]		
$\mathbf{VLEC}^n$	Properties[†]	IrVLC-IrURC-high	IrVLC-IrURC-low	IrVLC-URC-high
$\mathbf{VLEC}^1$	$(0.950, \ 2.412)$	$(0.000, 0.000)$	$(0.007, 0.004)$	$(0.000, 0.000)$
$\mathbf{VLEC}^2$	$(0.900, \ 2.518)$	$(0.000, 0.000)$	$(0.004, 0.002)$	$(0.000, 0.000)$
$\mathbf{VLEC}^3$	$(0.850, \ 2.617)$	$(\mathbf{0.225, 0.140})$	$(0.001, 0.001)$	$(0.000, 0.000)$
$\mathbf{VLEC}^4$	$(0.800, \ 2.710)$	$(0.036, 0.024)$	$(0.000, 0.000)$	$(0.000, 0.000)$
$\mathbf{VLEC}^5$	$(0.750, \ 2.793)$	$(0.044, 0.031)$	$(0.000, 0.000)$	$(\mathbf{0.361, 0.255})$
$\mathbf{VLEC}^6$	$(0.701, \ 2.922)$	$(0.000, 0.000)$	$(0.000, 0.000)$	$(0.000, 0.000)$
$\mathbf{VLEC}^7$	$(0.650, \ 2.981)$	$(0.074, 0.061)$	$(0.000, 0.000)$	$(0.024, 0.020)$
$\mathbf{VLEC}^8$	$(0.601, \ 4.364)$	$(0.000, 0.000)$	$(0.000, 0.000)$	$(0.000, 0.000)$
$\mathbf{VLEC}^9$	$(0.550, \ 4.628)$	$(\mathbf{0.162, 0.156})$	$(0.000, 0.000)$	$(0.000, 0.000)$
$\mathbf{VLEC}^{10}$	$(0.501, \ 4.807)$	$(0.000, 0.000)$	$(0.000, 0.000)$	$(0.000, 0.000)$
$\mathbf{VLEC}^{11}$	$(0.450, \ 5.869)$	$(0.000, 0.000)$	$(0.000, 0.000)$	$(\mathbf{0.605, 0.713})$
$\mathbf{VLEC}^{12}$	$(0.400, \ 6.573)$	$(0.040, 0.053)$	$(0.000, 0.000)$	$(0.009, 0.012)$
$\mathbf{VLEC}^{13}$	$(0.350, \ 8.475)$	$(0.000, 0.000)$	$(0.000, 0.000)$	$(0.000, 0.000)$
$\mathbf{VLEC}^{14}$	$(0.301, 10.384)$	$(0.016, 0.028)$	$(0.000, 0.000)$	$(0.000, 0.000)$
$\mathbf{VLEC}^{15}$	$(0.251, 12.449)$	$(0.019, 0.041)$	$(0.000, 0.000)$	$(0.000, 0.000)$
$\mathbf{VLEC}^{16}$	$(0.952, \ 2.394)$	$(0.009, 0.005)$	$(0.007, 0.004)$	$(0.000, 0.000)$
$\mathbf{VLEC}^{17}$	$(0.901, \ 2.373)$	$(0.000, 0.000)$	$(0.004, 0.002)$	$(0.000, 0.000)$
$\mathbf{VLEC}^{18}$	$(0.851, \ 2.357)$	$(0.000, 0.000)$	$(0.001, 0.001)$	$(0.000, 0.000)$
$\mathbf{VLEC}^{19}$	$(0.800, \ 2.312)$	$(0.000, 0.000)$	$(0.000, 0.000)$	$(0.000, 0.000)$
$\mathbf{VLEC}^{20}$	$(0.750, \ 2.308)$	$(0.000, 0.000)$	$(0.000, 0.000)$	$(0.000, 0.000)$
$\mathbf{VLEC}^{21}$	$(0.700, \ 2.275)$	$(0.000, 0.000)$	$(0.000, 0.000)$	$(0.000, 0.000)$
$\mathbf{VLEC}^{22}$	$(0.653, \ 2.269)$	$(0.052, 0.043)$	$(0.000, 0.000)$	$(0.000, 0.000)$
$\mathbf{VLEC}^{23}$	$(0.605, \ 2.257)$	$(0.000, 0.000)$	$(0.000, 0.000)$	$(0.000, 0.000)$
$\mathbf{VLEC}^{24}$	$(0.553, \ 2.225)$	$(0.000, 0.000)$	$(\mathbf{0.469, 0.450})$	$(0.000, 0.000)$
$\mathbf{VLEC}^{25}$	$(0.501, \ 2.133)$	$(0.000, 0.000)$	$(\mathbf{0.506, 0.536})$	$(0.000, 0.000)$
$\mathbf{VLEC}^{26}$	$(0.451, \ 2.133)$	$(0.000, 0.000)$	$(0.000, 0.000)$	$(0.000, 0.000)$
$\mathbf{VLEC}^{27}$	$(0.407, \ 2.113)$	$(\mathbf{0.321, 0.419})$	$(0.000, 0.000)$	$(0.000, 0.000)$
$\mathbf{VLEC}^{28}$	$(0.350, \ 2.155)$	$(0.000, 0.000)$	$(0.000, 0.000)$	$(0.000, 0.000)$
$\mathbf{VLEC}^{29}$	$(0.300, \ 2.123)$	$(0.000, 0.000)$	$(0.000, 0.000)$	$(0.000, 0.000)$
$\mathbf{VLEC}^{30}$	$(0.252, \ 2.117)$	$(0.000, 0.000)$	$(0.000, 0.000)$	$(0.000, 0.000)$

[†] The properties of each component VLEC codebook $\mathbf{VLEC}^n$ are provided using the format $(R(\mathbf{VLEC}^n), D(\mathbf{VLEC}^n))$.

[‡] The fractions are specified using the format (C^n, α^n).

Table 9.5: Fractions $\{\alpha^r\}_{r=1}^{10}$ of the interleaved transmission frame $\mathbf{u}'$ and the IrURC-encoded frame $\mathbf{v}$ that the 10 component URC codes $\{\mathbf{URC}^r\}_{r=1}^{10}$ encode in the $\eta = 2.12$ bits per channel symbol IrVLC-IrURC-high and IrVLC-IrURC-low arrangements

	α^r	
	IrVLC-	IrVLC-
	IrURC-	IrURC-
$\mathbf{URC}^r$	high	low
$\mathbf{URC}^1$	0.000	0.000
$\mathbf{URC}^2$	0.043	0.164
$\mathbf{URC}^3$	0.258	0.000
$\mathbf{URC}^4$	0.000	0.000
$\mathbf{URC}^5$	0.627	0.619
$\mathbf{URC}^6$	0.039	0.121
$\mathbf{URC}^7$	0.000	0.000
$\mathbf{URC}^8$	0.000	0.000
$\mathbf{URC}^9$	0.013	0.000
$\mathbf{URC}^{10}$	0.019	0.095

Table 9.6: Properties and composition of the VLEC codebook $\mathbf{VLEC}^{31}$

$\mathbf{VLEC}^n$	Properties[†]	Composition[‡]
$\mathbf{VLEC}^{31}$	(0.530, 4.665)	10,11,10,8,7,5,5,4,7,6,6,8,9,14,11,13,
		2131D83AD2B2367E0D4A040F03F00421F3

[†] The properties of the codebook are provided using the format $(R(\mathbf{VLEC}^{31}), D(\mathbf{VLEC}^{31}))$.

[‡] The composition of the codebook is specified by providing the $K = 16$ codeword lengths $\{I^{31,k}\}_{k=1}^K$, together with the hexadecimal representation of the ordered concatenation of the $K = 16$ VLEC codewords $\{\mathbf{VLEC}^{31,k}\}_{k=1}^K$ in the codebook.

The lowest E_b/N_0 values for which the $\eta = 2.12$ arrangements of the IrVLC-IrURC-high, IrVLC-IrURC-low, IrVLC-URC-high and VLC-URC parameterizations could achieve an open EXIT-chart tunnel [267] are provided in Table 9.1. Note that the open EXIT-chart tunnels obtained at these threshold E_b/N_0 values imply that iterative decoding convergence to an infinitesimally low probability of error can be achieved, provided that the iterative decoding trajectory approaches the inner and outer EXIT functions sufficiently closely.

As shown in Figure 9.9, the threshold E_b/N_0 values of the IrVLC-URC-high and VLC-URC arrangements are 0.83 dB and 1.69 dB from the channel's capacity bound of 4.28 dB, respectively. By contrast, the threshold E_b/N_0 values of the IrVLC-IrURC-high and IrVLC-IrURC-low arrangements are significantly closer, at just 0.33 dB from the channel's capacity bound. Note that this is lower than the 0.42 dB discrepancy of the IrVLC-URC parameterization in Chapter 8 that facilitates the 'nearest-capacity' operation, even though the IrVLC-IrURC-high and IrVLC-IrURC-low arrangements suffer from the effective throughput loss detailed in Section 9.5.3. By taking this effective throughput loss into account, we may observe in Figure 9.9 that the threshold E_b/N_0 values of the IrVLC-IrURC-high and IrVLC-IrURC-low arrangements are just 0.04 dB from the channel's attainable capacity bound of 4.57 dB, while those of the IrVLC-URC-high and VLC-URC arrangements are 0.54 dB and 1.4 dB away, respectively.

As predicted in Section 9.1, the improved near-capacity operation of the IrVLC-IrURC-high and IrVLC-IrURC-low arrangements may be explained by the increased degree of design freedom that is afforded by serially concatenating two irregular codecs. Additionally, our findings may be further explained with the aid of EXIT-chart analysis. The inner and inverted outer EXIT functions of the IrVLC-IrURC-high, IrVLC-IrURC-low, IrVLC-URC-high and VLC-URC arrangements are provided in Figure 9.10. Here, the inner EXIT functions were recorded for the threshold E_b/N_0 values of Table 9.1.

As may be observed in Figure 9.10, the inner and inverted outer EXIT functions of the IrVLC-IrURC-high and IrVLC-IrURC-low arrangements are nearly parallel, having open EXIT tunnels that are narrow all along their length. By contrast, the inner and inverted outer EXIT functions of the IrVLC-URC-high and VLC-URC arrangements are not parallel, having open EXIT tunnels that become narrow only in the vicinity of specific points along their length. This therefore explains the observed discrepancies between the threshold E_b/N_0 values and the channel's E_b/N_0 capacity bound, since these are proportional to the area between the inner and outer EXIT functions, according to the EXIT-chart area property [204].

Similarly to the observations of Section 8.5.4, much of the open EXIT-chart area between the inner and inverted outer EXIT functions of the IrVLC-URC-high and VLC-URC arrangements may be attributed to the URC EXIT function's high starting point along the I_e^i axis of the EXIT charts in Figure 9.10. In the case of the IrVLC-URC-high arrangement detailed in Table 9.4, this results in the activation of only the component VLEC codebooks $\{\mathbf{VLEC}^n\}_{n=1}^{15}$ having the 'S'-shaped inverted EXIT functions of Figure 9.5, which rise rapidly towards the URC EXIT function, emerging from a high starting point. As shown in Figure 9.5 and discussed in Section 9.5.1, these component VLEC codebooks are associated with a relatively high APP SISO decoding complexity. Note that owing to their slowly rising inverted EXIT functions, the lower complexity component IrVLC codebooks $\{\mathbf{VLEC}^n\}_{n=16}^{30}$ could not be activated in the IrVLC-URC-high arrangement. This explains why Section 9.5.3 was unable to employ the modified EXIT-chart matching algorithm of Section 9.2 to seek a reduced computational complexity during the design of an 'IrVLC-URC-low' parameterization.

In contrast, in the case of the IrVLC-IrURC-high and IrVLC-IrURC-low parameterizations, the presence of the component URC codes having the EXIT functions of Figure 9.7 that emerge from the $(0, 0)$ point of the EXIT chart facilitates the design of an IrURC scheme having an EXIT function that starts close to this point. Here, the joint EXIT matching algorithm is free to choose the degree to which the EXIT functions are 'S'-shaped. For this reason, an APP SISO IrVLC decoding complexity reduction could be attained by employing the modified EXIT-chart matching algorithm of Section 9.2 to invoke the lower-complexity component IrVLC codebooks that do not have 'S'-shaped EXIT functions during the design

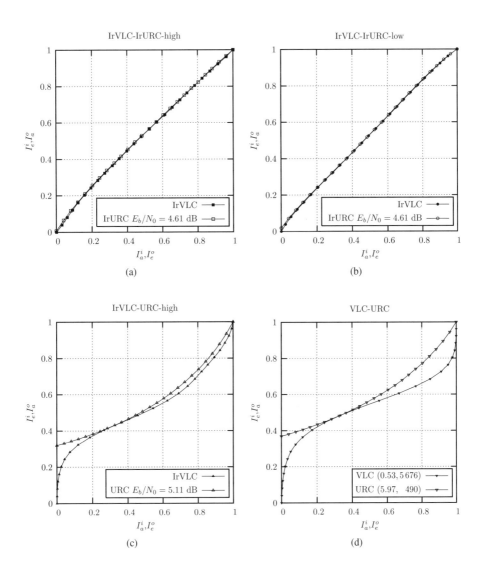

Figure 9.10: EXIT charts for the $\eta = 2.12$ bits per channel symbol arrangements of the IrVLC-IrURC parameterizations: (a) IrVLC-IrURC-high arrangement; (b) IrVLC-IrURC-low arrangement; (c) IrVLC-URC-high arrangement. (d) VLC-URC arrangement. The inner EXIT functions are provided for the threshold channel E_b/N_0 values, as specified in Table 9.1. ©IEEE [202] Maunder and Hanzo, 2009.

Table 9.7: Average number of ACS operations per source symbol performed during each APP SISO URC/IrURC decoding, APP SISO VLC/IrVLC decoding and VLC/IrVLC MAP sequence estimation iteration in the $\eta = 2.12$ bits per channel symbol arrangements of the IrVLC-IrURC-high, IrVLC-IrURC-low, IrVLC-URC-high and VLC-URC parameterizations. ©IEEE [202] Maunder and Hanzo, 2009.

Arrangement	APP SISO URC/IrURC decoding complexity	APP SISO VLC/IrVLC decoding complexity	VLC/IrVLC MAP sequence estimation complexity
IrVLC-IrURC-high	902	4 729	3 814
IrVLC-IrURC-low	1 004	3 120	2 640
IrVLC-URC-high	490	6 280	4 950
VLC-URC	490	5 676	4 489

of the IrVLC-IrURC-low parameterization. As described above, this results in the encoding of nearly 99% of the source symbols using low-complexity component VLEC codebooks from the set $\{\mathbf{VLEC}^n\}_{n=16}^{30}$ in the IrVLC-IrURC-low arrangement, compared with just over 38% in the IrVLC-IrURC-high parameterization, as shown in Table 9.4.

Let us now consider the computational complexity associated with APP SISO URC/IrURC decoding, APP SISO VLC/IrVLC decoding and VLC/IrVLC MAP sequence estimation in the $\eta = 2.12$ bits per channel symbol arrangements considered. These complexities were characterized by measuring the average number of ACS operations per source symbol performed during a single operation of the above-mentioned processes, as summarized in Table 9.7.

Observe in Table 9.7 that VLC/IrVLC MAP sequence estimation is associated with a lower computational complexity than is APP SISO VLC/IrVLC decoding in all arrangements, as may be expected. Furthermore, these processes are associated with lower computational complexities in the IrVLC-IrURC-high and IrVLC-IrURC-low arrangements than in the IrVLC-URC-high and VLC-URC arrangements. This may be explained by the IrVLC-IrURC-high and IrVLC-IrURC-low arrangements' activation of the component VLEC codebooks $\{\mathbf{VLEC}^n\}_{n=16}^{30}$ that are associated with relatively low computational complexities, as shown in Table 9.4 and Figure 9.5. Note that the APP SISO VLC/IrVLC decoding and VLC/IrVLC MAP sequence estimation complexities are particularly low in the case of the IrVLC-IrURC-low arrangement, which encodes nearly 99% of the source symbols using the low-complexity component VLEC codebooks $\{\mathbf{VLEC}^n\}_{n=16}^{30}$. In the case of the IrVLC-IrURC-high arrangement, the corresponding percentage is just over 38%, explaining its higher computational complexities.

As predicted above, the APP SISO URC/IrURC decoding complexity can be seen in Table 9.7 to be highest in the IrVLC-IrURC-low arrangement. This may be explained by this arrangement's employment of component URC codes having a relatively high coding memory of $L_{\mathrm{URC}} = 3$ to generate a significant percentage of the IrURC-encoded bits, as detailed above. The corresponding complexity can be seen to be slightly lower in the IrVLC-IrURC-low arrangement and is lowest in the IrVLC-URC-high and VLC-URC arrangements.

9.6 Simulation Results

In this section we characterize the achievable performance of the IrVLC-IrURC-high, the IrVLC-IrURC-low, the IrVLC-URC-high and the VLC-URC arrangements of Section 9.5.4, which employ an effective throughput of $\eta = 2.12$ bits per channel symbol. We characterize the use of each arrangement for transmission over 16QAM-based uncorrelated narrowband Rayleigh fading channels having a range of E_b/N_0 values above the capacity bound of 4.28 dB [300]. In each case two source symbol frame lengths were considered. More specifically, in one of the cases we simulated the transmission of 20 long source symbol frames. In this case, each source symbol frame $\mathbf{s}$ comprised $J^{\mathrm{sum}} = 140\,640$ randomly generated symbols, giving an average transmission frame $\mathbf{u}$ and interleaver length of $I^{\mathrm{sum}} = J^{\mathrm{sum}} E / R_{\mathrm{IrVLC}} \approx 1\,000\,000$ bits. By contrast, a shorter source symbol frame $\mathbf{s}$ length was employed in the other case, in which the transmission of 200 frames of $J^{\mathrm{sum}} = 14\,064$ symbols was simulated, corresponding to an average transmission frame $\mathbf{u}$ and interleaver length of $I^{\mathrm{sum}} \approx 100\,000$ bits.

In each of our arrangements considered, each activated component VLEC codebook $\mathbf{VLEC}^n$ is employed to encode the corresponding source symbol sub-frame $\mathbf{s}^n$, yielding a transmission sub-frame $\mathbf{u}^n$ comprising I^n bits, as described in Section 9.4.1. Since the lengths of the transmission sub-frames typically vary from frame to frame, it is necessary to explicitly convey their lengths as side information in order to facilitate their decoding in the receiver, as described in Section 9.4.1. The amount of side information required in each of the IrVLC-IrURC-high, IrVLC-IrURC-low, IrVLC-URC-high and VLC-URC arrangements may be determined as described in Section 8.5.5. The calculated number of bits per frame required to represent the side information is provided in Table 9.8 for each scheme and for transmission frame lengths of both $I^{\mathrm{sum}} \approx 1\,000\,000$ and $I^{\mathrm{sum}} \approx 100\,000$ bits. As suggested in Section 9.4.1, this error-sensitive side information may be protected by a low-rate block code in order to ensure its reliable transmission. Hence, for each arrangement considered, Table 9.8 also provides the fraction of the transmitted information that is constituted by the side information when it is protected by a $R_{\mathrm{rep}} = 1/3$-rate repetition code.

Observe in Table 9.8 that, when the transmission frame length I^{sum} is longer, a lower percentage of the transmitted information is constituted by the side information, which is protected by a $R_{\mathrm{rep}} = 1/3$-rate repetition code. Note that no more than 0.472% of the transmitted information was constituted by side information in any of the cases considered in Table 9.8. As with the fraction of 0.35% that was observed for the schemes of Section 8.5.5, we consider this to be negligible.

In each of our simulations, iterative decoding was continued until convergence was achieved and no perceivable further BER improvements could be attained. After each decoding iteration, we recorded the BER between the transmission frame $\mathbf{u}$ and the transmission frame estimate $\tilde{\mathbf{u}}$ that is obtained following the consideration of the *a priori* LLR frame $L_a^o(\mathbf{u})$ by the IrVLC/VLC MAP sequence estimator of Figure 9.3. Note that we opted for employing the BER associated with the transmission frame estimate $\tilde{\mathbf{u}}$ as our performance metric, rather than the SER associated with the source symbol frame estimate $\tilde{\mathbf{s}}$. This is because, owing to the nature of VLEC codes, synchronization can be occasionally lost during

Table 9.8: Number of side information bits required per frame and the percentage of the transmitted information that this represents when protected by a $R_{\text{rep}} = 1/3$-rate repetition code in the IrVLC-IrURC-high, the IrVLC-IrURC-low, the IrVLC-URC-high and the VLC-URC arrangements of Section 9.5.4, for transmission frame lengths of both $I^{\text{sum}} \approx 1\,000\,000$ and $I^{\text{sum}} \approx 100\,000$ bits

| | Side information bits and percentage | |
Arrangement	$I^{\text{sum}} \approx 1\,000\,000$	$I^{\text{sum}} \approx 100\,000$
IrVLC-IrURC-high	(206, 0.062%)	(158, 0.472%)
IrVLC-IrURC-low	(110, 0.033%)	(84, 0.251%)
IrVLC-URC-high	(74, 0.022%)	(54, 0.162%)
VLC-URC	(21, 0.006%)	(18, 0.054%)

the sequence estimation of some source symbol frame estimates $\tilde{\mathbf{s}}$, resulting in a particularly poor SER [179]. Hence, the transmission of a prohibitively high number of source symbol frames would have to be simulated in order that the SER could be averaged and employed as a reliable performance metric. For each BER value recorded during our simulations, we also noted the total number of ACS operations performed during IrVLC/VLC MAP sequence estimation and performed so far during the iterative IrVLC/VLC and IrURC/URC APP SISO decoding process. The corresponding plots of the BER attained following the achievement of iterative decoding convergence versus E_b/N_0 are provided for the considered simulations in Figure 9.11. Meanwhile, Figure 9.12 provides the corresponding plots of the average number of ACS operations required per source symbol in order to achieve a BER of 10^{-5} versus E_b/N_0.

As shown in Figure 9.11, the BER attained following the achievement of iterative decoding convergence reduces as the channel's E_b/N_0 value increases, for all cases considered. This may be explained by considering the EXIT-chart tunnels of Figure 9.10, which gradually open and become wider as the E_b/N_0 value is increased from the channel's capacity bound, allowing the iterative decoding trajectory to progress further, as explained in Section 4.3. Note that an open EXIT-chart tunnel implies that iterative decoding convergence to an infinitesimally low BER can be achieved, provided that the iterative decoding trajectory approaches the inner and outer EXIT functions sufficiently closely, as described in Section 4.3. However, it can be seen in Figure 9.11 that low BERs were not achieved at the threshold E_b/N_0 values, where the EXIT-chart tunnels open. This is imposed by the BCJR algorithm's assumption [48] that all correlation within the *a priori* LLR frames $L_a^o(\mathbf{u})$ and $L_a^i(\mathbf{u}')$ was successfully mitigated by the interleaver π_1 and deinterleaver π_1^{-1} of Figure 9.3. The interleaver's ability to mitigate this correlation is proportional to its length, but not even our long $I^{\text{sum}} \approx 1\,000\,000$-bit interleaver is able to mitigate all correlation in practice. As a result, the iterative decoding trajectory does not match perfectly with the inner and outer EXIT functions, and the tunnel must be further widened before the trajectory can reach the top right-hand corner of the EXIT chart, which is associated with an infinitesimally low BER, as described in Section 4.3. As may be predicted from this discussion, BERs of less than 10^{-5} could be obtained in Figure 9.11 at E_b/N_0 values that are closer to the threshold values in

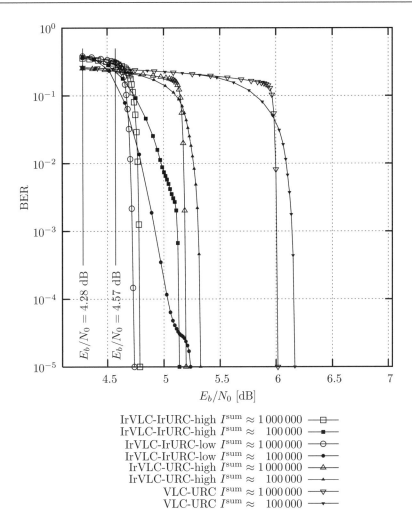

IrVLC-IrURC-high $I^{\mathrm{sum}} \approx 1\,000\,000$ —□—
IrVLC-IrURC-high $I^{\mathrm{sum}} \approx 100\,000$ —■—
IrVLC-IrURC-low $I^{\mathrm{sum}} \approx 1\,000\,000$ —○—
IrVLC-IrURC-low $I^{\mathrm{sum}} \approx 100\,000$ —●—
IrVLC-URC-high $I^{\mathrm{sum}} \approx 1\,000\,000$ —△—
IrVLC-URC-high $I^{\mathrm{sum}} \approx 100\,000$ —▲—
VLC-URC $I^{\mathrm{sum}} \approx 1\,000\,000$ —▽—
VLC-URC $I^{\mathrm{sum}} \approx 100\,000$ —▼—

Figure 9.11: Plots of the BER attained following the achievement of iterative decoding convergence versus E_b/N_0 for the IrVLC-IrURC-high, the IrVLC-IrURC-low, the IrVLC-URC-high and the VLC-URC arrangements of Section 9.5.4, when employing interleaver lengths of both $I^{\mathrm{sum}} \approx 1\,000\,000$ and $I^{\mathrm{sum}} \approx 100\,000$ bits. ©IEEE [202] Maunder and Hanzo, 2009.

the case of the arrangements employing an interleaver length of $I^{\mathrm{sum}} \approx 1\,000\,000$ bits than was possible by those employing a $I^{\mathrm{sum}} \approx 100\,000$-bit interleaver.

As shown in Figures 8.15 and 8.16, the BER attained following the achievement of iterative decoding convergence reduces as the channel's E_b/N_0 value increases, for all parameterizations considered. This may be explained by considering the EXIT-chart tunnel, which gradually opens and becomes wider as the E_b/N_0 value is increased from the

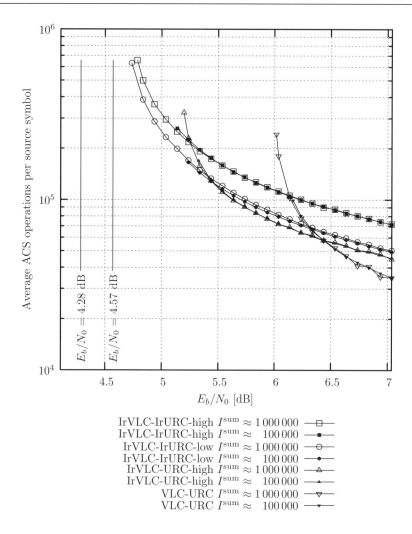

Figure 9.12: Average number of ACS operations per source symbol required to achieve a BER of 10^{-5} versus E_b/N_0 for the IrVLC-IrURC-high, the IrVLC-IrURC-low, the IrVLC-URC-high and the VLC-URC arrangements of Section 9.5.4, when employing interleaver lengths of both $I^{\text{sum}} \approx 1\,000\,000$ and $I^{\text{sum}} \approx 100\,000$ bits. ©IEEE [202] Maunder and Hanzo, 2009.

channel's capacity bound, allowing the iterative decoding trajectory to progress further, as explained in Section 4.3. Note that an open EXIT-chart tunnel implies that iterative decoding convergence to an infinitesimally low BER can be achieved, provided that the iterative decoding trajectory approaches the inner and outer EXIT functions sufficiently closely, as described in Section 4.3. However, it can be seen in Figures 8.15 and 8.16 that low BERs were not achieved at the threshold E_b/N_0 values, where the EXIT-chart tunnels open. This is because our relatively short 100 000-bit interleaver is unable to mitigate all of the correlation within the *a priori* LLR frames $L_a^o(\mathbf{u})$ and $L_a^i(\mathbf{u}')$, which the BCJR algorithm assumes to be uncorrelated [48]. As a result, the iterative decoding trajectory does not match perfectly with the inner and outer EXIT functions and the tunnel must be further widened before the trajectory can reach the top right-hand corner of the EXIT chart, which is associated with an infinitesimally low BER, as described in Section 4.3.

Observe in Figure 9.11 that, when using the longer average interleaver length of $I^{\text{sum}} \approx$ 1 000 000 bits, the IrVLC-IrURC-high and IrVLC-IrURC-low arrangements of Section 9.5.4 could achieve BERs of less than 10^{-5} for E_b/N_0 values in excess of 4.79 dB and 4.74 dB respectively. These values are 0.18 dB and 0.13 dB respectively, from the 4.61 dB threshold E_b/N_0 value provided in Table 9.1 for the IrVLC-IrURC-high and IrVLC-IrURC-low arrangements. By contrast, BERs of less than 10^{-5} could only be achieved by the IrVLC-IrURC-high and IrVLC-IrURC-low arrangements employing the shorter $I^{\text{sum}} \approx 100\,000$-bit interleaver for E_b/N_0 values in excess of 5.14 dB and 5.24 dB, respectively. These values correspond to significantly larger discrepancies of 0.53 dB and 0.63 dB from the threshold E_b/N_0 value respectively.

Note that smaller discrepancies between the threshold E_b/N_0 values and those at which BERs of less than 10^{-5} could be achieved in Figure 9.11 were observed for the IrVLC-URC-high and VLC-URC arrangements than for the IrVLC-IrURC-high and IrVLC-IrURC-low arrangements. When employing $I^{\text{sum}} \approx 1\,000\,000$- and $I^{\text{sum}} \approx 100\,000$-bit interleavers, the IrVLC-URC-high arrangement could achieve BERs of less than 10^{-5} for E_b/N_0 values in excess of 5.20 dB and 5.33 dB respectively, corresponding to discrepancies of 0.09 dB and 0.22 dB, respectively, from the threshold E_b/N_0 value of 5.11 dB. Similarly, the VLC-URC arrangement could achieve BERs of less than 10^{-5} for E_b/N_0 values in excess of 6.02 dB and 6.17 dB when employing $I^{\text{sum}} \approx 1\,000\,000$- and $I^{\text{sum}} \approx 100\,000$-bit interleavers, respectively, giving discrepancies from the threshold E_b/N_0 value of 5.97 dB of 0.05 dB and 0.2 dB, respectively.

The IrVLC-IrURC-high and IrVLC-IrURC-low arrangements' large discrepancies of up to 0.63 dB may be explained by their EXIT tunnels of Figure 9.10, which are narrow along their entire length. As a result, these arrangements are particularly sensitive to any narrowing of the EXIT chart tunnel that is owed to the presence of residual correlation within the *a priori* LLR frames $L_a^o(\mathbf{u})$ and $L_a^i(\mathbf{u}')$ of Figure 9.3. By contrast, the EXIT chart tunnels of the IrVLC-URC-high and VLC-URC arrangements shown in Figure 9.10 are narrow only at particular points along their length, resulting in smaller discrepancies of no more than 0.22 dB between the threshold E_b/N_0 values and those at which BERs of less than 10^{-5} could be achieved in Figure 9.11.

As shown in Figure 9.11, when employing the longer average interleaver length of $I^{\text{sum}} \approx$ 1 000 000 bits, the IrVLC-IrURC-high, the IrVLC-IrURC-low, the IrVLC-URC-high and the VLC-URC arrangements of Section 9.5.4 could achieve BERs of less than 10^{-5} for E_b/N_0 values in excess of limits that are 0.51 dB, 0.46 dB, 0.92 dB and 1.74 dB from the channel's capacity bound of 4.28 dB, respectively. Furthermore, these limits are just 0.22 dB, 0.17 dB, 0.63 dB and 1.45 dB, respectively, from the channel's attainable capacity bound of 4.57 dB.

Let us now comment on the iterative decoding computational complexity of the considered arrangements, as characterized in Figure 9.12. In all cases, a BER of 10^{-5} may be achieved at a reduced computational complexity as the E_b/N_0 value increases. This may be explained by the widening of the EXIT-chart tunnel as the E_b/N_0 value increases. As a result, fewer decoding iterations are required in order to reach the particular extrinsic mutual information that is associated with a BER of 10^{-5}. As may be expected, at high E_b/N_0 values the arrangements employing an interleaver length of $I^{\mathrm{sum}} \approx 100\,000$ bits are associated with a similar computational complexity to that of the corresponding arrangements employing an interleaver length of $I^{\mathrm{sum}} \approx 1\,000\,000$ bits.

Observe in Figure 9.12 that the IrVLC-IrURC-low arrangement has an iterative decoding complexity that is approximately 25% lower than that of the IrVLC-IrURC-high parameterization at high E_b/N_0 values. As predicted in Section 9.5.4, this is a benefit of the modification to the EXIT-chart matching algorithm of Section 9.2, which facilitated the joint EXIT-chart matching and the quest for a reduced computational complexity during the design of the IrVLC-IrURC-low parameterization.

Note in Figure 9.12 that for high E_b/N_0 values the IrVLC-URC-high and VLC-URC arrangements have a lower iterative decoding complexity than does the IrVLC-IrURC-low arrangement. This is because the corresponding EXIT-chart tunnels of the IrVLC-URC-high and VLC-URC arrangements seen in Figure 9.10 become narrow only towards the top right-hand corner of the EXIT chart. In contrast, the EXIT-chart tunnel of the IrVLC-IrURC-low arrangement is also narrow at the bottom left-hand corner of the EXIT chart seen in Figure 9.10b. Since a relatively high number of decoding iterations are required for maintaining convergence in these narrow EXIT-chart tunnel regions, the IrVLC-IrURC-low arrangement is associated with a relatively high computational complexity at high E_b/N_0 values.

Despite this, however, we may opt for the IrVLC-IrURC-low arrangement as our preferred scheme, since it facilitates the 'nearest-capacity' operation, achieving BERs of less than 10^{-5} for E_b/N_0 values in excess of a limit that is only 0.46 dB from the channel's E_b/N_0 capacity bound. Furthermore, the iterative decoding complexity of the preferred IrVLC-IrURC-low arrangement is only marginally higher than that of the IrVLC-URC-high arrangement at high E_b/N_0 values.

9.7 Summary and Conclusions

In Chapter 8 we observed that the inverted EXIT functions of outer irregular codecs can typically only be matched to the EXIT functions of regular inner codecs with limited accuracy. This is because inverted outer EXIT functions are constrained to starting from the $(0, 0)$ point of the EXIT chart, while the inner EXIT functions typically emerge from a relatively high point along the I_e^i axis of the EXIT chart, as described in Section 4.2. This results in a relatively large discrepancy between the channel's capacity bound and the threshold E_b/N_0 value at which an open EXIT-chart tunnel can be achieved. This is because the discrepancy is proportional to the relatively large area of the EXIT chart that is enclosed by the resultant tunnel, according to the EXIT-chart area property [204]. In the case of the IrVLC-URC arrangements of Section 8.5.4, the open EXIT-chart tunnels of Figure 8.13 could not be achieved within 0.42 dB of the BPSK-modulated uncorrelated narrowband Rayleigh fading channel's E_b/N_0 capacity bound. As always, these open EXIT-chart tunnels imply that iterative decoding convergence to an infinitesimally low probability of error can be achieved, provided that the iterative decoding trajectory approaches the inner and outer EXIT functions sufficiently closely, as described in Section 4.3.

The aforementioned findings of Chapter 8 motivated this chapter's introduction of a novel IrURC codec, having an EXIT function that may be shaped to emerge from a point on the EXIT chart's I_e^i axis that is closer to the inverted outer EXIT function's starting point of $(0, 0)$. In Section 9.3 we proposed a novel method for jointly matching the EXIT functions of two serially concatenated irregular codecs, seeking the highest coding rate that offers an open EXIT-chart tunnel. This joint EXIT-chart matching algorithm is able to exploit the increased degree of design freedom that is afforded by employing two irregular codecs in order to achieve an EXIT-chart tunnel that is narrow at all points along its length, facilitating near-capacity operation. Indeed, using this approach the open IrVLC-IrURC EXIT-chart tunnels of Figure 9.10 were obtained for threshold E_b/N_0 values in excess of a limit that was only 0.33 dB from the 16QAM-modulated uncorrelated narrowband Rayleigh fading channel's capacity bound. By contrast, an open EXIT-chart tunnel was only achieved for IrVLC-URC and VLC-URC benchmarks at significantly greater discrepancies of 0.83 dB and 1.69 dB, respectively, from the channel's E_b/N_0 capacity bound.

Note, however, that Section 9.5.3 showed that open EXIT charts could have been achieved at E_b/N_0 values closer to the capacity bound if the proposed scheme of Section 9.4 had not suffered from an effective throughput loss. In Section 9.5.2 this was attributed to the employment of a $M_{\mathrm{QAM}} = 16\mathrm{QAM}$ demodulator, offering a throughput of $\log_2(M_{\mathrm{QAM}}) = 4$ bits per channel symbol. This was chosen instead of the 1-bit-per-channel-symbol BPSK demodulator employed in Chapter 8, which did not suffer from an effective throughput loss. More specifically, the observed effective throughput loss may be attributed to the 'one-shot' operation of the $M_{\mathrm{QAM}} = 16\mathrm{QAM}$ demodulator in Figure 9.3. In this case, unlike that of the BPSK demodulator employed in Chapter 8, there is a benefit in employing iterative 16QAM demodulation and APP SISO URC/IrURC decoding, as evidenced by the EXIT chart of Figure 9.8. While this benefit was not exploited in the proposed scheme of Section 9.4, the resultant effective throughput loss was minimized by employing the Gray-coded 16QAM scheme of Figure 9.4, since this offers LLRs having a maximal mutual information in the absence of iterative demodulation [300], as shown in Figure 9.8.

In Section 9.5.3 the associated effective throughput loss was characterized and the channel's attainable E_b/N_0 capacity bound was defined. Furthermore, the IrVLC-IrURC threshold E_b/N_0 values corresponding to the open EXIT-chart tunnels of Figure 9.10 were found to be only 0.04 dB from the attainable capacity bound. This motivates the specific solution to the effective throughput loss that is proposed in Section 12.12, outlining our future work. This mitigates the effective throughput loss without requiring iterative demodulation, by relying on a symbol-based IrURC design that additionally facilitates irregular 16QAM transmission, employing a number of other constellation point mappings in addition to Gray coding.

In Section 9.6 we investigated the employment of both $I^{\mathrm{sum}} \approx 1\,000\,000$-bit and $I^{\mathrm{sum}} \approx 100\,000$-bit interleavers. We found that less side information was required to convey the length of the transmission sub-frames when the longer interleaver was employed, although, naturally, this was associated with a higher latency. Additionally, the longer interleaver was found to mitigate more of the correlation within the *a priori* LLR frames $L_a^o(\mathbf{u})$ and $L_a^i(\mathbf{u}')$ of Figure 9.3. Since the BCJR algorithm employed by the APP SISO decoders assumes that all *a priori* LLRs that can influence any particular decoding decision are uncorrelated [48], the longer interleaver was found to offer better matches between the iterative decoding trajectories as well as the inner and outer EXIT functions. As a result, BERs of 10^{-5} were facilitated at E_b/N_0 values that were closer to the capacity bound when longer interleavers were employed. Note, however, that not even the $I^{\text{sum}} \approx 1\,000\,000$-bit interleaver was sufficiently long to achieve BERs of 10^{-5} at the threshold E_b/N_0 values.

In Figure 9.11 we showed that an IrVLC-IrURC scheme employing a $I^{\text{sum}} \approx 1\,000\,000$-bit interleaver could achieve BERs of less than 10^{-5} for E_b/N_0 values in excess of a limit that is 0.46 dB from the channel's capacity bound. With the solution to the effective throughput loss of Section 12.12, we could expect this discrepancy to be reduced to equal the discrepancy of 0.17 dB from the attainable capacity bound. This is comparable with the 0.13 dB discrepancy demonstrated for Irregular Low-Density Parity Check (IrLDPC) codes [147,313] and superior to the 0.25 dB discrepancy found for irregular turbo codes [148].

Finally, in Section 9.2 we proposed a novel method for jointly performing EXIT-chart matching and seeking a reduced corresponding APP SISO decoding complexity. In Figure 9.12, an IrVLC-IrURC arrangement designed using this method was shown to impose a 25% lower complexity than that of a corresponding IrVLC-IrURC arrangement that was not designed to have a reduced complexity.

Part III

Applications of VLCs

Chapter **10**

Iteratively Decoded Variable-Length Space–Time Coded Modulation: Code Construction and Convergence Analysis[1]

S. X. Ng, University of Southampton, UK
M. Tao, Shanghai Jiao Tong University, China
J. Wang, L.-L. Yang and L. Hanzo

10.1 Introduction

In this chapter an Iteratively Decoded Variable-Length Space–Time Coded Modulation (VL-STCM-ID) scheme capable of simultaneously providing both coding and iteration gain as well as multiplexing and diversity gain is investigated. Non-binary unity-rate precoders are employed in order to assist the iterative decoding of the VL-STCM-ID scheme. The discrete-valued source symbols are first encoded into variable-length codewords that are spread to both the spatial and the temporal domains. Then the variable-length codewords are interleaved and fed to the precoded modulator. More explicitly, the proposed VL-STCM-ID arrangement is a jointly designed iteratively decoded scheme combining source coding, channel coding and modulation as well as spatial diversity/multiplexing. We demonstrate that, as expected, the higher the source correlation, the higher the achievable performance gain of the scheme becomes. Furthermore, the performance of the VL-STCM-ID scheme is about 14 dB better than that of the Fixed-Length STCM (FL-STCM) benchmark at a source symbol error ratio of 10^{-4}.

---

[1]Part of this chapter is based on [271] © IEEE (2007).

Near-Capacity Variable-Length Coding   Lajos Hanzo, Robert G. Maunder, Jin Wang and Lie-Liang Yang
© 2011 John Wiley & Sons, Ltd

To elaborate a little further, recall from Chapter 1 that Shannon's separation theorem stated that source coding and channel coding are best carried out in isolation [133]. However, this theorem was formulated in the context of potentially infinite-delay, lossless entropy coding and infinite-block-length channel coding. In practice, real-time wireless audio and video communications systems do not meet these ideal hypotheses. Explicitly, the source-encoded symbols often remain correlated, despite the lossy source encoder's efforts to remove all redundancy. Furthermore, they exhibit unequal error sensitivity. In these circumstances, it is often more efficient to use jointly designed source and channel encoders.

The wireless communication systems of future generations will be required to provide reliable transmissions at high data rates in order to offer a variety of multimedia services. Space–time coding schemes, which employ multiple transmitters and receivers, are among the most efficient techniques designed for providing high data rates by exploiting the high channel capacity potential of Multiple-Input Multiple-Output (MIMO) channels [314, 315]. More explicitly, the Bell Labs LAyered Space–Time architecture (BLAST) [316] was designed for providing full-spatial-multiplexing gain, while Space–Time Trellis Codes (STTCs) [317] were designed for providing the maximum attainable spatial-diversity gain.

As a further advance, in this chapter we investigate a jointly designed source coding and Space–Time Coded Modulation (STCM) scheme, where two-dimensional (2D) Variable-Length Codes (VLCs) are transmitted by exploiting both the spatial and the temporal domains. More specifically, the number of activated transmit antennas equals the number of modulated symbols in the corresponding VLC codeword in the spatial domain, where each VLC codeword is transmitted during a single symbol period. Hence, the transmission frame length is determined by the fixed number of source symbols, and therefore the proposed Variable-Length STCM (VL-STCM) scheme does not exhibit synchronization problems and does not require the transmission of side information for synchronization. Additionally, the associated source symbol correlation is exploited in order to achieve an increased product distance, hence resulting in an increased coding gain. Furthermore, the VL-STCM scheme advocated is capable of providing both multiplexing and diversity gains with the aid of multiple transmit antennas. Practical applications of the proposed scheme are related to the transmission of VLC-based MPEG-2-, 3- and 4-encoded video and audio sequences, for example. It is also possible to simply pack binary computer data into VLC-encoded symbols for the sake of removing any correlation amongst the symbols and for their near-capacity transmission.

Relevant work on the joint design of source coding and space–time coding can be found in [318] and [319], where the performance measure is based on the end-to-end analog source distortion. However, in this chapter we assume the presence of a discrete source, where the potentially analog source was quantized/discretized, before it was input to our VL-STCM encoder. Our objective is thus to minimize the error probability of the discrete source symbols. Furthermore, joint detection of conventional One-Dimensional (1D) VLC and STCM has been shown to approach the channel capacity in [320], although the related VLC schemes have to convey explicit side information regarding the total number of VLC-encoded bits or symbols per transmission frame.

On the other hand, it was shown in [321] that a binary Unity-Rate Code (URC) or precoder can be beneficially concatenated with Trellis-Coded Modulation (TCM) [322] for the sake of invoking iterative detection, and hence for attaining iteration gains. Since VL-STCM also belongs to the TCM family, we further develop the VL-STCM scheme for the sake of attaining additional iteration gains by introducing a novel non-binary URC between the variable-length space–time encoder and the modulator. The Iterative Decoding (ID) VL-STCM (VL-STCM-ID) scheme achieves a significant coding/iteration gain over both the non-iterative VL-STCM scheme and the Fixed-Length STCM (FL-STCM) benchmark.

The rest of the chapter is organized as follows. The overview of the space–time coding technique advocated is given in Section 10.2 and the 2D VLC design is outlined in Section 10.3. The description of the proposed VL-STCM and VL-STCM-ID schemes is presented in Sections 10.4 and 10.5 respectively. The convergence of the VL-STCM-ID scheme is analyzed in Section 10.6. In Section 10.7 the performance of the proposed schemes is discussed. Then the VL-STCM is combined with a higher-order modulation scheme in Section 10.8. Finally our conclusions are offered in Section 10.9.

10.2   Space–Time Coding Overview

Let us consider a MIMO system employing N_t transmit antennas and N_r receive antennas. The signal to be transmitted from transmit antenna m, $1 \leq m \leq N_t$, at the discrete time index t is denoted as $x_m[t]$. The signal received at antenna n, $1 \leq n \leq N_r$, and at time instant t can be modeled as

$$y_n[t] = \sqrt{E_s} \sum_{m=1}^{N_t} h_{n,m}[t] x_m[t] + w_n[t], \tag{10.1}$$

where E_s is the average energy of the signal constellation and $h_{n,m}[t]$ denotes the flat-fading channel coefficients between transmit antenna m and receive antenna n at time instant t, while $w_n[t]$ is the Additive White Gaussian Noise (AWGN) having zero mean and a variance of $N_0/2$ per dimension. The amplitude of the modulation constellation points is scaled by a factor of $\sqrt{E_s}$, so that the average energy of the constellation points becomes unity and the expected Signal-to-Noise Ratio (SNR) per receive antenna is given by $\gamma = N_t E_s / N_0$ [323]. Let us denote the transmission frame length as T symbol periods and define the space–time encoded codewords over T symbol periods as an ($N_t \times T$)-dimensional matrix $\mathbf{C}$ formed as

$$\mathbf{C} = \begin{bmatrix} c_1[1] & c_1[2] & \cdots & c_1[T] \\ c_2[1] & c_2[2] & \cdots & c_2[T] \\ \vdots & \vdots & \ddots & \vdots \\ c_{N_t}[1] & c_{N_t}[2] & \cdots & c_{N_t}[T] \end{bmatrix}, \tag{10.2}$$

where the elements of the tth column $\mathbf{c}[t] = [c_1[t] \ c_2[t] \cdots c_{N_t}[t]]^T$ are the space–time symbols transmitted at time instant t and the elements in the mth row $\mathbf{c_m} = [c_m[1] \ c_m[2] \cdots c_m[T]]$ are the space–time symbols transmitted from antenna m. The signal transmitted at time instant t from antenna m, which is denoted as $x_m[t]$ in Equation (10.1), is the modulated space–time symbol given by $x_m[t] = f(c_m[t])$ where $f(.)$ is the modulator's mapping function. The Pair-Wise Error Probability (PWEP) of erroneously detecting $\mathbf{E}$ instead of $\mathbf{C}$ is upper bounded at high SNRs by [317, 324]

$$p(\mathbf{C} \to \mathbf{E}) \leq \frac{1}{2} \left(\frac{E_s}{4N_0} \right)^{-E_H \cdot N_r} (E_P)^{-N_r}, \tag{10.3}$$

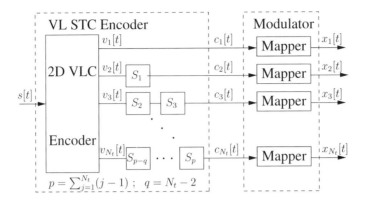

Figure 10.1: Block diagram of the VL-STCM transmitter. ©IEEE [271] Ng *et al.* 2007.

where E_H is referred to as the *effective Hamming distance*, which quantifies the transmitter-diversity order, and E_P is termed the *effective product distance* [317], which quantifies the coding advantage of a space–time code.

It was shown in [325] that a full-spatial-diversity STTC scheme having the minimum decoding complexity can be systematically designed based on two steps. The first step is to design a block code, while the second step is to transmit the block code diagonally across the space–time grid. The mechanism of the diagonal transmission across the space–time grid will be exemplified in Section 10.4 in the context of Figure 10.1. The Hamming distance and the product distance of a block code can be preserved when the block code is transmitted diagonally across the space–time grid. Hence, a full-spatial-diversity STTC scheme can be realized, when the Hamming distance of the block code used by the STTC scheme equals the number of transmitters. Based on the same principle, a joint source-coding and STTC scheme can be systematically constructed by first designing a 2D VLC and then transmitting the 2D VLC diagonally across the space–time grid. As mentioned above [325], this allows us to achieve a transmitter-diversity order[2], which is identical to the Hamming distance of the 2D VLC plus a coding advantage quantified by the product distance of the 2D VLC, as well as a multiplexing gain, provided that the number of possible source symbols N_s is higher than the number of modulation levels M.

Let us now commence our detailed discourse on the proposed VL-STCM-ID scheme in the following sections.

10.3 Two-Dimensional VLC Design

Consider as an example a source having $N_s = 8$ possible discrete values, and let the lth value be represented by a symbol $s^l = l$ for $l \in \{1, 2, \ldots, N_s\}$. Let us consider a source where the symbols emitted are independent of each other, but the symbol probability distribution is not uniform, and is given by

$$P(s^{l+1}) = 0.6P(s^l) = 0.6^l P(s^1) \, , \tag{10.4}$$

---

[2]The spatial-diversity order is the product of the transmitter-diversity order and the receiver-diversity order.

and $\sum_{l=1}^{N_s} P(s^l) = 1$. Hence, the source symbol s^1 has the highest occurrence probability of $P(s^1) = 0.4068$, and the source symbol s^8 has the lowest occurrence probability of $P(s^8) = 0.0114$. Note that a source is correlated when its entropy rate $\mathcal{H}(s)$ is smaller than $\log_2(N_s)$ [326]. For the independent source considered, the source entropy rate equals the source entropy $H(s)$, which is given by $\mathcal{H}(s) = H(s) = -\sum_{l=1}^{8} \log_2(P(s^l)) \cdot P(s^l) = 2.302$ bit. Since $\mathcal{H}(s) < \log_2(N_s)$, the source considered is a correlated source, where the higher the source correlation the smaller the source entropy rate. Let us now consider a 2D VLC codeword matrix, V_{VLC}, which encodes these $N_s = 8$ possible source symbols using $N_t = 3$ transmit antennas and BPSK modulation as follows:

$$
\mathbf{V}_{VLC} = \begin{bmatrix} \text{x} & 1 & \text{x} & 0 & \text{x} & 0 & 1 & 1 \\ \text{x} & \text{x} & 0 & \text{x} & 1 & 1 & 0 & 1 \\ 0 & \text{x} & \text{x} & 1 & 1 & \text{x} & 1 & 0 \end{bmatrix}, \tag{10.5}
$$

where each column of the (3×8)-dimensional matrix $\mathbf{V}_{VLC}$ corresponds to the specific VLC codeword conveying a particular source symbol, and the elements in the matrix denoted as '0' and '1' represent the BPSK symbols to be transmitted by the $N_t = 3$ transmit antennas, while 'x' represents 'no transmission.' 'No transmission' implies that the corresponding transmit antenna sends no signal. Let the lth source symbol s^l be encoded using the lth column of the $\mathbf{V}_{VLC}$ matrix seen in Equation (10.5). Hence, the source symbol s^1 is encoded into an N_t-element codeword using the first column of $\mathbf{V}_{VLC}$ in Equation (10.5), namely $[\text{x x } 0]^T$, where the first and second transmit antennas are in the 'no transmission' mode, while the third antenna transmits an 'active' symbol represented by the binary value '0'. If $L(s^l)$ is the number of 'active' symbols in the VLC codeword assigned to source symbol s^l, then we may define the average codeword length of the 2D VLC as

$$
L_{ave} = \sum_{l=1}^{N_s} P(s^l) L(s^l), \tag{10.6}
$$

where we have $L_{ave} = 1.233$ bits/VLC codeword for this system according to Equations (10.4) and (10.5). The corresponding BPSK signal mapper is characterized in Figure 10.2, where the 'no transmission' symbol is actually represented by the origin of the Euclidean space, i.e. we have $f(\text{x}) = 0$, where $f(.)$ is the mapping function. Since the 'no transmission' symbol is a zero-energy symbol, the amount of energy saving can be computed from

$$
A^2 = \frac{N_t}{L_{ave}}, \tag{10.7}
$$

where we have $A^2 = 3/1.233 = 2.433$, which is equivalent to $20 \log(A) = 3.86$ dB. Hence, more transmitted energy is saved, when there are more 'no transmission' symbols in a VLC codeword. Therefore, the columns of the matrix $\mathbf{V}_{VLC}$ in Equation (10.5), which are the VLC codewords, and the source symbols are specifically arranged so that the more frequently occurring source symbols are assigned to VLC codewords having more 'no transmission' components, in order to save transmit energy. The energy saved is then reallocated to the 'active' symbols for the sake of increasing their minimum Euclidean distance, as shown in Figure 10.2.

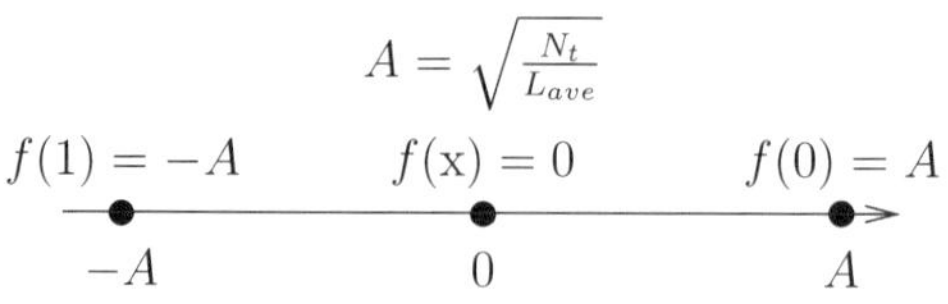

$$A = \sqrt{\frac{N_t}{L_{ave}}}$$

$$f(1) = -A \qquad\qquad f(\mathrm{x}) = 0 \qquad\qquad f(0) = A$$

$$-A \qquad\qquad\qquad 0 \qquad\qquad\qquad A$$

Figure 10.2: The signal mapper of the VL-STCM. ©IEEE [271] Ng *et al.* 2007.

Let us define the Hamming distance $E_{H\,min}$ as the number of different symbol positions of all the columns in the 2D VLC codeword matrix. Hence, we have $E_{H\,min} = 2$ for the 2D VLC codeword matrix in Equation (10.5). We further define the minimum product distance $E_{P\,min}$ as

$$E_{P\,min} = \min_{1 \le s < \tilde{s} \le N_s} \prod_{m \in \xi} |f(v_m) - f(\tilde{v}_m)|^2, \tag{10.8}$$

where ξ represents the set of VLC codeword component indices m satisfying the condition of $v_m \ne \tilde{v}_m$ for $1 \le m \le N_t$ and the VLC codeword $\mathbf{v} = [v_1 \; v_2 \ldots \; v_{N_t}]^T$ is defined in Figure 10.1 and Equation (10.5). The two VLC codewords conveying the source symbols s and $\tilde{s}$ are represented as $[v_1 \ldots v_{N_t}]^T$ and $[\tilde{v}_1 \ldots \tilde{v}_{N_t}]^T$, respectively, in Equation (10.8). We have $E_{P\,min} = 5.92$ based on Equation (10.5) and the signal mapper of Figure 10.2. Note that the mapper function $f(.)$ used in Equation (10.8) and portrayed in Figure 10.2 depends on the amount of energy saving. The amount of energy saving is given by Equation (10.7), where the numerator N_t is fixed and the denominator L_{ave} is given by Equation (10.6). As in the conventional 1D VLC, the higher the source correlation, the lower the average codeword length L_{ave} of the 2D VLC. A higher energy saving can be attained when the source is more correlated, due to a reduced average codeword length L_{ave}. In other words, the source correlation is converted into an increased minimum product distance, resulting in an increased coding gain when the source is more correlated. By contrast, the conventional 1D VLC exploits the source correlation in order to attain an increased compression ratio.

The design of the 2D VLC scheme is summarized in Algorithm 10.1.

Note that the search-space of Step 1 can be significantly reduced with the aid of the branch-and-bound algorithm of [327], where both $E_{H\,min}$ and $E_{P\,min}$ are used during the bounding operation. By contrast, the code search in [320, 325] also employs the branch-and-bound algorithm, but uses only $E_{P\,min}$ in the bounding operation. The 2D VLC matrix seen in Equation (10.5) was designed based on the above three steps. When $N_t = 3$ transmit antennas are employed for transmitting the N_t-element 2D VLC codewords denoted as $\mathbf{v} = [v_1 \ldots v_{N_t}]^T$ in Figure 10.1, we achieve a transmitter-diversity order of $E_{H\,min} = 2$, a coding gain quantified by $E_{P\,min} = 5.92$ and a spatial multiplexing gain quantified by $\log_2(N_s/M) = 2$, where $M = 2$ is the number of modulation levels of the original BPSK modulation and $N_s = 8$ is the number of source symbols. The throughput of the scheme is given by $\eta = \log_2(N_s) = 3$ bit/s/Hz and the Signal-to-Noise Ratio (SNR) per bit is given by $E_b/N_0 = \gamma/\eta$, where γ is the SNR per receive antenna.

Based on the above design principles, a range of 2D VLCs can be created for M-ary PSK modulation for $M > 2$, where the origin of the Euclidean space represents the 'no transmission' symbol.

Algorithm 10.1

1. Search for all possible VLC codeword matrices that have the maximum achievable minimum Hamming distance $E_{H\,\min}$ and product distance $E_{P\,\min}$ values, for each pair of the VLC codewords at a given N_s and N_t combination. Note that attaining a higher $E_{H\,\min}$ is given more weight than $E_{P\,\min}$, since $E_{H\,\min}$ is more dominant in the PWEP of Equation (10.3).

2. Rearrange the columns of the VLC codeword matrices in descending orders according to the number of 'no transmission' components. Assign the source symbols to the columns of the VLC codeword matrix, in descending orders according to the symbol probabilities.

3. Find the VLC codeword matrix that gives the shortest average codeword length, with the aid of Equation (10.6).

4. Assign the more frequently occurring source symbols to the VLC codewords having more 'no transmission' components, in the interest of saving transmission power.

10.4 VL-STCM Scheme

The block diagram of the VL-STCM transmitter is illustrated in Figure 10.1, which can be represented by two fundamental blocks, namely the Variable-Length Space–Time Code (VL-STC) encoder and the modulator. As seen in Figure 10.1, a VLC codeword $\mathbf{v}[t] = [v_1[t]\ v_2[t]\cdots v_{N_t}[t]]^T$ is assigned to each of the source symbols $s[t]$ generated by the source at time instant t, where we have $s[t] \in \{1, \ldots, N_s\}$ and N_s denotes the number of possible source symbols. Each of the VLC codewords $\mathbf{v}[t]$ seen in Figure 10.1 corresponds to one of the columns in the VLC matrix of Equation (10.5). Hence, the component $v_m[t]$ of the VLC codeword is represented by a symbol seen in the VLC matrix of Equation (10.5). As portrayed in Figure 10.1, the VLC codeword $\mathbf{v}[t]$ is transmitted diagonally across the space–time grid with the aid of appropriate-length shift registers, denoted as S_k in Figure 10.1, where we have $k \in \{1, 2, \ldots, \sum_{j=1}^{N_t}(j-1)\}$. As we can see from Figure 10.1, the codeword $\mathbf{v}[t] = [v_1[t]\ v_2[t]\cdots v_{N_t}[t]]^T$ is transmitted using N_t transmit antennas, where the mth element of each VLC codeword, for $1 \le m \le N_t$, is delayed by $(m-1)$ shift register cells before it is transmitted through the mth transmit antenna. Hence, the N_t components of each VLC codeword are transmitted on a diagonal of the space–time codeword matrix of Equation (10.2). Since the VLC codewords are encoded diagonally, the space–time coded symbol $c_m[t]$ transmitted by the mth antenna, $1 \le m \le N_t$, at a particular time-instant t is given by $c_m[t] = v_m[t - m + 1]$. Thus, for this specific case each element of the space–time codeword matrix of Equation (10.2) is given by $c_m[t] = v_m[t - m + 1]$, while the transmitted signal is given by

$$x_m[t] = f(c_m[t]) = f(v_m[t - m + 1]),\tag{10.9}$$

for $1 \le m \le N_t$. Note that originally there were only $N_s = 8$ legitimate 2D-VLC codewords in Equation (10.5). However, after these VLC codewords are diagonally mapped across the space–time grid using the shift registers shown in Figure 10.1, there is a total of $\bar{M}^{N_t} = 3^3 = 27$ legitimate space–time codewords[3], where $\bar{M}$ is the number of possible symbols in each

---

[3]A space–time codeword is defined as the N_t-element output word of the VL-STC encoder of Figure 10.1.

Table 10.1: The space–time codeword table. ©IEEE [271] Ng *et al.* 2007.

Index	$\mathbf{c}$	Index	$\mathbf{c}$	Index	$\mathbf{c}$
0	$[\,0\,\mathrm{x}\,\mathrm{x}\,]^T$	9	$[\,0\,0\,\mathrm{x}\,]^T$	18	$[\,0\,1\,\mathrm{x}\,]^T$
1	$[\,0\,\mathrm{x}\,1\,]^T$	10	$[\,0\,0\,1\,]^T$	19	$[\,0\,1\,1\,]^T$
2	$[\,0\,\mathrm{x}\,0\,]^T$	11	$[\,0\,0\,0\,]^T$	20	$[\,0\,1\,0\,]^T$
3	$[\,1\,\mathrm{x}\,\mathrm{x}\,]^T$	12	$[\,1\,0\,\mathrm{x}\,]^T$	21	$[\,1\,1\,\mathrm{x}\,]^T$
4	$[\,1\,\mathrm{x}\,1\,]^T$	13	$[\,1\,0\,1\,]^T$	22	$[\,1\,1\,1\,]^T$
5	$[\,1\,\mathrm{x}\,0\,]^T$	14	$[\,1\,0\,0\,]^T$	23	$[\,1\,1\,0\,]^T$
6	$[\,\mathrm{x}\,\mathrm{x}\,\mathrm{x}\,]^T$	15	$[\,\mathrm{x}\,0\,\mathrm{x}\,]^T$	24	$[\,\mathrm{x}\,1\,\mathrm{x}\,]^T$
7	$[\,\mathrm{x}\,\mathrm{x}\,1\,]^T$	16	$[\,\mathrm{x}\,0\,1\,]^T$	25	$[\,\mathrm{x}\,1\,1\,]^T$
8	$[\,\mathrm{x}\,\mathrm{x}\,0\,]^T$	17	$[\,\mathrm{x}\,0\,0\,]^T$	26	$[\,\mathrm{x}\,1\,0\,]^T$

position of the 2D VLC codewords. Note, however, that the number of legitimate space–time codewords may become lower than $\bar{M}^{N_t}$ when a different 2D VLC codeword matrix is employed.

The corresponding trellis diagram of the proposed VL-STC encoder is depicted in Figure 10.3. The $N_t = 3$-element space–time codeword seen in Figure 10.1 is given by $\mathbf{c} = [c_1\ c_2\ c_3]^T$. The relationship between the 27 codeword indices shown in Figure 10.3 and the space–time codeword $\mathbf{c}$ is defined in Table 10.1. The trellis states are defined by the contents of the shift register cells S_k shown in Figure 10.1, which are denoted by $\mathbf{S} = [S_1\ S_2\ S_3]$. For example, the state $\mathbf{S} = 0$ is denoted as $[\mathrm{x}\ 0\ 0]$ in the trellis diagram of Figure 10.3. Note that each shift register cell may hold $\bar{M} = 3$ possible values, namely $\{0, 1, \mathrm{x}\}$, in conjunction with the VL-STC encoder based on Equation (10.5). However, the number of legitimate trellis states can be less than $\bar{M}^p$, where $p = \sum_{j=1}^{N_t}(j - 1) = 3$ is the total number of shift registers. Hence, in our specific case there are only 24 legitimate trellis states out of the $\bar{M}^p = 27$ possible trellis states, which is a consequence of the constraints imposed by the 2D VLC of Equation (10.5). As we can see from Figure 10.3, there are always $N_s = 8$ diverging trellis branches from each state due to the eight possible source symbols. However, the number of converging trellis paths may vary from one state to another due to the variable length structure of the space–time codewords. Explicitly, there are six and nine trellis paths converging to state $\mathbf{S} = 0$ and state $\mathbf{S} = 1$, respectively, as seen in the trellis structure of Figure 10.3.

Similar to the full-spatial-diversity STTC scheme of [325], the above VL-STCM design has the minimum decoding complexity required for attaining the target transmitter-diversity and multiplexing gain. On one hand, it is possible to design a range of higher-complexity VL-STCM schemes in order to attain a higher coding gain. On the other hand, iterative decoding is well known for achieving near-channel-capacity performance with the aid of low-complexity constituent codes [328]. Hence, we will employ this minimum-complexity transmitter-diversity and multiplexing-based VL-STCM arrangement as one of the constituent codes in an iterative decoding scheme.

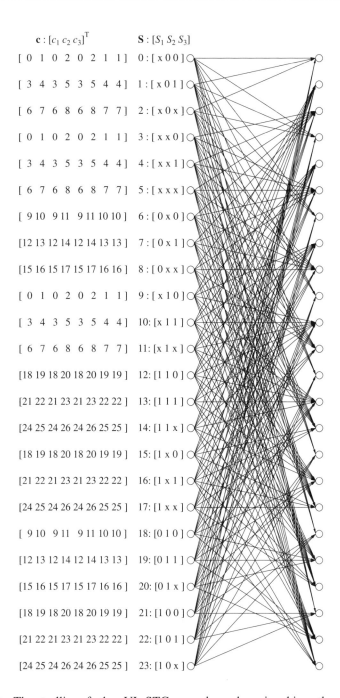

Figure 10.3: The trellis of the VL-STC encoder when invoking the 2D VLC of Equation (10.5). For the list of codewords see Table 10.1. ©IEEE [271] Ng *et al.* 2007.

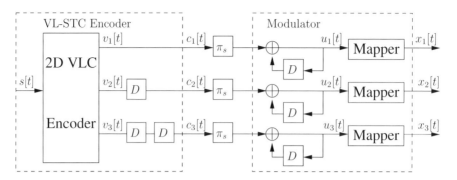

Figure 10.4: The VL-STCM-ID transmitter employing $N_t = 3$ transmit and $N_r = 2$ receive antennas, where π_s denotes a symbol interleaver. ©IEEE [271] Ng *et al.* 2007.

10.5   VL-STCM-ID Scheme

In order to invoke iterative detection and hence attain iteration gains as a benefit of the more meritoriously spread extrinsic information, we introduce a symbol-based random[4] interleaver and a non-binary URC for each of the $N_t = 3$ transmit antennas. The $N_t = 3$ parallel symbol-based interleavers were generated independently. As we can see from Figure 10.4, each element in the space–time codeword $c_m[t]$ for $m \in \{1, 2, 3\}$ is further interleaved and encoded by a non-binary URC, before being fed to the mapper. The convergence behavior of the iterative VL-STCM-ID decoder is analyzed in Section 10.6, where it is shown that the VL-STCM-ID scheme would be unable to converge at low SNRs, when the recursive feedback-assisted non-binary precoders or URCs are not used. In contrast, a simple single-cell non-binary URC invoked before the mapper of each transmit antenna would allow us to achieve a significant iteration gain. Hence, the VL-STC encoder was retained unaltered, but the original modulator was modified. The non-binary URC employs a modulo-$\bar{M}$ adder, where, again, $\bar{M} = 3$ is the number of different symbols in the VLC space–time codeword and we have $c_m[t] \in \{0, 1, x\}$. Accordingly, this single-cell URC possesses $\bar{M} = 3$ trellis states. Note that we represent the 'x' symbol using the number '2' during the modulo-$\bar{M}$ addition. Furthermore, the VL-STCM-ID scheme is very similar to the BICM-ID arrangement proposed in [329], where N parallel bit-based interleavers were used for interleaving the bits of the N-bit codeword before the modulator, in order to attain iteration gains. By contrast, the VL-STCM-ID scheme employs N_t parallel symbol-based interleavers for interleaving the symbols of the N_t-symbol space–time codeword before the modulator, in order to attain iteration gains with the aid of iterative decoding.

At the receiver, the symbol-based log-domain MAP algorithm [328] is used by both the VL-STC decoder and the URC decoder. The block diagram of the VL-STCM-ID receiver is depicted in Figure 10.5, where we denote the log-domain symbol probability of the VL-STC codeword c_m and the URC codeword u_m for the mth transmitter as $L^{i.m}_{(a,e)}(c)$ and $L^{i.m}_{(a,e)}(u)$, respectively. Furthermore, the subscripts a and e denote the *a priori* and extrinsic nature of the probabilities, while the superscript $i.m$ suggests that the probabilities belong to the

---

[4]The optimization of the interleaver is not considered in this chapter.

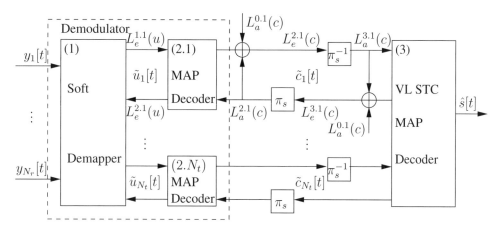

Figure 10.5: The VL-STCM-ID receiver for an $N_r \times N_t$ MIMO system. The notations $(\tilde{.})$ and $(\hat{.})$ indicate the extrinsic and *a priori* probability and the hard decision estimate of $(.)$, respectively. The notation $L^{i.m}_{(a,e)}(c, u)$ denotes the log-domain symbol probability of the VL-STC codeword c or the URC codeword u for the mth transmitter. The subscripts a and e denote the *a priori* and extrinsic nature of the probabilities while the superscript $i.m$ identifies that the probabilities belong to the ith stage decoder for the mth transmitter. Note that $i = 0$ means that the probabilities were calculated from the source symbol distribution. ©IEEE [271] Ng *et al.* 2007.

ith decoder stage of the mth transmitter. Note that $i = 0$ implies that the probabilities were calculated from the source symbol distribution.

The extrinsic probability of the URC codeword of transmit antenna m, namely $P_e(u_m[t])$, can be computed during each symbol period in the 'Soft Demapper' block of Figure 10.5. By dropping the time-related square bracket, we can compute $P_e(u_m)$ as

$$P_e(u_m = b) = \sum_{\mathbf{x} \in \chi(m,b)} \left(P(\mathbf{y}|\mathbf{x}) \prod_{\substack{j \neq m}}^{\text{all } j} P_a(u_j) \right), \tag{10.10}$$

$$m \in \{1, 2, \ldots, N_t\}, \ b \in \{0, 1, \mathbf{x}\}$$

where the subset $\chi(m, b)$ contains all the phasor combinations for the transmitted signal vector $\mathbf{x} = [x_1 \ x_2 \cdots x_{N_t}]^T$ where $x_m = f(u_m = b)$ holds, while $P(\mathbf{y}|\mathbf{x})$ is the Probability Density Function (PDF) of the MIMO Rayleigh fading channel given by

$$P(\mathbf{y}|\mathbf{x}) = \frac{1}{2\pi\sigma_n^2} \exp\left(\frac{-\|\mathbf{y} - \mathbf{Hx}\|^2}{2\sigma_n^2} \right), \tag{10.11}$$

and $\sigma_n^2 = N_0/2$ is the noise variance and $\mathbf{y}$ is the N_r-element complex received signal vector, while $\mathbf{H}$ is an $(N_r \times N_t)$-dimensional complex channel matrix during the time instant t. Furthermore, the *a priori* probability of u_m in Equation (10.10) is computed from the extrinsic log-domain probability of the mth URC MAP decoder as $P_a(u_m) = \exp(L_e^{2.m}(u))$, while the log-domain *a priori* probability of u_m for the mth URC MAP decoder is given by $L_e^{1.m}(u) = \ln(P_e(u_m))$. Hence, the soft demapper benefits from the *a priori* information of its input symbols u_m after the first iteration. Note that we employ the Jacobian logarithm [330] to compute Equation (10.10) in the log domain; hence there is no need for log domain to normal domain conversion. As we can see from Figure 10.5, each of the $N_t = 3$ URC MAP decoders seen inside the demodulator block benefits from the *a priori* information of its codeword $u_m[t]$ as well as from that of its input symbol $c_m[t]$, $m \in \{1, 2, \ldots, N_t\}$.

It is possible to attain some *a priori* probability for the N_t-element VL-STC codewords, **c**, (which also constitute the URC's input words), given the source symbol occurrence probability specified in Equation (10.4). Explicitly, the probability of the mth URC's input word c_m can be expressed as

$$P(c_m = d) = \sum_{l \in \mu(m,d)} P(s^l), \tag{10.12}$$

$$m \in \{1, 2, \ldots, N_t\}, \ d \in \{0, 1, x\},$$

where the subset $\mu(m, d)$ contains the specific indices of those columns in the VLC matrix, where the mth row element in that column equals d. Hence, we have $L_a^{0.m}(c) = \ln(P(c_m))$ as an additional *a priori* probability for symbol c_m during each iteration between the URC MAP decoder and the VL-STC MAP decoder, as shown in Figure 10.5. Note that $P(c_m)$ is directly computed from the source symbol occurrence probability $P(s^l)$; hence we do not use $P(s^l)$ again as the *a priori* probability of the VL-STC input word in the VL-STC MAP decoder, in order to avoid reusing the same information. Therefore, both the VL-STC and the URC MAP decoders benefit from the *a priori* or extrinsic information of the space–time codeword $\mathbf{c}[t] = [c_1[t] \ c_2[t] \ c_3[t]]^T$ received from each other as well as from the additional *a priori* information provided by the potentially different probability source symbols $s[t]$. A full iteration consists of a soft demapper operation, $N_t = 3$ URC MAP decoder operations and a VL-STC MAP decoder operation. For the non-iteratively decoded VL-STCM/FL-STCM, the soft demapper computes the *a priori* information of $\mathbf{c}[t] = [c_1[t] \ c_2[t] \ c_3[t]]^T$ and feeds it to the VL-STC/FL-STCM MAP decoder. Note that the VL-STC/FL-STC decoder of VL-STCM/FL-STCM also benefits from the *a priori* probability of its input word $s[t]$, which is given by the source symbol occurrence probability in Equation (10.4). Hence, as the source becomes correlated, the VL-STCM-ID, VL-STCM and FL-STCM schemes will benefit from the *a priori* probability of the source symbols. However, FL-STCM attains no energy savings.

10.6 Convergence Analysis

EXtrinsic Information Transfer (EXIT) charts designed for binary receivers [331] have been widely used for analyzing the convergence behavior of iterative-decoding-aided concatenated coding schemes. The *non-binary* EXIT charts were introduced on the basis of the multi-dimensional histogram computation of [332, 333]. However, the convergence analysis of the proposed three-stage VL-STCM-ID scheme requires the employment of novel three-dimensional (3D) non-binary EXIT charts, which evolved from the binary 3D EXIT charts used in [334, 335] for analyzing multiple concatenated codes.

To elaborate a little further, EXIT charts visualize the input and output characteristics of the constituent MAP decoders in terms of the mutual information transfer between the input sequence and the *a priori* information at the input, as well as between the input sequence and the extrinsic information at the output of the constituent decoder. Hence, there are two steps in generating an EXIT chart. Firstly, we have to model the *a priori* probabilities of the input sequence and then feed it to the decoder. Secondly, we have to compute the mutual information of the extrinsic probabilities at the output of the decoder. Let us now model the *a priori* probabilities of the VL-STC codeword, $\mathbf{c} = [c_1\ c_2\ \cdots\ c_{N_t}]^T$.

Let us assume that the input symbol of one of the URC encoders, denoted as c, is transmitted across an AWGN channel using the $\mathcal{M} = 3$-phasor mapper shown in Figure 10.2, and the received signal is given by $y = x + n$, where n is the AWGN noise having a zero mean and a variance of $\bar{\sigma}_n^2$. Furthermore, we have $x = f(c)$, where $f(.)$ is the mapper function portrayed in Figure 10.2. Since $f(.)$ is a memoryless function, the probability of occurrence for x is the same as that of c. Hence, we have $P(x) = P(c)$, which can be expressed from Equation (10.12). At a given probability of occurrence for x, the mutual information between x and y can be formulated as

$$I(x, y) = \sum_{i=1}^{\mathcal{M}} \int_y P(x_i, y) \log_2 \left(\frac{P(x_i, y)}{P(x_i)P(y)} \right) dy,$$
$$= H(x) - H(x|y), \tag{10.13}$$

where $H(x)$ is the entropy of x, given by

$$H(x) = -\sum_{i=1}^{\mathcal{M}} P(x_i) \log_2(P(x_i)), \tag{10.14}$$

and $H(x|y)$ is the conditional entropy of x given y, which can be expressed as

$$H(x|y) = -\sum_{i=1}^{\mathcal{M}} P(x_i) E\left[\log_2 \left(\sum_{j=1}^{\mathcal{M}} \frac{P(x_j)}{P(x_i)} \exp(\Psi_{i,j}) \right) \right], \tag{10.15}$$

where $\exp(\Psi_{i,j}) = P(y|x_i)/P(y|x_j)$ and $P(y|x)$ is the conditional Gaussian PDF, while the exponent $\Psi_{i,j}$ is given by

$$\Psi_{i,j} = \frac{-|x_i - x_j + n|^2 + |n|^2}{2\bar{\sigma}_n^2}. \tag{10.16}$$

The expectation term $E[.]$ in Equation (10.15) is taken over different representations of the AWGN noise n.

Let us now denote the *a priori* information of c as $I_A(c) = I(x, y)$. Note that $I(x, y)$ is monotonically decreasing with respect to $\bar{\sigma}_n^2$ and we can simplify Equation (10.13) to a form where $I(x, y)$ is expressed as a function $J(.)$ of $\bar{\sigma}_n^2$, i.e. as $I(x, y) = J(\bar{\sigma}_n^2)$. Hence, at a given I_A value we can find the corresponding noise variance with the aid of the inverse function $\bar{\sigma}_n^2 = J^{-1}(I_A(c))$. Then we can generate a noise sample n' having a variance of $\bar{\sigma}_n^2$. Consequently, we can produce $y' = x + n'$, where again $x = f(c)$ represents the mapper function portrayed in Figure 10.2 and c is the actual input symbol of the particular URC encoder. Finally, we can generate the *a priori* symbol probabilities for $P_a(c)$ using the conditional Gaussian PDF:

$$P_a(c) = \frac{1}{2\pi\bar{\sigma}_n^2} \exp\left(\frac{-|y' - f(c)|^2}{2\bar{\sigma}_n^2} \right), \qquad (10.17)$$

for $c \in \{0, 1, \mathrm{x}\}$. Then we feed these symbol probabilities to the corresponding MAP decoder. Note that the above method can be used for any symbol-interleaved serially concatenated coding schemes, where the symbol probabilities are directly created for a given I_A value.

Next, we compute the mutual information of the extrinsic symbol probabilities $I_E(c)$ of the VL-STC or URC decoder output using the method proposed in [332] for the symbol c. Finally, the mutual information for the N_t-element VL-STC codeword $\mathbf{c} = [c_1\ c_2 \cdots c_{N_t}]^T$ is expressed as the sum of mutual information valid for its symbol components c_m, $m \in \{1, 2, \ldots, N_t\}$. Hence we have $I_A(\mathbf{c}) = \sum_{m=1}^{N_t} I_A(c_m)$ and $I_E(\mathbf{c}) = \sum_{m=1}^{N_t} I_E(c_m)$. We also compute the mutual information for the URC codeword $\mathbf{u}$ based on the same procedure. However, the experimentally evaluated histogram of $\mathbf{u}$ exhibited a near-uniform distribution (in other words, it was associated with equiprobable symbols) when the symbol interleaver length was sufficiently high. This is because the URC employed can be viewed as an accumulator.

The 3D EXIT charts and the actual iterative decoding trajectories for the VL-STCM-ID scheme having $N_t = 3$ and $N_r = 2$ when using an uncorrelated source are shown in Figures 10.6 and 10.7 respectively. Let us denote the three axes of the 3D EXIT charts using the letters x, y and z, while $I_A(*)$ and $I_E(*)$ denote the *a priori* and extrinsic information for (*), respectively, where (*) is either the VL-STC MAP decoder (VL-STC) or the URC MAP decoder (URC) or, alternatively, the soft demapper (Demod). As we can see from Figure 10.5, each of the URC MAP decoders takes (provides) the *a priori* (extrinsic) probabilities of its input word c and output word (or codeword) u as the input (output). Hence, the mutual information of the input word and output word of the URC decoder will be represented by $I_{(A,E)}^i(\text{URC})$ and $I_{(A,E)}^o(\text{URC})$, respectively. Each of the N_t symbol interleavers shown in Figure 10.4 has a length of 10 000 symbols.

The EXIT plane marked with triangles in Figure 10.6 was computed based on the extrinsic probabilities of the soft demapper $L_e^{1.m}(u)$, for $m \in \{1, 2, 3\}$, at the given values in the x and y axes at $E_b/N_0 = 4$ dB. The other EXIT plane marked with dashed lines in Figure 10.6 is SNR-independent and was plotted based on the URC decoders' output word extrinsic probabilities $L_e^{2.m}(u)$, for $m \in \{1, 2, 3\}$, at the given values in the y and z axes. The decoding trajectory, which was determined based on the extrinsic probabilities $L_e^{1.m}(\mathbf{u})$, $L_e^{2.m}(u)$ and $L_e^{3.m}(c)$ for $m \in \{1, 2, 3\}$ as in Figure 10.5, is under the EXIT plane marked with triangles and on the left of the EXIT plane marked with dashed lines in Figure 10.6.

Figure 10.7 depicts the 3D EXIT charts, when the x axis of Figure 10.6 is changed from $I_E^o(\text{URC})$ to $I_E^i(\text{URC})$. More explicitly, the vertical EXIT plane seen in Figure 10.7, which

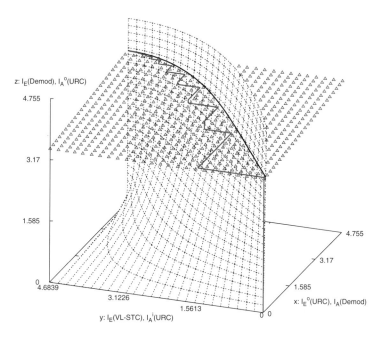

Figure 10.6: The 3D EXIT charts for the VL-STCM-ID scheme having $N_t = 3$ and $N_r = 2$, when using an uncorrelated source. The iterative trajectory is computed at $E_b/N_0 = 4$ dB. ©IEEE [271] Ng *et al.* 2007.

is independent of the z axis and the SNR, was computed based on the VL-STC decoder's extrinsic probabilities $L_e^{3.m}(c)$, for $m \in \{1, 2, 3\}$, at the given values in the x and z axes. The slanted EXIT plane shown in Figure 10.7 is also SNR independent and was computed based on the URC decoders' input word extrinsic probabilities $L_e^{2.m}(c)$, for $m \in \{1, 2, 3\}$, at the given values in the y and z axes. The stepwise-linear iterative decoding trajectory displayed in Figure 10.7 was plotted based on the extrinsic probabilities $L_e^{1.m}(u)$, $L_e^{2.m}(c)$ and $L_e^{3.m}(c)$ for $m \in \{1, 2, 3\}$.

According to [335], the intersection of the planes in Figure 10.6 represents the points of convergence between the soft demapper and the URC decoder. The intersection points, where we have $I_A(\text{Demod}) = I_E^o(\text{URC})$ at the corresponding $I_A^i(\text{URC})$ and $I_A^o(\text{URC})$ values, are shown in Figure 10.6 as a solid line. The corresponding values of $I_E^i(\text{URC})$ are shown in Figure 10.7 as a solid line on the slanted EXIT plane. Similarly, the intersection of the EXIT planes seen in Figure 10.7 represents the points of convergence between the URC decoder and the VL-STC decoder. Hence, projecting these two intersection curves onto $z = 0$ in Figure 10.7 gives us the equivalent 2D EXIT chart seen in Figure 10.8. Therefore, the 3D EXIT charts generated for multiple concatenated codes can be projected onto an equivalent 2D EXIT chart [335].

More specifically, we can observe an open tunnel between the EXIT curves of the VL-STC and URC schemes in the 2D EXIT charts of Figure 10.8 at $E_b/N_0 = 4$ dB, which indicates that decoding convergence can be achieved. However, the URC EXIT curve is

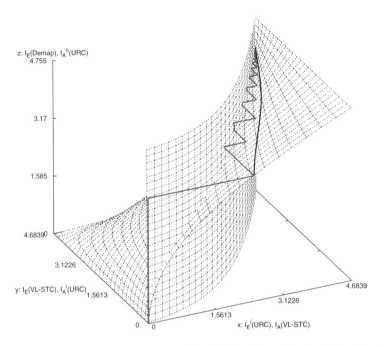

Figure 10.7: The 3D EXIT charts for the VL-STCM-ID scheme having $N_t = 3$ and $N_r = 2$, when using an uncorrelated source. The iterative trajectory is computed at $E_b/N_0 = 4$ dB. ©IEEE [271] Ng *et al.* 2007.

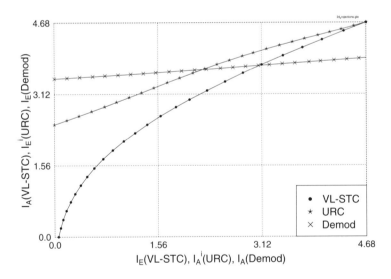

Figure 10.8: The 2D EXIT charts projection for the VL-STCM-ID scheme having $N_t = 3$ and $N_r = 2$, when using an uncorrelated source at $E_b/N_0 = 4$ dB. ©IEEE [271] Ng *et al.* 2007.

SNR dependent, and by reducing the E_b/N_0 values the angle between the two curves at the top right-hand corner would be further reduced, and hence the open tunnel would become closed, preventing decoding convergence. Therefore, a better URC code may be designed by ensuring that the EXIT curve exhibits a wider angle with respect to the VL-STC EXIT curve. Note, furthermore, that the EXIT curve generated for the soft demapper is also depicted in Figure 10.8 at $E_b/N_0 = 4$ dB, where it is flat and it intersects with the VL-STC EXIT curve, before the maximum value of 4.68 bits is reached. Hence, decoding convergence cannot be achieved at $E_b/N_0 = 4$ dB, when the URC was not invoked between the soft demapper and the VL-STC. Furthermore, at $I_E(\text{VL-STC}) = 0$ we have $I_E^i(\text{URC}) < I_E(\text{Demod})$. Therefore, before any iteration feedback is exploited, the VL-STCM-ID scheme would not outperform its VL-STCM counterpart. It was also concluded in [334] that for a three-stage serially concatenated system a unity-rate recursive encoder, such as the URC used, should be employed at the intermediate stage in order to achieve optimal decoding convergence.

Figure 10.9 shows the 3D EXIT charts and the corresponding convergence curve of the soft demapper and the URC decoder as well as the actual iterative decoding trajectory for the VL-STCM-ID scheme having $N_t = 3$ and $N_r = 2$ when using the correlated source defined in Equation (10.4). As the source becomes correlated, the entropy of the codeword c_m for $m \in \{1, 2, 3\}$ reduces. Hence, the maximum values for the x and y axes in Figure 10.9 are smaller than those in Figure 10.7. However, the open spatial segment of the 3D space between the two EXIT planes becomes wider when the source is correlated, since the decoders exploit the additional *a priori* probabilities given by Equation (10.12).

The convergence curve of the soft demapper and the URC decoder is projected as a dashed line onto $I_E(\text{Demod}) = 0$ in Figure 10.9. Similarly, the projection of the intersection line between the VL-STC and URC EXIT planes is represented by the curve lying on the vertical EXIT plane at $I_E(\text{Demod}) = 0$. As can be seen from Figure 10.9 at $I_E(\text{Demod}) = 0$, an open tunnel exists between the two projection curves at $E_b/N_0 = 3$ dB. Hence, the iterative decoder converged at $E_b/N_0 = 3$ dB, i.e. at a 1 dB lower value, when employing the correlated source instead of the uncorrelated source. Again, the decoding convergence of VL-STCM-ID is limited by the angle between the two projection curves at the convergence point when using the correlated source. A better URC may be designed for an earlier convergence with the aid of the 3D and 2D EXIT charts, but we leave this issue for future research.

According to the MIMO channel capacity formula derived for the Discrete-input Continuous-output Memoryless Channel (DCMC) in [336], the DCMC capacity for the $N_r = 2$ and $N_t = 3$ MIMO scheme employing the signal mapper seen in Figure 10.2 is $E_b/N_0 = 1.25$ dB at a bandwidth efficiency of 3 bit/s/Hz. Hence, the performance of the VL-STCM-ID scheme is about 2.75 dB and 1.75 dB away from the MIMO channel capacity.

10.7   Simulation Results

Let us introduce a Fixed Length (FL) STCM (FL-STCM) scheme as our benchmark, where the FL codeword matrix is given by

$$\mathbf{V}_{FLC} = \begin{bmatrix} 0 & 0 & 0 & 0 & 1 & 1 & 1 & 1 \\ 0 & 0 & 1 & 1 & 0 & 0 & 1 & 1 \\ 0 & 1 & 0 & 1 & 0 & 1 & 0 & 1 \end{bmatrix}. \tag{10.18}$$

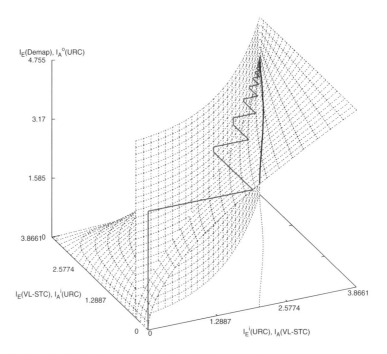

Figure 10.9: The 3D EXIT charts for the VL-STCM-ID scheme having $N_t = 3$ and $N_r = 2$, when using the correlated source defined in Equation (10.4). The iterative trajectory is computed at $E_b/N_0 = 3$ dB. ©IEEE [271] Ng *et al.* 2007.

The FL-STCM transmitter obeys the schematic of Figure 10.1, except that it employs the V_{FLC} of Equation (10.18). For the FL-STCM, the minimum Hamming distance and product distance are 1 and 4, respectively. It attains the same multiplexing gain as that of the VL-STCM or VL-STCM-ID arrangements. Note that it is possible to create an iterative FL-STCM-ID scheme by replacing the VL-STC encoder in Figure 10.4 with the FL-STC encoder. However, the EXIT curve of the FL-STC scheme of Equation (10.18) was found to be too flat for attaining any iteration gain due to its unity minimum Hamming distance. Let us now evaluate the performance of the VL-STCM, VL-STCM-ID and FL-STCM schemes in terms of their source Symbol-Error Ratio (SER) versus the E_b/N_0 ratio. Again we have $E_b/N_0 = \gamma/\eta$, where γ is the SNR per receive antenna and $\eta = \log_2(N_s) = 3$ bit/s/Hz is the effective information throughput.

Figure 10.10 depicts the SER versus E_b/N_0 performance of the VL-STCM, VL-STCM-ID and FL-STCM schemes, when communicating over uncorrelated Rayleigh fading channels using BPSK, three transmitters, two receivers and a block/interleaver length of 10 000 symbols. As expected, the VL-STCM arrangement attains a higher gain when the source is correlated compared with FL-STCM. However, the FL-STCM benchmark also benefits from the probability-related *a priori* information of the source symbols as the source becomes correlated. Both the FL-STCM and VL-STCM-ID schemes also manage to achieve some coding gain, since their decoder benefits from the probability-related *a priori*

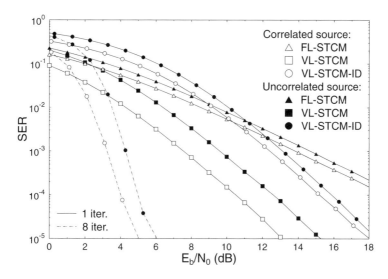

Figure 10.10: SER versus E_b/N_0 performance of the VL-STCM, VL-STCM-ID and FL-STCM schemes, when communicating over uncorrelated Rayleigh fading channels using BPSK, $N_t = 3$ and $N_r = 2$. ©IEEE [271] Ng *et al.* 2007.

information of the source symbols. By contrast, the coding gain attained as a benefit of transmitting correlated source symbols increases as the number of iterations invoked by the VL-STCM-ID scheme increases. Although the VL-STCM-ID arrangement performs worse than VL-STCM during the first iteration, the performance of VL-STCM-ID at SER = 10^{-4} after the eighth iteration is approximately 6.5 (15) dB and 7.5 (14.6) dB better than that of the VL-STCM (FL-STCM) scheme, when employing correlated and uncorrelated sources respectively. The price of using the VL-STCM-ID scheme for attaining a near-channel-capacity performance is the associated higher decoding and interleaver delay as well as the increased decoding complexity. Hence, for a delay-sensitive and complexity-constrained system, using VL-STCM is a better choice. By contrast, for a system that requires a higher performance and can afford a higher delay and complexity, VL-STCM-ID should be employed.

10.8   Non-Binary VL-STCM

In this section, we will show that our VLC-STCM design can be extended to higher-order modulation schemes for attaining a higher throughput. More specifically, we will design a *non-binary* VL-STCM scheme using 2D *non-binary* VLCs. Similar to the binary VL-STCM scheme, the number of activated transmit antennas in the non-binary VL-STCM encoder equals the number of non-binary symbols of the corresponding VLC codeword in the spatial domain, where each VLC codeword is transmitted during a single symbol period. Hence, the transmission frame length is determined by the fixed number of source symbols and therefore this non-binary VL-STCM scheme does not exhibit synchronisation problems and does not require the transmission of side information.

10.8.1 Code Design

$$
\underset{\xrightarrow{\;s[t]\;}}{
\begin{array}{l}
V_{VLC} = \\[2pt]
\begin{bmatrix}
\text{x x } 3\ 2\ 0\ 1\ \text{x x x } 2\ 0\ 1\ 0\ 3\ 3\ 2\ 0\ 0 \\[4pt]
\text{x } 0\ \text{x } 3\ \text{x x } 2\ 1\ 3\ \text{x } 1\ 2\ 0\ 0\ 1\ 0\ 3\ 2 \\[4pt]
0\ \text{x x x } 3\ 2\ 3\ 1\ 2\ 1\ \text{x x } 0\ 3\ 2\ 2\ 1\ 2
\end{bmatrix}
\begin{array}{l}
\xrightarrow{\;v_1[t]\;} \\[4pt]
\xrightarrow{\;v_2[t]\;} \\[4pt]
\xrightarrow{\;v_3[t]\;}
\end{array}
\end{array}}
$$

Figure 10.11: The $N_s = 18$ possible VLC codewords $\mathbf{v}[t]$ of the codeword matrix in the 2D non-binary VLC V_{VLC} employing 4PSK modulation, which signals the $N_s = 18$ possible source symbols with the aid of using $N_t = 3$ transmit antennas.

The block diagram of the non-binary VL-STCM transmitter is similar to Figure 10.1 except that a 2D non-binary VLC encoder is used. An example of the 2D non-binary VLC codeword matrix, V_{VLC}, is depicted in Figure 10.11 when there are $N_s = 18$ possible source symbols and $N_t = 3$ transmit antennas. More explicitly, each column of the (3×18)-dimensional matrix V_{VLC} corresponds to one codeword and the elements in the matrix represent the 4PSK symbols to be transmitted by the different antennas, while 'x' represents 'no transmission'. 'No transmission' implies that the corresponding transmitter antenna sends no signal. The VLC codeword $\mathbf{v}[t]$ corresponding to source symbol $s[t] = j$, $j = 1, 2, \ldots, 18$, at time instant t is given by the jth column of the matrix V_{VLC}. Hence, each of the $N_t = 3$ components of $\mathbf{v}[t]$ may assume one of the following values $v_i[t] \in \{\text{x}, 0, 1, 2, 3\}$, where $i \in \{1, 2, \ldots, N_t\}$ when 4PSK modulation is employed[5]. Note that more transmission energy is saved, which may be reallocated to 'active' symbols, when there are more 'no transmission' symbols in a VLC codeword $\mathbf{v}[t]$. Therefore, the columns of the matrix V_{VLC}, which are the VLC codewords, and the source symbols are specifically arranged, so that the more frequently occurring source symbols are assigned to VLC codewords having more 'no transmission' components, in order to save transmit energy.

The signal mapper blocks of the non-binary VL-STCM scheme are characterized in Figure 10.12. Again, the 'no transmission' symbol is represented by the origin of the Euclidean space. Furthermore, the amount of energy saving can be computed from Eq. (10.7) as before. where L_{ave} is the average codeword length of the VLC codeword matrix V_{VLC}. As an example, we use the $N_s = 18$ source symbol probabilities of $P(s) = \{0.1776, 0.158, 0.1548, 0.1188, 0.1162, 0.077, 0.0682, 0.0388, 0.0374, 0.0182, 0.017, 0.0068, 0.0066, 0.002, 0.0016, 0.0007, 0.0002, 0.0001\}$. Hence, the average codeword length is given by $L_{ave} = \sum_{s=1}^{N_s} P(s)L(s) = 1.5208$ for the V_{VLC} matrix of Figure 10.11, where $L(s)$ is the number of 'active' transmitted symbols in the VLC codeword assigned to source symbol s according to Figure 10.11. Hence, in this specific example a power saving of $20 \log_{10}(\sqrt{3/1.5208}) = 2.9505$ dB can be obtained. Note however that the more correlated the source, the lower the average codeword length, and hence the higher the power saving. In order to normalize the average energy of the multiple transmitter-based modulation constellations to unity, we have to scale the 4PSK constellation of each transmitters such that the radius of the 4PSK constellation becomes $A = \sqrt{3/1.5208} = 1.4045$.

---

[5]At this early phase of the research we set aside the related crest-factor problems for future work.

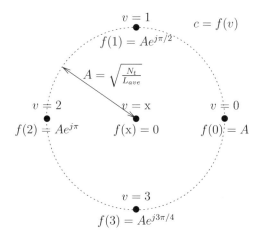

Figure 10.12: The signal mapper of the non-binary VL-STCM.

As we can see from Figures. 10.1 and 10.11, each of the 18 non-binary VLC codewords $\mathbf{v}[t] = [v_1[t] \; v_2[t] \; v_3[t]]^T$ is transmitted using $N_t = 3$ transmitters, where the mth element of each VLC codeword, for $1 \leq m \leq N_t$, is delayed by $(m-1)$ shift register cells, before it is transmitted through the mth transmit antenna. Hence, the N_t number of components of each VLC codeword are transmitted on a diagonal of the spacetime codeword matrix of Equation (10.2). Since the VLC codewords are transmitted diagonally, the transmitted signal at the mth antenna, for $1 \leq m \leq N_t$, at a particular time-instant t is given by $x_m[t] = f(v_m[t-m+1])$. Hence, for the spacetime codeword matrix of Equation (10.2) can be written as:

$$
\mathbf{C} = \begin{bmatrix}
\cdots & f(v_1[t]) & f(v_1[t+1]) & \cdots & f(v_1[t+N_t-1]) & \cdots \\
\cdots & f(v_2[t-1]) & f(v_2[t]) & \cdots & f(v_2[t+N_t-2]) & \cdots \\
\vdots & \vdots & \vdots & \ddots & \vdots & \vdots \\
\cdots & f(v_{N_t}[t-N_t+1]) & f(v_{N_t}[t-N_t+2]) & \cdots & f(v_{N_t}[t]) & \cdots
\end{bmatrix}.
$$

It can be shown based on [325] that the minimum Hamming distance E_H of the VL-STCM scheme is given by:

$$
E_{H \min} = d_H^{VLC}, \tag{10.19}
$$

where d_H^{VLC} is the *symbol*-based minimum Hamming distance of the 2D VLC employed. The *symbol*-based Hamming distance between any two VLC codewords, which are selected from the columns of the codeword matrix V_{VLC} of Figure 10.11, is given by the number of different symbols - rather than bits - between the two VLC codewords. For example, the *symbol*-based Hamming distance between the first and the last columns of V_{VLC} shown in Figure 10.11 is three, where the 'no transmission' symbol 'x' is included as the fifth symbol. The minimum product distance $E_{P \min}$ is given by Equation (10.8).

In the example of Figure 10.11 we have $E_{H \min} = d_H^{VLC} = 2$ and $E_{P \min} = 3.89$, which were found to be the maximum achievable values at $N_s = 18$ and $N_t = 3$ by our exhaustive computer search for the best solution. Furthermore, the effective throughput of the VL-STCM scheme is given by $\log_2(N_s) = 4.17$ bits per symbol (bps), which is significantly higher than $\log_2(4) = 2$ bps in the case of classic 4PSK modulation. Therefore, firstly a transmit diversity order of $E_{H \min} = 2$ was attained. Secondly, a coding advantage directly related to the value of $E_{P \min} = 3.89$ was achieved. Thirdly, a multiplexing gain can also be attained for $N_s > M$, where M is the number of modulation levels in the original modulation scheme. Additionally, the correlation of the source only affects the specific mapping of the different-probability source symbols to the 2D VLC codewords of Figure 10.11, but not the design of the 2D VLC itself. Therefore, the higher the source correlation, the lower the average codeword length L_{ave}. Hence, a higher power saving can be obtained, when the source is more correlated, since the amount of energy saving is given by Equation (10.7), where N_t is fixed.

10.8.2 Benchmarker

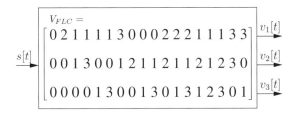

Figure 10.13: The codeword matrix of the 2D non-binary FLC C_{FLC} employing 4PSK modulation, having $N_s = 18$ number of possible source symbols and using $N_t = 3$ transmit antennas.

In order to benchmark the proposed VL-STCM scheme of Figure 10.11, a Fixed-Length STCM (FL-STCM) benchmarker was created. The codeword matrix of the Fixed Length Code (FLC) is shown in Figure 10.13, where again, 4PSK modulation was employed in conjunction with $N_s = 18$ and $N_t = 3$. In the FLC all N_t transmit antennas are active at all time instants, hence there are only $M = 4$ phasors in the modulation constellation and no power saving is achieved due to the absence of 'no transmission' symbols. Similar to the VLC scheme of Figure 10.11, this was designed by exhaustive computer search, evaluating Equations. 10.19 and 10.8 for all possible combinations of the 4PSK symbols.

At $N_s = 18$, we have $E_{H \min} = 1$ and $E_{P \min} = 2$, which are the maximum achievable values for the FLC at a given combination of $N_s = 18$ and $N_t = 3$. However, these values are significantly smaller than the $E_{H \min}$ and $E_{P \min}$ of the 2D VLC. Nonetheless, both FL-STCM and VL-STCM achieve the same multiplexing gain of $\log_2(N_s/M) = 2.17$ due to having the same average throughput of $\log_2(N_s) = 4.17$ bps when compared to the 4PSK ($M = 4$) modulation.

10.8.3 Decoding

At the receiver, both the Viterbi and the MAP algorithms may be used by the trellis-based decoder. The trellis states are defined by the contents of the shift register cells D shown in Figure 10.1. Note that each shift register cell may hold five possible values, namely $\{x, 0, 1, 2, 3\}$ for the non-binary VL-STCM and four 4PSK values for the non-binary FL-STCM scheme. However, not all consecutive state sequences are possible due to the constraint imposed by the 2D VLC of Figure 10.11 or by the FLC code of Figure 10.13. More specifically, there are 90 and 44 possible trellis states in the trellis of the VL-STCM and

FL-STCM schemes, respectively. Hence, the decoding complexity of the VL-STCM based on the V_{VLC} matrix of Figure 10.11 is about twice that of the FL-STCM based on the C_{FLC} matrix of Figure 10.13. The log-domain symbol metrics can be computed based on the PDF of the channel as:

$$Pr\left(\mathbf{y}[t] \mid \mathbf{x}[t]\right) = -\frac{1}{N_0} \sum_{n=1}^{N_r} \left| y_n[t] - \sqrt{E_s} \sum_{m=1}^{N_t} h_{n,m}[t] x_m[t] \right|^2, \qquad (10.20)$$

where $\mathbf{y}[t] = [y_1[t] \ \cdots \ y_{N_r}[t]]$ is the received signal vector at time-instant t, $\mathbf{x}[t] = [x_1[t] \ \cdots \ x_{N_t}[t]]$ is the N_t-element space-time coded symbol and $x_m[t] = f(v_m[t - m + 1])$. When the different source symbols' probability of occurrence $P(d)$ is different, it can be used as the *a priori* probability for assisting the trellis decoder. Hence, the log-domain trellis branch transition metrics of the transition from the previous state S^{-1} to the current state S may be computed as:

$$\gamma(S^{-1}, S) \quad = \quad Pr\left(\mathbf{y}[t] \mid \mathbf{x}[t]\right) + Pr(s), \qquad (10.21)$$

where $\mathbf{x}[t]$ and s are the associated codeword and source symbol for that trellis branch, while $Pr\left(\mathbf{y}[t] \mid \mathbf{x}[t]\right)$ is given by Equation (10.20) and $Pr(s) = \ln P(s)$ is the log-domain *a priori* probability of the source symbol s.

It is worth mentioning that it is possible to design a VL-STCM scheme, which has a higher minimum product distance, $E_{P\,min}$, at the cost of a higher decoding complexity. More explicitly, the design of the proposed VL-STCM is based on the search for an optimum non-binary 2D VLC matrix V_{VLC}, as shown in Figure 10.11, and its trellis is based on the shift-registers shown in Figure 10.1. By contrast, a VL-STCM scheme having a higher $E_{P\,min}$ can be designed based on the search of an optimum space-time codeword matrix $\mathbf{X}$, given by Equation (10.2), using a trellis that has a higher number of trellis states.

10.8.4   Simulation results

Let us now evaluate the performance of both the VL-STCM and FL-STCM schemes in terms of their source SER versus the SNR per bit, which is given by $E_b/N_0 = \gamma/\eta$, where γ is the SNR per receive antenna and $\eta = \log_2(18) = 4.17$ bps is the effective information throughput.

Figure 10.14 depicts the SER versus E_b/N_0 performance of the VL-STCM and FL-STCM schemes, when communicating over uncorrelated Rayleigh fading channels using 4PSK, three transmitters and two receivers. Note that when the source is uncorrelated, we have $L_{ave} = 2.1667$ and $A_0 = \sqrt{\frac{N_t}{L_{ave}}} = 1.38$ for the VL-STCM scheme, hence the power saving of VL-STCM is given by $20 \log_{10}(A_0) = 1.4133$ dB and the effective product distance is given by $E_{P\,min} = 1.92$. By contrast, the results for the correlated source investigated are based on the source symbol probabilities $P(s)$ defined in Section 10.8.1. The upper bounds of the PWEP defined in Equation (10.3) are also shown in Figure 10.14 for comparison with the simulation results. As we can see from Figure 10.14, the performance of VL-STCM is about 1.7 dB better, when the source is correlated as compared to when the source is uncorrelated. This is due to the higher achievable $E_{P\,min}$ value, when the source is correlated. Observe in Figure 10.14 that the PWEP upper bounds of the VL-STCM are close to the simulation results. By contrast, the performance of the FL-STCM is about 1.1 dB better, when the

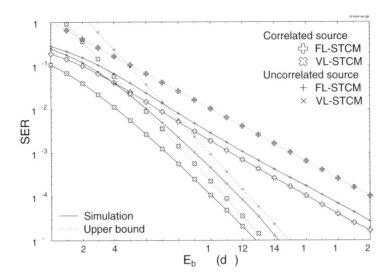

Figure 10.14: SER versus E_b/N_0 performance of the VL-STCM and FL-STCM schemes, when communicating over uncorrelated Rayleigh fading channels using 4PSK, $N_t = 3$ and $N_r = 2$.

source is correlated compared to the uncorrelated source. This is because the branch transition metrics in Equation (10.21) benefit from the different source symbol probabilities, when the source is correlated.

By comparing the simulation performance of VL-STCM and FL-STCM at SER $= 10^{-4}$, the performance of VL-STCM is approximately 6 dB and 5.4 dB better than that of FL-STCM, when employing correlated and uncorrelated sources, respectively. Again, this coding gain is achieved at the cost of approximately doubling the decoding complexity from 44 to 90 trellis states.

10.9 Conclusions

An iteratively decoded variable-length space–time coded modulation design was proposed. The joint design of source-coding, space–time coded modulation and iterative decoding was shown to achieve both spatial diversity and multiplexing gain, as well as coding and iteration gains at the same time. The variable-length structure of the individual codewords mapped to the maximum of N_t transmit antennas imposes no synchronization and error propagation problems. The convergence properties of the proposed VL-STCM-ID were analyzed using 3D symbol-based EXIT charts as well as 2D EXIT chart projections. A significant iteration gain was achieved by the VL-STCM-ID scheme, which hence outperformed both the non-iterative VL-STCM scheme and the FL-STCM benchmark with the aid of N_t unity-rate recursive feedback precoders. The VL-STCM-ID scheme attains a near-MIMO channel capacity performance.

The proposed VL-STCM design may also be employed in the context of higher-order modulation schemes by first designing a two-dimensional VLC mapping high-order modulation constellations to the N_t transmit antennas. A non-binary VLC-STCM was designed and analyzed in Section 10.8, where an overall performance gain of 6 dB was attained by the non-binary VL-STCM scheme over the non-binary FL-STCM benchmarker scheme at SER=10^{-4} when aiming for a throughput of 4.17 bps. The non-binary VL-STCM may also be extended to a non-binary VL-STCM-ID scheme for the sake of attaining further iteration gains. Our future research will incorporate both explicit channel coding and real-time multimedia source codecs.

Chapter 11

Iterative Detection of Three-Stage Concatenated IrVLC FFH-MFSK[1]

Dr Sohail Ahmed, National University of Sciences and Technology (NUST), Pakistan
R. G. Maunder, L.-L. Yang and L. Hanzo

11.1 Introduction

Fast Frequency Hopping (FFH) M-ary Frequency Shift Keying (MFSK) has been shown to efficiently combat Partial Band Noise Jamming (PBNJ) [338]. The source of PBNJ may be intentional, as in the case of military communication systems [339] where enemy jammers try to disrupt the friendly communication. But it may be caused by microwave sources and lighting devices operating adjacent to a Frequency Hopping (FH) network. The military communication systems usually operate using a wide frequency band. Similarly, a number of commercial FH networks utilize the Industrial Scientific Medical band. Thus both types of communication systems are susceptible to partial-band or full-band noise interference imposed by intentional or unintentional jammers. By transmitting every symbol using multiple hops, FFH systems benefit from both time and frequency diversity, which assists in mitigating the detrimental effects of PBNJ. Furthermore, in the FFH-MFSK receiver, a suitable diversity combining method may be invoked to further suppress the effects of PBNJ. We employ clipped combining [338] for this purpose. This is an effective, low-complexity combining scheme dispensing with estimation of channel gains, which would be required in order to invoke maximal ratio combining [340]. Furthermore, as a benefit of noncoherent detection, MFSK-based schemes are eminently suitable for communication in channels characterized by fast fading, which typically prevents effective channel estimation.

Although uncoded FFH schemes have been analyzed when communicating in channels contaminated by jamming [341–343], coded FFH schemes employing Soft Decision Decoding (SDD) have attracted little attention. The concept of extrinsic information exchange between an M-ary orthogonal demodulator and a Soft-Input Soft-Output (SISO) decoder

---

[1]Part of this chapter is based on [337] © IEEE (2009).

has been investigated in [344, 345]. In [345], further insights have been provided with the aid of EXtrinsic Information Transfer (EXIT) chart analysis. EXIT charts constitute useful semi-analytical tools, which were proposed by ten Brink [346] for analyzing the convergence behavior of iteratively decoded schemes. This powerful tool, however, has not been applied to FH-assisted schemes. Fiebig and Robertson [347] employed SDD-assisted convolutional, turbo and Reed Solomon codes, although no attempt was made to exploit the decoder's soft output. Park and Lee [348] extended the work reported in [347] by employing Iterative Decoding (ID). Again, in the open literature, little attention has been devoted to the investigation of SISO decoding in noncoherent MFSK-based schemes, and in particular to the FFH-MFSK arrangement. The reasons for this trend might include the difficulty of deriving soft information from the received signal, as well as the challenge of the subsequent exploitation of the soft information forwarded by the SISO decoder to the demodulator.

By contrast, a number of useful tools have been employed for the analysis as well as for the performance enhancement of iteratively decoded coherently modulated schemes. A powerful technique for improving the iterative gain of iteratively decoded schemes is constituted by precoders, which improve the extrinsic information exchange between the channel decoder and the demodulator [349,350]. The precoder imposes memory upon the channel, thus rendering it recursive.[2] Hence, SISO systems have been shown to benefit substantially from employment of unity-rate precoders without reducing the effective throughput of the system [351].

EXIT-chart analysis [346] has opened avenues towards the investigation and performance improvement of iteratively decoded communication systems. A further significant contribution made in the field of ID with the aid of EXIT charts is the introduction of Irregular Convolutional Codes (IrCCs) [149]. An IrCC consists of multiple sub-codes, each having a different code rate, which are used to encode specific fractions of the source data block. By optimally choosing the code rates of the IrCC component codes and the associated fractions of the input block, the convergence of the ID process may be achieved at low Signal-to-Noise Ratios (SNRs). In analogy with IrCCs [149], we introduce the novel family of so-called Irregular Variable-Length Codes (IrVLCs) [197] by employing a number of component VLC codebooks having different coding rates [179] for encoding particular fractions of the input source symbol stream. With the aid of EXIT charts [346], the appropriate lengths of these fractions may be chosen in order to shape the inverted EXIT curve of the IrVLC codec to ensure that it does not cross the EXIT curve of the inner channel codec. In this way, an open EXIT chart tunnel may be created even at low SNR values.

Our aim in this contribution is to employ the classical tools of ID to achieve significant performance improvement in FFH-MFSK systems. Some such tools, for example SISO decoding, unity-rate precoding and EXIT-chart analysis, have been successfully applied to coherent schemes but have attracted little attention as far as the noncoherent schemes are concerned. The open problems solved in this chapter are as follows.

1. We investigate for the first time an IrVLC-coded and precoder-assisted noncoherently detected FFH-MFSK system communicating in an uncorrelated Rayleigh fading channel contaminated by PBNJ.

2. The soft metrics passed by the noncoherent FFH-MFSK demodulator to the decoder are derived, assuming uncorrelated Rayleigh fading channels.

---

[2]To be recursive in this context implies that the channel has an infinite impulse response.

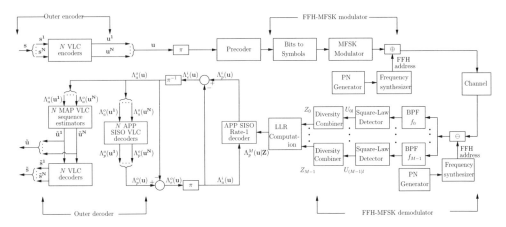

Figure 11.1: Schematic of the IrVLC- and VLC-based schemes employed in conjunction with FFH-MFSK. In the IrVLC coded scheme $N = 16$, whilst $N = 1$ in the VLC-coded scheme. ©IEEE [198] Ahmed *et al.* 2009.

3. By employing EXIT charts, we investigate the serial concatenation of the FFH-MFSK demodulator, unity-rate decoder and IrVLC outer decoder in order to attain a good performance even at low SNR values.

4. We contrast the two-stage ID between the inner rate-1 decoder and the IrVLC outer decoder with the three-stage extrinsic information exchange amongst the demodulator, inner decoder and outer decoder.

The proposed system might find fruitful application in FFH-MFSK-based ad hoc or cellular networks transmitting, for example, VLC-compressed H.264/MPEG-4 video or audio signals. The VLC compression may also provide useful data rate and bandwidth efficiency gains in pure data transmission by exploiting the potential latent correlation of bits.

The rest of this chapter is structured as follows. In Section 11.2 the system considered is described, and the concept of joint source and channel coding is elaborated. In Section 11.3 the soft metrics are derived, and the ID process is discussed with emphasis on the EXIT characteristics of the precoder and the demodulator. In Section 11.4 the choice of IrVLC code parameters is discussed with the aid of EXIT charts, and our Bit-Error Ratio (BER) results are discussed. Finally, in Section 11.5, we present our conclusions.

11.2 System Overview

In this section we consider two serially concatenated and iteratively decoded joint source and channel coding schemes. Specifically, we consider an IrVLC codec and an equivalent regular VLC-based benchmark in this role. We refer to these schemes as the IrVLC and the VLC-FFH-MFSK arrangements, respectively. The schematic that is common to both of these schemes is shown in Figure 11.1.

11.2.1 Joint Source and Channel Coding

The schemes considered are designed for facilitating the low-BER transmission of source symbol values over an uncorrelated Rayleigh fading channel. We consider $K = 16$-ary source symbol values that have the probabilities of occurrence that result from the Lloyd–Max quantization [164] of independent Laplacian distributed source samples. More explicitly, we consider the four-bit Lloyd–Max quantization of a Gaussian source. Note that these occurrence probabilities vary by more than an order of magnitude, between 0.0023 and 0.1616. These probabilities correspond to entropy or average information values between 2.63 bits and 8.74 bits, motivating the application of VLC and giving an overall source entropy of $E = 3.47$ bits/VLC symbol.

In the transmitter of Figure 11.1, the source symbol frame s to be transmitted is partitioned into J four-bit source symbols corresponding to $K = 16$-ary values of $\{s_j\}_{j=1}^{J} \in [1 \ldots K]$. These four-bit source symbols are then decomposed into N component streams $\{\mathbf{s}^n\}_{n=1}^{N}$ to be protected by the N different-rate IrVLC component codes, where we opted for $N = 16$ in the case of the IrVLC-FFH-MFSK scheme and $N = 1$ in the case of the VLC-based benchmark scheme. The number of symbols in the source symbol frame s that are decomposed into the source symbol frame component $\mathbf{s}^n$ is specified as J^n, where we have $J^1 = J$ in the case of the VLC-based scheme. By contrast, in the case of the IrVLC-based scheme, the specific values of $\{J^n\}_{n=1}^{N}$ may be chosen in order to shape the inverted EXIT curve of the IrVLC codec so that it does not cross the EXIT curve of the precoder, as will be detailed in Section 11.4.

Each of the N source symbol frame components $\{\mathbf{s}^n\}_{n=1}^{N}$ is VLC-encoded using the corresponding codebook from the set of N VLC codebooks $\{\mathbf{VLC}^n\}_{n=1}^{N}$, having a range of coding rates $\{R^n\}_{n=1}^{N} \in [0, 1]$ satisfying

$$\sum_{n=1}^{N} \alpha_n R^n = R, \tag{11.1}$$

where $\alpha_n = J^n/J$ is the particular fraction of the source symbol frame coded by the nth sub-code and R is the average code rate of the VLC or the IrVLC scheme. The specific source symbols having the value of $k \in [1 \ldots K]$ and encoded by the specific VLC codebook $\mathbf{VLC}^n$ are represented by the codeword $\mathbf{VLC}^{n,k}$, which has a length of $I^{n,k}$ bits. The J^n VLC codewords that represent the J^n source symbols in the source symbol frame component $\mathbf{s}^n$ are concatenated to provide the transmission frame component $\mathbf{u}^n = \{\mathbf{VLC}^{n,s_{j^n}^n}\}_{j^n=1}^{J^n}$.

Owing to the variable length of the VLC codewords, the number of bits comprised by each transmission frame component $\mathbf{u}^n$ will typically vary from frame to frame. In order to facilitate the VLC decoding of each transmission frame component $\mathbf{u}^n$, it is necessary to explicitly convey the exact number of bits $I^n = \sum_{j^n=1}^{J^n} I^{n,s_{j^n}^n}$ to the receiver with the aid of side information. Furthermore, this highly error-sensitive side information must be reliably protected against transmission errors. This may be achieved using a low-rate block code or repetition code, for example. For the sake of avoiding obfuscating details, this is not shown in Figure 11.1.

11.2.2 FFH-MFSK Modulation

The N transmission frame components $\{\mathbf{u}^n\}_{n=1}^{N}$ encoded by the different IrVLC component codes are concatenated at the transmitter, as shown in Figure 11.1. The resultant transmission frame u has a length of $\sum_{n=1}^{N} I^n$ bits. Following the bit interleaver π, the binary transmission frame u is precoded and the precoded bits are converted to M-ary symbols [347], which are transmitted by the FFH-MFSK modulator of Figure 11.1. To elaborate, in the FFH-MFSK transmitter the M-ary symbols are mapped to M frequency tones [352]. The particular MFSK tone chosen for transmission modulates a carrier generated by a frequency synthesizer, which is controlled by the L-tuple FFH sequence generated by a Pseudo-Noise (PN) generator, where L is the number of frequency hops per symbol. Hence, the transmitted FFH signal hops from one frequency to another in a pseudorandom fashion over a wide bandwidth. The signal transmitted during the ith symbol duration of T_s may be expressed as

$$s(t) = \sum_{l=0}^{L-1} \sqrt{2PP_{T_h}}(t - iT_s - lT_h) \cos\{2\pi[f_m + f_l]t + \phi\}, \tag{11.2}$$

where P is the power of the transmitted signal; $P_{T_h}(t)$ denotes a rectangular signaling waveform associated with one hop duration of $T_h = T_s/L$; $f_m, m = 0, 1, \ldots, M - 1$ is the MFSK tone's frequency and f_l is the lth frequency, $l = 0, 1, \ldots, L - 1$, in the frequency hopping sequence during the time interval $lT_h \le t < (l + 1)T_h$. Furthermore, in Equation (11.2), ϕ represents the cumulative phase induced by the FFH-MFSK modulation.

11.2.3 The Channel

The channel is assumed to be a frequency-flat Rayleigh fading medium for each of the transmitted frequencies. Furthermore, we assume that the separation between the adjacent frequencies is higher than the coherence bandwidth of the channel. The separation between adjacent frequency tones is assumed to be $R_h = 1/T_h$, which also represents the bandwidth occupied by a single FFH-MFSK tone. Therefore, all FFH tones conveying the same symbol experience independent fading.

The transmitted signal is also corrupted by Additive White Gaussian Noise (AWGN) and a PBNJ signal [341, 352], having single-sided power spectral densities of N_0 and N_J respectively. We assume that the PBNJ signal jams a fraction $0 \le \rho \le 1$ of the total spread spectrum bandwidth W_{ss}. We also assume that the PBNJ signal is contiguous and hence that all the M FSK tones of a particular band are jammed if the jamming signal is present in that band. Thus, the probability that a band or a tone is jammed is given by ρ, while the probability that the band is not jammed is $(1 - \rho)$.

11.2.4 FFH-MFSK Demodulation

The receiver schematic is also shown in Figure 11.1, where the frequency de-hopper, which is identical to and aligned with the frequency hopper of the transmitter, de-spreads the received signal by exploiting the knowledge of the transmitter's unique FFH address. The demodulator of Figure 11.1 consists of M branches, each corresponding to a single MFSK tone and consisting of a square-law detector [352] as well as a diversity combiner, which performs clipping followed by linear combining of the signals received in all hops. The process of clipping is expressed by [338, 353]

$$f(U_{ml}) = \begin{cases} C, & \text{if } U_{ml} \ge C \\ U_{ml}, & \text{otherwise} \end{cases} \tag{11.3}$$

$$m = 0, 1, \ldots, M - 1, l = 0, 1, \ldots, L - 1,$$

where U_{ml} represents the square-law detector's output seen in Figure 11.1, corresponding to the mth tone in the lth hop, and C represents an appropriately chosen clipping threshold. The decision variable recorded after clipped combining is given by

$$Z_m = \sum_{l=0}^{L-1} f(U_{ml}), \quad m = 0, 1, \ldots, M-1. \tag{11.4}$$

At the output of the diversity combiner, the corresponding symbol probabilities and Logarithmic Likelihood Ratios (LLRs) are computed, as explained in the following section, and these are then fed to the unity-rate decoder of Figure 11.1.

11.3 Iterative Decoding

In this section we discuss how soft information is derived from the channel's output observations and how ID is carried out by exchanging extrinsic information between the demodulator, the unity-rate decoder and the outer decoder seen in Figure 11.1.

11.3.1 Derivation of Soft Information

The LLR output by the demodulator corresponding to the ith bit b_i (given that $\mathbf{Z} = [Z_0, Z_1, \ldots, Z_{M-1}]$, which represents the set of M diversity combiner outputs of Figure 11.1 is received) may be expressed as [354]

$$\Lambda_p^M(i|\mathbf{Z}) = \log\left[\frac{P_p^M(b_i = 1|\mathbf{Z})}{P_p^M(b_i = 0|\mathbf{Z})}\right], \tag{11.5}$$

where $P(\cdot)$ denotes the probability of an event, the superscript M indicates the MFSK-based demodulator, and the subscript p denotes *a posteriori* probability. In order to compute the LLRs, we need the probability that the mth symbol s_m was transmitted, $m = 0, \ldots, M-1$, given that the signal $\mathbf{Z}$, which represents the set of M outputs of the diversity combiners of Figure 11.1, is received. This probability is given by

$$P_p^M(s_m|\mathbf{Z}) = \frac{p(\mathbf{Z}|s_m)P(s_m)}{p(\mathbf{Z})}, \tag{11.6}$$

where $p(\mathbf{Z}|s_m)$ is the Probability Density Function (PDF) of the received signal $\mathbf{Z}$, given that s_m is transmitted. Furthermore, $P(s_m)$ is the *a priori* probability of the symbol s_m, while $p(\mathbf{Z}) = \sum_{m=0}^{M-1} p(\mathbf{Z}|s_m)P(s_m)$ is the probability of receiving the signal set $\mathbf{Z}$, which is independent of s_m, $m = 0, \ldots, M-1$, for a given $\mathbf{Z}$. Moreover, for equiprobable symbols, we have $P(s_m) = 1/M$. Hence, the PDF $p(\mathbf{Z}|s_m)$ uniquely and unambiguously describes the statistics required for estimating the probability $P(s_m|\mathbf{Z})$. For independent fading of the M tones, the PDF $p(\mathbf{Z}|s_m)$ is given by

$$p(\mathbf{Z}|s_m) = f_{Z_m}(x_m|s_m) \prod_{n=0,n\neq m}^{M-1} f_{Z_n}(x_n|s_m), \tag{11.7}$$

where $f_{Z_n}(x_n|s_m)$ represents the PDF of the nth diversity combiner output, $n = 0, 1, \ldots,$ $M - 1$, given that s_m is transmitted.

In order to derive the PDFs of the diversity combiner outputs, we first consider the PDFs of the square-law detector outputs before diversity combining, as seen in Figure 11.1. We derive the soft information from the channel's output observations assuming a somewhat simplistic but tractable interference-free channel model. This is because we aim to derive an expression for the soft information that may be computed using the parameters that may be either known or readily estimated at the receiver. By contrast, if the soft information is derived by considering the presence of PBNJ, the desired expression will be a function of parameters such as the signal-to-jammer ratio and the PBNJ duty factor ρ, which are not known at the receiver. Moreover, although clipped combining is employed in the receiver, we perform the forthcoming analysis assuming linear combining, using no clipping at the receiver. This assumption has been stipulated in order to further simplify the analysis, and is supported by the observation that clipping is an operation performed to reduce effects of PBNJ [338]. Hence our analysis may result in suboptimal soft information, but we nevertheless demonstrate in Section 11.4 that valuable performance improvements can be achieved using this approach.

Assuming that s_m is transmitted in the lth hop, it can be readily shown that for independent Rayleigh fading the PDF of the noise-normalized square-law detector output U_{ml}, seen in Figure 11.1, may be expressed as [352]

$$f_{U_{ml}}(y_m|s_m) = \frac{1}{1 + \gamma_h} e^{-y_m/(1+\gamma_h)}, \quad y_m \geq 0, \tag{11.8}$$

where $\gamma_h = bRE_b/(N_0 L)$ is the SNR per hop, E_b is the transmitted energy per bit and $b = \log_2 M$ is the number of bits per symbol.

Similarly, for all the interfering or non-signal tones associated with $n = 0, 1, \ldots, M - 1$, $n \neq m$, the corresponding PDF is given by

$$f_{U_{nl}}(y_n|s_m) = e^{-y_n}, \quad y_n \geq 0. \tag{11.9}$$

From Equation (11.8), using the Characteristic Function approach [352][3], we can derive the PDF of the linear combiner's output Z_m seen in Figure 11.1, which can be expressed as

$$f_{Z_m}(x_m|s_m) = \frac{x_m^{L-1}}{(1 + \gamma_h)^L \Gamma(L)} e^{-x_m/(1+\gamma_h)}. \tag{11.10}$$

Similarly, for all the non-signal tones, $n = 0, 1, \ldots, M - 1$, $n \neq m$ we have

$$f_{Z_n}(x_n|s_m) = \frac{x_n^{L-1}}{\Gamma(L)} e^{-x_n}. \tag{11.11}$$

---

[3]The Characteristic Function is the Fourier transform of the PDF, and the fact that the Characteristic Function of a sum of random variables is the product of their individual Characteristic Functions [352] has been exploited here.

Inserting Equations (11.10) and (11.11) in Equation (11.7) and after further simplifications, we have

$$p(\mathbf{Z}|s_m) = \left[\frac{1}{(1+\gamma_h)^L} \frac{1}{\Gamma^M(L)} \prod_{n=0}^{M-1} x_n^{L-1} e^{-x_n} \right] \exp\left(\frac{x_m \gamma_h}{1+\gamma_h} \right). \tag{11.12}$$

We can see in Equation (11.12) that all the terms except the last exponential term are common for any of the m symbols, $m = 0, 1, \ldots, M-1$. Since the computation of the LLRs requires the logarithm of the bit probabilities, we consider the common terms as a normalization factor and express the normalized probability $p(\mathbf{Z}|s_m)$ as

$$p(\mathbf{Z}|s_m) = \exp\left(\frac{x_m \gamma_h}{1+\gamma_h} \right). \tag{11.13}$$

Upon inserting Equation (11.13) in Equation (11.6), we can derive the corresponding symbol probabilities. The resultant bit probabilities can be derived from the symbol probabilities, assuming the bits-to-symbol mapping of [347] using the relation of [344, 347]

$$P_p^M(b_i = 0|\mathbf{Z}) = \sum_{t=0}^{M/2-1} P_p^M(s_m|\mathbf{Z}), \quad m = t + \left\lfloor \frac{t}{2^i} \right\rfloor 2^i, \tag{11.14}$$

where $\lfloor x \rfloor$ represents the largest integer less than or equal to x. Finally, the LLRs Λ_p^M are computed from the bit probabilities using Equation (11.5) and after subtracting the *a priori* LLRs, the extrinsic LLRs may be fed to the decoder.

In the foregoing analysis, we noted that the knowledge of the received SNR is required to compute symbol probabilities from Equation (11.13). In noncoherent systems, especially in FFH-MFSK, accurate estimation of the channel during a hop interval that is typically a fraction of the symbol interval may not be practicable. However, the SNR may be estimated to a reasonable degree of accuracy by computing a running average of the received AWGN power over a sufficiently large number, say 20, of received samples. We assume in this contribution that the received SNR is accurately known at the receiver.

11.3.2 Two-Stage and Three-Stage Concatenated Schemes

Following our derivation of the soft information from the received signal, two types of serial concatenation schemes are possible. In the first case, depicted in Figure 11.1, the *A Posteriori* Probability (APP) SISO unity-rate decoder and the outer decoder perform ID, both invoking the Bahl–Cocke–Jelinek–Raviv (BCJR) algorithm [48] using bit-based trellises [180]. The extrinsic soft information is iteratively exchanged between the rate-1 decoder and the VLC decoding stages in the form of LLRs for the sake of assisting each other's operation, as detailed in [275]. We refer to this configuration of the ID as the *two-stage scheme*. Alternatively, the system may be modified so that the FFH-MFSK demodulator, the unity-rate inner decoder and the IrVLC outer decoder exchange their extrinsic Mutual Information (MI), as shown in Figure 11.2. We refer to this arrangement as the *three-stage scheme*. The three-stage scheme requires an additional interleaver between the precoder and the FFH-MFSK modulator of Figure 11.2.

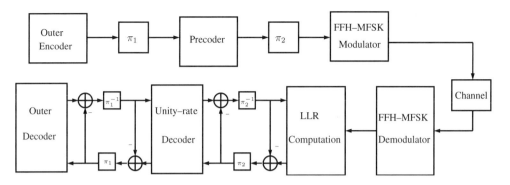

Figure 11.2: Schematic of the system of Figure 11.1, modified for a three-stage serial concatenation of the demodulator, unity-rate decoder and outer VLC/IrVLC decoder. The FFH-MFSK modulator and demodulator and the outer encoder and decoder are as shown in Figure 11.1. ©IEEE [198] Ahmed *et al.* 2009.

11.3.3 EXIT-Chart Analysis

Let us now study the MI characteristics of the FFH-MFSK demodulator using EXIT curves [346]. All MI measurements were made using the histogram-method-based approximation of the true distribution [346]. Note that in Figure 11.1, $\Lambda(\cdot)$ denotes the LLRs of the bits concerned, where the superscript i indicates the inner decoder (or demodulator), while o corresponds to outer decoding. Additionally, a subscript denotes the dedicated role of the LLRs, with a, p and e indicating *a priori*, *a posteriori* and extrinsic information, respectively. Moreover, unless otherwise stated, we employ the following parameter values: source symbol frame length of $J = 70\,000$, code rate of 0.5, MFSK modulation order of $M = 16$ and FFH diversity order of $L = 3$. Finally, we assume optimum clipping thresholds, defined in Equation (11.3), for all simulations.

In Figure 11.3, the MI transfer characteristics of the FFH-MFSK demodulator recorded in a Rayleigh fading channel are shown for various M values, assuming $L = 3$ and $E_b/N_0 = 8$ dB. In order to portray these EXIT curves in the correct perspective, the inverted EXIT curve of a half-rate Recursive Systematic Convolutional (RSC) decoder characterized by the octal generator polynomial of $(7, 5)$ is also shown. Figure 11.3 demonstrates that the FFH-MFSK demodulator yields low-gradient EXIT curves for all values of the modulation order M shown. When the SNR is sufficiently high, the EXIT curves can be shifted upward in the EXIT plane, and hence an arbitrarily low BER may be achieved. However, this would be achieved at the cost of having a large area between the demodulator's and the RSC decoder's EXIT curves, implying that the scheme operates far from capacity.

In order to generate an inner decoder EXIT curve that is more conducive to fruitful extrinsic information exchange with the outer IrVLC decoder, we employ a precoder, characterized by the generator polynomial of $1/(1 + D)$ [350], between the IrVLC encoder and the FFH-MFSK modulator, as shown in Figure 11.1. The MI transfer characteristics of the unity-rate decoder recorded for various values of precoder memory are shown in Figure 11.4. It can be seen that the precoder renders the channel to appear recursive [350], and hence has steeper EXIT curves than does the stand-alone demodulator. Furthermore, as the precoder's

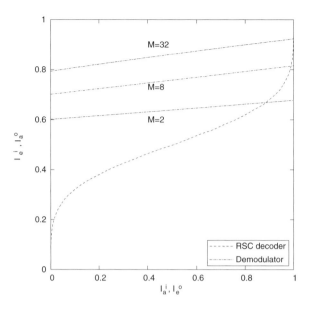

Figure 11.3: EXIT characteristics of FFH-MFSK demodulator in interference-free Rayleigh fading channel, assuming $E_b/N_0 = 8$ dB and various M values.

memory is increased, the EXIT curves become steeper. Moreover, in contrast with those of the demodulator, the rate-1 decoder's EXIT curves do indeed reach the $(I_E, I_A) = (1, 1)$ point in Figure 11.4, implying that the precoder allows the iterative decoding to converge to an arbitrarily low BER. In our subsequent analysis, we opt for a memory-3 precoder.

Let us now employ the EXIT-chart technique in order to investigate the three-stage MI information exchange amongst all the three components of the receiver; namely, the FFH-MFSK demodulator, the inner unity-rate decoder and the outer IrVLC decoder. Ideally, this would require a 3D EXIT chart depicting the evolution of the MI at the output of all the three serially concatenated components. However, a simpler and almost equally effective method of investigating this three-way MI exchange was proposed in [355], which implies treating the FFH-MFSK demodulator and the unity-rate inner decoder as a single module. This effectively allows us to analyze the three-stage concatenation as a two-stage one using 2D EXIT charts. We will refer to this module as the *combined module*. The EXIT characteristics of the combined module are depicted in Figure 11.5 for various values of the precoder memory at $E_b/N_0 = 4.8$ dB. We note that the combined module yields an EXIT curve similar to that of the unity-rate decoder seen in Figure 11.4, although at a much lower SNR value. This indicates that our three-stage concatenated scheme may yield arbitrarily low BERs at even lower SNR values. We also observe in Figure 11.5 that both precoder memory 1 and memory 2 yield equally attractive EXIT gradients, while a precoder memory higher than 2 results in an EXIT curve that starts at a very low I_E value on the y axis and consequently is unsuitable for accurate matching with the outer IrVLC decoder. Thus, for the three-stage scheme, we employ a precoder of memory 1.

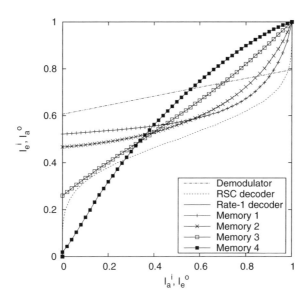

Figure 11.4: EXIT characteristics of the rate-1 decoder in interference-free Rayleigh fading channel, assuming $E_b/N_0 = 6.5$dB and various values of the precoder memory. ©IEEE [198] Ahmed *et al.* 2009.

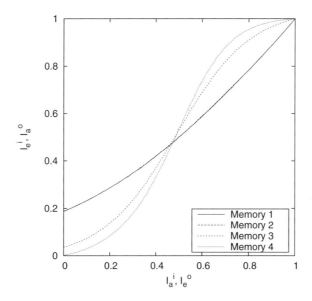

Figure 11.5: EXIT characteristics of the combined module in interference-free Rayleigh fading channel, assuming $E_b/N_0 = 4.8$ dB and various values of the precoder memory.

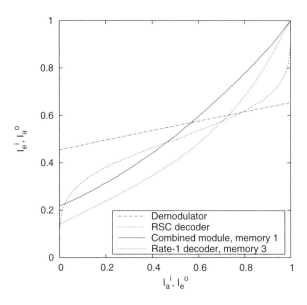

Figure 11.6: EXIT characteristics of the demodulator, the rate-1 decoder and the combined module in Rayleigh fading channel contaminated by PBNJ, assuming $E_b/N_0 = 6$ dB, $E_b/N_J = 10$ dB and $\rho = 0.1$.

In Figure 11.6 we further compare the EXIT characteristics of the three types of inner modules considered, namely the stand-alone FFH-MFSK demodulator, the stand-alone rate-1 decoder and the combined module of demodulator plus the rate-1 decoder, in a channel contaminated by PBNJ. In this figure, we have assumed $E_b/N_0 = 6$ dB, $E_b/N_J = 10$ dB and a jamming duty factor of $\rho = 0.1$. We observe that the combined module retains its superiority in this type of channel as well, and, although its EXIT curve has a similar gradient to that of the rate-1 decoder, its EXIT curve emerges from a higher point in the EXIT plane, indicating that the combined module would allow satisfactory communication at lower SNR values. The superiority of the combined module over the stand-alone rate-1 decoder stems from the fact that when employing the combined module extrinsic information is exchanged between the two SISO modules, i.e. the demodulator and the rate-1 decoder, thus enabling optimum exploitation of the soft information.

Note that our combined module invokes a single iteration of extrinsic information exchange between the demodulator and the unity-rate decoder. We have also investigated the use of more than one iteration between these two blocks, but it was found that no additional benefit is achieved. Hence, when the three-stage scheme of Figure 11.2 is employed, each ID iteration involves one iteration between the demodulator and the unity-rate decoder followed by one iteration between the unity-rate inner decoder and the outer decoder.

11.3.4 VLC Decoding

Since N separate VLC encoders are employed in the IrVLC-FFH-MFSK transmitter, N separate VLC decoders are employed in the corresponding receiver seen in Figure 11.1.

In parallel with the composition of the bit-based transmission frame $\mathbf{u}$ from N VLC components, the *a priori* LLRs $\Lambda_a^o(\mathbf{u})$ are decomposed into N components, as shown in Figure 11.1. This is achieved with the aid of the explicit side information that we assume for conveying the number of bits I^n in each transmission frame component $\mathbf{u^n}$. Each of the N VLC decoders is provided with the *a priori* LLR sub-frame $\Lambda_a^o(\mathbf{u^n})$ and in response it generates the *a posteriori* LLR sub-frame $\Lambda_p^o(\mathbf{u^n})$, $n \in [1 \ldots N]$. These *a posteriori* LLR sub-frames are concatenated in order to provide the *a posteriori* LLR frame $\Lambda_p^o(\mathbf{u})$, as shown in Figure 11.1.

During the final decoding iteration, N bit-based MAP VLC sequence estimation processes are invoked instead of single-class APP SISO VLC decoding, as shown in Figure 11.1. In this case, each transmission frame component $\mathbf{u^n}$ is estimated from the corresponding *a priori* LLR frame component $\Lambda_a^o(\mathbf{u^n})$. The resultant transmission frame component estimates $\tilde{\mathbf{u}}^n$ of Figure 11.1 may be concatenated to provide the transmission frame estimate $\tilde{\mathbf{u}}$. Additionally, the transmission frame component estimates $\tilde{\mathbf{u}}^n$ may be VLC decoded to provide the source symbol frame component estimates $\tilde{\mathbf{s}}^n$. Since this process cannot exploit the knowledge that the source symbol frame component $\mathbf{s}^n$ comprises J^n source symbols, the resultant source symbol frame component estimate $\tilde{\mathbf{s}}^n$ may not contain J^n source symbols. For the sake of preventing the resultant loss of synchronization that this would imply, source symbol estimates are appropriately removed from, or appended to the end of, each source symbol component estimate $\tilde{\mathbf{s}}^n$, so that they each comprise the required J^n number of source symbol estimates. These adjusted source symbol component estimates $\tilde{\mathbf{s}}^n$ may be concatenated to generate the source symbol frame estimate $\tilde{\mathbf{s}}$, as shown in Figure 11.1.

In the next section we detail the design of our IrVLC scheme and characterize the IrVLC- and VLC-based schemes with the aid of EXIT-chart analysis.

11.4   System Parameter Design and Results

As described in Section 11.2, we opted for employing $N = 16$ component VLC codebooks $\{\mathbf{VLC}^n\}_{n=1}^N$ having approximately equally spaced coding rates in the range of $[0.2, 0.95]$ in the IrVLC-FFH-MFSK schemes of Figures 11.1 and 11.2. In each case, we employ a Variable-Length Error-Correcting (VLEC) codebook [179] that is tailored to the source symbol values' probabilities of occurrence described in Section 11.2. By contrast, in the VLC scheme we employ just $N = 1$ VLC codebook, which is identical to the VLC codebook $\mathbf{VLC}^{10}$ of the IrVLC scheme, having a coding rate of $R = 0.5$. Note that this coding rate results in an average interleaver length of $J \cdot E/R$ bits.

In Figure 11.7 we provide the inverted EXIT curves that characterize the bit-based APP SISO VLC decoding of the above-mentioned VLC codebooks, together with the rate-1 decoder's EXIT curves at E_b/N_0 values of 6.6 and 6.9 dB, assuming an uncorrelated Rayleigh fading channel contaminated by PBNJ characterized by $E_b/N_J = 10$ dB and $\rho = 0.1$. All the EXIT curves were generated using uncorrelated Gaussian distributed *a priori* LLRs, based on the assumption that the transmission frame's bits have equiprobable logical values. This is justified because we employ a long interleaver length and because the entropy of the VLC encoded bits was found to be only negligibly different from unity for all the VLC codebooks considered.

Figure 11.7 also shows the inverted EXIT curve of the IrVLC scheme. This was obtained as the appropriately weighted superposition of the $N = 16$ component VLC codebooks'

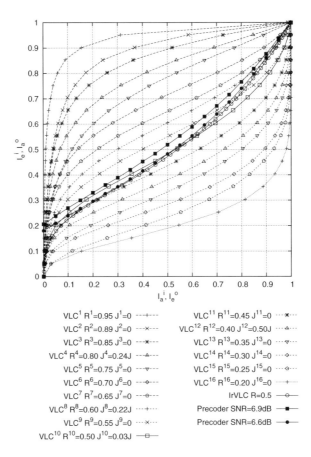

Figure 11.7: Inverted VLC EXIT curves and rate-1 decoder EXIT curves, assuming $E_b/N_J = 10$ dB and $\rho = 0.1$, in uncorrelated Rayleigh fading channel. ©IEEE [198] Ahmed *et al.* 2009.

inverted EXIT curves, where the weight applied to the inverted EXIT curve of the component VLC codebook **VLC^n** is proportional to the specific number of source symbols employed for encoding J^n [149]. Using the approach of [149], the values of $\{J^n\}_{n=1}^{N}$ given in Figure 11.7 were designed so that the IrVLC coding rate matches that of our regular VLC scheme, namely 0.5. Furthermore, we ensured that the inverted IrVLC EXIT curve did not cross the rate-1 decoder's EXIT curve at E_b/N_0 of 6.6 dB. We note that only four of the 16 VLC components were indeed activated by the algorithm of [149] in order to encode a nonzero number of source symbols. As shown in Figure 11.7, the presence of the resultant open EXIT-chart tunnel implies that an infinitesimally low BER may be achieved by the IrVLC-FFH-MFSK scheme for E_b/N_0 values above 6.6 dB. In contrast, an open EXIT-chart tunnel is not afforded for E_b/N_0 values below 6.9 dB in the case of the benchmark VLC-based scheme.

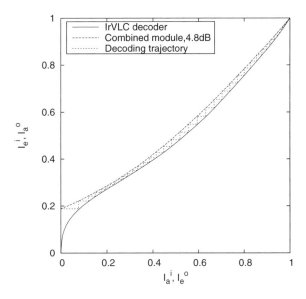

Figure 11.8: EXIT curves of the IrVLC decoder and the combined module, and decoding trajectory, assuming an interference-free, uncorrelated Rayleigh fading channel.

Analogously to the IrVLC design of Figure 11.7, we have also designed IrVLC codes for both the two-stage and the three-stage ID schemes, assuming various jamming scenarios in Rayleigh fading channels. The design parameter values are listed in Table 11.1, where the relevant minimum values of E_b/N_0 or E_b/N_J at which an open convergence tunnel is formed are shown alongside the specific fractions of the source symbol frame encoded by each component code of the IrVLC scheme given by $\alpha_n = J_n/J$. For both the IrVLC and the VLC schemes, theoretical as well as the actual values are shown. The theoretical values imply those predicted by the EXIT-chart analysis, while the actual values are those achieved in the Monte Carlo simulations.

The validity of our EXIT-chart analysis was verified by carrying out the ID process of Figure 11.1 and recording the MI values at the output of the inner and outer decoders, as seen in Figure 11.8. We observe that the ID trajectory more or less follows the EXIT curves of the inner and outer IrVLC decoders.

Let us now focus our attention on the BER performance of the proposed system as well as on that of the benchmark in the context of the two-stage and the three-stage schemes considered. Note that the BER depicted in Figures 11.9 to 11.10 corresponds to the encoded bits of the transmission frame $\mathbf{u}$ and the corresponding received frame $\tilde{\mathbf{u}}$. The performance predictions of Figure 11.7 are verified for the two-stage scheme by the BER versus E_b/N_0 results of Figure 11.9, where we assume $E_b/N_J = 10$ dB and $\rho = 0.1$ when the transmitted signal encounters PBNJ. Also shown is the BER performance of the VLC scheme dispensing with the precoder and thus invoking ID between the demodulator and the VLC decoder. We observe that both the precoded arrangements, namely the VLC- and the IrVLC-based schemes, yield significantly superior performance compared with the system

Table 11.1: List of parameter values for the IrVLC codes used in the two-stage as well as the three-stage concatenated schemes under various channel conditions, with the code rates of the IrVLC's component codes given by [0.95 0.89 0.85 0.8 0.75 0.7 0.65 0.6 0.55 0.5 0.45 0.4 0.35 0.3 0.25 0.2]

| Channel | Scheme | Minimum value at which convergence is achieved | | | | $\alpha_n = J_n/J$ values |
| | | IrVLC | | VLC | | |
		Theory	Actual	Theory	Actual	
Interference-free	2-stage	$\frac{E_b}{N_0} = 5.7$ dB	$\frac{E_b}{N_0} = 5.7$ dB	$\frac{E_b}{N_0} = 5.9$ dB	$\frac{E_b}{N_0} = 5.9$ dB	[0,0,0,0.18,0,0.12,0, 0.18,0,0,0,0.52,0,0,0,0]
	3-stage	$\frac{E_b}{N_0} = 4.7$ dB	$\frac{E_b}{N_0} = 4.8$ dB	$\frac{E_b}{N_0} = 5.0$ dB	$\frac{E_b}{N_0} = 5.1$ dB	[0,0,0,0.26,0,0.24,0,0, 0,0,0.34,0.13,0.03,0,0]
$\frac{E_b}{N_J} = 10$ dB $\rho = 0.1$	2-stage	$\frac{E_b}{N_0} = 6.6$ dB	$\frac{E_b}{N_0} = 6.7$ dB	$\frac{E_b}{N_0} = 6.9$ dB	$\frac{E_b}{N_0} = 7$ dB	[0,0,0.24,0,0,0.22, 0,0.03,0,0.5,0,0,0]
	3-stage	$\frac{E_b}{N_0} = 5.8$ dB	$\frac{E_b}{N_0} = 5.9$ dB	$\frac{E_b}{N_0} = 6.1$ dB	$\frac{E_b}{N_0} = 6.2$ dB	[0,0,0,0.18,0,0.27,0,0, 0,0,0.55,0,0,0]
$\frac{E_b}{N_J} = 10$ dB $\rho = 0.5$	2-stage	$\frac{E_b}{N_J} = 9.4$ dB	$\frac{E_b}{N_J} = 10.4$ dB	$\frac{E_b}{N_J} = 9.8$ dB	$\frac{E_b}{N_J} = 10.7$ dB	[0,0,0,0.08,0,0.16,0.11, 0,0.5,0,0,0,0]
	3-stage	$\frac{E_b}{N_J} = 6.8$ dB	$\frac{E_b}{N_J} = 7.5$ dB	$\frac{E_b}{N_J} = 7.9$ dB	$\frac{E_b}{N_J} = 8.7$ dB	[0,0,0,0.13,0,0.32,0,0, 0,0,0.55,0,0,0]

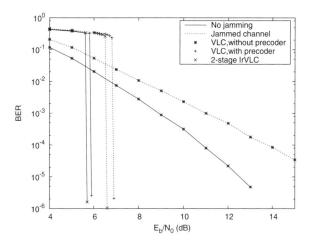

Figure 11.9: BER versus SNR performance of the two-stage VLC- and IrVLC-based schemes in both jammed and interference-free uncorrelated Rayleigh fading channels. $E_b/N_J = 10$ dB and $\rho = 0.1$ are assumed in the jammed channel.

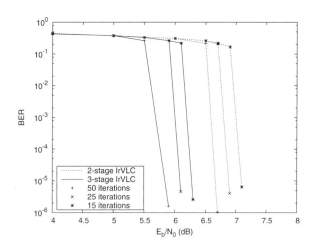

Figure 11.10: BER versus SNR performance of the two-stage and three-stage IrVLC-based schemes in jammed, uncorrelated Rayleigh fading channels for various numbers of decoding iterations, assuming $E_b/N_J = 10$ dB and $\rho = 0.1$. ©IEEE [198] Ahmed *et al.* 2009.

operating without the precoder, attaining a E_b/N_0 gain of approximately 7 dB at a BER of 10^{-4} in the jammed channel. Furthermore, we observe that the IrVLC scheme achieves an arbitrarily low BER at E_b/N_0 values in excess of 5.7 dB in interference-free channels and at 6.6 dB in the channels contaminated by PBNJ, as predicted by the EXIT-chart of Figure 11.7. The corresponding E_b/N_0 values for the VLC-based scheme are 5.9 and 6.9 dB respectively.

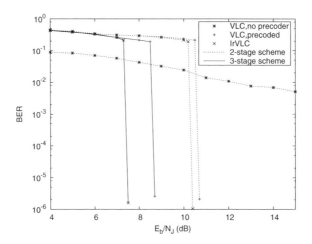

Figure 11.11: BER versus signal-to-jammer ratio performance of the two-stage and three-stage VLC- and IrVLC-based schemes in jammed, uncorrelated Rayleigh fading channels, assuming $E_b/N_0 = 10$ dB and $\rho = 0.5$. ©IEEE [198] Ahmed *et al.* 2009.

In Figure 11.11 we provide the BER versus E_b/N_J performance comparison of both the two-stage and the three-stage schemes, assuming $E_b/N_0 = 10$ dB and $\rho = 0.5$. We observe that both the VLC- and the IrVLC-based schemes result in superior performance compared with the system operating without the precoder, which encounters an error floor. We also note that the three-stage IrVLC scheme yields a further improvement of nearly 1.1 dB over the two-stage IrVLC. This performance gain confirms the EXIT-chart prediction of Figure 11.6. However, the performance gain achieved by employing three-stage concatenation comes at the cost of an increased complexity, since an additional interleaver is required in the three-stage scheme as shown in Figure 11.2. Moreover, each decoding iteration of the three-stage scheme involves the additional extrinsic information exchange with the demodulator as well.

An increased complexity is also imposed by the increased number of decoding iterations, which is a natural consequence of operating at a lower SNR. This is especially true in the case of the IrVLC scheme, where typically a narrower EXIT tunnel exists between the EXIT curves of the outer and the inner decoders, as seen in Figure 11.7. In Figures 11.9 and 11.11, up to 80 iterations are needed for achieving convergence to $(I_E, I_A) = (1, 1)$ when employing the IrVLC-based scheme. The effect of reducing the number of iterations on both the two-stage and the three-stage schemes is depicted in Figure 11.10, where the BER versus E_b/N_0 performance of the IrVLC schemes considered is shown. We observe that the number of decoding iterations can be significantly reduced by raising the E_b/N_0 value by as little as 0.2 to 0.4 dB.

Finally, we observe in Figures 11.9 to 11.10 and Table 11.1 that the actual minimum value of E_b/N_0 or E_b/N_J at which convergence is achieved is different from the theoretical prediction when jamming is stronger. Note that under Rayleigh fading the PBNJ becomes more detrimental as its duty factor ρ approaches unity [356, 357]. This also explains the error floor suffered by the system without precoder, as seen in Figure 11.11. Thus, as seen in

Table 11.1, when $\rho = 0.5$ is assumed the actual and theoretical convergence threshold values differ by up to 1 dB. The reason for this observation is that the PBNJ induces correlation in the received signal, which cannot be entirely eliminated even by using very large interleaver lengths. Thus, as the jamming gets more detrimental, so does the effect of correlation.

11.5   Conclusion

We have investigated the serial concatenation of IrvLC coding with a FFH-MFSK modem operating in a Rayleigh fading channel, when the transmitted signal was corrupted by PBNJ. Using EXIT-chart analysis, we found that a stand-alone FFH-MFSK demodulator does not promise significant iterative gains, motivating us to employ a unity-rate precoder for the sake of making the channel appear recursive. Furthermore, the IrVLC code was designed in such a way that the inverted EXIT curve of the IrVLC decoder matches the EXIT curve of the inner decoder. In this way, an open EXIT-chart tunnel may be created even at low SNR values, providing source-correlation-dependent additional performance gains of up to 1.1 dB over the VLC-based scheme. Since the employment of the VLC involves non-identical occurrence probabilities for the source symbols, it is not possible to provide a comparison of the proposed scheme with the state of the art in the context of coded FFH-MFSK, which traditionally employs equiprobable source symbols or bits. However, we have provided a comparison of the IrVLC scheme with a VLC scheme dispensing with the precoder; consequently we noted that the precoder-aided schemes yield a E_b/N_0 gain in excess of 7 dB over the system dispensing with the precoder, which suffers from an error floor when jamming is severe.

Moreover, we demonstrated that the three-stage concatenation involving the demodulator, the rate-1 decoder and the outer IrVLC decoder yields superior performance compared with the two-stage concatenation of the rate-1 decoder and the outer decoder. However, the three-stage scheme imposes a high complexity owing to the use of an additional interleaver, and because a three-stage decoding iteration entails higher complexity. In contrast, we found that a precoder of memory 1 is more suitable for the three-stage IrVLC scheme, while the two-stage scheme requires a precoder of memory 3, and thus its decoder imposes a higher complexity.

In conclusion, the precoder-aided FFH-MFSK-VLC scheme constitutes a moderate-complexity design option, which can be employed in systems communicating through channels contaminated by PBNJ for transmission of joint source- and channel-encoded audio or video signals. If a higher complexity can be afforded, the IrVLC-based scheme offers additional performance improvements.

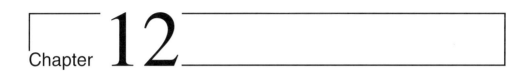

Chapter 12

Conclusions and Future Research

12.1 Chapter 1: Introduction

This chapter constituted the general background of our studies throughout the book. More specifically, a brief overview of the literature of source encoding and soft source decoding was presented in Section 1.1.1. Then the development of iterative decoding techniques and their convergence analysis was described in Section 1.1.3. Furthermore, as a special case of iterative decoding, joint source–channel decoding was introduced and the main contributions to the open literature were summarized in Section 1.1.2. Finally, the organization of the book was described in Section 1.4, while our novel contributions were highlighted in Section 1.5.

12.2 Chapter 2: Information Theory Basics

In this chapter we focussed our attention on the basic Shannonian information transmission scheme and highlighted the differences between Shannon's theory, valid for ideal source and channel codecs as well as for Gaussian channels, and its ramifications for Rayleigh channels. We also argued that practical finite-delay source codecs cannot operate at transmission rates as low as the entropy of the source. However, these codecs do not have to operate losslessly, since perceptually unobjectionable distortions can be tolerated. This allows us to reduce the associated bit rate.

Since wireless channels exhibit bursty error statistics, the error bursts can only be randomized with the aid of infinite-length channel interleavers, which are not appropriate for real-time interactive multimedia communications. Although with the advent of high-delay turbo channel codecs it is possible to operate near the Shannonian performance limits over Gaussian channels, over bursty and dispersive channels different information-theoretical channel capacity limits apply.

We considered the entropy of information sources both with and without memory and highlighted a number of algorithms, such as the Shannon–Fano, the Huffman and run-length coding algorithms, designed for the efficient encoding of sources exhibiting memory. This was followed by considering the transmission of information over noise-contaminated channels, leading to Shannon's channel coding theorem. Our discussions continued by

Near-Capacity Variable-Length Coding   Lajos Hanzo, Robert G. Maunder, Jin Wang and Lie-Liang Yang
© 2011 John Wiley & Sons, Ltd

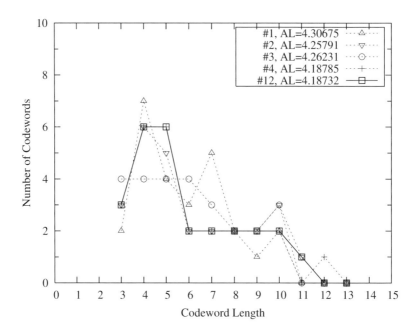

Figure 12.1: Evolution of the RVLC codeword length histograms. The RVLC is designed for the English alphabet, and its detailed construction process is described in Section 3.3.4.2. The codeword length distribution is optimized via a number of iterations for the sake of reducing the average codeword length.

considering the capacity of communications channels in the context of the Shannon–Hartley law. The chapter was concluded by considering the ramifications of Shannon's messages for wireless channels.

12.3   Chapter 3: Sources and Source Codes

Chapter 3 commenced with the description of general source models, among which a memoryless source model having a known finite alphabet, such as that described in Section 3.2, was used throughout the monograph. Then various source codes such as Huffman codes, RVLCs and VLEC codes were introduced in Section 3.3, along with their construction methods. An important contribution of this chapter is that a generic algorithm was presented for the construction of efficient RVLCs and VLEC codes. The philosophy of our proposed algorithm is that we first construct an initial RVLC or VLEC code using existing methods such as those described in [21, 23, 30], and then we optimize the codeword length distribution of the resultant code length by length. For example, Figure 12.1 shows the evolution of the codeword length histograms of the RVLC designed for the English alphabet in Section 3.3.4.2. After 12 iterations of optimization, the best codeword length distribution is found, resulting in a RVLC having the lowest average codeword length of $AL = 4.18732$.

Consequently, as shown in Tables 3.3, 3.4 and 3.5, for a variety of memoryless sources the proposed algorithm was capable of generating RVLCs of higher code efficiency and/or shorter maximum codeword length than the algorithms previously disseminated in the literature. Furthermore, as seen from Tables 3.9 and 3.10, the proposed algorithm was also capable of constructing VLEC codes having similar code efficiency to those generated by the existing algorithm [29], but incurring a significantly lower complexity.

In Section 3.4 various VLC decoding methods were presented. First, the source information, such as the number of bits or symbols in the transmitted frames, and the constraints imposed by a source code and formulated in terms of the corresponding codebook, were translated into a trellis representation, such as the symbol-based trellis described in Section 3.4.1.1 or the bit-based trellis described in Section 3.4.1.1. Then MAP/ML sequence estimation or MAP decoding could be performed, as introduced in Sections 3.4.2.1 and 3.4.2.2 respectively. It was shown in Section 3.4 that trellis-based soft decoding provides an effective way of capitalizing on the available information as much as possible. In general, the more information is utilized, the better the performance. This information can be explicit, such as the transmission frame length information, or implicit, such as the code constraint of a VLC. For example, soft decoding generally outperforms hard decoding, and the symbol-level trellis-based decoding outperforms the bit-level trellis-based decoding. Furthermore, as expected, VLCs having higher free distances outperform VLCs having lower free distances at the price of a reduced system throughput. Figure 12.2 provides some quantitative results, summarizing the conclusions of Section 3.4. It can be seen from Figure 12.2a that soft-decision decoding significantly outperforms hard-decision decoding and the attainable E_b/N_0 gain improves upon increasing the VLC's free distance. Moreover, as seen from Figure 12.2b, the performance of soft ML decoding improves upon increasing the free distance of the VLC used.

12.4   Chapter 4: Iterative Source–Channel Decoding

Chapter 4 provided an investigation of iterative source–channel decoding techniques. In this chapter the source code, the channel code and the ISI channel were viewed as a serially concatenated system. Hence, iterative decoding could be performed, provided that the source decoder, the channel decoder and the channel equalizer designed for the ISI channel were all SISO modules.

This chapter commenced with an overview of various concatenated schemes, as described in Section 4.1. Then a SISO APP decoding algorithm was introduced in Section 4.2.1. This algorithm provides a general description of any trellis-based APP decoding/detection scheme, which can be applied to source decoding, channel decoding and channel equalization. Hence it constitutes the core module of iterative decoding schemes.

EXIT charts were introduced in Section 4.2.2. The mutual information between the data bits at the transmitter and the soft values at the receiver was used for characterizing the decoding behavior of a SISO APP module, resulting in the so-called EXIT functions. A histogram-based algorithm and its simplified version were introduced in Section 4.2.2.3 in order to evaluate the EXIT functions of a SISO APP module, followed by several examples of typical EXIT functions of SISO APP modules embedded in different positions of a concatenation scheme.

Given the EXIT characteristics of the constituent modules of a concatenated scheme, we may either predict or explain its convergence behavior. This was carried out for iterative source–channel decoding for transmission over non-dispersive AWGN channels

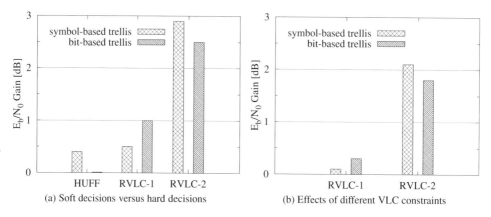

(a) Soft decisions versus hard decisions

(b) Effects of different VLC constraints

Figure 12.2: Comparison of the various VLC decoding schemes investigated in Section 3.4. Figure 12.2a compares the performance of soft-decision- and hard-decision-decoding-based schemes, where the E_b/N_0 gain is defined as the difference of the minimum E_b/N_0 values required for achieving a SER of 10^{-5} for transmission over AWGN channels, when using ML decoding. Figure 12.2b demonstrates the effects of different VLC free distances, $d_f = 1$ (RVLC-1) and $d_f = 2$ (RVLC-2). The Huffman code (HUFF)-based scheme is used as a benchmark, where the E_b/N_0 gain is defined as the difference of the minimum E_b/N_0 values required for achieving a SER of 10^{-5} for transmission over AWGN channels when using soft ML decoding.

in Section 4.3 and for transmission over dispersive AWGN channels in Section 4.4. In the scenario of non-dispersive AWGN channels, it was shown in Figure 4.15 that the free distance of the source code has to be larger than $d_f = 2$ in order that the iterative decoding scheme becomes capable of converging to the perfect mutual-information point of $(1, 1)$, which implies attaining infinitesimally low SERs. Furthermore, it was shown in Figures 4.17–4.24 that given a specific channel code, the system's convergence threshold decreases upon increasing the free distance of the source code, resulting in an improved SER performance. Figure 12.3 serves as a summary of our main results provided in Section 4.3. It is worth noting that when the free distance of the VLC code is increased from $d_f = 1$ to $d_f = 2$, i.e. when using the code RVLC-2 instead of the code HUFF or RVLC-1, the system's throughput is only slightly decreased, but a significant E_b/N_0 gain is attained. Further increasing the free distance will continue to increase the attainable E_b/N_0 gain, while incurring a considerable loss of throughput.

In the scenario of dispersive channels, it was shown by both our EXIT-chart analysis and our Monte Carlo simulations provided in Section 4.4.2 that the redundancy in the source codes is capable of effectively eliminating the ISI imposed by the channel, provided that channel equalization and source decoding are performed jointly and iteratively. Furthermore, the higher the free distance of the source code, the closer the SER performance approaches the SER bound of non-dispersive AWGN channels.

Additionally, in Section 4.4.3 precoding was shown to be an effective way of 'modifying' the EXIT characteristic of a channel equalizer. Most importantly, in conjunction with

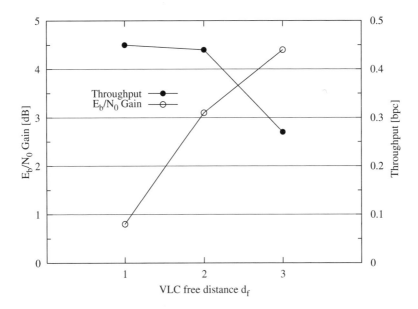

Figure 12.3: Free distance versus E_b/N_0 gain and throughput, where the E_b/N_0 gain is based on the minimum SNR value required for achieving a SER of 10^{-4} for transmission over AWGN channels, and the scheme using the Huffman code (HUFF) is used as a benchmark. The system model is described in Figure 4.14, where the transmitter is constituted by a VLC encoder and a convolutional encoder, and the receiver is constituted by an APP convolutional decoder as well as an APP VLC decoder, which performs channel decoding and source decoding iteratively.

precoding the EXIT function of a channel equalizer becomes capable of reaching the point of $(I_A = 1,\ I_E = 1)$ as shown in Figure 4.35, which is critical for avoiding potential error floors at the receiver's output. It was demonstrated in Figures 4.36–4.39 that the choice of the precoder depends both on the EXIT characteristics of the channel equalizer and those of the source decoder, so that these two are matched to each other, hence achieving the lowest possible E_b/N_0 convergence threshold.

Figure 12.4 summarizes the main results of Section 4.4. It can be seen from Figure 12.4 that the SER performance both of the scheme using RVLC-2 and of that using VLEC-3 can be improved when using appropriate precoders. However, although the precoder of $1 + D^2$ is optimal for the scheme using RVLC-2, the precoder of $1 + D$ constitutes a better choice for the scheme using VLEC-3.

Finally, the performance of a three-stage iterative receiver was evaluated in Section 4.4.4. The receiver of Figure 4.41 consists of a channel equalizer, a channel decoder and a source decoder, where the extrinsic information is exchanged among all the three SISO modules, which hence constitutes a joint source–channel decoding and equalization scheme. It was shown in Figure 4.42 that by exploiting the source redundancy in the iterative decoding process, the system's performance was improved by 2 dB in terms of the E_b/N_0 values

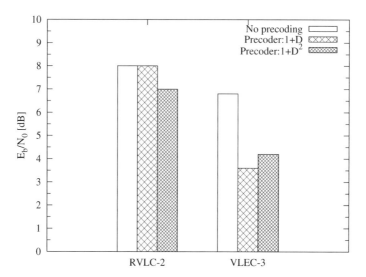

Figure 12.4: The effects of precoding and those of VLC free distances of $d_f = 2$ for the RVLC-2 and $d_f = 3$ for the VLEC-3 schemes on the attainable SER performance, when communicating over dispersive AWGN channels, where the E_b/N_0 value is the minimum SNR value required for achieving a SER of 10^{-4}. The system model is described in Figures 4.28 and 4.29, where the transmitter is constituted by a VLC encoder as well as a precoder if precoding is employed, and the receiver is constituted by an APP channel equalizer as well as an APP VLC decoder, which performs channel equalization and source decoding iteratively.

required for achieving the same SER, when compared with the separate source–channel decoding scheme. The convergence behavior of this scheme was analyzed using EXIT charts in Section 5.7 after we introduced the convergence analysis technique for multi-stage concatenated schemes in Chapter 5.

12.5 Chapter 5: Three-Stage Serially Concatenated Turbo Equalization

Chapter 5 investigated the design of the three-stage serially concatenated turbo MMSE equalization scheme seen in Figure 5.4, which consisted of an inner channel equalizer, a unity-rate recursive intermediate channel code and an outer channel code. Firstly, a brief introduction to SISO MMSE equalization was offered in Section 5.2, followed by an example of conventional two-stage turbo equalization in Section 5.3. The main body of this chapter focussed on the optimization of three-stage turbo equalization schemes by using EXIT-chart analysis.

With the aid of the EXIT modules as proposed in Figure 4.10 of Section 4.2.2, 3D EXIT-chart analysis may be simplified to 2D EXIT analysis as shown in Figures 5.7, 5.8 and 5.9 of Section 5.4.2. It was also shown in Figure 5.8 of Section 5.4.2.1 that, by employing a unity-rate recursive convolutional code as the intermediate constituent code, the three-stage scheme becomes capable of converging to the perfect mutual information point.

Moreover, the outer constituent code was optimized in Section 5.4.2.2 for achieving the lowest possible E_b/N_0 convergence threshold. Interestingly, it was observed in Figure 5.8 that relatively weak codes having short memories resulted in a lower convergence threshold than did strong codes having long memories.

Additionally, the activation order of the component decoders was optimized in Section 5.4.2.3 for achieving the convergence at the lowest possible E_b/N_0 value, while maintaining a low decoding complexity. It was found in Table 5.2 that by invoking the outer and intermediate decoder of Figure 5.4 more frequently the total number of decoder activations is reduced, resulting in a decreased decoding complexity.

The BER performance of the optimized scheme was evaluated in Section 5.4.3, which verified the EXIT-chart analysis provided in Section 5.4.2. The iterative decoding process was visualized using both 3D and 2D EXIT charts as shown in Figures 5.12–5.15 of Section 5.4.4. Furthermore, the effects of different interleaver block lengths were discussed in Figure 5.16 of Section 5.4.5. Generally, the longer the interleaver length, the more closely the simulated performance matches the EXIT-chart analysis. It was found in Figure 5.16 that an interleaver length of the order of 10^5 bits is sufficiently high for achieving a good match with the decoding trajectory recorded. Figure 12.5 provided some quantitative results summarized from Section 5.4.5. It can be seen from Figure 12.5 that when the interleaver depth is increased from $L = 10^3$ bits to $L = 10^4$ bits, a significant coding gain may be attained. Further increasing the interleaver depth to $L = 10^5$ bits, however, results in a marginal increase of the coding gain. Naturally, the attainable iteration gain is increased upon increasing the interleaver depth.

In Section 5.5 the maximum achievable information rate of the three-stage turbo equalization scheme of Figure 5.4 was analyzed. Then an IRCC was invoked as the outer constituent code, whose EXIT function was optimized for matching that of the combined module of the inner channel equalizer and the intermediate channel decoder, so that the EXIT-tunnel area between these two EXIT functions was minimized. The Monte Carlo simulation results provided in Figure 5.23 of Section 5.5.4 show that the performance of the resultant scheme is only 0.5 dB away from the channel capacity.

Finally, the employment of non-unity-rate intermediate codes was also considered in Section 5.6. It was shown in Figure 5.27 that, as expected, the maximum achievable information rate of such schemes was reduced in comparison with that of the schemes using unity-rate intermediate codes. In contrast, the E_b/N_0 convergence threshold may be decreased when only regular convolutional codes are used. A number of optimized serially concatenated codes were obtained, and listed in Table 5.6.

As a summary, Figure 12.6 compares the distance to capacity for the various MMSE turbo equalization schemes discussed in Chapter 5.

In Part II of this book we introduced the novel concept of Irregular Variable Length Coding (IrVLC) and investigated its applications, characteristics and performance in the context of wireless telecommunications. As discussed throughout Part II, IrVLCs encode various components of the source signal with different sets of binary codewords, having a range of appropriately selected lengths. Three particular applications of IrVLCs were investigated in this book, namely joint source and channel coding, EXtrinsic Information

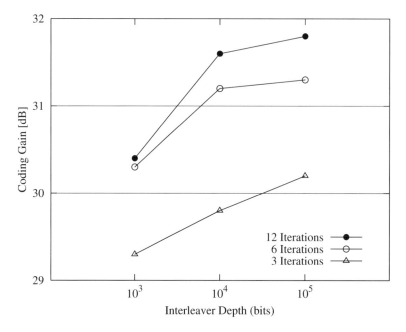

Figure 12.5: Achievable coding gains at a BER of 10^{-4} for the three-stage turbo equalization scheme of Figure 5.10 using different interleaver depths. The turbo equalization scheme is constituted by a RSC$(2, 1, 2)$ code as the outer code, a unity-rate RSC$(1, 1, 2)$ code as the intermediate code and an inner MMSE equalizer as described in Section 5.4.

Transfer (EXIT)-chart matching and Unequal Error Protection (UEP). These are detailed in the following sections, together with a discussion of our future work.

12.6   Chapter 6: Joint Source and Channel Coding

In Chapter 6 we exemplified the application of IrVLCs for the joint source and channel coding of video information. This application was motivated by the observation that Shannon's source and channel coding separation theorem [133] is invalid in the context of practical video transmission. While source and channel coding can be performed in isolation without imposing any performance loss, if the assumptions discussed in Section 6.1 apply, these conditions are not fulfilled in the case of practical video transmission. We therefore proposed the novel joint source and channel coding scheme of Section 6.2, which employs both Variable-Dimension Vector Quantization (VDVQ) [290] as a special case of Vector Quantization (VQ) [170] and the Reversible Variable-Length Code (RVLC) [276] class of Variable-Length Codes (VLCs).

Here, the employment of VDVQ tiles having a range of dimensions facilitates the efficient representation of both large areas of the video frame that have a low luminance-variance and small areas of high variance, as exemplified in Figure 6.5. Additionally, the employment of RVLC codewords having various lengths facilitates the representation of more frequently occurring VDVQ tiles with the aid of shorter codewords, giving a reduced average

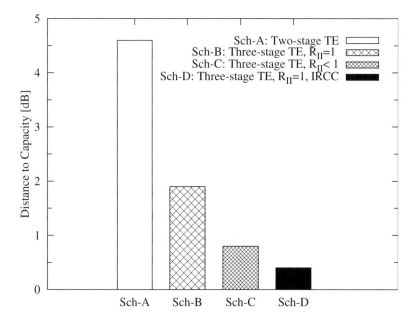

Figure 12.6: Distance to capacity for the various MMSE turbo equalization schemes of Chapter 5, where Sch-A represents the conventional two-stage turbo equalization scheme of Figure 5.1. Sch-B, Sch-C and Sch-D denote the same three-stage turbo equalization scheme of Figure 5.4, but differ in the channel codes used. Sch-B employs a unity-rate $RSC(1, 1, 2)$ code as the intermediate code and a $RSC(2, 1, 2)$ code as the outer code. Sch-C uses a SCC of SCC-A2, described in Table 5.6, which is constituted by a rate-3/4 $RSC(3, 4, 2)$ code as the intermediate code and a $RSC(2, 3, 3)$ code as the outer code. Sch-D employs the same unity-rate $RSC(1, 1, 2)$ code used in Sch-B as the intermediate code, while using the IRCC described in Section 5.5.3 as the outer code.

codeword length and providing source coding. Furthermore, channel coding is provided by the redundancy that is inherent in the RVLC codewords [276], facilitating an error-correction capability during RVLC decoding. The VDVQ/RVLC video codec advocated therefore employs a joint source and channel coding philosophy.

In Section 6.3.3 we imposed a number of constraints governing the allocation of the VDVQ tiles and RVLC codebooks in order to represent the various components of the video source frame. More specifically, these code constraints enforced the legitimate tessellation of the VDVQ tiles having a range of dimensions and ensured that the various fractions of the source video frame were encoded using the same number of bits. Since the set of RVLC codewords that can be employed during video encoding varies depending on which component of the source video frame is being encoded, the VDVQ/RVLC video codec can be said to employ IrVLCs.

In the VDVQ/RVLC video codec, the complete set of the above-mentioned code constraints was described by the novel trellis structure of Section 6.3.4, which is reminiscent of a symbol-based VLC trellis [181]. Hence, the employment of this trellis structure facilitated the consideration of all legitimate transmission frame permutations. This fact was exploited in order to perform novel Minimum Mean Square Error (MMSE) VDVQ/RVLC encoding using a variation of the Viterbi algorithm [193], as described in Section 6.4.

Additionally, the employment of the trellis structure during VDVQ/RVLC decoding was shown to guarantee the recovery of legitimate – although not necessarily error-free – video information in Section 6.5. This ensured that useful video information was never discarded, unlike in the conventional video decoders of [273, 274], where a single transmission error may render an entire video frame invalid. A novel modification of the Bahl–Cocke–Jelinek–Raviv (BCJR) algorithm [48] was employed during *A Posteriori* Probability (APP) Soft-Input Soft-Output (SISO) VDVQ/RVLC decoding in order to facilitate the iterative exchange [278] of extrinsic information with a serially concatenated APP SISO Trellis-Coded Modulation (TCM) [277] decoder, as well as to facilitate the soft MMSE reconstruction of the video sequence. Since the VDVQ/RVLC trellis structure describes the complete set of VDVQ/RVLC-induced code constraints, all of the associated redundancy was beneficially exploited with the aid of the modified BCJR algorithm.

In Section 6.6 the serially concatenated and iteratively decoded VDVQ/RVLC-TCM scheme of Section 6.2 was shown to outperform two suitably designed separate source- and channel-coding benchmarks. This was attributed to the benefits of the VDVQ/RVLC codec described above, which were realized owing to the joint source- and channel-coding philosophy adopted. Indeed, Figure 6.12 shows that the VDVQ/RVLC-TCM scheme was capable of achieving subjectively pleasing video reconstructions having a Peak Signal-to-Noise Ratio (PSNR) of 29.5 dB at a channel Signal-to-Noise Ratio (SNR) that is 1.1 dB lower than that of the VQ-based benchmark [273] and 1.6 dB lower than that of the MPEG-4 [158]-based benchmark [274].

12.7   Chapters 7–9: EXIT-Chart Matching

In Chapters 7–9 we considered the application of IrVLCs for EXIT-chart matching. This was motivated by the fact that an open EXIT-chart tunnel was only created for the VDVQ/RVLC-TCM scheme of Section 6.2 if the Rayleigh fading channel SNR was in excess of a threshold that was 1.29 dB higher than the channel's SNR capacity bound, as shown in Figure 12.1. Note that, as described in Section 4.3, an infinitesimally low probability of decoding error can only be achieved if the EXIT-chart tunnel is open and if the iterative decoding trajectory approaches the inner and outer EXIT functions sufficiently closely to facilitate iterative decoding convergence to the $(1, 1)$ point of the EXIT chart. Hence, operation closer than 1.29 dB from the channel's capacity bound was prevented for the VDVQ/RVLC-TCM scheme, as shown in Figure 6.12. Note that similar discrepancies of 1 dB were obtained for the SBVLC-TCM and BBVLC-TCM schemes of Section 7.3.2, as shown in Figure 12.1. Like the VDVQ/RVLC-TCM scheme of Section 6.2, the SBVLC-TCM and BBVLC-TCM schemes employed the serial concatenation and iterative decoding of a VLC-based outer codec with a TCM inner codec, and were not designed using EXIT-chart matching. Furthermore, Figure 12.1 shows that a similar discrepancy of 1.4 dB between the threshold E_b/N_0 value and the channel's attainable capacity bound was obtained for the VLC-URC scheme of Section 9.5.4, which employs Unity-Rate Coding (URC) for the inner codec instead of TCM. Instead of the capacity bound, the channel's *attainable* capacity bound is considered in this case, since it is this that imposes the fundamental limit on the VLC-URC scheme's operation, as described in Section 9.5.3. This is justified, since we will propose a solution to the associated effective throughput loss in Section 12.12, when outlining our future work. The corresponding EXIT chart obtained for the VLC-URC scheme of Section 9.5.4 was provided in Figure 9.10, together with those of the other schemes introduced in Section 9.5.4, and these are repeated for convenience in Figure 12.7.

In Section 6.6 we observed that open EXIT-chart tunnels could have been created for channel SNRs that are closer to the channel's capacity bound if the inverted VDVQ/RVLC EXIT function of Figure 6.11 had offered a better match with the TCM scheme's EXIT function. More specifically, this would have enabled the EXIT-chart tunnel to remain open and be further narrowed as the channel SNR was reduced towards the channel's capacity bound. The described observation of Section 6.6 may be explained by the area property of EXIT charts [204], which states that the EXIT-chart area enclosed by the threshold EXIT-chart tunnel is commensurate with the discrepancy between the channel's capacity bound and the threshold SNR.

Hence, in Section 7.3.2 we demonstrated that the inverted EXIT function of an outer IrVLC codec can be shaped to match with an inner EXIT function. Here, the IrVLC scheme generated particular fractions of the IrVLC-encoded transmission frame using different component VLC codebooks of either the RVLC or the Variable-Length Error-Correction (VLEC) [179] class. We showed that the inverted EXIT function of the corresponding APP SISO IrVLC decoder depends on the specifically chosen fractions of the IrVLC-encoded transmission frame that are generated by each component VLC codebook. More explicitly, the inverted IrVLC EXIT function may be obtained using Equation (7.4), which employs the described fractions as weights during the averaging of the component VLC codebooks' inverted EXIT functions.

Section 7.3.2 showed that the EXIT-chart matching algorithm of [149] may be employed to design specific parameterizations of the SBIrVLC-TCM and BBIrVLC-TCM schemes detailed in Section 7.2. Here, the algorithm of [149] was employed to shape the inverted IrVLC EXIT functions to match the EXIT function of the serially concatenated TCM codec. This facilitated the creation of open EXIT-chart tunnels at channel E_b/N_0 values in excess of a threshold that is 0.5 dB from the channel's capacity bound, as shown in Table 12.1. This is equal to the 0.5 dB discrepancy shown in Table 12.1 that was obtained when matching the inverted EXIT function of an Irregular Convolutional Code (IrCC) [306] to the TCM EXIT function during the parameterization of the Huffman-IrCC-TCM scheme of Section 7.4.1.

Furthermore, Table 12.1 shows that the open EXIT-chart tunnel of Figure 12.7b was achieved at a similar E_b/N_0 discrepancy of 0.54 dB from the channel's attainable capacity bound for the IrVLC-URC-high arrangement detailed in Section 9.5.4. Note that this scheme employed a serial concatenation of an IrVLC outer codec and a URC inner codec. A URC inner codec was also employed by the IrVLC-URC scheme of Section 8.4. Discrepancies of 0.42 dB and 0.7 dB are shown in Table 12.1 for parameterizations of this scheme that employed IrVLC coding rates of 0.55 and 0.85 respectively. This suggests that an improved EXIT-chart matching was achieved when employing lower IrVLC coding rates, resulting in open EXIT-chart tunnels at channel E_b/N_0 values that are closer to the channel's capacity bound, as shown in Figures 8.12 and 9.9.

Table 12.1: Iterative decoding performance and complexity of the various schemes considered in Chapters 6–9

Chapter	Scheme	Outer codec Coding rate	Outer codec EXIT matched	Inner codec Coding rate	Inner codec EXIT matched	Modem	Capacity bound – threshold E_b/N_0	Interleaver length [bits]	Capacity bound – operating E_b/N_0	ACS complexity at 2 dB from capacity bound
6	VDVQ/RVLC-TCM	0.667	No	0.75	No	SP 16QAM	1.29 dB	1 485 / 74 250	3.04 dB / 1.79 dB	N/A / N/A
7	SBVLC-TCM	0.52	No	0.75	No	SP 16QAM	1.00 dB	217 500	1.10 dB	3.5×10^5
	BBVLC-TCM	0.52	No	0.75	No	SP 16QAM	1.00 dB	217 500	1.10 dB	4.7×10^4
	SBIrVLC-TCM	0.52	Yes	0.75	No	SP 16QAM	0.50 dB	217 500	0.60 dB	1.2×10^6
	BBIrVLC-TCM	0.52	Yes	0.75	No	SP 16QAM	0.50 dB	217 500	0.60 dB	8.1×10^4
	Huffman-IrCC-TCM	0.52	Yes	0.75	No	SP 16QAM	0.50 dB	217 500	0.80 dB	4.3×10^4
8	IrVLC-URC †	0.55	Yes	1	No	BPSK	0.42 dB	100 000	1.41 dB	6.0×10^4
		0.85	Yes	1	No	BPSK	0.70 dB	100 000	1.15 dB	3.1×10^4
9	VLC-URC	0.53	No	1	No	Gray-coded 16QAM	1.40 dB ‡	100 000 / 1 000 000	1.60 dB ‡ / 1.45 dB ‡	5.0×10^4 ‡ / 5.0×10^4 ‡
	IrVLC-URC-high	0.53	Yes	1	No	Gray-coded 16QAM	0.54 dB ‡	100 000 / 1 000 000	0.76 dB ‡ / 0.63 dB ‡	5.6×10^4 ‡ / 5.6×10^4 ‡
	IrVLC-IrURC-high	0.53	Yes	1	Yes	Gray-coded 16QAM	0.04 dB ‡	100 000 / 1 000 000	0.57 dB ‡ / 0.22 dB ‡	8.6×10^4 ‡ / 8.6×10^4 ‡
	IrVLC-IrURC-low	0.53	Yes	1	Yes	Gray-coded 16QAM	0.04 dB ‡	100 000 / 1 000 000	0.67 dB ‡ / 0.17 dB ‡	6.1×10^4 ‡ / 6.1×10^4 ‡

† The IrVLC comprises the component VLEC codebooks $\{\text{VLEC}^n\}_{n=12}^{22}$ of Table 8.6, which were designed using the GA of Section 8.3.

‡ The channel's *attainable* E_b/N_0 capacity bound is employed.

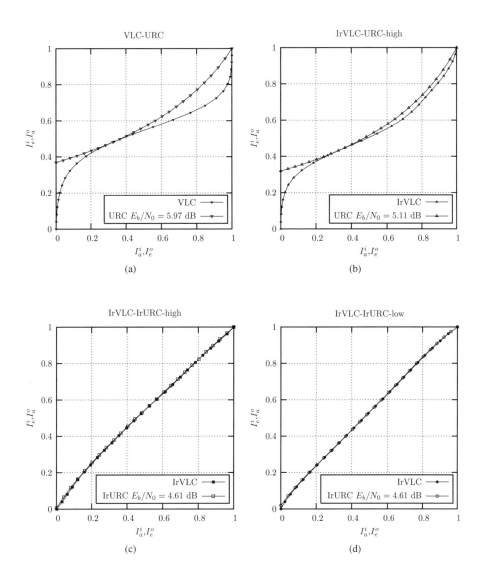

Figure 12.7: EXIT charts for the schemes of Section 9.5.4: (a) VLC-URC arrangement; (b) IrVLC-URC-high arrangement; (c) IrVLC-IrURC-high arrangement; (d) IrVLC-IrURC-low arrangement. The inner EXIT functions are provided for the threshold channel E_b/N_0 values, as specified in Table 9.1.

Owing to the aforementioned benefits of EXIT-chart matching, the observed discrepancies in the range of 0.42 dB–0.7 dB are lower than those obtained when EXIT-chart matching was not employed, which are in the range of 1 dB–1.4 dB, as described above.

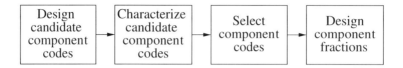

Figure 12.8: Conventional irregular coding design process.

12.8 Chapter 8: GA-Aided Design of Irregular VLC Components

Chapter 8 showed that our ability to perform EXIT-chart matching and to achieve an open EXIT-chart tunnel at E_b/N_0 values that are close to the channel's capacity bound is commensurate with the degree of diversity in the shapes exhibited by the inverted EXIT functions of the component VLC codebook suite. For this reason, the conventional irregular coding design process strives towards obtaining a component VLC codebook suite having a wide variety of inverted EXIT functions, as shown in Figure 12.8. The component VLEC codebooks employed by the IrVLC schemes of Chapters 7 and 8 were designed using Algorithm 3.3 of Section 3.3. As discussed in Section 8.1, this algorithm attempts to design VLEC codebooks having maximal coding rates that satisfy particular specified distance criteria. However, this algorithm does not facilitate the direct control or prediction of the inverted EXIT function shapes that correspond to the designed VLEC codebooks. Hence, in the conventional irregular coding design process depicted in Figure 12.8, a significant amount of 'trial-and-error'-based human interaction is required. This involves the design of a high number of candidate component VLEC codebooks, the characterization of their inverted EXIT functions and the selection of a suite having a wide variety of inverted EXIT functions, as exemplified in Chapter 7.

The trial-and-error efforts required to design a suite of IrVLC component codebooks using Algorithm 3.3 of Section 3.3 motivated the design of a novel Genetic Algorithm (GA) for generating the VLEC codebooks of Section 8.3. Unlike Algorithm 3.3 of Section 3.3, this GA was shown to facilitate the direct control and prediction of the inverted EXIT function shapes that result for the designed VLEC codebooks, eliminating the trial-and-error efforts in the irregular coding design process. While maintaining desirable VLEC-encoded bit entropies and IrVLC decoding complexities, the GA of Section 8.3 seeks VLEC codebooks having arbitrary coding rates and Real-Valued Free Distance Metrics (RV-FDMs).

This novel RV-FDM was proposed in Section 8.2 as an alternative to the Integer-Valued Free Distance (IV-FD) lower bound of [179] for the characterization of a VLEC codebook's error-correction capability. Like the IV-FD lower bound of [179], the RV-FDM considers the minimum number of differing bits in any pair of equal-length legitimate VLEC-encoded bit sequences, characterizing the probability of occurrence for the most likely undetectable transmission error scenario, as described in Section 8.1. However, *unlike* the IV-FD lower bound, the RV-FDM of Section 8.2 also considers how susceptible the VLEC-encoded bits are to this transmission error scenario. As a result, the RV-FDM exists within the real domain, allowing the comparison of the error-correction capabilities of two VLEC codebooks having equal IV-FD lower bounds. This facilitates its employment within the objective function of the novel GA proposed in Section 8.3.

In Section 8.2, we showed that a VLEC codebook's RV-FDM affects the number of inflection points appearing in the corresponding inverted EXIT function. More specifically, we showed that high RV-FDMs are associated with 'S'-shaped inverted EXIT functions having up to two points of inflection, whilst low RV-FDMs result in inverted EXIT functions having no more than one point of inflection. Furthermore, we showed that the inverted EXIT function of a VLEC codebook will reach the top right-hand corner of the EXIT chart if its RV-FDM is at least equal to two [298]. These findings complement the property [204] that the area below an inverted VLEC EXIT function equals the corresponding coding rate. Therefore, since the inverted VLEC EXIT function shape of a VLEC codebook depends on both its coding rate and its RV-FDM, the GA of Section 8.3 facilitates the direct control and prediction of the inverted EXIT function shapes that result for the designed VLEC codebooks.

The employment of both the novel GA of Section 8.3 and Algorithm 3.3 of Section 3.3 to design suites of IrVLC component codebooks was investigated in Section 8.5. The suite of component VLEC codebooks designed in Section 8.5.1 by our novel GA had the wide variety of inverted EXIT functions shown in Figure 8.8. This was obtained by seeking component VLEC codebooks having a wide variety of coding rates and RV-FDMs. In some cases, high RV-FDMs were sought, resulting in 'S'-shaped inverted EXIT functions having up to two points of inflection, whilst low RV-FDMs were sought for the remaining component VLEC codebooks, which were associated with inverted EXIT functions having no more than one point of inflection. Here, we found that more extreme RV-FDMs could be obtained for VLEC codebooks having lower coding rates. This may be explained by the higher degree of design freedom that is facilitated for lower coding rates owing to the longer codewords that this implies.

When the novel GA of Section 8.3 was employed to design component VLEC codebooks, the trial-and-error approach was not employed. Similarly, no trial and error was employed when Algorithm 3.3 of Section 3.3 was used, facilitating a fair comparison. Instead, a different IV-FD lower bound was sought for each component VLEC codebook designed using Algorithm 3.3 of Section 3.3. However, the resultant component VLEC codebooks were found to have relatively high RV-FDMs and only a limited variety of coding rates, resulting in the limited variety of 'S'-shaped inverted EXIT functions shown in Figure 8.7.

Owing to its employment of a wider variety of coding rates, and as a benefit of its both high and low RV-FDMs, the suite of component VLEC codebooks designed by our novel GA in Section 8.5.1 was found to be more suitable for use in EXIT-chart matching than that designed using Algorithm 3.3 of Section 3.3. More specifically, open EXIT-chart tunnels could be created for the IrVLC-URC scheme of Section 8.4 at channel E_b/N_0 values within 1 dB of the Rayleigh fading channel's capacity bound for a wide range of effective throughputs, when employing the suite of component VLEC codebooks generated using our GA, as shown in Figure 8.12. In contrast, open EXIT-chart tunnels could only be achieved when employing the suite designed by Algorithm 3.3 of Section 3.3 for a limited range of effective throughputs and within a significantly higher margin of 4.4 dB from the E_b/N_0 capacity bound. This confirmed the observation that our ability to perform EXIT-chart matching depends on how much variety is exhibited within the inverted EXIT functions of the suite of component VLEC codebooks.

However, regardless of the component VLEC codebook suite employed, we observed in Section 8.5.4 that the inverted IrVLC EXIT function can only be matched to the EXIT functions of a regular inner codec with limited accuracy. This is because inverted outer EXIT functions are constrained to starting from the $(0, 0)$ point of the EXIT chart, while the inner EXIT functions typically emerge from a relatively high point along the I_e^i axis of the EXIT chart, as described in Section 4.2. As a result, we cannot create an arbitrarily narrow open EXIT-chart tunnel. Instead, a lower bound is imposed upon the enclosed EXIT-chart area and, hence, upon the discrepancy between the threshold E_b/N_0 value and the channel's capacity bound, owing to the area property of EXIT charts [204].

12.9   Chapter 9: Joint EXIT-Chart Matching of IRVLCs and IRURCs

The above-mentioned findings motivated the introduction of novel Irregular Unity-Rate Codes (IrURCs) in Chapter 9, which encode different fractions of the transmission frame using different component URCs, having various EXIT functions. In analogy with those of IrVLCs and IrCCs, IrURC EXIT functions may be shaped by specifically selecting the fraction of the transmission frame that is encoded by each component URC. In this way, the IrURC EXIT function may be shaped to emerge from a point on the EXIT chart's I_e^i axis that is closer to the inverted outer EXIT function's starting point of $(0, 0)$.

The serial concatenation and iterative decoding of an IrVLC outer codec with an IrURC inner codec was demonstrated in Section 9.4. Here, the IrVLC's suite of component VLEC codebooks was designed using the GA of Section 8.3 in order to generate the required diversity of inverted EXIT function shapes shown in Figure 9.5 and repeated for convenience in Figure 12.9. In contrast, the EXIT functions shown in Figure 9.7 and repeated for convenience in Figure 12.10 were obtained by selecting the IrURC's suite of component URCs from a large number of candidates, as described in Section 9.5.2. In Section 9.3

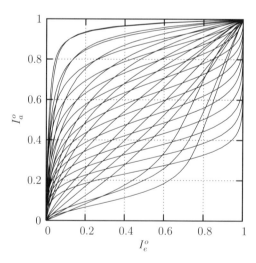

Figure 12.9: Inverted EXIT functions for the component VLEC codebooks employed by the IrVLC-IrURC-high and IrVLC-IrURC-low arrangements of Section 9.5.4.

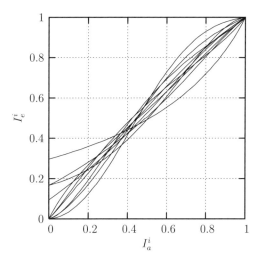

Figure 12.10: EXIT functions corresponding to a Gray-coded 16QAM-modulated Rayleigh fading channel SNR of 8 dB for the component URC codes employed by the IrVLC-IrURC-high and IrVLC-IrURC-low arrangements of Section 9.5.4.

we proposed a novel method for jointly matching the EXIT functions of the two serially concatenated irregular codecs. This method iteratively applies the EXIT-chart matching algorithm of [149] to alternately match the outer EXIT function to the inner and vice versa, simultaneously seeking the highest coding rate that offers an open EXIT-chart tunnel. Note however, that the novel modification of Section 9.2 was required in order to allow the EXIT-chart matching of the IrURC EXIT function, since all component URCs have the same unity coding rate. The joint EXIT-chart matching algorithm of Section 9.3 was shown to be able to exploit the increased degree of design freedom that is afforded by employing two irregular codecs in order to create an EXIT-chart tunnel that is narrow at all points along its length. This facilitated the creation of the marginally open EXIT-chart tunnels shown in Figures 12.7c and 12.7d for the IrVLC-IrURC-high and IrVLC-IrURC-low arrangements of Section 9.5.4, respectively. Owing to the area property of EXIT charts, these were obtained at E_b/N_0 values that were just 0.04 dB from the channel's attainable capacity bound, as shown in Figure 12.1.

Note that an open EXIT-chart tunnel implies that iterative decoding convergence to an infinitesimally low probability of error can be achieved, provided that the iterative decoding trajectory approaches the inner and outer EXIT functions sufficiently closely, as described in Section 4.3. However, throughout this monograph we found that high-quality reconstructions could not be achieved at the threshold E_b/N_0 values, where the EXIT-chart tunnels open. This is due to the BCJR algorithm's assumption [48] that all correlation within the LLR frames exchanged by the APP SISO decoders is successfully mitigated by the intermediary interleavers. If this is not the case, the iterative decoding trajectory will not match perfectly with the inner and outer EXIT functions and the tunnel must be further widened before the trajectory can reach the top right-hand corner of the EXIT chart, which is associated with an infinitesimally low probability of error, as described in Section 4.3. Since the interleaver's

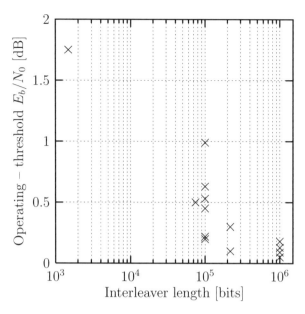

Figure 12.11: Scatter plot of the interleaver lengths provided in Table 12.1 and the corresponding discrepancies between the E_b/N_0 value at which EXIT chart tunnel opens and that at which it is sufficiently widened to facilitate a high reconstruction quality.

ability to mitigate the correlation is proportional to its length, longer interleavers can be expected to yield lower discrepancies between the E_b/N_0 value at which the EXIT-chart tunnel opens and that at which it is sufficiently widened to facilitate a high reconstruction quality. Indeed, this relationship may be observed in Figure 12.11, which provides a scatter plot of the discrepancies and interleaver lengths given in Table 12.1, as will be detailed in our forthcoming discussions.

In Section 6.6 we had characterized the discrepancy between the threshold E_b/N_0 value at which an open EXIT-chart tunnel could be created for the VDVQ/RVLC-TCM video transmission scheme and the operating E_b/N_0 value at which it could achieve a high-quality reconstruction having a Peak Signal-to-Noise Ratio (PSNR) of 29.5 dB. This discrepancy was found to be 1.75 dB when the interleaver length was equal to that of a single encoded video frame, namely 1 485 bits, as shown in Table 12.1. In contrast, when 50 encoded video frames were concatenated to give an interleaver length of 74 250 bits, the discrepancy was reduced to just 0.5 dB, facilitating operation at 1.79 dB from the channel's E_b/N_0 capacity bound. However, in Section 6.6 this scheme was shown to incur a 5 s latency, since the video frame rate was 10 fps and because all 50 frames must be received before they can be deinterleaved.

The discrepancy between the threshold E_b/N_0 value at which an open EXIT-chart tunnel could be created for the arrangements of Section 9.5.4 and the operating E_b/N_0 value at which they could achieve a BER of 10^{-5} was characterized in Section 9.6. When a 100 000-bit interleaver was employed, the discrepancies for the VLC-URC and IrVLC-URC-high arrangements were found to be 0.2 dB and 0.22 dB respectively, as shown in Table 12.1.

However, these discrepancies were reduced to 0.05 dB and 0.09 dB respectively when we employed a longer interleaver, having a length of 1 000 000 bits. Larger discrepancies were observed for the IrVLC-IrURC-high and IrVLC-IrURC-low arrangements, owing to their narrow EXIT-chart tunnels, as discussed in Section 9.6. These were 0.53 dB and 0.67 dB respectively when the 100 000-bit interleaver was employed, as compared with 0.18 dB and 0.13 dB respectively when the 1 000 000-bit interleaver was employed. Note that the IrVLC-IrURC-low arrangement using the 1 000 000-bit interleaver could achieve a BER of less than 10^{-5} for E_b/N_0 in excess of a limit that was just 0.17 dB from the channel's attainable capacity bound, as shown in Table 12.1. This is comparable with the 0.13 dB discrepancy demonstrated for Irregular Low-Density Parity Check (IrLDPC) codes [147,313] and superior to the 0.25 dB discrepancy found for irregular turbo codes [148].

In Section 7.4 we showed that the SBVLC-, BBVLC-, SBIrVLC- and BBIrVLC-TCM schemes using a 217 500-bit interleaver could achieve a high-quality source sample reconstruction SNR of 20 dB for E_b/N_0 values in excess of a limit that was 0.1 dB from the threshold at which an open EXIT-chart tunnel was created, as shown in Table 12.1. However, in the case of the Huffman-IrCC-TCM scheme of Section 7.4.1, the corresponding discrepancy was equal to the higher value of 0.3 dB. This was explained in Section 7.4.3 by the relatively high sensitivity of the APP SISO IrCC decoder to any residual correlation within the iteratively exchanged LLRs that was insufficiently mitigated by the 217 500-bit interleaver. This resulted in the poor correlation between the iterative decoding trajectory and the inverted IrCC EXIT function. More specifically, we concluded that an APP SISO IrCC decoder's sensitivity to this correlation increases if its coding rate is reduced or if, in particular, we increase its coding memory. Hence, the high sensitivity of the Huffman-IrCC-TCM scheme's APP SISO IrCC decoder was attributed to its relatively high coding memory of 4.

Section 7.4.3 also concluded that an APP SISO VLC decoder's sensitivity to the aforementioned correlation depends only on its coding rate. Indeed, in Section 8.6 the APP SISO IrVLC decoder's sensitivity to this extrinsic information correlation was found to increase as the IrVLC coding rate was reduced. As shown in Table 12.1, the IrVLC-URC scheme of Section 8.4 using a 100 000-bit interleaver and an IrVLC coding rate of 0.85 could achieve a BER of less than 10^{-5} for E_b/N_0 values in excess of a limit that was 0.45 dB from the threshold at which an open EXIT-chart tunnel was created. However, this discrepancy grew to 0.99 dB when the IrVLC coding rate was reduced to 0.55, as shown in Table 12.1.

Throughout this book we have considered receivers in which soft information is iteratively exchanged between APP SISO decoders and in which a final hard decision is made by a MAP sequence estimator. These components of the receiver apply the BCJR algorithm [48] and the Viterbi algorithm [193] to suitably designed trellises [141, 180]. These require only Add, Compare and Select (ACS) operations if all calculations are performed in the logarithmic probability domain and if a lookup table is employed for correcting the Jacobian approximation [296]. Since each individual ACS operation requires the same resources in a systolic-array-based chip, the number of ACS operations performed by a receiver may be employed to characterize the complexity/area/speed trade-off required for its implementation.

In Section 7.4 we introduced the novel plot of Figure 7.12 for characterizing the iterative decoding complexity of a receiver. This plot provides the average number of ACS operations required per source symbol to achieve particular reconstruction qualities as a function of the channel's E_b/N_0 value. This plot, as well as those of Figures 7.13, 8.17, 8.18 and 9.12, showed that particular reconstruction qualities can be achieved with lower complexities as the channel's E_b/N_0 value is increased. This may be explained by the associated widening of the open EXIT-chart tunnel, requiring fewer decoding iterations for the iterative decoding trajectory to reach the particular point on the EXIT chart that is associated with the reconstruction quality considered.

Additionally, Figures 7.12 and 7.13 showed that lower complexities may be maintained, provided that lower reconstruction qualities can be tolerated, since fewer decoding iterations are required for the iterative decoding trajectory to reach the particular point on the EXIT chart that is associated with a lower reconstruction quality. However, Section 7.4.5 observed that, in the approach of iterative decoding convergence, large reconstruction quality gains are obtained for relatively small amounts of additional computational complexity. We concluded that if the channel E_b/N_0 value is sufficiently high as to create an open EXIT-chart tunnel, then we can typically justify the computational complexity required for the iterative decoding trajectory to reach the $(1, 1)$ point of the EXIT chart, owing to the infinitesimally low probability of error that results.

In Sections 7.4.5, 8.6 and 9.5.4 we showed that outer APP SISO decoding and MAP sequence estimation are typically associated with significantly higher computational complexities than is inner APP SISO decoding. In the most extreme case considered in this volume, the outer decoders of the SBVLC- and SBIrVLC-TCM schemes of Section 7.2 accounted for about 97% of the ACS operations employed per source sample. In contrast, the outer decoders of the Huffman-IrCC-TCM scheme of Section 7.4.1 were responsible for about 60% of the iterative decoding complexity, in the most balanced case considered.

In Table 12.1, we provide the average number of ACS operations required per source symbol to achieve high-quality reconstructions at an E_b/N_0 value that is 2 dB from the channel's capacity bound for each of the schemes considered in Sections 7.2 and 8.5.4. Additionally, for the schemes of Section 9.6, the complexity at an E_b/N_0 value that is 2 dB from the channel's *attainable* capacity bound is provided in Table 12.1. Furthermore, Figure 12.12 plots the complexities of the aforementioned schemes for a range of E_b/N_0 discrepancies from the capacity bounds. While the complexities shown in Table 12.1 and Figure 12.12 for the schemes of Section 7.2 are associated with obtaining a high-quality source sample reconstruction SNR of 20 dB, those provided for the schemes of Sections 8.5.4 and 9.6 are associated with achieving a BER of 10^{-5}. The comparison of the described complexities is fair, since each of the schemes considered in Chapters 7–9 facilitates the transmission of 16-ary source symbols over an uncorrelated narrowband Rayleigh fading channel. Furthermore, in all cases the source symbols have the probabilities of occurrence that result from the Lloyd–Max quantization [164, 165] of Gaussian distributed source samples, as described in Section 7.2.1.

Note that Figure 12.12 illustrates the discrepancies between the channel's appropriate capacity bounds and the E_b/N_0 values above which the schemes considered in Chapters 7–9 may achieve high-quality reconstructions, confirming the discrepancies provided in Table 12.1. Furthermore, at high discrepancies from the channel's E_b/N_0 capacity bounds, Figure 12.12 shows that similar iterative decoding complexities may be observed for the BBVLC-, BBIrVLC- and Huffman-IrCC-TCM schemes of Section 7.2 as well as for each

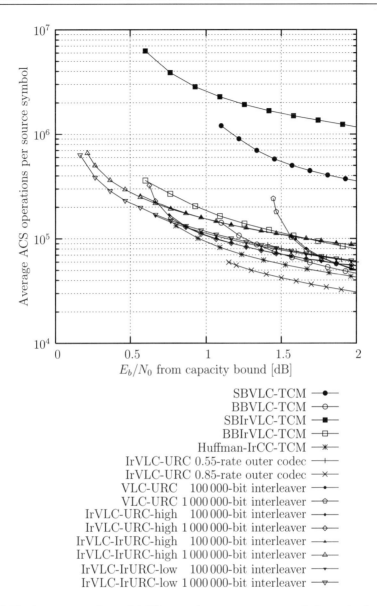

Figure 12.12: Average number of ACS operations per source symbol required to achieve high-quality reconstructions at a range of E_b/N_0 discrepancies from the appropriate capacity bounds for the schemes of Chapters 7–9.

scheme introduced in Sections 8.5.4 and 9.6. Indeed, the corresponding ACS counts provided in Table 12.1 can be seen to have similar values in the range of $[3.1 \times 10^4, 8.6 \times 10^4]$. This similarity may be explained because all of these schemes employ bit-based trellises [141,180] as the basis of their outer APP SISO decoders and MAP sequence estimators. In contrast, the

SBVLC- and SBIrVLC-TCM schemes of Section 7.2 employed the symbol-based VLC trellis of [181] as the basis of their APP SISO decoders. For these schemes, Table 12.1 provides ACS operation counts of 3.5×10^5 and 1.2×10^6 respectively, which are significantly higher than those provided for the schemes employing bit-based trellises, as illustrated in Figure 12.12. This increased complexity may be attributed to the number of trellis transitions that are employed in symbol-based VLC trellises, which is typically significantly higher than the number employed in their bit-based equivalents, as described in Section 3.4.

The computational complexities provided in Figure 12.12 for the IrVLC-URC arrangement of Section 8.5.4 having an IrVLC coding rate of 0.85 can be seen to be lower than that associated with an IrVLC coding rate of 0.55. This may be explained by the higher number of bit-based trellis transitions that are required to represent the longer codewords of the 0.55-rate IrVLC, as discussed in Section 8.6. For this reason, we can expect lower computational complexities to be associated with VLCs and IrVLCs having higher coding rates in general. This is confirmed by the average numbers of ACS operations performed per source symbol that are provided for VLEC codebooks having various coding rates in Figures 8.7–8.10 and 9.5. These figures additionally show that VLEC codebooks having relatively low RV-FDMs are also associated with low computational complexities. This was exploited during the design of the IrVLC-IrURC-low arrangement of Section 9.5.4. More specifically, the novel modification of the EXIT-chart matching algorithm [149] of Section 9.2 was employed jointly to perform EXIT-chart matching while seeking a reduced IrVLC computational complexity by invoking component VLEC codebooks having a low RV-FDM. As a result, in Table 12.1 the computational complexity of the IrVLC-IrURC-low arrangement can be seen to be 25% lower than that of the IrVLC-IrURC-high arrangement, which was designed without seeking a reduced IrVLC computational complexity. Note that a reduced computational complexity could not be achieved when the IrVLC's EXIT function was matched to that of a regular URC, as discussed in Section 9.5.4. This was found to be because, unlike the 'S'-shaped inverted EXIT functions of the component VLEC codebooks having a high RV-FDM, those associated with a low RV-FDM do not rise rapidly enough to match with the URC EXIT function, which starts from a high point along the EXIT chart's I_e^i axis.

In Section 7.2.1 we showed that the number of transitions employed by a symbol-based VLC trellis, and hence its computational complexity and memory requirement, scales with the square of the number of source symbols that it simultaneously decodes. For this reason, the total computational complexity and memory requirement can be reduced by decomposing each source symbol frame into sub-frames, which are decoded separately. However, owing to the nature of VLCs, the lengths of VLC-encoded transmission sub-frames typically vary from frame to frame. In order to facilitate their decoding in the receiver, the transmitter must convey the lengths of the sub-frames as explicit side information, which should be protected using a low-rate channel code owing to its error-sensitive nature. Hence, the choice of how many sub-frames to employ is a trade-off between the amount of side information required and the computational complexity, as well as the memory requirements per source symbol. Note that the complexities provided in Table 12.1 and Figure 12.12 for the SBVLC- and SBIrVLC-TCM schemes of Section 7.2 are therefore specific to the particular considered case, in which each source symbol sub-frame comprised 100 symbols. In this arrangement, the required side information was found to account for 4% of the total information conveyed in Section 7.3.

In contrast, Section 7.2.1 showed that the number of transitions per source symbol employed by a bit-based trellis is independent of the number of source symbols that it simultaneously decodes. Hence, the total computational complexity and memory requirement *cannot* be reduced by decomposing each source symbol frame into sub-frames in this case. However, the memory required to decode each source symbol sub-frame will be reduced if more sub-frames are employed. If the sub-frames are decoded sequentially, this memory can be reused for each sub-frame and a lower-cost implementation will result. By contrast, the amount of memory required by an implementation that decodes all sub-frames concurrently will not be affected by the number of sub-frames employed.

In the schemes of Sections 8.5.4 and 9.6, the amount of side information required was significantly reduced by employing just a single source symbol sub-frame for each IrVLC component code. Using this approach, we found that less side information was required when a longer interleaver was employed, as described in Section 9.6. Indeed, the required side information was reduced to just 0.006% of the total information conveyed when a 1 000 000-bit interleaver was employed.

12.10   Chapter 10: Iteratively Decoded VLC Space–Time Coded Modulation

In this chapter an iteratively decoded variable-length space–time coded modulation design was proposed. The joint design of source coding, space–time coded modulation and iterative decoding was shown to achieve both spatial diversity and multiplexing gain, as well as coding and iteration gains at the same time. The variable-length structure of the individual codewords mapped to the maximum of N_t transmit antennas imposes no synchronization and error-propagation problems. The convergence properties of the proposed VL-STCM-ID scheme were analyzed using 3D symbol-based EXIT charts as well as 2D EXIT-chart projections. A significant iteration gain was achieved by the VL-STCM-ID scheme, which hence outperformed both the non-iterative VL-STCM scheme and the FL-STCM benchmark with the aid of N_t unity-rate recursive feedback precoders. The VL-STCM-ID scheme attains a near-MIMO channel capacity performance.

12.11   Chapter 11: Iterative Detection of Three-Stage Concatenated IrVLC FFH-MFSK

In this chapter we investigated a serially concatenated IrvLC/FFH-MFSK transceiver operating in a Rayleigh fading channel, when the transmitted signal was also corrupted by PBNJ. Our EXIT-chart analysis demonstrated that a two-stage concatenated FFH-MFSK requires the employment of an additional unity-rate precoder for the sake of making the channel appear recursive. In order to ensure near-capacity operation, the IrVLC codec was specifically designed to ensure that the inverted EXIT curve of the IrVLC decoder matches the EXIT curve of the inner decoder. In this way, an open EXIT-chart tunnel may be created even at low SNR values, providing source-correlation-dependent additional performance gains of up to 1.1 dB over the regular VLC-based benchmark scheme. Since the employment of the VLC involves non-identical occurrence probabilities for the source symbols, it is not possible to provide a comparison of the proposed scheme with the state of the art in the context of coded FFH-MFSK, which traditionally employs equiprobable source symbols or bits. However, we have provided a comparison of the IrVLC scheme with a VLC scheme dispensing with the precoder; consequently we noted that the precoder-aided schemes yield an E_b/N_0 gain in excess of 7 dB over the system dispensing with the precoder, which suffers from an error floor when jamming is severe.

Moreover, we demonstrated that the three-stage concatenation involving the demodulator, the rate-1 decoder and the outer IrVLC decoder yields superior performance compared with the two-stage concatenation of the rate-1 decoder and the outer decoder. Naturally, the three-stage scheme imposes a higher complexity. By contrast, we found that a precoder of memory 1 is more suitable for the three-stage IrVLC scheme, while the two-stage scheme requires a precoder of memory 3; thus the memory-3 rate-1 decoder imposes a somewhat higher complexity than does its memory-1 counterpart.

In conclusion, the precoder-aided FFH-MFSK-VLC scheme constitutes a moderate-complexity design option, which can be employed in systems communicating through channels contaminated by PBNJ for transmission of joint source and channel encoded audio or video signals. If a higher complexity can be afforded, the IrVLC-based scheme offers additional performance improvements. In our future work we will investigate more sophisticated three-stage iterative decoding, exchanging extrinsic information amongst the demodulator, the rate-1 decoder and the outer decoder.

12.12 Future Work

As shown in Table 12.1, the schemes of Sections 6.2 and 7.2 employed a $R_{\text{TCM}} = 3/4$-rate TCM inner codec together with Set Partitioned (SP) $M_{\text{TCM}} = 16$-ary Quadrature Amplitude Modulation (16QAM) [277] to facilitate transmissions over an uncorrelated narrowband Rayleigh fading channel. However, in these schemes the maximum effective throughput is limited to $R_{\text{TCM}} \cdot \log_2(M_{\text{TCM}}) = 3$ bits per channel use. Owing to the less-than-unity TCM coding rate, an effective throughput loss occurs for high channel E_b/N_0 values, where the capacity of the 16QAM modulated channel will exceed the maximum effective throughput of 3 bits per channel use and will approach $\log_2(M_{\text{TCM}}) = 4$ bits per channel use.

This motivated the employment of an inner URC codec in Section 8.4, which used $M_{\text{BPSK}} = 2$ Binary Phase-Shift Keying (BPSK) [300] to facilitate transmissions over an uncorrelated narrowband Rayleigh fading channel, as shown in Figure 12.1. Here, the maximum effective throughput was equal to the maximum capacity of $\log_2(M_{\text{BPSK}}) = 1$ bit per channel use and no effective throughput loss was incurred. Indeed, the areas beneath the URC EXIT functions provided in Figure 8.13 were found to be equal to the corresponding channel capacities, as predicted by the area property of EXIT charts [204].

In the scheme of Section 9.4 we opted for employing a URC-based inner codec together with $M_{\text{QAM}} = 16$QAM instead of BPSK, since this facilitates a higher maximum effective throughput of $\log_2(M_{\text{QAM}}) = 4$ bits per channel use. In Section 9.5.2 we showed that the receiver of Figure 9.3 would benefit from the iterative extrinsic information exchange of the 16QAM demodulator and the inner APP SISO decoder. However, for the sake of obtaining an implementational and computational complexity saving, the receiver of Figure 9.3 employed only the 'one-shot' activation of the 16QAM demodulator. However, as a result, when multiplied by $\log_2(M_{\text{QAM}}) = 4$ the average area beneath the URC EXIT functions exemplified in Figure 9.7 did not equal the corresponding channel capacities.

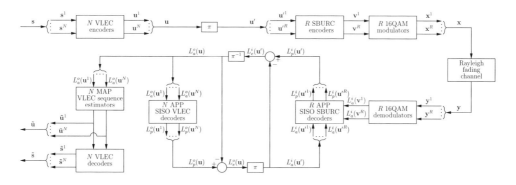

Figure 12.13: Schematic of the IrVLC-SBIrURC scheme.

In Section 9.5.3 we defined the attainable capacity of a channel having a particular E_b/N_0 value as being equal to the average area beneath the corresponding URC EXIT functions, multiplied by $\log_2(M_{QAM}) = 4$. We showed that the channel's attainable capacity represents an upper bound to the maximum effective throughput for which an open EXIT-chart tunnel can be achieved. This is because a scheme's effective throughput may be approximated by multiplying the area beneath the inverted outer EXIT function by $\log_2(M_{QAM}) = 4$ [204]. Since this area must be lower than that beneath the inner EXIT function in order for an open EXIT-chart tunnel to be facilitated, iterative decoding convergence to an infinitesimally low probability of error is prevented when the effective throughput is higher than the channel's attainable capacity. The discrepancy between the channel's capacity and its *attainable* capacity therefore imposes an effective throughput loss.

Section 9.5.2 showed that this effective throughput loss was minimized by employing Gray-coded 16QAM [311], since the corresponding EXIT function of Figure 9.8 is optimized to emerge from the highest possible point on the EXIT chart's I_e^m axis. As shown in Figure 9.9, the effective throughput loss resulted in a discrepancy of 0.29 dB between the channel's E_b/N_0 capacity bound and its *attainable* capacity bound. Hence, the 0.17 dB discrepancy between the channel's attainable capacity bound and the E_b/N_0 value at which the IrVLC-IrURC-low arrangement could achieve a BER of 10^{-5} that is shown in Table 12.1 represents a 0.46 dB discrepancy from the channel's capacity bound.

In this section we propose a method for mitigating the effective throughput loss of the IrVLC-IrURC scheme detailed in Section 9.4. However, this solution does not employ iterative extrinsic information exchange between the 16QAM demodulator and the inner APP SISO decoder. Instead, the benefit of iterative demodulation is mitigated by replacing the bit-based IrURC inner codec of Figure 9.3 with a Symbol-Based Irregular Unity-Rate Code (SBIrURC). Unlike a bit-based IrURC, this SBIrURC can directly employ the symbol probabilities obtained for the demodulator's $M_{QAM} = 16$ constellation points without first converting them into sets of $\log_2(M_{QAM}) = 4$ bit probabilities. We refer to this proposed solution as the IrVLC-SBIrURC scheme and Figure 12.13 provides its schematic, which is reminiscent of the IrVLC-IrURC schematic provided in Figure 9.3.

In the IrVLC-SBIrURC scheme of Figure 12.13, IrVLC encoding, APP SISO decoding and MAP sequence estimation are performed in exactly the same way as in the IrVLC-IrURC scheme of Figure 9.3. Furthermore, the source symbol frame $\mathbf{s}$, the transmission frame $\mathbf{u}$ and the LLR frames $L_a^o(\mathbf{u})$ as well as $L_p^o(\mathbf{u})$ are composed of N sub-frames, as in the IrVLC-IrURC scheme of Figure 9.3. Likewise, the interleaved transmission frame $\mathbf{u}'$ and the LLR frames $L_a^i(\mathbf{u}')$ as well as $L_p^i(\mathbf{u}')$ are composed of R sub-frames, as before. Additionally, iterative decoding is performed as in the IrVLC-IrURC receiver of Figure 9.3, with the subtraction of the *a priori* LLR frames from the *a posteriori* LLR frames and the interleaving π of the resultant extrinsic LLR frames, as shown in Figure 12.13. Finally, as in the IrVLC-IrURC scheme of Figure 9.3, the outer and inner EXIT functions of the IrVLC-SBIrURC scheme may be shaped by specifically selecting the fractions, $\{\alpha^n\}_{n=1}^N$ and $\{\alpha^r\}_{r=1}^R$ of the frames $\mathbf{u}$ and $\mathbf{u}'$ that are composed by the sub-frames $\{\mathbf{u}^n\}_{n=1}^N$ and $\{\mathbf{u}'^r\}_{r=1}^R$ respectively. The IrVLC-SBIrURC scheme of Figure 12.13 differs from the IrVLC-IrURC scheme of Figure 9.3 in terms of the operation of the irregular inner codec and the modem.

In the IrVLC-SBIrURC transmitter of Figure 12.13, each interleaved transmission sub-frame $\mathbf{u}'^r$ is decomposed into sets of $\log_2(M_{\text{QAM}}) = 4$ consecutive bits, which are converted into $M_{\text{QAM}} = 16$-ary symbol values. In analogy with the IrVLC-IrURC scheme of Figure 9.3, the $M_{\text{QAM}} = 16$-ary symbol values corresponding to each interleaved transmission sub-frame $\mathbf{u}'^r$ are encoded using a Symbol-Based Unity-Rate Code (SBURC) having a different symbol-based Linear Feedback Shift Register (LFSR) design. For example, these LFSRs could employ the designs of Figure 9.6 if they were modified to employ modulo-16 additions and memory elements. Following SBURC encoding in the IrVLC-SBIrURC transmitter, the sub-frame $\mathbf{v}^r$ of SBURC-encoded $M_{\text{QAM}} = 16$-ary symbol values is obtained, as shown in Figure 12.13.

In the IrVLC-SBIrURC transmitter of Figure 12.13, the $M_{\text{QAM}} = 16$-ary symbol values of each SBURC-encoded sub-frame $\mathbf{v}^r$ are mapped to $M_{\text{QAM}} = 16$QAM constellation points in order to generate the corresponding channel input symbol sub-frame $\mathbf{x}^r$. However, a different mapping scheme may be employed for each SBURC-encoded sub-frame $\mathbf{v}^r$, facilitating irregular modulation, as shown in Figure 12.13. Suitable $M_{\text{QAM}} = 16$QAM mapping schemes include Gray coding [300], SP [277], Modified Set Partitioning (MSP) [358], the mixed mapping of [358], the Maximum Squared Euclidean Weight (MSEW) mapping of [359] and the $M16^a$ and $M16^r$ mappings of [360].

Following modulation, the resultant channel input symbol sub-frames $\{\mathbf{x}^r\}_{r=1}^R$ are concatenated in order to obtain the channel input symbol frame $\mathbf{x}$. This is transmitted over an uncorrelated narrowband Rayleigh fading channel and received as the channel output symbol frame $\mathbf{y}$, as shown in Figure 12.13. In the IrVLC-SBIrURC receiver of Figure 12.13, the channel output symbol frame $\mathbf{y}$ is decomposed into R sub-frames $\{\mathbf{y}^r\}_{r=1}^R$, each of which is interpreted by a different $M_{\text{QAM}} = 16$QAM demodulator. More specifically, for each channel output symbol, the demodulators determine the probability that the corresponding channel input symbol conveyed each of the $M_{\text{QAM}} = 16$ constellation points. Following this, the $M_{\text{QAM}} = 16$ probabilities associated with each channel output symbol in the sub-frame $\mathbf{y}^r$ are provided as *a priori* information to the corresponding APP SISO SBURC decoder in the form of the logarithmic *a priori* probability sub-frame $L_a^i(\mathbf{v}^r)$, as shown in Figure 12.13.

In the IrVLC-SBIrURC receiver of Figure 12.13, each *a priori* LLR sub-frame $L_a^i(\mathbf{u}'^r)$ is decomposed into sets of $\log_2(M_{\text{QAM}}) = 4$ consecutive LLRs, which are converted into sets of $M_{\text{QAM}} = 16$ log-APPs in a manner similar to that of TCM [277]. The TCM symbol-based trellis [277] is employed to interpret these log-APPs, together with those of the log-APP sub-frame $L_a^i(\mathbf{v}^r)$ provided by the demodulator. Here, the BCJR algorithm [48] is employed to determine sets of $M_{\mathrm{QAM}} = 16$ *a posteriori* log-APPs, which are converted into sets of $\log_2(M_{\mathrm{QAM}}) = 4$ LLRs for the *a posteriori* LLR sub-frame $L_p^i(\mathbf{u}'^r)$.

In addition to mitigating the effective throughput loss of the IrVLC-IrURC scheme of Figure 9.3, the IrVLC-SBIrURC scheme facilitates a higher degree of design freedom, owing to its employment of irregular modulation. While a different EXIT function may be obtained for each component URC code in the IrVLC-IrURC scheme of Figure 9.3, the IrVLC-SBIrURC scheme benefits from a different EXIT function for each *combination* of the component URC code and component 16QAM mapping. As a result, a greater variety of inner EXIT function shapes can be obtained, facilitating the improved joint matching of the inner and outer EXIT functions, as described in Section 9.3.

Note, however, that APP SISO SBURC decoders are associated with a significantly higher computational complexity than are their bit-based equivalents, owing to the significantly higher number of trellis transitions that they employ [277]. For example, a $M_{\mathrm{QAM}} = 16$-ary SBURC employing just one memory element in its LFSR is associated with a trellis that employs $M_{\mathrm{QAM}} = 16$ transitions from each of $M_{\mathrm{QAM}} = 16$ states for each set of $\log_2(M_{\mathrm{QAM}}) = 4$ bits. By contrast, the equivalent bit-based URC employs two transitions from each of two states for each bit. We can therefore expect the APP SISO decoder of the described SBURC to have a 16 times higher complexity than that of the equivalent bit-based URC. With reference to Table 9.7, we may observe that a 16 times increase in the inner APP SISO decoder's complexity would cause it to eclipse that of the outer APP SISO decoder and dominate the iterative decoding complexity. This could be countered, however, by employing the novel modification to the EXIT-chart matching algorithm [149] of Section 9.2 for the sake of jointly performing EXIT chart matching and seeking a reduced SBIrURC computational complexity.

In the light of these discussions, our future work will consider the design and characterization of the IrVLC-SBIrURC scheme of Figure 9.3.

12.13   Closing Remarks

Throughout this book we have introduced novel IrVLC-aided wireless telecommunication schemes and methodologies for their design, in the pursuit of near-capacity operation. In Chapter 6 we developed a scheme without making any particular effort to facilitate its near-capacity operation. Here, EXIT-chart analysis was only employed in order to quantify how close to capacity the scheme could operate. By contrast, EXIT-chart analysis was employed as an integral part of the design process in Chapter 7. More specifically, EXIT-chart matching was employed to shape the IrVLC EXIT function to match that of the serially concatenated inner codec and hence to facilitate near-capacity operation. Further gains were achieved in Chapter 8 by challenging the conventional irregular coding design process of Figure 12.8. Instead of selecting a suite of IrVLC components having a wide variety of EXIT function shapes from a set of many candidates, a suite was directly designed using the RV-FDM of Section 8.2 and the GA of Section 8.3. In Chapter 9 we invoked an irregular inner codec to complement the IrVLC, facilitating a higher degree of design freedom. This was exploited by the joint EXIT-chart matching algorithm of Section 9.3 in order to match the IrVLC and inner EXIT functions to each other, facilitating even 'nearer-to-capacity' operation. Finally, in Section 12.12 outlining our future work, we proposed a method for mitigating the effective throughput loss that was associated with the scheme of Chapter 9, as well as for facilitating the employment of irregular modulation and for providing an even higher degree of design freedom. With these benefits, we may expect to achieve 'very-near-capacity' operation.

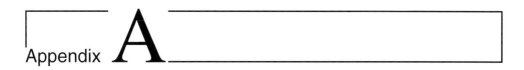

VLC Construction Algorithms

A.1 First RVLC Construction Algorithm

We now introduce the MRG-based algorithm proposed in [21] by slightly paraphrasing it as follows. Continuing the list of definitions in Section 3.3.1, we add:

Definition 12 (Minimum Repetition Gap (MRG) [21]) A codeword $\mathbf{c}$ having k bits is represented as $c_1 c_2 \ldots c_k$. We form a new bit pattern $\mathbf{b}$ by concatenating $\mathbf{c}$ with a k-bit pattern $\mathbf{x}$ yielding the pattern $\mathbf{b} = \mathbf{cx}$. Every bit in $\mathbf{x}$ can be viewed as '0' or '1' in the following calculation. The minimum repetition gap g of $\mathbf{c}$ may be defined as the minimal-length interval that can be created for which it is valid that all the bits in $\mathbf{b}$ are repeated every g bits, ie. appear to be repeated every g bits. This may be formulated mathematically as:

$$\min_{1 \leq g \leq k} \{b_i = b_j\}, \text{ for all } 1 \leq i, j \leq (g + k) \ s.t. \ i \equiv j \pmod{g}.$$

For example, given a codeword $\mathbf{c} = 010$, we have $k = 3$. The resultant $2k$-bit long concatenated pattern is formulated as $\mathbf{b} = 010x_1x_2x_3$. We can opt, for example, for setting $\mathbf{x} = 101$, resulting in $b_1 = b_3 = b_5$ (i.e. $x_2) = 0$ and $b_2 = b_4(x_1) = b_6(x_3) = 1$. Therefore, by observing the resultant six-bit pattern of 010101 we infer that the two-bit pattern of 01 repeats itself. Hence we conclude that g is equal to 2 in this case. Alternatively, we may opt for $\mathbf{x} = 010$, in which case we have a six-bit pattern of 010010 emerging, and hence we infer that the three-bit pattern of 010 repeats itself. Therefore we conclude that g is equal to 3 in this case. Thus we have $\mathrm{MRG} = \min\{2, 3\} = 2$.

The RVLC construction method of [21] is based on the following conjecture. For any two candidate codewords $\mathbf{c}_1$ and $\mathbf{c}_2$ at level l, let g_1 and g_2 denote the MRGs of $\mathbf{c}_1$ and $\mathbf{c}_2$, respectively. $V(\mathbf{c}_1, k)$ and $V(\mathbf{c}_2, k)$ denote the sets of candidate codewords at level k that violate the affix condition for $\mathbf{c}_1$ and $\mathbf{c}_2$. If $g_1 \leq g_2$, then $|V(\mathbf{c}_1, k)| \leq |V(\mathbf{c}_2, k)|$ for $k \geq l$.

Therefore, if we always choose codewords with the smallest MRGs while satisfying the affix condition, we can make the set of usable candidate codewords for the next levels as large as possible.

Near-Capacity Variable-Length Coding   Lajos Hanzo, Robert G. Maunder, Jin Wang and Lie-Liang Yang
© 2011 John Wiley & Sons, Ltd

Let $U = \{u_1, \ldots, u_M\}$ be a M-ary i.i.d. information source with the probability mass function $P_u = \{p_1, \ldots, p_M\}$, $p_1 \geq \cdots \geq p_M$. The RVLC construction algorithm of [21] is then summarized as Algorithm A.1.

Algorithm A.1

Step 1: Construct a Huffman code C_H, which maps the source symbols into binary codewords $\mathbf{w}_1, \ldots, \mathbf{w}_M$ of length $l_1, \ldots, l_M$, $L_{\min} = l_1, \leq \cdots \leq l_M = L_{\max}$. Denote the bit length vector of C_H by $(n_H(1), \ldots, n_H(L_{\max}))$, where $n_H(i)$ represents the number of Huffman codewords having length i. Initialize the bit length vector of the target asymmetrical RVLC, $n(i)$, by $n_H(i)$.

Step 2: Calculate the total number of available candidate codewords of a full binary tree at level i, denoted by $n_a(i)$, and the MRG for each candidate codeword, and then arrange these codewords based on the increasing order of MRGs. If $n(i) \leq n_a(i)$, assign $n(i)$ codewords. Otherwise, assign all the available codewords, and the remaining $(n(i) - n_a(i))$ codewords are added to $n(i+1)$ so as to be assigned at the next stage, i.e.

$$n(i+1) := n(i+1) + n(i) - n_a(i), \quad n(i) := n_a(i).$$

Step 3: Step 2 is repeated until every source symbol has been assigned a codeword.

A.2   Second RVLC Construction Algorithm

In [23] Lakovic and Villasenor improved the above algorithm by establishing the formal relationship between the number of available RVLC codewords of any length and the structure of shorter RVLC codewords that had been selected, and pointed out that the MRG metric was a special case of this relationship. We now re-derive this relationship.

Let a binary codeword $\mathbf{w}$ of length i be denoted as $\mathbf{w} = (w_1 w_2 \cdots w_i)$. Define the prefix set $P_j(\mathbf{w})$ as the set of length-j codewords, $j > i$, which are prefixed by $\mathbf{w}$, and define the suffix set $S_j(\mathbf{w})$ as the set of length-j codewords, $j > i$, which are suffixed by $\mathbf{w}$. It is obvious that

$$|P_j(\mathbf{w})| = |S_j(\mathbf{w})| = 2^{(j-i)}.$$

Assume that $W = \{\mathbf{w}_1, \mathbf{w}_2, \ldots, \mathbf{w}_m\}$ is a set of RVLC codewords of lengths $l_1 \leq l_2 \leq \cdots \leq l_m$. Denote with $n_a(j)$ the number of codewords of length j, $j > l_m$, which are neither prefixed nor suffixed by any codeword from W. Since the total number of length-j codewords is equal to 2^j, it holds that

$$n_a(j) = 2^j - \left| \bigcup_{\mathbf{w}_k \in W} P_j(\mathbf{w}_k) \right| - \left| \bigcup_{\mathbf{w}_k \in W} S_j(\mathbf{w}_k) \right| + \left| \bigcup_{\mathbf{w}_k, \mathbf{w}_h \in W} P_j(\mathbf{w}_k) \cap S_j(\mathbf{w}_h) \right|.$$

For RVLCs all the prefix sets, as well as all the suffix sets, are disjoint, and therefore

$$n_a(j) = 2^j - \sum_{\mathbf{w}_k \in W} (|P_j(\mathbf{w}_k)| + |S_j(\mathbf{w}_k)|) + \sum_{\mathbf{w}_k \in W} \sum_{\mathbf{w}_h \in W} |P_j(\mathbf{w}_k) \cap S_j(\mathbf{w}_h)|$$

$$= 2^j - \sum_{k=1}^{m} 2^{j-l_k+1} + \sum_{\mathbf{w}_k \in W} \sum_{\mathbf{w}_h \in W} |P_j(\mathbf{w}_k) \cap S_j(\mathbf{w}_h)|.$$

Now define the affix set $A_j(\mathbf{w}_k, \mathbf{w}_h)$ as the set of codewords of length j, $j > \max(l_k, l_h)$, prefixed by $\mathbf{w}_k$ and suffixed by $\mathbf{w}_h$:

$$A_j(\mathbf{w}_k, \mathbf{w}_h) = P_j(\mathbf{w}_k) \cap S_j(\mathbf{w}_h).$$

Obviously, $n_a(j)$ depends on the cardinalities of affix sets $A_j(\mathbf{w}_k, \mathbf{w}_h)$, i.e., on the affix index of the set W, defined as

$$a_j(W) = \sum_{\mathbf{w}_k \in W} \sum_{\mathbf{w}_h \in W} |A_j(\mathbf{w}_k, \mathbf{w}_h)|.$$

The cardinalities of affix sets $A_j(\mathbf{w}_k, \mathbf{w}_h)$ are given as follows [27]. For $\max(l_k, l_h) < j < l_k + l_h$,

$$|A_j(\mathbf{w}_k, \mathbf{w}_h)| = \begin{cases} 1, & \text{if } (w_{k_{j-l_h+1}} w_{k_{j-l_h+2}} \cdots w_{k_{l_k}}) = (w_{h_1} w_{h_2} \cdots w_{h_{l_k+l_h-j}}) \\ 0, & \text{otherwise} \end{cases}$$

while for $l_k + l_h \leq j$,

$$|A_j(\mathbf{w}_k, \mathbf{w}_h)| = 2^{j-l_k-l_h}.$$

With the above formulas, the MRG can be interpreted as follows. The MRG of a length-i codeword $\mathbf{w}$, denoted as $g(\mathbf{w})$, is equal to the minimal positive integer g such that $|A_{i+g}(w, w)| > 0$. Algorithm A.1 [21] selects the codewords with the lowest $g(\mathbf{w})$, which contributes to the increase in $n_a(i + g)$, with the lowest g first, because $n_a(i + g)$ depends on $|A_{i+g}(\mathbf{w}, \mathbf{w})|$. However, this algorithm does not consider the elements $|A_{i+g}(\mathbf{w}_k, \mathbf{w}_h)|$, $\mathbf{w}_k \neq \mathbf{w}_h$, which have a significant impact on $n_a(i + g)$, and consequently on the coding efficiency. Therefore, Lajovic and Villasenor [23] proposed a new construction algorithm, shown here as Algorithm A.2, that differed from Algorithm A.1

Algorithm A.2 can generate RVLCs with smaller average code lengths than Algorithm A.1; however, the complexity is much higher. At each level i, with $n_a(i) \geq n(i)$, Algorithm A.1 considers $n_a(i)$ MRGs, while Algorithm A.2 considers $\binom{n_a(i)}{n(i)}$ affix indices.

A.3 Greedy Algorithm and Majority Voting Algorithm (MVA)

Both the greedy algorithm and the MVA may be applied for the construction of a fixed-length code of length n having a given block distance of d from a given set of codewords that may not necessarily contain all possible $n$tuples. Given a set of codewords W, the greedy algorithm initialises the target codeword set C with an arbitrary codeword in W, and then selects the first available codeword in W that is at a distance of at least d from

Algorithm A.2

Step 1 and Step 3: Same as Step 1 and Step 3 of Algorithm A.1.

Step 2: If $n(i) \leq n_a(i)$, $n(i)$ codewords should be assigned. There are $\binom{n_a(i)}{n(i)}$ candidate sets to be selected. Calculate $n_a(i+1)$ for all possible selections, and select the set with the maximal $n_a(i+1)$. If there are several candidate sets that yield the same $n_a(i+1)$, choose the set that yields the highest $n_a(i+2)$. If $n(i) > n_a(i)$, assign all the available codewords, and the remaining $(n(i) - n_a(i))$ codewords are added to $n(i+1)$ so as to be assigned at the next stage, i.e.

$$n(i+1) = n(i+1) + n(i) - n_a(i), \quad n(i) = n_a(i).$$

all the existing codewords in the set C. This procedure is repeated until no more legitimate codewords are available in the set W. This algorithm is easy to implement; however, the number of codewords in the resultant codeword set may be far from the maximum achievable value, since it ignores the effect that the codewords selected at the current stage will limit the number of legitimate codewords at the next stage. By contrast, the MVA is usually capable of generating larger codeword sets by 'looking' a step forward. Its procedure can be described as in Algorithm A.3 [30].

Algorithm A.3

Step 1: Initialize the target code C to the null code (i.e. no codewords selected).

Step 2: For each codeword in W determine the number of remaining codewords in W that are at least distance d from the given codeword.

Step 3: Choose the codeword that has got the maximum number of other codewords that satisfy the minimum distance requirement. If there is more than one such codeword, then choose an arbitrary one. Add this codeword to the target code C.

Step 4: Remove from W all those codewords that do not satisfy the minimum distance requirement to the chosen codeword.

Step 5: Repeat Steps 2–4 with the new set W. Each time, the selected codeword is added to the code C. The algorithm ends when W is empty.

The main disadvantage of Algorithm A.3 is that as the number of codewords in W increases it becomes too computationally expensive, and the greedy algorithm will then be preferred.

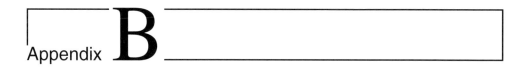

SISO VLC Decoder

In Chapter 3 we derive the MAP VLC decoder as an inner code, which processes the channel output directly. In the following we derive the APP VLC decoder as an outer code, which processes the extrinsic output of the inner code. We derive the decoding procedure based on the SISO APP module introduced in [254], which was a slightly modified version of the BCJR algorithm [48] and was originally designed for convolutional codes.

The SISO module is a four-terminal device having two inputs and two outputs, as shown in Figure B.1. It accepts as its inputs $P(\mathbf{u}; \mathbf{I})$ and $P(\mathbf{c}; \mathbf{I})$, namely the probability distributions of the information symbols $\mathbf{u}$ and code symbols $\mathbf{c}$ labeling the edges of the code trellis, and forms the resultant outputs, $P(\mathbf{u}; \mathbf{O})$ and $P(\mathbf{c}; \mathbf{O})$, which constitute an update of these distributions based upon the code constraints. These outputs represent the extrinsic information.

Following the notation of [254], the extrinsic information is calculated as follows. At time k the output probability distribution is evaluated as

$$\tilde{P}_k(c:O) = \tilde{H}_k \sum_{e:c(e)=c} \alpha_{k-1}(s^S(e))\beta_k(s^E(e))\gamma_k(e), \qquad \text{(B.1)}$$

where e represents a branch of the trellis, and $c(e)$, $s^S(e)$, and $s^E(e)$ are, respectively, the output symbol, the starting state and the terminating state of the branch e, while $\tilde{H}_k$ is a normalizing factor that ensures maintaining $\tilde{P}_k(0; O) + \tilde{P}_k(1; O) = 1$. The quantities $\alpha_k(\cdot)$ and $beta_k(\cdot)$ are obtained through the forward and backward recursions [254], respectively,

Figure B.1: A SISO module.

Near-Capacity Variable-Length Coding Lajos Hanzo, Robert G. Maunder, Jin Wang and Lie-Liang Yang
© 2011 John Wiley & Sons, Ltd

which can be expressed as

$$\alpha_k(s) = \sum_{e:s^E(e)=s} \alpha_{k-1}[s^S(e)]\gamma_k(e), \tag{B.2}$$

$$\beta_k(s) = \sum_{e:s^S(e)=s} \beta_{k+1}[s^S(e)]\gamma_{k+1}(e), \tag{B.3}$$

in conjunction with the initial values of $\alpha_0(s) = 0$ and $\beta_N(s) = 0$ for all states except for the root state, since the trellis always starts and ends at the root state.

In order to incorporate the source information, the branch transition probability in [254] is modified as follows:

$$\gamma_k(e) = P_k(c(e); I) \times P(e), \tag{B.4}$$

where $P(e) \triangleq P(c(e), s^E(e)|s^S(e))$ is the source *a priori* information associated with the branch transition probability. It is time invariant and determined by the source distribution, which can be calculated with the aid of the state transition probabilities as [249]

$$P(c(e), s^E(e)|s^S(e)) = \frac{\sum_{i \in g(s^E(e))} p_i}{\sum_{j \in g(s^S(e))} p_j}, \tag{B.5}$$

where $g(n)$ are all the codeword indices associated with node n in the code tree.

Therefore, the extrinsic information can be extracted by excluding the input probability of $P_k(c; I)$ from the output probability, yielding

$$P_k(c; O) = \frac{\tilde{P}_k(c; O)}{P_k(c; I)}$$

$$= H_k \sum_{e:c(e)=c} \alpha_{k-1}(s^S(e))\beta_k(s^E(e))P(e), \tag{B.6}$$

where, again, H_k is a normalization factor. To reduce the computational complexity, the above multiplicative algorithm can be converted to an additive form in the log domain. The logarithms of the above quantities are defined as follows:

$$\pi_k(c; I) \triangleq \ln[P_k(c; I)] \tag{B.7}$$

$$\pi_k(c; O) \triangleq \ln[P_k(c; O)] \tag{B.8}$$

$$A_k(s) \triangleq \ln[\alpha_k(s)] \tag{B.9}$$

$$B_k(s) \triangleq \ln[\beta_k(s)] \tag{B.10}$$

$$\Gamma_k(e) \triangleq \ln[\gamma_k(e)]. \tag{B.11}$$

With the aid of these definitions, the SISO algorithm defined by Equations (B.2), (B.3) and (B.6) can be correspondingly described as follows:

$$\pi_k(c; O) = \ln\left[\sum_{e:c(e)=c} \exp\{A_{k-1}[s^S(e)] + \Gamma_k(e) + B_k[s^E(e)]\}\right] + h_k \qquad \text{(B.12)}$$

where h_k is a normalization constant. The quantities $A_k(\cdot)$ and $B_k(\cdot)$ can be obtained through the forward and backward recursions, respectively, as follows:

$$A_k(s) = \ln\left[\sum_{e:s^E(e)=s} \exp\{A_{k-1}[s^S(e)] + \Gamma_k[e]\}\right] \qquad \text{(B.13)}$$

$$B_k(s) = \ln\left[\sum_{e:s^S(e)=s} \exp\{B_{k+1}[s^S(e)] + \Gamma_{k+1}[e]\}\right] \qquad \text{(B.14)}$$

with the initial values

$$A_0(s) = \begin{cases} 0 & s \text{ is the root node} \\ -\infty & \text{otherwise} \end{cases}$$

$$B_N(s) = \begin{cases} 0 & s \text{ is the root node} \\ -\infty & \text{otherwise.} \end{cases}$$

For simplicity, we use LLRs instead of probabilities themselves. Therefore, the output extrinsic information in the form of LLRs can be computed as

$$L(c_k; O) = \pi_k(c = 0; O) - \pi_k(c = 1; O)$$

$$= \ln\left[\sum_{e:c(e)=0} \exp\{A_{k-1}[s^S(e)] + \Gamma_k[e] + B_k[s^E(e)]\}\right]$$

$$- \ln\left[\sum_{e:c(e)=1} \exp\{A_{k-1}[s^S(e)] + \Gamma_k[e] + B_k[s^E(e)]\}\right]. \qquad \text{(B.15)}$$

Correspondingly, the *a priori* probability $P_k(c; I)$ may also be calculated from the *a priori* LLR $L(c_k; I)$ as

$$P_k(c; I) = \frac{e^{\frac{c_k}{2}L(c_k;I)}}{e^{-\frac{1}{2}L(c_k;I)} + e^{\frac{1}{2}L(c_k;I)}}$$

$$= \eta_k \, e^{\frac{c_k}{2}L(c_k;I)} \qquad \text{(B.16)}$$

where η_k is independent of the value of c_k, and can be canceled out. Hence, the logarithm of the *a priori* probability may be calculated as

$$\pi_k(c_k; I) = \frac{c_k}{2} L(c_k; I). \qquad \text{(B.17)}$$

Appendix C

APP Channel Equalization

The APP equalizer outputs the channel information, which is expressed as $\{L_f(d_k)\}$, associated with all data bits, given N_c received samples hosted by the vector $\mathbf{y} = y_1^{N_c}$ as well as the *a priori* information $\{L_b(d_k)\}$ corresponding to all data bits. The feed-forward information $L_f(d_k)$ corresponding to bit d_k is given by

$$L_f(d_k) \triangleq \ln \frac{P(d_k = +1|\mathbf{y})}{P(d_k = -1|\mathbf{y})} - \ln \frac{P(d_k = +1)}{P(d_k = -1)}$$
$$= L(d_k|\mathbf{y}) - L_b(d_k) \qquad (\text{C.1})$$

where $L(d_k|\mathbf{y}) = \ln[P(d_k = +1|\mathbf{y})/P(d_k = -1|\mathbf{y})]$ represents the *a posteriori* Logarithmic Likelihood Ratio (LLR) of the data bit d_k, which may be computed using the algorithm described below by following the approaches outlined in [69, 70, 72, 99].

The *a posteriori* LLR of the data bit d_k can be expressed in conjunction with the channel states s_{k-1} and s_k as

$$L(d_k|\mathbf{y}) = \ln \frac{P(d_k = +1|\mathbf{y})}{P(d_k = -1|\mathbf{y})}$$
$$= \ln \frac{\sum_{\forall (i,j) \in \mathcal{B}:\, d_{i,j}=+1} P(s_k = q_j;\, s_{k-1} = q_i|y_1^{N_c})}{\sum_{\forall (i,j) \in \mathcal{B}:\, d_{i,j}=-1} P(s_k = q_j;\, s_{k-1} = q_i|y_1^{N_c})}$$
$$= \ln \frac{\sum_{\forall (i,j) \in \mathcal{B}:\, d_{i,j}=+1} P(s_k = q_j;\, s_{k-1} = q_i;\, y_1^{N_c})}{\sum_{\forall (i,j) \in \mathcal{B}:\, d_{i,j}=-1} P(s_k = q_j;\, s_{k-1} = q_i;\, y_1^{N_c})}. \qquad (\text{C.2})$$

The sequence $y_1^{N_c}$ involved in the computation of the probability $P(s_{k-1}, s_k, y_1^{N_c})$ can be written in a more detailed form as $P(s_{k-1}, s_k, y_1^{k-1}, y_k, y_{k+1}^{N_c})$. Applying the chain rule for joint probabilities, i.e., exploiting that we have $P(a, b) = P(a)P(b|a)$, this expression

Near-Capacity Variable-Length Coding Lajos Hanzo, Robert G. Maunder, Jin Wang and Lie-Liang Yang
© 2011 John Wiley & Sons, Ltd

yields the following decomposition of $P(s_{k-1}, s_k, y_1^{N_c})$:

$$P(s_{k-1}, s_k, y_1^{N_c}) = P(s_{k-1}, y_1^{k-1})P(s_k, y_k|s_{k-1})P(y_{k+1}^{N_c}|s_k). \tag{C.3}$$

Let us introduce the following definitions:

$$\alpha_k(s) \triangleq P(s_k = s, y_1^k), \tag{C.4}$$

$$\beta_k(s) \triangleq P(y_{k+1}^{N_c}|s_k), \tag{C.5}$$

$$\gamma_k(s_{k-1}, s_k) \triangleq P(s_k, y_k|s_{k-1}). \tag{C.6}$$

Then according to [48] the term $\alpha_k(s)$ can be computed via the recursion

$$\alpha_k(s) = \sum_{\forall\, s' \in \mathcal{S}} \alpha_{k-1}(s')\gamma_k(s', s) \tag{C.7}$$

associated with the initial values of $\alpha_0(q_0) = 1$ and $\alpha_0(q_i) = 0$ for $i \neq 0$. The term $\beta_k(s)$ can be computed via the recursion

$$\beta_k(s) = \sum_{\forall\, s' \in \mathcal{S}} \beta_{k+1}(s')\gamma_{k+1}(s, s'), \tag{C.8}$$

in conjunction with the initial values of $\beta_{N_c}(q_0) = 1$ and $\beta_{N_c}(q_i) = 0$ for $i \neq 0$. The term $\gamma_k(s_{k-1}, s_k)$ represents the transition probability from state s_{k-1} to state s_k, which can be further decomposed into

$$\gamma_k(s_{k-1}, s_k) = P(s_k|s_{k-1}) \cdot p(y_k|s_{k-1}, s_k). \tag{C.9}$$

The transition probability $\gamma_k(s_{k-1} = q_i, s_k = q_j)$ is zero if the index pair (i, j) is not in $\mathcal{B}$. For pairs (i, j) within $\mathcal{B}$, the probability $P(s_k = q_j|s_{k-1} = q_i)$ is governed by the corresponding input symbol $d_{i,j}$, and $p(y_k|s_{k-1}, s_k)$ is governed by the corresponding output symbol $\nu_{i,j}$, which can be described as

$$\gamma_k(s_{k-1} = q_i, s_k = q_j) = \begin{cases} P(d_k = d_{i,j}) \cdot p(y_k|\nu_k = \nu_{i,j}), & (i, j) \in \mathcal{B} \\ 0 & (i, j) \notin \mathcal{B}. \end{cases} \tag{C.10}$$

Given the AWGN channel of Equation (4.49), it follows that $p(y_k|\nu_k)$ is given by

$$p(y_k|\nu_k) = \frac{1}{\sqrt{2\pi\sigma^2}} \exp\left(-\frac{(y_k - \nu_k)^2}{2\sigma^2}\right). \tag{C.11}$$

From Equation (C.2) and from the decomposition of $P(s_{k-1}, s_k, y_1^{N_c})$ given in Equation (C.3), it follows that

$$L(d_k|y_1^{N_c}) = \ln \frac{\sum_{\forall (i,j) \in \mathcal{B}:\, d_{i,j}=+1} \alpha_{k-1}(q_i) \cdot \gamma_k(q_i, q_j) \cdot \beta_k(q_j)}{\sum_{\forall (i,j) \in \mathcal{B}:\, d_{i,j}=-1} \alpha_{k-1}(q_i) \cdot \gamma_k(q_i, q_j) \cdot \beta_k(q_j)}. \tag{C.12}$$

As shown in Equation (C.10), once the channel output samples are available the corresponding values of $\gamma_k(s', s)$ can be computed and stored. With the aid of these values, both $\alpha_k(s)$ and $\beta_k(s)$ of Equations (C.7) and (C.8) can be computed, and, finally, the *a posteriori* LLR of Equation (C.12) can be obtained. For the sake of reducing the associated complexity, these computations can be performed in the log domain, as follows.

Let us introduce the following notations:

$$A_k(s) \triangleq \ln[\alpha_k(s)], \tag{C.13}$$

$$B_k(s) \triangleq \ln[\beta_k(s)], \tag{C.14}$$

$$\Gamma_k(s', s) \triangleq \ln[\gamma_k(s', s)]. \tag{C.15}$$

Consequently, $A_k(s)$ can be formed iteratively by:

1. **Initialization**:

$$A_0(0) = 0 \quad \text{and} \quad A_0(s \neq 0) = -\infty; \tag{C.16}$$

2. **Forward Recursion**:

$$A_k(s) = \ln\left[\sum_{\forall\, s' \in \mathcal{S}} \exp\{A_{k-1}(s') + \Gamma_k(s', s)\}\right], \quad k = 1, \ldots, N_c. \tag{C.17}$$

$B_k(s)$ of (C.14) can be solved similarly, by:

1. **Initialization**:

$$B_{N_c}(0) = 0 \quad \text{and} \quad A_{N_c}(s \neq 0) = -\infty; \tag{C.18}$$

2. **Backward Recursion**:

$$B_k(s) = \ln\left[\sum_{\forall\, s' \in \mathcal{S}} \exp\{B_{k+1}(s') + \Gamma_{k+1}(s, s')\}\right], \quad k = N_c - 1, \ldots, 0. \tag{C.19}$$

Furthermore, given the L-values of the encoded bits d_k, i.e., given the *a priori* information of $L(d_k)$, and given the channel observations y_k, $\Gamma_k(s', s)$ seen in Equation (C.15) can be calculated as follows:

$$\Gamma_k(q_i, q_j) = \begin{cases} \dfrac{1}{2} d_{i,j} L(d_k) - \dfrac{(y_k - \nu_k)^2}{2\sigma^2}, & (i, j) \in \mathcal{B} \\ 0 & (i, j) \notin \mathcal{B} \end{cases} \quad k = 1, \ldots, N_c. \tag{C.20}$$

Finally, the *a posteriori* LLR of Equation (C.12) can be obtained as follows:

$$L(d_k | y_1^{N_c}) = \ln \sum_{\forall (i,j) \in \mathcal{B}:\, d_{i,j} = +1} \exp\{A_{k-1}(q_i) + \Gamma_k(q_i, q_j) + B_k(q_j)\}$$

$$- \ln \sum_{\forall (i,j) \in \mathcal{B}:\, d_{i,j} = -1} \exp\{A_{k-1}(q_i) + \Gamma_k(q_i, q_j) + B_k(q_j)\}$$

$$k = 1, \ldots, N_c. \tag{C.21}$$

The above calculations entail the evaluation of the logarithm of a sum of exponentials. These calculations may be carried out using either either the MAX-Log approximation or the Jacobian logarithm employing a look-up table [250].

Bibliography

[1] C. E. Shannon, 'A mathematical theory of communication,' *Bell Systems Technical Journal*, vol. 27, pp. 379–423, 623–656, 1948.

[2] L. Hanzo, P. J. Cherriman and J. Streit, *Wireless Video Communications: Second to Third Generation Systems and Beyond*. IEEE Press, 2001.

[3] F. I. Alajaji, N. C. Phamdo and T. E. Fuja, 'Channel codes that exploit the residual redundancy in CELP-encoded speech,' *IEEE Transactions on Speech Audio Processing*, vol. 4, pp. 325–336, September 1996.

[4] K. Sayood and J. C. Borkenhagen, 'Use of residual redundancy in the design of joint source/channel coders,' *IEEE Transactions on Communications*, vol. 39, pp. 838–846, June 1991.

[5] K. Sayood, F. Liu and J. D. Gibson, 'A constrained joint source/channel coder design,' *IEEE Journal in Selected Areas of Communications*, vol. 12, pp. 1584–1593, December 1994.

[6] R. E. V. Dyck and D. J. Miller, 'Transport of wireless video using separate, concatenated, and joint source-channel coding,' *Proceedings of IEEE*, vol. 87, no. 10, pp. 1734–1750, October 1999.

[7] C. Guillemot and P. Siohan, 'Joint source-channel decoding of variable-length codes with soft information: A survey,' *EURASIP Journal on Applied Signal Processing*, pp. 906–927, June 2005.

[8] J. Proakis, *Digital Communications*, 4th ed. New York: McGraw-Hill, 2001.

[9] D. A. Huffman, 'A method for the construction of minimum redundancy codes,' *Proceedings of Institute of Radio Engineering*, vol. 40, pp. 1098–1101, September 1952.

[10] J. J. Rissanen, 'Arithmetic codings as number representations,' *Acta Polytechnica Scandinavica*, vol. 31, pp. 44–51, 1979.

[11] J. C. Kieffer, 'A survey of the theory of source coding with a fidelity criterion,' *IEEE Transactions on Information Theory*, pp. 1473–1490, September 1993.

[12] Y. Takishima, M. Wada and H. Murakami, 'Reversible Variable Length Codes,' *IEEE Transactions on Communications*, vol. 43, no. 2/3/4, pp. 158–162, Feb./Mar./Apr. 1995.

[13] J. Wen and J. D. Villasenor, 'A class of reversible variable length codes for robust image and video coding,' in *Proceedings of IEEE International Conference on Image Processing*, vol. 2, (Lausanne, Switzerland), October 1997, pp. 65–68.

[14] ——, 'Reversible variable length codes for efficient and robust image and video coding,' in *Data Compression Conference (DCC'98)* , Snowbird, USA, 1998, pp. 471–480.

[15] S. W. Golomb, 'Run-length encodings,' *IEEE Transactions on Information Theory*, pp. 1473–1490, July 1966.

[16] R. F. Rice, 'Some practical universal noiseless coding techniques,' *Technical Report JPL-79-22, Jet Propulsion Laboratory, Pasadena, California*, pp. 79–22, March 1979.

[17] J. Teuhola, 'A compression method for clustered bit-vectors,' *Information Processing Letters*, pp. 308–311, October 1978.

[18] 'Video coding for low bit rate communication,' *ITU-T Recommendation H.263*, 1998.

[19] 'Generic coding of audio-visual objects,' *ISO/IEC 14496-2 MPEG-4*, 1999.

[20] C. Tsai and J. Wu, 'Modified symmetrical reversible variable-length code and its theoretical bounds,' *IEEE Transactions on Information Theory*, vol. 47, no. 6, pp. 2543–2548, September 2001.

[21] ——, 'On constructing the Huffman-code-based reversible variable-length codes,' *IEEE Transactions on Communications*, vol. 49, no. 9, pp. 1506–1509, September 2001.

[22] K. Lakovic and J. Villasenor, 'On design of error-correcting reversible variable length codes,' *IEEE Communications Letters*, vol. 6, no. 8, pp. 337–339, August 2002.

[23] ——, 'An algorithm for construction of efficient fix-free codes,' *IEEE Communications Letters*, vol. 7, no. 8, pp. 391–393, August 2003.

[24] C. Lin, Y. Chuang and J. Wu, 'Generic construction algorithms for symmetric and asymmetric RVLCs,' in *Proceedings of IEEE International Conference on Computational Science(ICCS)*, Amsterdam, The Netherlands, April 2002, pp. 968–972.

[25] Z. Kukorelly and K. Zeger, 'New binary fix-free codes with Kraft sum 3/4,' in *IEEE International Symposium on Information Theory*, Lausanne, Switzerland, June 2002, p. 178.

[26] S. Yekhanin, 'Sufficient conditions for existence of fix-free codes,' in *IEEE International Symposium on Information Theory*, Washington, USA, June 2001, p. 284.

[27] C. Ye and R. W. Yeung, 'Some basic properties of fix-free codes,' *IEEE Transactions on Information Theory*, vol. 47, no. 1, pp. 72–87, January 2001.

[28] V. Buttigieg and P. G. Farrell, 'On variable-length error-correcting codes,' in *IEEE International Symposium on Information Theory (ISIT'94)*, Trondheim, Norway, June–July 1994, p. 507.

[29] ——, 'Variable-length error-correcting codes,' *IEE Proc.-Commun.*, vol. 147, no. 4, pp. 211–215, August 2000.

[30] V. Buttigieg, 'Variable-length error-correcting codes,' Ph.D. dissertation, University of Manchester, Department of Electrical Engineering, 1995.

[31] C. Lamy and J. Paccaut, 'Optimised constructions for variable-length error correcting codes,' in *Proceedings of IEEE Information Theory Workshop (ITW2003)*, Paris, France, March 31–April 4 2003, pp. 183–186.

[32] J. Wang, L.-L. Yang and L. Hanzo, 'Iterative Construction of Reversible Variable Length Codes and Variable Length Error Correcting Codes,' *IEEE Communications Letters*, pp. 671–673, August 2004.

[33] M. Schindler, 'Practical Huffman Coding. [Online],' Available: http://www.compressconsult.com/huffman/.

[34] C. Fogg, 'Survey of software and hardware VLC architectures,' in *SPIE Proceedings on Image and Video Compression Conference*, May 1994, pp. 29–37.

[35] P. C. Tseng, Y. C. Chang, Y. W. Huang, H. C. Fang, C. T. Huang and L. G. Chen, 'Advances in hardware architectures for image and video coding-A survey,' *Proceedings of IEEE*, vol. 93, no. 1, pp. 184–197, January 2005.

[36] M. Park and D. J. Miller, 'Decoding entropy-coded symbols over noisy channels by MAP sequence estimation for asynchronous HMMs,' in *Proceedings of Conference on Information Sciences and Systems*, Princeton, NJ, USA, March 1998, pp. 477–482.

[37] ——, 'Joint source-channel decoding for variable-length encoded data by exact and approximate MAP sequence estimation,' in *Proceedings of IEEE International Conference on Acoustics, Speech, Signal Processing (ICASSP'99)*, Phoenix, AZ, USA, March 1999, pp. 2451–2454.

[38] N. Demir and K. Sayood, 'Joint source/channel coding for variable length codes,' in *Proceedings of Data Compression Conference (DCC'98)*, Snowbird, UT, USA, March-April 1998, pp. 139–148.

[39] D. J. Miller and M. Park, 'A sequence-based approximate MMSE decoder for source coding over noisy channels using discrete hidden Markov models,' *IEEE Transactions on Communications*, vol. 46, no. 2, pp. 222–231, 1998.

[40] R. Bauer and J. Hagenauer, 'Symbol by Symbol MAP Decoding of Variable Length Codes,' in *Proceedings of 3rd International ITG Conference on Source and Channel Coding*, Munich, Germany, January 2000, pp. 111–116.

[41] V. B. Balakirsky, 'Joint Source-Channel Coding with Variable Length Codes,' in *Proceedings of 1997 IEEE International Symposium on Information Theory(ISIT'97)*, Ulm Germany, July 1997, p. 419.

[42] R. Bauer and J. Hagenauer, 'On Variable Length Codes for Iterative Source/Channel Decoding,' in *Proceedings of IEEE Data Compression Conference (DCC)*, Snowbird, USA, April 2001, pp. 273–282.

[43] J. Kliewer and R. Thobaben, 'Combining FEC and optimal soft-input source decoding for the reliable transmission of correlated variable-length encoded signals,' in *Proceedings of CoIEEE Data Compression Conference*, Snowbird, UT, USA, April 2002, pp. 83–91.

[44] R. Thobaben and J. Kliewer, 'Robust decoding of variable-length encoded Markov sources using a three-dimensional trellis,' *IEEE Communications Letters*, vol. 7, no. 7, pp. 320–322, July 2003.

[45] ——, 'Low complexity iterative joint source-channel decoding for variable-length encoded Markov sources,' *IEEE Transactions on Communications*, vol. 53, no. 12, pp. 2054–2064, December 2005.

[46] A. Guyader, E. Fabre, C. Guillemot and M. Robert, 'Joint source-channel turbo decoding of entropy-coded sources,' *IEEE Journal of Selected Areas in Communications*, vol. 19, no. 9, pp. 1680–1696, September 2001.

[47] A. J. Viterbi, 'Error Bounds for Convolutional Codes and an Asymptotically Optimum Decoding Algorithm,' *IEEE Transactions on Information Theory*, vol. IT-13, pp. 260–269, April 1967.

[48] L. Bahl, J. Cocke, F. Jelinek and J. Raviv, 'Optimal decoding of linear codes for minimizing symbol error rate (Corresp.),' *IEEE Transactions on Information Theory*, vol. 20, no. 2, pp. 284–287, March 1974.

[49] M. Bystrom, S. Kaiser and A. Kopansky, 'Soft source decoding with applications,' *IEEE Transaction on Circuits and Systems for Video Technology*, vol. 11, no. 10, pp. 1108–1120, October 2001.

[50] L. Perros-Meilhac and C. Lamy, 'Huffman tree-based metric derivation for a low-complexity sequential soft VLC decoding,' in *Proceedings of IEEE International Conference on Communications (ICC'02)*, New York, NY, USA, April–May 2002, pp. 783–787.

[51] P. Sweeney, *Error Control Coding: From Theory to Practice*. John Wiley & Sons, 2002.

[52] A. H. Murad and T. E. Fuja, 'Joint source-channel decoding of variable-length encoded sources,' in *Proceedings of Information Theory Workshop (ITW'98)*, Killarney, Ireland, June 1998, pp. 94–95.

[53] J. Hagenauer, 'The turbo principle: Tutorial introduction and state of the art,' in *Proceedings of the 1st International Symposium on Turbo Codes and Related Topics*, Brest, France, 1997, pp. 1–12.

[54] N. Gortz, 'On the iterative approximation of optimal joint source-channel decoding,' *IEEE Journal on Selected Areas in Communications*, vol. 19, no. 9, pp. 1662–1670, September 2001.

[55] J. Wen and J. D. Villasenor, 'Utilizing soft information in decoding of variable length codes,' in *Data Compression Conference (DCC'99)*, Snowbird, USA, 1999, pp. 131–139.

[56] M. Adrat, U. von Agris and P. Vary, 'Convergence behavior of iterative source-channel decoding,' in *IEEE International Conference on Acoustics, Speech, Signal Processing (ICASSP'03)*, Hong Kong, China, April 2003, pp. 269–272.

[57] J. Kliewer and R. Thobaben, 'Parallel concatenated joint source-channel coding,' *Electronics Letters*, vol. 39, no. 23, pp. 1664–1666, 2003.

[58] J. Hagenauer, 'Source-controlled channel decoding,' *IEEE Transactions on Communications*, vol. 43, no. 9, pp. 2449–2457, September 1995.

[59] R. Steele and L. Hanzo, Eds., *Mobile Radio Communications: Second and Third Generation Cellular and WATM Systems*, 2nd ed. New York, USA: IEEE Press – John Wiley & Sons, 1999.

[60] J. Garcia-Frias and J. D. Villasenor, 'Joint turbo decoding and estimation of hidden Markov sources,' *IEEE Journal on Selected Areas in Communications*, vol. 19, no. 9, pp. 1671–1679, September 2001.

[61] G.-C. Zhu, F. Alajaji, J. Bajcsy and P. Mitran, 'Non-systematic turbo codes for non-uniform i.i.d. sources over AWGN channels,' in *Proceedings of Conference on Information Sciences and Systems (CISS'02)*, Princeton, NJ, USA, March 2002.

[62] A. Elbaz, R. Pyndiah, B. Solaiman and O. Ait Sab, 'Iterative decoding of product codes with a priori information over a Gaussian channel for still image transmission,' in *Proceedings of IEEE Global Telecommunications Conference (GLOBECOM'99)*, Rio de Janeiro, Brazil, December 1999.

[63] Z. Peng, Y.-F. Huang and D. J. Costello, 'Turbo codes for image transmission-a joint channel and source decoding approach,' *IEEE Journal on Selected Areas in Communications*, vol. 18, no. 6, pp. 868–879, 2000.

[64] L. Guivarch, J.-C. Carlach and P. Siohan, 'Joint source-channel soft decoding of Huffman codes with turbo-codes,' in *Proceedings of IEEE Data Compression Conference (DCC'00)*, Snowbird, Utah, USA, March 2000, pp. 8–92.

[65] M. Jeanne, J.-C. Carlach and P. Siohan, 'Joint source-channel decoding of variable-length codes for convolutional codes and turbo codes,' *IEEE Transactions on Communications*, vol. 53, no. 1, pp. 10–15, 2005.

[66] L. Guivarch, J.-C. Carlach and P. Siohan, 'Low complexity soft decoding of Huffman encoded Markov sources using turbo-codes,' in *Proceedings of IEEE 7th International Conference on Telecommunications (ICT'00)*, Acapulco, Mexico, May 2000, pp. 872–876.

[67] X. Jaspar and L. Vandendorpe, 'Three SISO modules joint source-channel turbo decoding of variable length coded images,' in *Proceedings of 5th International ITG Conference on Source and Channel Coding (SCC'04)*, Erlangen, Germany, January 2004, pp. 279–286.

[68] N. Phamdo and N. Farvardin, 'Optimal detection of discrete Markov sources over discrete memoryless channels – applications to combined source-channel coding,' *IEEE Transactions on Information Theory*, vol. 40, no. 1, pp. 186–193, 1994.

[69] C. Berrou, A. Glavieux and P. Thitimajshima, 'Near-Shannon limit error-correcting coding and decoding: Turbo-codes,' in *Proceedings of IEEE International Conference on Communications (ICC'93)*, Geneva, Switzerland, May 1993, pp. 1064–1070.

[70] L. Hanzo, T. H. Liew and B. L. Yeap, *Turbo Coding, Turbo Equalisation and Space-time Coding for Transmission over Fading Channels*. IEEE Press, 2002.

[71] S. Benedetto, D. Divsalar, G. Montorsi and F. Pollara, 'Serial concatenation of interleaved codes: performance analysis, design, and iterative decoding,' *IEEE Transactions on Information Theory*, vol. 44, no. 3, pp. 909–926, May 1998.

[72] J. Hagenauer, E. Offer and L. Papke, 'Iterative decoding of binary block and convolutional codes,' *IEEE Transactions on Information Theory*, vol. 42, pp. 429–445, Mar. 1996.

[73] R. Pyndiah, 'Near-optimum decoding of product codes: block turbo codes,' *IEEE Transactions on Communications*, vol. 46, no. 8, pp. 1003–1010, August 1998.

[74] P. Sweeney and S. Wesemeyer, 'Iterative soft-decision decoding of linear block codes,' *IEE Proceedings of Communications*, vol. 147, no. 3, pp. 133–136, June 2000.

[75] D. Divsalar, H. Jin and R. McEliece, 'Coding theorems for 'turbo-like' codes,' in *Proceedings of 36th Allerton Conference on Communication, Control and Computing*, Allerton, IL, USA, September 1998, pp. 201–210.

[76] D. J. C. MacKay and R. M. Neal, 'Near Shannon limit performance of Low Density Parity Check Codes,' *Electronics Letters*, vol. 32, no. 18, pp. 1645–1646, 1996.

[77] T. Richardson and R. Urbanke, 'The capacity of LDPC codes under message passing decoding,' *IEEE Transactions on Information Theory*, vol. 47, pp. 595–618, February 2001.

[78] T. Richardson, M. A. Shokrollahi and R. Urbanke, 'Design of capacity approaching low density parity-check codes,' *IEEE Transactions on Information Theory* , vol. 47, pp. 619–637, February 2001.

[79] R. G. Gallager, 'Low-density parity-check codes,' *IRE Transactions on Information Theory*, vol. 8, no. 1, pp. 21–28, 1962.

[80] M. Tüchler and J. Hagenauer, 'EXIT charts of irregular codes,' in *Proceedings of Conference on Information Science and Systems [CDROM]*, Princeton University, 20-22 March 2002.

[81] M. Tüchler, 'Design of serially concatenated systems depending on the blocklength,' *IEEE Transactions on Communications*, vol. 52, no. 2, pp. 209–218, February 2004.

[82] C. Douillard, M. Jezequel, C. Berrou, A. Picart, P. Didier and A. Glavieux, 'Iterative correction of intersymbol interference: Turbo equalization,' *European Transactions on Telecommunications*, vol. 6, pp. 507–511, Sept.–Oct 1995.

[83] X. Li and J. A. Ritcey, 'Bit-interleaved coded modulation with iterative decoding,' in *Proceedings of International Conference on Communications*, Vancouver, Canada, Juen 1999, pp. 858–863.

[84] S. ten Brink, J. Speidel and R. Yan, 'Iterative demapping and decoding for multilevel modulation,' in *Proceedings of IEEE Global Telecommunications Conference*, 1998, pp. 858–863.

[85] S. Benedetto, D. Divsalar, G. Montorsi and F. Pollara, 'Serial concatenated trellis coded modulation with iterative decoding,' in *Proceedings of IEEE International Symposium on Information Theory*, 29 June–4 July 1997, p. 8.

[86] D. Divsalar, S. Dolinar and F. Pollara, 'Serial turbo trellis coded modulation with rate-1 inner code,' in *Proceedings of International Symposium on Information Theory*, Sorrento, Italy, June 2000, p. 194.

[87] M. Moher, 'An iterative multiuser decoder for near-capacity communications,' *IEEE Transactions on Communications*, vol. 46, no. 7, pp. 870–880, July 1998.

[88] M. C. Reed, C. B. Schlegel, P. Alexander and J. Asenstorfer, 'Iterative multiuser detection for DS-CDMA with FEC: Near single user performance,' *IEEE Transactions on Communications*, vol. 46, no. 7, pp. 1693–1699, July 1998.

[89] A. Nayak, J. Barry and S. McLaughlin, 'Joint timing recovery and turbo equalization for coded partial response channels,' *IEEE Transactions on Magnetics*, vol. 38, no. 5, September 2002.

[90] X. Wang and H. Poor, 'Iterative (turbo) soft interference cancellation and decoding for coded CDMA,' *IEEE Transactions on Communications*, vol. 47, no. 10, pp. 1046–1061, July 1999.

[91] J. Pearl, *Probabilistic Reasoning in Intelligent Systems: Networks of Plausible Inference*. San Mateo, CA, USA: Morgan Kaufmann Publishers, 1988.

[92] R. J. McEliece, D. J. C. MacKay and J.-F. Cheng, 'Turbo decoding as an instance of pearl's "belief propagation" algorithm,' *IEEE Journal on Selected Areas in Communications*, vol. 16, no. 2, pp. 140–152, 1998.

[93] F. R. Kschischang, B. J. Frey and H.-A. Loeliger, 'Factor graphs and the sum-product algorithm,' *IEEE Transactions on Information Theory*, vol. 47, no. 2, pp. 498–519, 2001.

[94] H.-A. Loeliger, 'An introduction to factor graphs,' *IEEE Signal Processing Magazine*, pp. 28–41, January 2004.

[95] F. R. Kschischang and B. J. Frey, 'Iterative decoding of compound codes by probability propagation in graphical models,' *IEEE Journal on Selected Areas in Communications*, vol. 16, no. 2, pp. 219–230, 1998.

[96] E. Biglieri, 'Receivers for coded signals: a unified view based on factor graph,' in *Proceedings of the Fifth International Conference on Information, Communications and Signal Processing*, Bangkok, Thailand, 6–9 December 2005, p. 97.

[97] J. Hagenauer, 'The EXIT chart – Introduction to extrinsic information transfer in iterative processing,' in *Proceedings of 12th European Signal Processing Conference (EUSIPCO)*, September 2004, pp. 1541–1548.

[98] S. ten Brink, 'Convergence behaviour of iteratively decoded parallel concatenated codes,' *IEEE Transactions on Communications* , pp. 1727–1737, Oct. 2001.

[99] S. Benedetto and G. Montorsi, 'Unveiling turbo codes: some results on parallel concatenated coding schemes,' *IEEE Transactions on Information Theory*, vol. 42, pp. 409–428, Mar. 1996.

[100] D. Divsalar and R. J. McEliece, 'Effective free distance of turbo codes,' *Electronics Letters*, vol. 32, no. 5, pp. 445–446, February 1996.

[101] L. C. Perez, J. Seghers and D. J. Costello, 'A distance spectrum interpretation of turbo codes,' *IEEE Transactions on Information Theory*, vol. 42, pp. 1698–1709, November 1996.

[102] T. Duman and R. Salehi, 'New performance bounds for turbo codes,' in *Proceedings of IEEE Global Telecommunications Conference (GLOBECOM'97)*, Phoenix, AZ, USA, November 1997, pp. 634–638.

[103] S. Chung, T. Richardson and R. Urbanke, 'Analysis of sum-product decoding of low-density parity-check codes using a Gaussian approximation,' *IEEE Transactions on Information Theory*, vol. 47, pp. 657–670, February 2001.

[104] D. Divsalar, S. Dolinar and F. Pollara, 'Low complexity turbo-like codes,' in *Proceedings of the 2nd International Symposium on Turbo Codes and Related Topics*, Brest, France, September 2000, pp. 73–80.

[105] K. R. Narayanan, 'Effect of precoding on the convergence of turbo equalization for partial response channels,' *IEEE Journal on Selected Areas in Communications*, vol. 19, no. 4, pp. 686–698, April 2001.

[106] S. ten Brink, 'Convergence of iterative decoding,' *Electronics Letters*, vol. 35, pp. 806–808, 1999.

[107] M. Tüchler, S. ten Brink and J. Hagenauer, 'Measures for tracing convergence of iterative decoding algorithms,' in *Proceedings of 4th IEEE/ITG Conference on Source and Channel Coding*, Berlin, Germany, January 2002, pp. 53–60.

[108] S. Kullback, *Information Theory and Statistics*. Dover Publications Inc., New York, 1968.

[109] Q. Luo and P. Sweeney, 'Method to analyse convergence of turbo codes,' *Electronics Letters*, vol. 41, no. 13, pp. 757–758, June 2005.

[110] M. Moher and T. A. Gulliver., 'Cross-entropy and iterative decoding,' *IEEE Transactions on Information Theory*, vol. 44, no. 7, pp. 3097–3104, November 1998.

[111] S. ten Brink, 'Iterative decoding trajectories of parallel concatenated codes,' in *Proceedings of the 3rd IEEE/ITG Conference on Source and Channel Coding*, Munich, Germany, January 2000, pp. 75–80.

[112] ——, 'Design of serially concatenated codes based on iterative decoding convergence,' in *Proceedings of the 2nd International Symposium on Turbo Codes and Related Topics*, Brest, France, September 2000, pp. 319–322.

[113] M. Tüchler, R. Koetter and A. C. Singer, 'Turbo equalization: Principles and new results,' *IEEE Transactions on Communications*, vol. 50, pp. 754–767, May 2002.

[114] F. Schreckenbach and G. Bauch, 'Measures for tracing convergence of iterative decoding algorithms,' in *Proceedings of 12th European Signal Processing Conference (EUSIPCO)*, Vienna, Austria, September 2004, pp. 1557–1560.

[115] S. ten Brink, 'Designing iterative decoding schemes with the extrinsic information transfer chart,' *AEÜ International Journal of Electronics and Communications*, vol. 54, no. 6, pp. 389–398, November 2000.

[116] A. Ashikhmin, G. Kramer and S. ten Brink, 'Extrinsic information transfer functions: model and erasure channel properties,' *IEEE Transactions on Information Theory*, vol. 50, no. 11, pp. 2657–2673, November 2004.

[117] S. ten Brink, G. Kramer and A. Ashikhmin, 'Design of Low-Density Parity-Check Codes for modulation and detection,' *IEEE Transactions on Communications*, vol. 52, no. 4, pp. 670–678, April 2004.

[118] S. ten Brink and G. Kramer, 'Design of repeat-accumulate codes for iterative detection and decoding,' *IEEE Transactions on Signal Processing*, vol. 51, no. 11, pp. 2764–2772, November 2003.

[119] S. ten Brink, 'Convergence of multi-dimensional iterative decoding schemes,' in *Proceedings of Thirty-Fifth Asilomar Conference on Signals, Systems and Computers*, Pacific Grove, CA, October/November 2001, pp. 270–274.

[120] M. Tüchler, 'Convergence prediction for iterative decoding of threefold concatenated systems,' in *Proceedings of IEEE Global Telecommunications Conference (GLOBECOM'02)*, Taipei, China, 17–21, November 2002, pp. 1358–1362.

[121] F. Brännström, L. K. Rasmussen and A. J. Grant, 'Optimal scheduling for multiple serially concatenated codes,' in *International Symposium on Turbo Codes and Related Topics*, Brest, France, September 2003, pp. 383–386.

[122] ——, 'Convergence analysis and optimal scheduling for multiple concatenated codes,' *IEEE Transactions on Information Theory*, vol. 51, no. 9, pp. 3354–3364, September 2005.

[123] A. Grant, 'Convergence of non-binary iterative decoding,' in *Proceedings of IEEE Global Telecommunications Conference (GLOBECOM'01)*, San Antonio TX, USA, November 2001, pp. 668–670.

[124] J. Kliewer, S. X. Ng and L. Hanzo, 'On the computation of EXIT characteristics for symbol-based iterative decoding,' in *Proceedings of the 4th International Symposium on Turbo Codes (ISTC'06) in Connection with the 6th International ITG-Conference on Source and Channel Coding*, Munich, Germany, 3–7 April 2006.

[125] G. D. Forney, *Concatenated Codes*. Cambridge, MA: MIT Press, 1966.

[126] D. Divsalar and F. Pollara, 'Multiple turbo codes for deep space communications,' Jet Propulsion Laboratory,' TDA Progress Report 42–121, May 1995.

[127] S. Benedetto and G. Montorsi, 'Serial concatenation of block and convolutional codes,' *Electronics Letters*, vol. 32, no. 10, pp. 887–888, May 1996.

[128] ——, 'Iterative decoding of serially concatenated convolutional codes,' *Electronics Letters*, vol. 32, no. 13, pp. 1186–1187, June 1996.

[129] G. Ungerboeck, 'Channel coding with multilevel/phase signals,' *IEEE Transactions on Information Theory*, vol. IT-28, pp. 55–67, 1982.

[130] S. Benedetto, D. Divsalar, G. Montorsi and F. Pollara, 'Analysis, design, and iterative decoding of double serially concatenated codes with interleavers,' *IEEE Journal on Selected Areas in Communications*, vol. 16, pp. 231–244, February 1998.

[131] M. Tüchler, A. C. Singer and R. Koetter, 'Minimum mean squared error equalization using a priori information,' *IEEE Transactions on Signal Processing*, vol. 50, no. 3, pp. 673–682, March 2002.

[132] I. Land, Peter A. Hoeher and S. Gligorevic, 'Computation of symbol-wise mutual information in transmission systems with LogAPP decoders and application to EXIT charts,' in *Proceedings of International ITG Conference on Source and Channel Coding*, Erlangen, Germany, 2004, pp. 195–202.

[133] C. E. Shannon, 'The mathematical theory of communication,' *Bell System Technical Journal*, vol. 27, pp. 379–423, July 1948.

[134] D. J. Costello and G. D. Forney, 'Channel coding: The road to channel capacity,' *Proceedings of the IEEE*, vol. 95, no. 6, pp. 1150–1177, June 2007.

[135] I. Jacobs and E. Berlekamp, 'A lower bound to the distribution of computation for sequential decoding,' *IEEE Transactions on Information Theory*, vol. 13, no. 2, pp. 167–174, April 1967.

[136] J. Wozencraft, 'Sequential decoding for reliable communication,' *IRE National Convention Record*, vol. 5, no. 2, pp. 11–25, August 1957.

[137] J. M. Wozencraft and B. Reiffen, *Sequential Decoding*. Cambridge, MA: MIT Press, 1961.

[138] J. Massey, *Threshold Decoding*. Cambridge, MA: MIT Press, 1961.

[139] R. Fano, 'A heuristic discussion of probabilistic decoding,' *IEEE Transactions on Information Theory*, vol. 9, no. 2, pp. 64–74, April 1963.

[140] P. Elias, 'Coding for noisy channels,' *IRE International Convention Record*, vol. 3, no. 4, pp. 37–46, 1955.

[141] G. D. Forney, 'Review of random tree codes,' NASA Ames Research Center, Moffett Field, CA, USA, Tech. Rep. NASA CR73176, December 1967.

[142] C. Berrou, A. Glavieux, and P. Thitimajshima, 'Near Shannon limit error-correcting coding and decoding: Turbo-codes (1),' in *Proceedings of the IEEE International Conference on Communications*, vol. 2, Geneva, May 1993, pp. 1064–1070.

[143] R. G. Gallager, 'Low-density parity-check codes,' *IRE Transactions on Information Theory*, vol. 8, no. 1, pp. 21–28, January 1962.

[144] ——, *Low Density Parity Check Codes*. Cambridge, MA: MIT Press, 1963.

[145] D. J. C. MacKay and R. M. Neal, 'Near Shannon limit performance of low density parity check codes,' *Electronics Letters*, vol. 32, no. 18, pp. 457–458, August 1996.

[146] R. Tanner, 'A recursive approach to low complexity codes,' *IEEE Transactions on Information Theory*, vol. 27, no. 5, pp. 533–547, September 1981.

[147] M. Luby, M. Mitzenmacher, A. Shokrollahi, D. Spielman and V. Stemann, 'Practical loss-resilient codes,' in *Proceedings of the ACM Symposium on Theory of Computing*, El Paso, TX, USA, May 1997, pp. 150–159.

[148] B. J. Frey and D. J. C. MacKay, 'Irregular turbo-like codes,' in *Proceedings of the International Symposium on Turbo Codes*, Brest, France, September 2000, pp. 67–72.

[149] M. Tüchler and J. Hagenauer, 'EXIT charts of irregular codes,' in *Proceedings of the Conference on Information Sciences and Systems*, Princeton, NJ, USA, March 2002, pp. 748–753.

[150] A. J. Viterbi and J. K. Omura, *Principles of Digital Communications and Coding*. New York, NY, USA: McGraw-Hill, 1979.

[151] L. Hanzo, P. J. Cherriman, and J. Streit, *Video Compression and Communications: H.261, H.263, H.264, MPEG4 and Proprietary Codecs for HSDPA-Style Adaptive Turbo-Transceivers*. John Wiley and IEEE Press, September 2007.

[152] L. Hanzo, F. C. A. Somerville and J. P. Woodard, *Voice and Audio Compression for Wireless Communications*. John Wiley and IEEE Press, 2007.

[153] N. Ahmed, T. Natarajan and K. R. Rao, 'Discrete cosine transform,' *IEEE Transactions on Computers*, vol. C-23, pp. 90–93, January 1974.

[154] A. N. Netravali and J. D. Robbins, 'Motion-compensated television coding. I,' *Bell System Technical Journal*, vol. 58, pp. 631–670, March 1979.

[155] D. A. Huffman, 'A method for the construction of minimum-redundancy codes,' *Proceedings of the IRE*, vol. 40, pp. 1098–1101, September 1952.

[156] *Information Technology – Coding of Moving Pictures and Associated Audio for Digital Storage Media at up to About 1.5 Mbit/s*, ISO/IEC Std. 11 172, June 1996.

[157] *ISO/IEC CD 13818 MPEG 2 International Standard, Information Technology, Generic Coding of Moving Video and Associated Audio Information, Parts 1–3*, International Standards Organization.

[158] *Information Technology – Coding of Audio-visual Objects*, ISO/IEC Std. 14 496.

[159] *Video Codec for Audiovisual Services at $p \times 64$ kbit/s*, ITU-T Std. H.261, March 1993.

[160] *Video Coding for Low Bit Rate Communication*, ITU-T Std. H.263, January 2005.

[161] *Advanced Video Coding for Generic Audiovisual Services*, ITU-T Std. H.264, March 2005.

[162] J. L. Massey, 'Joint source and channel coding,' in *Communication Systems and Random Process Theory*, J. K. Skwirzynski, Ed. Amsterdam, The Netherlands: Sijthoff and Noordhoff, December 1978, pp. 279–293.

[163] R. E. van Dyck and D. J. Miller, 'Transport of wireless video using separate, concatenated, and joint source-channel coding,' *Proceedings of the IEEE*, vol. 87, no. 10, pp. 1734–1750, October 1999.

[164] S. Lloyd, 'Least squares quantization in PCM,' *IEEE Transactions on Information Theory*, vol. 28, no. 2, pp. 129–137, March 1982.

[165] J. Max, 'Quantizing for minimum distortion,' *IRE Transactions on Information Theory*, vol. 6, no. 1, pp. 7–12, March 1960.

[166] A. Kurtenbach and P. Wintz, 'Quantizing for noisy channels,' *IEEE Transactions on Communications*, vol. 17, no. 2, pp. 291–302, April 1969.

[167] N. Farvardin and V. Vaishampayan, 'Optimal quantizer design for noisy channels: An approach to combined source-channel coding,' *IEEE Transactions on Information Theory*, vol. 33, no. 6, pp. 827–838, November 1987.

[168] H. Kumazawa, M. Kasahara, and T. Namekawa, 'A construction of vector quantizers for noisy channels,' *Electronics and Engineering in Japan*, vol. 67-B, no. 4, pp. 39–47, January 1984.

[169] N. Farvardin, 'A study of vector quantization for noisy channels,' *IEEE Transactions on Information Theory*, vol. 36, no. 4, pp. 799–809, July 1990.

[170] Y. Linde, A. Buzo, and R. Gray, 'An algorithm for vector quantizer design,' *IEEE Transactions on Communications*, vol. 28, no. 1, pp. 84–95, January 1980.

[171] K. Sayood and J. C. Borkenhagen, 'Use of residual redundancy in the design of joint source/channel coders,' *IEEE Transactions on Communications*, vol. 39, no. 6, pp. 838–846, June 1991.

[172] N. Phamdo and N. Farvardin, 'Optimal detection of discrete Markov sources over discretememoryless channels – applications to combined source-channel coding,' *IEEE Transactions on Information Theory*, vol. 40, no. 1, pp. 186–193, January 1994.

[173] T. Fingscheidt and P. Vary, 'Softbit speech decoding: A new approach to error concealment,' *IEEE Transactions on Speech and Audio Processing*, vol. 9, no. 3, pp. 240–251, March 2001.

[174] J. Kliewer and R. Thobaben, 'Iterative joint source-channel decoding of variable-length codes using residual source redundancy,' *IEEE Transactions on Wireless Communications*, vol. 4, no. 3, pp. 919–929, May 2005.

[175] F. Lahouti and A. K. Khandani, 'Soft reconstruction of speech in the presence of noise and packet loss,' *IEEE Transactions on Audio, Speech and Language Processing*, vol. 15, no. 1, pp. 44–56, January 2007.

[176] R. Thobaben and J. Kliewer, 'Low-complexity iterative joint source-channel decoding for variable-length encoded Markov sources,' *IEEE Transactions on Communications*, vol. 53, no. 12, pp. 2054–2064, December 2005.

[177] D. J. Miller and M. Park, 'A sequence-based approximate MMSE decoder for source coding over noisy channels using discrete hidden Markov models,' *IEEE Transactions on Communications*, vol. 46, no. 2, pp. 222–231, February 1998.

[178] M. W. Marcellin and T. R. Fischer, 'Trellis coded quantization of memoryless and Gauss-Markov sources,' *IEEE Transactions on Communications*, vol. 38, no. 1, pp. 82–93, January 1990.

[179] V. Buttigieg and P. G. Farrell, 'Variable-length error-correcting codes,' *IEE Proceedings on Communications*, vol. 147, no. 4, pp. 211–215, August 2000.

[180] V. B. Balakirsky, 'Joint source-channel coding with variable length codes,' in *Proceedings of the IEEE International Symposium on Information Theory*, Ulm, Germany, June 1997, p. 419.

[181] R. Bauer and J. Hagenauer, 'Symbol by symbol MAP decoding of variable length codes,' in *Proceedings of the ITG Conference on Source and Channel Coding*, Munich, Germany, January 2000, pp. 111–116.

[182] M. Park and D. J. Miller, 'Decoding entropy-coded symbols over noisy channels by MAP sequence estimation for asynchronous HMMs,' in *Proceedings of the Conference on Information Sciences and Systems*, Princeton, NJ, USA, March 1998, pp. 477–482.

[183] M. Bystrom, S. Kaiser and A. Kopansky, 'Soft source decoding with applications,' *IEEE Transactions on Circuits and Systems for Video Technology*, vol. 11, no. 10, pp. 1108–1120, October 2001.

[184] N. Görtz, 'A generalized framework for iterative source-channel decoding,' *Annals of Telecommunications*, pp. 435–446, July/August 2001.

[185] J. Hagenauer and N. Görtz, 'The turbo principle in joint source-channel coding,' in *Proceedings of the IEEE Information Theory Workshop*, Paris, France, March 2003, pp. 275–278.

[186] X. Jaspar, C. Guillemot and L. Vandendorpe, 'Joint source-channel turbo techniques for discrete-valued sources: From theory to practice,' *Proceedings of the IEEE*, vol. 95, no. 6, pp. 1345–1361, June 2007.

[187] J. Modestino and D. Daut, 'Combined source-channel coding of images,' *IEEE Transactions on Communications*, vol. 27, no. 11, pp. 1644–1659, November 1979.

[188] M. Grangetto, B. Scanavino, G. Olmo and S. Benedetto, 'Iterative decoding of serially concatenated arithmetic and channel codes with JPEG 2000 applications,' *IEEE Transactions on Image Processing*, vol. 16, no. 6, pp. 1557–1567, June 2007.

[189] Q. Qu, Y. Pei and J. W. Modestino, 'An adaptive motion-based unequal error protection approach for real-time video transport over wireless IP networks,' *IEEE Transactions on Multimedia*, vol. 8, no. 5, pp. 1033–1044, October 2006.

[190] Y. C. Chang, S. W. Lee, and R. Komiya, 'A low-complexity unequal error protection of H.264/AVC video using adaptive hierarchical QAM,' *IEEE Transactions on Consumer Electronics*, vol. 52, no. 4, pp. 1153–1158, November 2006.

[191] T. Gan, L. Gan, and K.-K. Ma, 'Reducing video-quality fluctuations for streaming scalable video using unequal error protection, retransmission, and interleaving,' *IEEE Transactions on Image Processing*, vol. 15, no. 4, pp. 819–832, April 2006.

[192] R. Bauer and J. Hagenauer, 'Iterative source/channel-decoding using reversible variable length codes,' in *Proceedings of the IEEE Data Compression Conference*, Snowbird, UT, USA, March 2000, pp. 93–102.

[193] A. Viterbi, 'Error bounds for convolutional codes and an asymptotically optimum decoding algorithm,' *IEEE Transactions on Information Theory*, vol. 13, no. 2, pp. 260–269, April 1967.

[194] R. G. Maunder, J. Kliewer, S. X. Ng, J. Wang, L.-L. Yang and L. Hanzo, 'Joint iterative decoding of trellis-based VQ and TCM,' *IEEE Transactions on Wireless Communications*, vol. 6, no. 4, pp. 1327–1336, April 2007.

[195] ——, 'Iterative joint video and channel decoding in a trellis-based vector-quantized video codec and trellis-coded modulation aided wireless videophone,' in *Proceedings of the IEEE Vehicular Technology Conference*, Dallas, TX, USA, September 2005, pp. 922–926.

[196] R. G. Maunder, J. Wang, S. X. Ng, L.-L. Yang and L. Hanzo, 'On the performance and complexity of irregular variable length codes for near-capacity joint source and channel coding,' *IEEE Transactions on Wireless Communications*, vol. 7, no. 4, pp. 1338–1347, 2008.

[197] ——, 'Iteratively decoded irregular variable length coding and trellis coded modulation,' in *Proceedings of the IEEE Workshop on Signal Processing Systems*, Shanghai, China, October 2007, pp. 222–227.

[198] S. Ahmed, R. G. Maunder, L.-L. Yang and L. Hanzo, 'Iterative detection of unity-rate precoded FFH-MFSK and irregular variable length coding,' in *IEEE Transactions on Vehicular Technology*, vol. 58, no. 7, September 2009, pp. 3765–3770.

[199] M. El-Hajjar, R. G. Maunder, O. Alamri, S. X. Ng and L. Hanzo, 'Iteratively decoded irregular variable length coding and sphere-packing modulation-aided differential space-time spreading,' in *Proceedings of the IEEE Vehicular Technology Conference*, Baltimore, MD, USA, September 2007.

[200] R. G. Maunder and L. Hanzo, 'Genetic algorithm aided design of variable length error correcting codes for arbitrary coding rate and error correction capability,' *IEEE Transactions on Communications*, vol. 57, no. 5, pp. 1290–1297, 2009.

[201] ——, 'Genetic algorithm aided design of near-capacity irregular variable length codes,' *IEEE Wireless Communications and Networking Conference, WCNC 2008.* Las Vegas, NV, USA, March 2008, 1256–1260.

[202] ——, 'Near-capacity irregular variable length coding and irregular unity rate coding,' *IEEE Transactions on Wireless Communications*, vol. 8, no. 11, pp. 5500–5507, 2009.

[203] ——, 'Concatenated irregular variable length coding and irregular unity rate coding,' *IEEE International Conference on Communications,* Beijing, China, May 2008, pp. 795–799.

[204] A. Ashikhmin, G. Kramer and S. ten Brink, 'Code rate and the area under extrinsic information transfer curves,' in *Proceedings of the IEEE International Symposium on Information Theory*, Lausanne, Switzerland, June 2002, p. 115.

[205] V. Buttigieg, 'Variable-Length Error-Correcting Codes,' Ph.D. dissertation, Department of Electronic Engineering, University of Manchester, Manchester, UK, 1995.

[206] C. Lamy and J. Paccaut, 'Optimised constructions for variable-length error correcting codes,' in *Proceedings of the IEEE Information Theory Workshop*, Paris, France, March 2003, pp. 183–186.

[207] J. Wang, L.-L. Yang and L. Hanzo, 'Iterative construction of reversible variable-length codes and variable-length error-correcting codes,' *IEEE Communications Letters*, vol. 8, no. 11, pp. 671–673, November 2004.

[208] R. Y. Tee, R. G. Maunder, J. Wang and L. Hanzo, 'Capacity approaching irregular bit-interleaved coded modulation,' submitted to *IEEE Vehicular Technology Conference,* Marina Bay, Singapore, May 2008.

[209] J. Wang, L.-L. Yang and L. Hanzo, 'Iterative channel equalization, channel decoding and source decoding,' in *Proceedings of the 61st IEEE Vehicular Technology Conference*, Sweden, May 2005.

[210] L.-L. Yang, J. Wang, R. Maunder and L. Hanzo, 'Iterative channel equalization and source decoding for vector quantization sources,' in *Proceedings of the 63rd IEEE Vehicular Technology Conference*, Melbourne, Australia, 7–10 May 2006.

[211] J. Wang, L.-L. Yang and L. Hanzo, 'Combined serially concatenated codes and MMSE equalisation: an EXIT chart aided perspective,' in *Proceedings of the 64th IEEE Vehicular Technology Conference [CDROM]*, Montréal, Canada, 25–28 September 2006.

[212] A. G. Lillie, A. R. Nix and J. P. McGeehan, 'Performance and design of a reduced complexity iterative equalizer for precoded ISI channels,' in *IEEE Vehicular Technology Conference*, Orlando, FL, USA, October 2003, pp. 299–303.

[213] J. Wang, S. X. Ng, A. Wolfgang, L.-L. Yang, S. Chen and L. Hanzo, 'Near-capacity three-stage MMSE turbo equalization using irregular convolutional codes,' in *Proceedings of the 4th International Symposium on Turbo Codes (ISTC'06) in Connection with the 6th International ITG-Conference on Source and Channel Coding*, Munich, Germany, 3–7 April 2006.

[214] O. Alamri, J. Wang, S. X. Ng, L.-L. Yang and L. Hanzo, 'Near-capacity three-stage turbo detection of irregular convolutional coded joint sphere-packing modulation and space-time coding,' *IEEE Transactions on Communications*, vol. 57, no. 5, pp. 1486–1495, 2009.

[215] C. Shannon, 'A mathematical theory of communication – part I,' *Bell Systems Technical Journal*, pp. 379–405, 1948.

[216] ——, 'A mathematical theory of communication – part II,' *Bell Systems Technical Journal*, pp. 405–423, 1948.

[217] ——, 'A mathematical theory of communication – part III,' *Bell Systems Technical Journal*, vol. 27, pp. 623–656, 1948.

[218] H. Nyquist, 'Certain factors affecting telegraph speed,' *Bell System Technical Journal*, p. 617, April 1928.

[219] R. Hartley, 'Transmission of information,' *Bell Systems Technical Journal*, p. 535, 1928.

[220] N. Abramson, *Information Theory and Coding*. New York: McGraw-Hill, 1975.

[221] A. Carlson, *Communication Systems*. New York: McGraw-Hill, 1975.

[222] H. Raemer, *Statistical Communication Theory and Applications*. Englewood Cliffs, NJ: Prentice-Hall, 1969.

[223] P. Ferenczy, *Telecommunications Theory, in Hungarian: Hirkozleselmelet*. Budapest, Hungary: Tankonyvkiado, 1972.

[224] K. Shanmugam, *Digital and Analog Communications Systems*. New York: John Wiley and Sons, 1979.

[225] C. Shannon, 'Communication in the presence of noise,' *Proceedings of the I.R.E.*, vol. 37, pp. 10–22, January 1949.

[226] ——, 'Probability of error for optimal codes in a Gaussian channel,' *Bell Systems Technical Journal*, vol. 38, pp. 611–656, 1959.

[227] A. Jain, *Fundamentals of Digital Image Processing*. Englewood Cliffs, NJ: Prentice-Hall, 1989.

[228] A. Hey and R. Allen, Eds., *R.P. Feynman: Feynman Lectures on Computation*. Reading, MA: Addison-Wesley, 1996.

[229] C. Berrou, A. Glavieux and P. Thitimajshima, 'Near Shannon Limit Error-Correcting Coding and Decoding: Turbo Codes,' in *Proceedings of the International Conference on Communications*, Geneva, Switzerland, May 1993, pp. 1064–1070.

[230] L. Hanzo, W. Webb and T. Keller, *Single- and Multi-Carrier Quadrature Amplitude Modulation: Principles and Applications for Personal Communications, WLANs and Broadcasting*. IEEE Press, 2000.

[231] J. Hagenauer, 'Quellengesteuerte kanalcodierung fuer sprach- und tonuebertragung im mobilfunk,' *Aachener Kolloquium Signaltheorie*, pp. 67–76, 23–25 March 1994.

[232] A. Viterbi, 'Wireless digital communications: A view based on three lessons learned,' *IEEE Communications Magazine*, pp. 33–36, September 1991.

[233] J. B. MacQueen, 'Some methods for classification and analysis of multivariate observations,' in *Proceedings of the Berkeley Symposium on Mathematical Statistics and Probability*, CA, USA, March 1966, pp. 281–297.

[234] B. McMillan, 'Two inequalities implied by unique decipherability,' *IRE Transactions on Information Theory*, vol. IT-2, pp. 115–116, Dec. 1956.

[235] D. J. C. MacKay, *Information Theory, Inference, and Learning Algorithms*. Cambridge University Press, Cambridge, 2003.

[236] L. G. Kraft, 'A device for quantizing, grouping, and coding of amplitude modulated pulses,' Master's thesis, Electrical Engineering Department, M.I.T., Mar. 1949.

[237] E. S. Schwartz and B. Kallick, 'Generating a canonical prefix encoding,' *Communications of ACM*, vol. 7, no. 3, pp. 166–169, Mar. 1964.

[238] C. Lamy and F. Bergot, 'Lower bounds on the existence of binary error-correcting variable-length codes,' in *Proceedings of IEEE Information Theory Workshop*, Paris, France, 31 March–4 April 2003, pp. 300–303.

[239] W. E. Hartnett, Ed., *Foundations of Coding Theory*. Dordrecht, Holland: D. Reidling Publishing Co., 1974.

[240] E. J. Dunscombe, 'Some applications of mathematics to coding theory,' Ph.D. dissertation, Royal Holloway and Bedford New College, University of London, 1988.

[241] M. A. Bernard and B. D. Sharma, 'Variable length perfect codes,' *Journal of Information & Optimization Science*, vol. 13, pp. 143–151, January 1992.

[242] A. Escott, 'A new performance measure for a class of two-length error correcting codes,' in *Codes and Cyphers: Cryptography and Coding IV*, P.G.Farrell, Ed. Essex, England: Institute of Mathematics and its Applications, 1995, pp. 167–181.

[243] 'Information technology – Digital compression and coding of continuous-tone still images– requirements and guidelines,' *ITU-T T.81 (ISO/IEC 10918-1)*, 1992.

[244] 'Information technology – JPEG2000,' *ISO/IEC 15444-1 (ITU-T T.800)*, 2000.

[245] 'Coding of moving pictures and associated audio for digital storage media at up to 1.5 Mbps,' *ISO/IEC 11172 MPEG-1*, 1993.

[246] 'Generic coding of moving pictures and associated audio information,' *ISO/IEC 13818 MPEG-2*, 1994.

[247] A. J. Viterbi, 'Wireless digital communications: A view based on three lessons learnt,' *IEEE Communications Magazine*, pp. 33–36, September 1991.

[248] G. D. Forney, 'The Viterbi algorithm,' *Proceedings of the IEEE*, vol. 61, no. 3, pp. 268–278, March 1973.

[249] W. Xiang, S. Pietrobon and S. Barbulescu, 'Iterative decoding of JPEG coded images with channel coding,' in *IEEE International Conference on Telecommunications*, Tahiti, French, Feb. 2003, pp. 1356–1360.

[250] P. Robertson, E. Villebrun and P. Hoeher, 'A comparison of optimal and sub-optimal MAP decoding algorithms operating in the log domain,' in *Proceedings of IEEE International Conference on Communications*, Seattle, WA, June 1995, pp. 1009–1013.

[251] T. Okuda, E. Tanaka and T. Kasai, 'A method for the correction of garbled words based on the Levenshtein metric,' *IEEE Transactions on Computers*, vol. C-25, no. 2, pp. 172–176, February 1976.

[252] M. Adrat and P. Vary, 'Iterative source-channel decoding: Improved system design using EXIT charts,' *EURASIP Journal on Applied Signal Processing (Special Issue: TURBO Processing)*, no. 6, pp. 928–941, May 2005.

[253] D. Divsalar and F. Pollara, 'Hybrid concatenated codes and iterative decoding,' Jet Propulsion Laboratory,' TDA Progress Report 42-130, August 1997.

[254] S. Benedetto, D. Divsalar, G. Montorsi and F. Pollara, 'A soft-input soft-output APP module for iterative decoding of concatenated codes,' *IEEE Communications Letters*, vol. 1, no. 1, pp. 22–24, January 1997.

[255] ——, 'Soft-input soft-output modules for the construction and distributed iterative decoding of code networks,' *European Transactions on Telecommunications*, vol. 9, pp. 155–172, March 1998.

[256] S. Huettinger and J. Huber, 'Design of multiple turbo codes with transfer characteristics of component codes,' in *Proceedings of Conference on Information Sciences and Systems*, March 2002.

[257] D. Raphaeli and Y. Zarai, 'Combined turbo equalization and turbo decoding,' *IEEE Communications Letters*, pp. 107–109, April 1998.

[258] W. E. Ryan, L. L. McPheters and S. W. McLaughlin, 'Combined turbo coding and turbo equalization for PR4-equalized Lorentzian channels,' in *Proceedings of Conference on Information Sciences and Systems*, March 1998, pp. 489–493.

[259] J. Kliewer, S. X. Ng and L. Hanzo, 'Efficient computation of EXIT functions for non-binary iterative decoding,' *IEEE Transactions on Communications (Letter)*, vol. 54, no. 12, pp. 2133–2136, 2006.

[260] L. Hanzo, C.H. Wong and M.S. Yee, *Adaptive Wireless Transceivers: Turbo-Coded, Space-Time Coded TDMA, CDMA and OFDM Systems*. Chichester, UK: John Wiley-IEEE Press, 2002.

[261] A. Glavieux, C. Laot and J. Labat, 'Turbo equalization over a frequency selective channel,' in *Proceedings of International Symposium on Turbo Codes*, Brest, France, 1997, pp. 96–102.

[262] D. Raphaeli, 'Combined turbo equalization and turbo decoding,' *IEEE Communications Letters*, vol. 2, no. 4, pp. 107–109, April 1998.

[263] I. Lee, 'The effect of precoder on serially concatenated coding systems with an ISI channel,' *IEEE Transactions on Communications*, vol. 49, no. 7, pp. 1168–1175, July 2001.

[264] W. Ryan, 'Performance of high rate turbo codes on a PR4 equalized magnetic recording channel,' in *IEEE International Conference on Communications (ICC)*, Atlanta, GA, USA, June 1998, pp. 947–951.

[265] T. Souvignier, A. Friedman, M. Oberg, P. H. Siegel, R. E. Swanson and J. Wolf, 'Turbo decoding for PR4: Parallel versus serial concatenation,' in *IEEE International Conference on Communications (ICC)*, Vancouver, Canada, June 1999, pp. 1638–1642.

[266] D. N. Doan and K. R. Narayanan, 'Some new results on the design of codes for inter-symbol interference channels based on convergence of turbo equalization,' in *IEEE International Conference on Communications (ICC'02)*, Anchorage Alaska, USA, 2002, pp. 1873–1877.

[267] A. Ashikhmin, G. Kramer and S. ten Brink, 'Extrinsic information transfer functions: Model and erasure channel properties,' *IEEE Transactions on Information Theory*, vol. 50, no. 11, pp. 2657–2673, November 2004.

[268] M. Adrat, J. Brauers, T. Clevorn and P. Vary, 'The EXIT-characteristic of softbit-source decoders,' *IEEE Communications Letters*, vol. 9, no. 6, pp. 540–542, June 2005.

[269] J. G. Proakis, *Digital Communications*. McGraw-Hill, 1983.

[270] H. Jin, A. Khandekar and R. McEliece, 'Irregular repeat-accumulate codes,' in *Proceedings of 2nd International Symposium on Turbo Codes and Related Topics*, Brest, France, September 2000, pp. 1–8.

[271] S. X. Ng, J. Wang, M. Tao, L.-L. Yang and L. Hanzo, 'Iteratively decoded variable-length space-time coded modulation: code construction and convergence analysis,' *IEEE Transactions on Wireless Communications*, vol. 6, no. 5, pp. 1953–1963, 2007.

[272] S. X. Ng, J. Y. Chung and L. Hanzo, 'Turbo-detected unequal protection MPEG-4 wireless video telephony using multi level coding, trellis coded modulation and space-time trellis coding,' *IEE Proceedings on Communications*, vol. 152, no. 6, pp. 1116–1124, December 2005.

[273] S. X. Ng, R. G. Maunder, J. Wang, L.-L. Yang and L. Hanzo, 'Joint iterative-detection of reversible variable-length coded constant bit rate vector-quantized video and coded modulation,' in *Proceedings of the European Signal Processing Conference*, Vienna, Austria, September 2004, pp. 2231–2234.

[274] S. X. Ng, J. Y. Chung, F. Guo and L. Hanzo, 'A turbo-detection aided serially concatenated MPEG-4/TCM videophone transceiver,' in *Proceedings of the IEEE Vehicular Technology Conference*, vol. 4, Los Angeles, CA, USA, September 2004, pp. 2606–2610.

[275] S. Benedetto and G. Montorsi, 'Serial concatenation of block and convolutional codes,' *Electronics Letters*, vol. 32, no. 10, pp. 887–888, May 1996.

[276] Y. Takishima, M. Wada and H. Murakami, 'Reversible variable length codes,' *IEEE Transactions on Communications*, vol. 43, no. 234, pp. 158–162, February 1995.

[277] G. Ungerboeck, 'Channel coding with multilevel/phase signals,' *IEEE Transactions on Information Theory*, vol. 28, no. 1, pp. 55–67, January 1982.

[278] S. Benedetto and G. Montorsi, 'Iterative decoding of serially concatenated convolutional codes,' *Electronics Letters*, vol. 32, no. 13, pp. 1186–1188, June 1996.

[279] C. Bergeron and C. Lamy-Bergot, 'Soft-input decoding of variable-length codes applied to the H.264 standard,' in *Proceedings of the IEEE Workshop on Multimedia Signal Processing*, Siena, Italy, September 2004, pp. 87–90.

[280] K. P. Subbalakshmi and Q. Chen, 'Joint source-channel decoding for MPEG-4 coded video over wireless channels,' in *Proceeding of the IASTED International Conference on Wireless and Optical Communications*, Banff National Park, Canada, July 2002, pp. 617–622.

[281] Q. Chen and K. P. Subbalakshmi, 'Trellis decoding for MPEG-4 streams over wireless channels,' in *Proceedings of SPIE Electronic Imaging: Image and Video Communications and Processing*, Santa Clara, CA, USA, January 2003, pp. 810–819.

[282] X. F. Ma and W. E. Lynch, 'Iterative joint source-channel decoding using turbo codes for MPEG-4 video transmission,' in *Proceedings of the IEEE International Conference on Acoustics, Speech, and Signal Processing*, vol. 4, Montreal, Quebec, Canada, May 2004, pp. 657–660.

[283] Q. Chen and K. P. Subbalakshmi, 'Joint source-channel decoding for MPEG-4 video transmission over wireless channels,' *IEEE Journal on Selected Areas in Communications*, vol. 21, no. 10, pp. 1780–1789, December 2003.

[284] Y. Wang and S. Yu, 'Joint source-channel decoding for H.264 coded video stream,' *IEEE Transactions on Consumer Electronics*, vol. 51, no. 4, pp. 1273–1276, November 2005.

[285] Q. Chen and K. P. Subbalakshmi, 'An integrated joint source-channel decoder for MPEG-4 coded video,' *IEEE Vehicular Technology Conference*, vol. 1, pp. 347–351, October 2003.

[286] K. Lakovic and J. Villasenor, 'Combining variable length codes and turbo codes,' in *Proceedings of the IEEE Vehicular Technology Conference*, vol. 4, Birmingham, AL, USA, May 2002, pp. 1719–1723.

[287] M. Grangetto, B. Scanavino, and G. Olmo, 'Joint source-channel iterative decoding of arithmetic codes,' in *Proceedings of the IEEE International Conference on Communications*, vol. 2, Paris, France, June 2004, pp. 886–890.

[288] H. Nguyen and P. Duhamel, 'Iterative joint source-channel decoding of variable length encoded video sequences exploiting source semantics,' in *Proceedings of the International Conference on Image Processing*, vol. 5, Singapore, October 2004, pp. 3221–3224.

[289] H. Xiao and B. Vucetic, 'Channel optimized vector quantization with soft input decoding,' in *Proceedings of the International Symposium on Signal Processing and Its Applications*, vol. 2, Gold Coast, Queensland, Australia, August 1996, pp. 501–504.

[290] D. J. Vaisey and A. Gersho, 'Variable block-size image coding,' in *Proceedings of the IEEE International Conference on Acoustics, Speech, and Signal Processing*, vol. 12, Dallas, TX, USA, April 1987, pp. 1051–1054.

[291] A. Makur and K. P. Subbalakshmi, 'Variable dimension VQ encoding and codebook design,' *IEEE Transactions on Communications*, vol. 45, no. 8, pp. 897–899, August 1997.

[292] J. Hagenauer, E. Offer, and L. Papke, 'Iterative decoding of binary block and convolutional codes,' *IEEE Transactions on Information Theory*, vol. 42, no. 2, pp. 429–445, March 1996.

[293] S. X. Ng and L. Hanzo, 'Space-time IQ-interleaved TCM and TTCM for AWGN and Rayleigh fading channels,' *Electronics Letters*, vol. 38, no. 24, pp. 1553–1555, 2002.

[294] S. ten Brink, 'Convergence of iterative decoding,' *Electronics Letters*, vol. 35, no. 10, pp. 806–808, May 1999.

[295] P. Robertson, E. Villebrun and P. Hoeher, 'A comparison of optimal and sub-optimal MAP decoding algorithms operating in the log domain,' in *Proceedings of the IEEE International Conference on Communications*, vol. 2, Seattle, WA, USA, June 1995, pp. 1009–1013.

[296] L. Hanzo, T. H. Liew and B. L. Yeap, *Turbo Coding, Turbo Equalisation and Space Time Coding for Transmission over Wireless Channels*. Chichester, UK: Wiley, 2002.

[297] V. Franz and J. B. Anderson, 'Concatenated decoding with a reduced-search BCJR algorithm,' *IEEE Journal on Selected Areas in Communications*, vol. 16, no. 2, pp. 186–195, February 1998.

[298] J. Kliewer, N. Görtz and A. Mertins, 'Iterative source-channel decoding with Markov random field source models,' *IEEE Transactions on Signal Processing*, vol. 54, no. 10, pp. 3688–3701, October 2006.

[299] J. Kliewer, A. Huebner and D. J. Costello, 'On the achievable extrinsic information of inner decoders in serial concatenation,' in *Proceedings of the IEEE International Symposium on Information Theory*, Seattle, WA, USA, July 2006, pp. 2680–2684.

[300] L. Hanzo, S. X. Ng, T. Keller and W. Webb, *Quadrature Amplitude Modulation*. Chichester, UK: Wiley, 2004.

[301] R. Bauer and J. Hagenauer, 'On variable length codes for iterative source/channel decoding,' in *Proceedings of the IEEE Data Compression Conference*, Snowbird, UT, USA, March 2001, pp. 273–282.

[302] R. C. Bose and D. K. Ray-Chaudhuri, 'On a class of error-correcting binary group codes,' *Information and Control*, vol. 3, pp. 68–79, March 1960.

[303] A. Hocquenghem, 'Codes correcteurs d'erreurs,' *Chiffres*, vol. 2, pp. 147–156, September 1959.

[304] B. Masnick and J. Wolf, 'On linear unequal error protection codes,' *IEEE Transactions on Information Theory*, vol. 13, no. 4, pp. 600–607, October 1967.

[305] M. Rosenblatt, 'A central limit theorem and a strong mixing condition,' *Proceedings of the National Academy of Sciences*, no. 42, pp. 43–47, 1956.

[306] M. Tüchler, 'Design of serially concatenated systems depending on the block length,' *IEEE Transactions on Communications*, vol. 52, no. 2, pp. 209–218, February 2004.

[307] D. E. Goldberg, *Genetic Algorithms in Search, Optimization and Machine Learning*. Addison-Wesley, 1989.

[308] D. Divsalar, S. Dolinar and F. Pollara, 'Serial concatenated trellis coded modulation with rate-1 inner code,' in *Proceedings of the IEEE Global Telecommunications Conference*, vol. 2, San Francisco, CA, USA, November 2000, pp. 777–782.

[309] G. Golub and C. van Loan, *Matrix Computations*. Baltimore: Johns Hopkins University Press, 1996.

[310] F. Gray, US Patent 2 632 058, March, 1953.

[311] G. Caire, G. Taricco and E. Biglieri, 'Bit-interleaved coded modulation,' *IEEE Transactions on Information Theory*, vol. 44, no. 3, pp. 927–946, May 1998.

[312] S. ten Brink, 'Code characteristic matching for iterative decoding of serially concatenated codes,' *Annals of Telecommunications*, vol. 56, no. 7–8, pp. 394–408, July–August 2001.

[313] T. J. Richardson, M. A. Shokrollahi and R. L. Urbanke, 'Design of capacity-approaching irregular low-density parity-check codes,' *IEEE Transactions on Information Theory*, vol. 47, no. 2, pp. 619–637, February 2001.

[314] G. Foschini Jr. and M. Gans, 'On limits of wireless communication in a fading environment when using multiple antennas,' *Wireless Personal Communications*, vol. 6, no. 3, pp. 311–335, March 1998.

[315] E. Telatar, 'Capacity of multi-antenna Gaussian channels,' *European Transactions on Telecommunication*, vol. 10, no. 6, pp. 585–595, Nov–Dec 1999.

[316] G. J. Foschini, Jr., 'Layered space-time architecture for wireless communication in a fading environment when using multi-element antennas,' *Bell Labs Technical Journal*, pp. 41–59, 1996.

[317] V. Tarokh, N. Seshadri and A. R. Calderbank, 'Space-time codes for high rate wireless communication: Performance analysis and code construction,' *IEEE Transactions on Information Theory*, vol. 44, no. 2, pp. 744–765, March 1998.

[318] S. Lin, A. Stefanov and Y. Wang, 'Joint source and space-time block coding for MIMO video communications,' in *IEEE Vehicular Technology Conference*, Los Angeles, CA, Sept 2004, pp. 2508–2512.

[319] T. Holliday and A. Goldsmith, 'Joint source and channel coding for MIMO systems,' in *Proceedings of Allerton Conference on Communication, Control, and Computing*, Oct. 2004.

[320] S. X. Ng, F. Guo and L. Hanzo, 'Iterative detection of diagonal block space time trellis codes, TCM and reversible variable length codes for transmission over Rayleigh fading channels,' in *IEEE Vehicular Technology Conference*, Los Angeles, CA, USA, 26–29 September 2004, pp. 1348–1352.

[321] D. Divsalar, S. Dolinar and F. Pollara, 'Serial concatenated trellis coded modulation with rate-1 inner code,' in *GLOBECOM '00*, San Francisco, CA, Nov–Dec 2000, pp. 777–782.

[322] G. Ungerböck, 'Channel coding with multilevel/phase signals,' *IEEE Transactions on Information Theory*, vol. 28, pp. 55–67, January 1982.

[323] A. F. Naguib, V. Tarokh, N. Seshadri and A. R. Calderbank, 'A space-time coding modem for high-data-rate wireless communications,' *IEEE Journal on Selected Areas in Communications*, vol. 16, no. 8, pp. 1459–1478, October 1998.

[324] D. Divsalar and M. K. Simon, 'Trellis coded modulation for 4800-9600 bits/s transmission over a fading mobile satellite channel,' *IEEE Journal on Selected Areas in Communications*, vol. 5, no. 2, pp. 162–175, February 1987.

[325] M. Tao and R. S. Cheng, 'Diagonal block space-time code design for diversity and coding advantage over flat rayleigh fading channels,' *IEEE Transactions on Signal Processing*, no. 4, pp. 1012–1020, April 2004.

[326] T. M. Cover and J. A. Thomas, *Elements of Information Theory*. New York, USA: John Wiley IEEE Press, 1991.

[327] G. L. Nemhauser, A. H. G. Rinnooy Kan and M. J. Todd, *Optimization*. Amsterdam, The Netherlands: North-Holland, 1989.

[328] L. Hanzo, T. H. Liew and B. L. Yeap, *Turbo Coding, Turbo Equalisation and Space Time Coding for Transmission over Wireless Channels*. New York, USA: John Wiley IEEE Press, 2002.

[329] X. Li and J. A. Ritcey, 'Bit-interleaved coded modulation with iterative decoding using soft feedback,' *IEE Electronics Letters*, vol. 34, pp. 942–943, May 1998.

[330] P. Robertson, E. Villebrun and P. Höher, 'A comparison of optimal and sub-optimal MAP decoding algorithms operating in log domain,' in *Proceedings of the International Conference on Communications*, June 1995, pp. 1009–1013.

[331] S. ten Brink, 'Convergence behaviour of iteratively decoded parallel concatenated codes,' *IEEE Transactions on Communications*, vol. 49, no. 10, pp. 1727–1737, October 2001.

[332] H. Chen and A. Haimovich, 'EXIT charts for turbo trellis-coded modulation,' *IEEE Communications Letters*, vol. 8, no. 11, pp. 668–670, November 2004.

[333] A. Grant, 'Convergence of non-binary iterative decoding,' in *Proceedings of the IEEE Global Telecommunications Conference (GLOBECOM)*, San Antonio TX, USA, November 2001, pp. 1058–1062.

[334] M. Tuchler, 'Convergence prediction for iterative decoding of threefold concatenated systems,' in *GLOBECOM '02*, Taipei, Taiwan, 17–21 November 2002, pp. 1358–1362.

[335] F. Brannstrom, L. K. Rasmussen and A. J. Grant, 'Convergence analysis and optimal scheduling for multiple concatenated codes,' *IEEE Transaction on Information Theory*, pp. 3354–3364, Sept 2005.

[336] S. X. Ng and L. Hanzo, 'On the MIMO channel capacity of multi-dimensional signal sets,' *IEEE Transactions on Vehicular Technology*, vol. 55, no. 2, pp. 528–536, 2006.

[337] S. Ahmed, R. G. Maunder, L.-L. Yang and L. Hanzo, 'Iterative detection of unity-rate precoded FFH-MFSK and irregular variable-length coding,' in *IEEE transactions on Vehicular Technology*, vol. 58, no. 7, 2009, pp. 3765–3770.

[338] K. C. Teh, A. C. Kot and K. H. Li, 'FFT-based clipper receiver for fast frequency-hopping spread-spectrum system,' in *Proceedings of the IEEE Symposium on Circuits and Systems*, vol. 4, May-June 1998, pp. 305–308.

[339] R. E. Ziemer and R. L. Peterson, *Digital Communications and Spread Spectrum Systems*. New York: Mcmillan Inc., 1985.

[340] V. A. Aalo and J. Zhang, 'Performance analysis of maximal ratio combining in the presence of multiple equal-power cochannel interferers in a Nakagami fading channel,' *IEEE Transactions on Vehicular Technology*, vol. 50, no. 2, pp. 497–503, March 2001.

[341] J. S. Lee, L. E. Miller and Y. K. Kim, 'Probability of error analyses of a BFSK frequency-hopping system with diversity under partial-band jamming interference-Part II: Performance of square-law nonlinear combining soft decision receivers,' *IEEE Transaction on Communications*, vol. COM-32, no. 12, pp. 1243–1250, December 1984.

[342] C.-L. Chang and T.-M. Tu, 'Performance analysis of FFH/BFSK product-combining receiver with partial-band jamming over independent Rician fading channels,' *IEEE Transactions on Wireless Communications*, vol. 4, no. 6, pp. 2629–2635, November 2005.

[343] R. C. Robertson and T. T. Ha, 'Error probabilities of fast frequency-hopped MFSK receiver with noise-normalization combining in a fading channel with partial-band interference,' *IEEE Transactions on Communications*, vol. 40, no. 2, pp. 404–412, February 1992.

[344] P. C. P. Liang and W. E. Stark, 'Algorithm for joint decoding of turbo codes and M-ary orthogonal modulation,' in *IEEE International Symposium on Information Theory, 2000*, June 2000, p. 191.

[345] M. C. Valenti, E. Hueffmeier, B. Bogusch and J. Fryer, 'Towards the capacity of noncoherent orthogonal modulation: BICM-ID for turbo coded NFSK,' in *IEEE Conference on Military Communications, 2004. MILCOM 2004*, vol. 3, October–November 2004, pp. 1549–1555.

[346] S. ten Brink, 'Convergence of iterative decoding,' *IEEE Transactions on Communications*, vol. 49, no. 10, pp. 1727–1737, October 2001.

[347] U. C. Fiebig, 'Soft-decision and erasure decoding in fast frequency-hopping systems with convolutional, turbo, and Reed-Solomon codes,' *IEEE Transactions on Communications*, vol. 47, no. 11, pp. 1646–1654, November 1999.

[348] D. Park and B. G. Lee, 'Iterative decoding in convolutionally and turbo coded MFSK/FH-SSMA systems,' in *IEEE International Conference on Communications, ICC 2001*, vol. 9, June 2001, pp. 2784–2788.

[349] W. E. Ryan, 'Concatenated codes for class IV partial response channels,' *IEEE Transactions on Communications*, vol. 49, no. 3, pp. 445–454, March 2001.

[350] R. Y. S. Tee, S. X. Ng and L. Hanzo, 'Precoder-aided iterative detection assisted multilevel coding and three-dimensional EXIT-chart analysis,' in *IEEE Wireless Communications and Networking Conference, 2006. WCNC 2006*, vol. 3, April 2006, pp. 1322–1326.

[351] D. Divsalar, S. Dolinar and F. Pollara, 'Serial concatenated trellis coded modulation with rate-1 inner code,' in *IEEE Global Telecommunications Conference, 2000. GLOBECOM '00.*, vol. 2, November–December 2000, pp. 777–782.

[352] J. G. Proakis, *Digital Communications*. Singapore: McGraw-Hill, 2001.

[353] S. Ahmed, S. X. Ng, L.-L. Yang and L. Hanzo, 'Iterative decoding and soft interference cancellation in fast frequency hopping multiuser system using clipped combining,' in *Proceedings of the IEEE Wireless Communications and Networking Conference, WCNC 2007*, March 2007, pp. 723–728.

[354] L. Hanzo, T. H. Liew and B. L. Yeap, *Turbo Coding, Turbo Equalisation and Space-Time Coding for Transmission over Fading Channels*. England: John Wiley and Sons, England, 2002.

[355] J. Wang, S. X. Ng, A. Wolfgang, L.-L. Yang, S. Chen and L. Hanzo, 'Near-capacity three-stage MMSE turbo equalization using irregular convolutional codes,' in *Proceedings of the International Symposium on Turbo Codes*, Munich, Germany, April 2006, electronic publication.

[356] P. J. Crepeau, 'Performance of FH/BFSK with generalized fading in worst-case partial-band Gaussian interference,' *IEEE Transactions on Selected Areas of Communications*, vol. 8, no. 5, pp. 884–886, June 1990.

[357] G. Huo and M. S. Aluoini, 'Another look at the BER performance of FFH/BFSK with product combining over partial-band jammed Rayleigh fading channels,' *IEEE Transactions on Vehicular Technology*, vol. 50, no. 5, pp. 1203–1215, September 2001.

[358] A. Chindapol and J. A. Ritcey, 'Design, analysis, and performance evaluation for BICM-ID with square QAM constellations in Rayleigh fading channels,' *IEEE Journal on Selected Areas in Communications*, vol. 19, no. 5, pp. 944–957, May 2001.

[359] J. Tan and G. L. Stuber, 'Analysis and design of interleaver mappings for iteratively decoded BICM,' in *Proceedings of the IEEE International Conference on Communications*, vol. 3, New York, NY, USA, April 2002, pp. 1403–1407.

[360] F. Schreckenbach, N. Görtz, J. Hagenauer and G. Bauch, 'Optimized symbol mappings for bit-interleaved coded modulation with iterative decoding,' in *Proceedings of the IEEE Global Telecommunications Conference*, vol. 6, San Francisco, CA, USA, December 2003, pp. 3316–3320.

Glossary

16QAM	16-ary Quadrature Amplitude Modulation
1D	One-Dimensional
3G	Third-Generation (network)
ACF	Autocorrelation Function
ACS	Add, Compare and Select
ADC	Analog-to-Digital Converter
ALU	Arithmetic and Logic Unit
APP	*A Posteriori* Probability
AR	Autoregressive
ARVLC	Asymmetric Reversible Variable-Length Coding
AWGN	Additive White Gaussian Noise
B-F	Buttigieg-Farrell
BB	Bit-Based
BCD	Binary-Coded Decimal
BCH	Bose–Chaudhuri–Hocquenghem
BCJR	Bahl–Cocke–Jelinek–Raviv
BEC	Binary Erasure Channel
BER	Bit Error Ratio
BICM	Bit-Interleaved Coded Modulation
BiCM-ID	Bit-interleaved Coded Modulation using Iterative Decoding
BLAST	Bell Labs LAyered Space-Time architecture
bpc	bits per channel use
BPSK	Binary Phase Shift Keying
BSC	Binary Symmetric Channel
BTC	Block Turbo Code
CABAC	Context Adaptive Binary Arithmetic Coding
CAVLC	Context Adaptive Variable Length Coding
CC	Convolutional Coding
CDF	Cumulative Density Function

CDMA	Code-Division Multiple Access
CIR	Channel Impulse Response
COVQ	Channel-Optimized Vector Quantization
CSI	Channel State Information
DCMC	Discrete-input Continuous-output Modulated Channel
DCT	Discrete Cosine Transform
DMS	Discrete Memoryless Source
DRI	Decoder Reliability Information
EWVLC	Even Weight Variable Length Coding
EXIT	EXtrinsic Information Transfer
FD	Frame Difference
FDM	Free Distance Metric
FDMA	Frequency Division Multiple Access
FEC	Forward Error Correction
FFH	Fast Frequency Hopping
FH	Frequency Hopping
FIR	Finite-length Impulse Response
FL-STCM	Fixed-Length Space-Time Coded Modulation
FLC	Fixed-Length Coding
GA	Genetic Algorithm
GSM	Global System of Mobile telecommunication
HA	Heuristic Algorithm
hi-fi	high fidelity
HISO	Hard-In Hard-Out
HMM	Hidden Markov Model
i.i.d.	independent and identically distributed
ID	Iterative Decoding
IIR	Infinite Impulse Response
IQ	In-phase Quadrature-phase
IRA	Irregular Repeat-Accumulate
IrCC	Irregular Convolutional Coding
IrLDPC	Irregular Low Density Parity Check
IrURC	Irregular Unity Rate Coding
IrVLC	Irregular Variable Length Coding
ISCD	Iterative Source-Channel Decoding
ISI	InterSymbol Interference
IV-FD	Integer-Valued Free Distance
JSCD	Joint Source-Channel Decoding
LBG	Linde–Buzo–Gray
LDPC	Low-Density Parity-Check (codes)

LFSR	Linear Feedback Shift Register
LLR	Logarithmic Likelihood-Ratio
log-APP	Logarithmic A Posteriori Probability
LTI	Linear Time-Invariant
MAP	Maximum *A posteriori* Probability
MB	Macro-Block
MC	Motion Compensation
MCE	Minimum Cross-Entropy
MFSK	M-ary Frequency Shift Keying
MI	Mutual Information
MIMO	Multiple-In Multiple-Out
ML	Maximum Likelihood
MLSE	Maximum Likelihood Sequence Estimation
MMSE	Minimum Mean-Squared-Error
MPEG	Motion Picture Experts Group
MRG	Minimum Repetition Gap
MSEW	Maximum Squared Euclidean Weight
MSP	Modified Set Partitioning
MSSL	Maximum Symmetric-Suffix Length
MUD	Multi-User Detection
MVA	Majority Voting Algorithm
NSC	Non-Systematic Convolutional (code)
PBNJ	Partial Band Noise Jamming
PCC	Parallel Concatenated ('turbo') Code
PDF	Probability Distribution Function
PM	Path Metric
PN	Pseudo-Noise
PSD	Power Spectral Density
PSK	Phase Shift Keying
PSNR	Peak Signal to Noise Ratio
PWEP	Pair-Wise Error Probability
QAM	Quadrature Amplitude Modulation
QCIF	Quarter Common Intermediate Format
RA	Repeat-Accumulate (codes)
rhs	right-hand side
RLC	Run-Length Coding
RMS	Root Mean Squared
RSC	Recursive Systematic Convolutional (code)
RV-FDM	Real-Valued Free Distance Metric
RVLC	Reversible Variable-Length Code
SB	Symbol-Based
SBIrURC	Symbol-Based Irregular Unity-Rate Code

SBURC	Symbol-Based Unity-Rate Code
SCC	Serially Concatenated Code
SCCD	Source-Controlled Channel Decoding
SDD	Soft Decision Decoding
SDI	Soft-Decision Information
SE	Sequence Estimation
SER	Symbol Error Ratio
SIHO	Soft-In Hard-Out
SISO	Soft-In Soft-Out
SNR	Signal to Noise Ratio
SOBIT	SOft BIT
SOVA	Soft Output Viterbi Algorithm
SP	Set Partitioning
SQNR	Signal to Quantization Noise Ratio
SRVLC	Symmetric Reversible Variable-Length Coding
SSI	Source Significance Information
STCM	Space-Time Coded Modulation
STP	State Transition Probability
STTC	Space-Time Trellis Code
TCM	Trellis Coded Modulation
TCQ	Trellis Coded Quantization
TDMA	Time Division Multiple Access
TE	Turbo Equalization
TTCM	Turbo Trellis-Coded Modulation
UEP	Unequal Error Protection
URC	Unity Rate Coding
VA	Viterbi Algorithm
VB	Video Block
VDVQ	Variable Dimension Vector Quantization
VL-coding	Variable-Length Coding
VL-STC	Variable-Length Space-Time Code
VL-STCM-ID	Iteratively Decoded Variable-Length Space-Time Coded Modulation
VLC	Variable-Length Code
VLEC	Variable-Length Error Correction
VLSI	Very-Large-Scale Integration
VQ	Vector Quantization
WGN	White Gaussian Noise
WN	White Noise

Subject Index

Additive white Gaussian noise (AWGN), 29, 32
Additive white Gaussian noise channel, 27
Aging, 29
AR model properties, 49–50
area property, 208
asymmetric RVLC, 101
Autoregressive model, 49
average codeword length, 98
Average conveyed mutual information per symbol, 67
Average error entropy, 68
Average information loss, 68
Average joint information, 67
Average loss of information per symbol, 69

backward recursion, 125, 148
Bahl–Cocke–Jelinek–Raviv (BCJR) algorithm, **248**, 266, 305, 347, 447
Bayes's rule, 56–59, 117
Binary Phase Shift Keying (BPSK), 307, 444
Binary symmetric channel example, 55–56

canonical Huffman code, 100
Canterbury Corpus, 104
Capacity of continuous channels, 75–82
Capacity of discrete channels, 70–72
Channel, 27
channel capacity, **11**, 319, 357
    E_b/N_0 bound, 253, 269, 321, 358
    loss, 294, 357, 445
channel coding, **11**
Channel coding theorem, 73
Channel encoder, 27
Channel impairments, 26
channel-optimized source coding, **16**
Coaxial cable, 30
code constraint, **242**, 245, 260, 292
Code efficiency, 96

Coding efficiency, 37
coding rate, 73, 251, **260**, 284, 292, 342, 431, 439, 442
combined module, 194
Communications channels, 28
Communications system design considerations, 26
computational complexity, 256, 266, *283*, 322, *333*, 343, *368*, 439, *441*
convergence distance, 97
Cumulative density function, 26
curve-fitting problem, 212
cutoff rate, **11**

divergence distance, 97

effective throughput, 252, 269, 319, 357
entropy, **11**, 34–37, 93, 251, **260**, 292, 293, 346
Entropy of analog sources, 76
Entropy of sources exhibiting memory, 43–45
Equivocation, 63, 69
Error entropy, 63, 64
Error entropy via imperfect channels, 63–70
error propagation, 100
error shoulder, 184
EXIT tunnel, 164
extended code, 98
EXtrinsic Information Transfer (EXIT)
    chart, *254, 272, 277, 300, 314, 324, 353, 355, 356, 363, 433, 436, 437*
    decoding trajectory, *254*, 279, 330, 366, 369, 437
    matching, 276, 319, 342, **344**, 356, **430**
    tunnel, 253, 274, 276, 278, 319, 330, 357, 366, 367, 431
    function, 253, 271, 276, 291, 296, 339

first-order Markov model, 51, 94
fix-free, 101

forward recursion, 124, 148
Four-state Markov model for a two-bit quantizer, 48
frame differencing, **235**
free distance, 98, **292**
    lower bound, 251, 268, **293**
    metric, **295**, 349, 434
Free-space, 30

Gaussian, 31
Gaussian noise, 29–32
Generating model sources, 49–51
Genetic Algorithm (GA), **300**, 310, 349, 359, 434
Group delay, 29

Hamming distance, 96
Hamming weight, 96
hard information, 232, 268, 297
histogram, 154
Huffman code, 99
Huffman coding, 2, 39–43, 242, 276
    codebook, *265*

Ideal communications system, 82
Information loss, 62
Information loss via imperfect channels, 62–63
Information of a source, 32–33
Information properties, 33
Information sources, 25
Information theory, 25–84
instantaneously decodable, 99
inter-frame redundancy, **235**
interleaver, 27, 238, 266, **280**, 307, 327, 347, 365, 437
intra-frame redundancy, **235**
Irregular Convolutional Coding (IrCC), 276
irregular trellis, **244**, *244*, 246, 248, 429
    path, **245**
    state, **245**
    transition, **245**
Irregular Unity Rate Coding (IrURC), 345, 436
Irregular Variable Length Codes (IrVLCs), **11**
Irregular Variable Length Coding (IrVLC), 233, 262, 303, 345
Issues in information theory, 25–28
iterative decoding, **237**, **266**, **307**, **347**
    convergence, 251, 269, 283, 318, 354, 365

Jacobian logarithm, 127
Joint information, 66
joint source and channel coding, **16**, 234, 262, 306, 345, **428**

Kraft inequality, 99

latency, 251, 438
Levenshtein distance, 131

Linde–Buzo–Gray (LBG) algorithm, 242
Log-MAP algorithm, 250, 266, 306, 325, 343
Logarithmic Likelihood Ratio (LLR), 248, 266, 297, 307, 347, 366
loss of synchronization, 100
lossless compression, **235**
lossy compression, **235**, 263

macro-block, 240
MAP/ML Sequence Estimation, 5
Markov model, 43
Markov processes, 94
Max-Log approximation, 127
maximum achievable information rate, 210
Maximum entropy of a binary source, 34–36
Maximum entropy of a q-ary source, 36–37
McMillan inequality, 99
memoryless source, 93
Microwave channels, 29
minimum block distance, 96
minimum convergence distance, 97
minimum divergence distance, 97
Minimum Mean Square Error (MMSE) Decoding, 5
Modulator, 27
mutual information, 59–61, 150
Mutual information example, 61–62

N-state Markov model for discrete sources exhibiting memory, 45
Noise, 29
Noise – natural, 30
Noise – synthetic, 30
non-singular, 99

Optical fiber, 29

Peak Signal-to-Noise Ratio (PSNR), 254
Power spectral density, 26
Practical evaluation of the Shannon–Hartley law, 79–82
Practical Gaussian channels, 30
prefix codes, 99
Probability density function, 26

Quadrature Amplitude Modulation (QAM), 252, 269, 347
quantization, 25, **92**, 262, 309, 345

Rayleigh fading channel, 252, 262, 307, 345, 347
redundancy, **11**, 243, 260, 292
residual redundancy, **16**
Reversible Variable Length Coding (RVLC), 233, 260
    codebook, **241**, *241*
    decoding, **248**
    encoding, *243*, **246**
RLC efficiency, 52–54

Run-length coding, 51–54
Run-length coding principle, 51–52
RVLCs for the English alphabet, 104

Sampling, 25
separation theorem, 1
sequence estimation, 116, 267, 278, 297, 308, 349
serial concatenation, 237, 266, 305
Shannon's channel coding theorem, 73–75
Shannon's message for wireless channels, 82–84
Shannon–Fano coding, 37–39
Shannon–Hartley law, 79
side information, 266, 267, **270**, 307, **328**, 346, 365,
    365, 442
Signal-to-noise ratio, 26
Skin effect, 30
soft information, 249, 268
    a posteriori, 267, 278, 308
    a priori, 238, 266, 308
    a posteriori, 238, **249**, 348
    a priori, **248**, 348
    extrinsic, 238, 266, 308, 347
    mutual information, 253, **271**, 299, 354
Soft-In Soft-Out (SISO) decoding, 237, 266, 307,
    347
source and channel coding separation theorem, **15**,
    231, 428
Source coding, 1, **11**, 37–41, 95
Source encoder, 27, 37
Speech systems, 82
subcode, 210
Symbol-by-Symbol MAP Decoding, 5
symmetric RVLC, 101

T-BCJR algorithm, 250
Thermal noise, 30
Time Division Multiple Access (TDMA), 26
Transducer, 26
Transfer function, 29

Transmission via discrete channels, 54–70
Trellis Coded Modulation (TCM), 237, 262
trellis encoder, 146
turbo principle, 143
turbo-cliff, 9
Twisted pairs, 29
Two-state Markov model example, 46–48
Two-state Markov model for discrete sources
    exhibiting memory, 43–45

Unequal Error Protection (UEP), **17**, 261
uniquely decodable, 99
Unity-Rate Coding (URC), 294, 350
unity-rate intermediate code, 214

Variable Dimension Vector Quantization (VDVQ),
    237
    codebook, **241**, *241*
    decoding, **248**
    encoding, *243*, **246**
Variable Length Coding (VLC), 262
    average codeword length, **260**, 292
Variable-Length Codes (VLCs), **11**
Variable-Length Error-Correction (VLEC) coding,
    260, 291, 349
    bit-based trellis, 266, 295, 306, 347
    codebook, *299, 310, 312, 351, 352, 361*
    symbol-based trellis, 266
Vector Quantization (VQ), **93**
video
    block, 239
    coding, 232
    compression, **234**
    information, 232
Viterbi algorithm, 246
Viterbi's three lessons, 112
VLECs for the English Alphabet, 111

Waveguides, 29

Author Index

Aalo [340], 401
Adrat [268], 205
Ahmed [153], 15, 232, 234, 261
Ahmed [198], 20, 403, 409, 411, 414, 418
Ahmed [353], 406
Alamri [199], 20
Aluoini [357], 419
Anderson [297], 250
Ashikhmin [204], 20, 205, 207, 208, 253, 292, 305,
    323, 325, 334, 335, 340, 344, 355, 357,
    362, 370, 431, 435, 436, 444, 445
Ashikhmin [267], 205, 207, 208, 253, 259, 274, 291,
    340, 354, 361

Bahl [48], 5, 9, 13, 19, 116, 122, 146, 148, 233, 238,
    248, 266, 279, 305, 330, 343, 347, 366,
    368, 371, 408, 430, 437, 439, 447, 453,
    458
Balakirsky [180], 16, 232, 262, 266, 288, 295, 306,
    347, 408, 439, 441
Bauch [360], 446
Bauer [181], 16, 19, 233, 245, 262, 264, 266, 282,
    284, 288, 295, 306, 429, 442
Bauer [192], 19
Bauer [301], 256, 282
Benedetto [188], 17
Benedetto [275], 231, 232, 237, 259, 261, 262, 266,
    291, 294, 305, 307, 339, 408
Benedetto [278], 231, 232, 238, 259, 261, 262, 266,
    291, 294, 305, 307, 339, 430
Bergeron [279], 232
Berlekamp [135], 11
Berrou [142], 13, 16
Biglieri [311], 347, 445
Bogusch [345], 402
Borkenhagen [171], 16
Bose [302], 260
Brauers [268], 205

Brink [204], 20, 205, 207, 208, 253, 292, 305, 323,
    325, 334, 335, 340, 344, 355, 357, 362,
    370, 431, 435, 436, 444, 445
Brink [267], 205, 207, 208, 253, 259, 274, 291, 340,
    354, 361
Brink [294], 238, 253, 271, 291, 313, 322, 325, 339,
    341, 350, 354
Brink [312], 354
Brink [346], 402, 409
Buttigieg [179], 16, 260, 291–293, 296, 315, 339,
    366, 402, 413, 431, 434
Buttigieg [205], 21, 293, 294
Buzo [170], 16, 93, 231, 232, 242, 428
Bystrom [183], 16

Caire [311], 347, 445
Chang [190], 17, 20, 261
Chang [342], 401
Chen [280], 232
Chen [281], 232
Chen [283], 232
Chen [285], 232
Chen [332], 386, 388
Chen [355], 410
Cherriman [151], 15, 17, 231, 232, 234–236, 250,
    254, 259, 261, 262
Chindapol [358], 446
Chung [272], 222
Chung [274], 231–234, 255, 256, 430
Clevorn [268], 205
Cocke [48], 5, 9, 13, 19, 116, 122, 146, 148, 233,
    238, 248, 266, 279, 305, 330, 343, 347,
    366, 368, 371, 408, 430, 437, 439, 447,
    453, 458
Costello [134], 11
Costello [299], 251, 253, 269, 271, 318, 354
Crepeau [356], 419

Daut [187], 17
Divsalar [308], 305, 340, 345
Divsalar [351], 402
Dolinar [308], 305, 340, 345
Dolinar [351], 402
Duhamel [288], 232, 233
Dyck [163], 16

El-Hajjar [199], 20
Elias [140], 11, 259, 260, 280

Fano [139], 11
Farrell [179], 16, 260, 291–293, 296, 315, 339, 366,
        402, 413, 431, 434
Farvardin [167], 16, 17
Farvardin [169], 16, 232, 242
Farvardin [172], 16
Fiebig [347], 402, 405, 408
Fingscheidt [173], 16
Fischer [178], 16
Forney [134], 11
Forney [141], 13, 306, 347, 350, 359, 439, 441
Franz [297], 250
Frey [148], 13, 372, 439
Fryer [345], 402

Görtz [184], 16
Görtz [185], 16, 232
Görtz [298], 251, 253, 269, 271, 293, 314, 350, 435
Görtz [360], 446
Gallager [143], 13
Gallager [144], 13
Gan [191], 17, 20, 261
Gersho [290], 232, 234, 237, 241, 428
Glavieux [142], 13, 16
Goldberg [307], 293, 294, 300
Golub [309], 343
Grangetto [188], 17
Grangetto [287], 232
Grant [333], 386
Gray [170], 16, 93, 231, 232, 242, 428
Gray [310], 347
Guillemot [186], 16
Guo [274], 231–234, 255, 256, 430

Ha [343], 401
Hagenauer [149], 13, 20, 21, 259, 274, 276, 287, 291,
        319–321, 335, 339–344, 356–358, 402,
        414, 431, 437, 442, 447
Hagenauer [181], 16, 19, 233, 245, 262, 264, 266,
        282, 284, 288, 295, 306, 429, 442
Hagenauer [185], 16, 232
Hagenauer [192], 19
Hagenauer [292], 233, 237, 248, 253, 266, 307, 339,
        341, 343, 347
Hagenauer [301], 256, 282

Hagenauer [360], 446
Haimovich [332], 386, 388
Hanzo [151], 15, 17, 231, 232, 234–236, 250, 254,
        259, 261, 262
Hanzo [152], 15, 259, 262
Hanzo [194], 20, 231, 234, 235, 239, 241, 244, 247,
        254, 255
Hanzo [195], 20
Hanzo [196], 20, 263, 272, 278, 283
Hanzo [197], 20, 402
Hanzo [198], 20, 403, 409, 411, 414, 418
Hanzo [199], 20
Hanzo [200], 20, 21, 291, 297, 300, 305, 306, 314
Hanzo [201], 20, 21
Hanzo [202], 20, 21, 259, 339, 344, 345, 353, 355,
        356, 358, 363, 364, 367, 368
Hanzo [203], 20, 21
Hanzo [207], 21
Hanzo [208], 21
Hanzo [271], 222
Hanzo [272], 222
Hanzo [273], 231–234, 255, 256, 430
Hanzo [274], 231–234, 255, 256, 430
Hanzo [293], 234, 237, 238, 252, 269
Hanzo [296], 250, 262, 266, 267, 282, 306, 308, 313,
        322, 325, 343, 347, 355, 439
Hanzo [300], 253, 269, 305, 347, 355–358, 365, 371,
        444, 446
Hanzo [350], 402, 410
Hanzo [353], 406
Hanzo [354], 406
Hanzo [355], 410
Hocquenghem [303], 260
Hoeher [295], 250
Huebner [299], 251, 253, 269, 271, 318, 354
Hueffmeier [345], 402
Huffman [155], 16, 232, 242, 259, 260, 265
Huo [357], 419

Jacobs [135], 11
Jaspar [186], 16
Jelinek [48], 5, 9, 13, 19, 116, 122, 146, 148, 233,
        238, 248, 266, 279, 305, 330, 343, 347,
        366, 368, 371, 408, 430, 437, 439, 447,
        453, 458

Kaiser [183], 16
Kasahara [168], 16, 232, 242
Keller [300], 253, 269, 305, 347, 355–358, 365, 371,
        444, 446
Khandani [175], 16
Kim [341], 401, 405
Kliewer [174], 16, 233, 248, 249
Kliewer [176], 16
Kliewer [194], 20, 231, 234, 235, 239, 241, 244, 247,
        254, 255
Kliewer [195], 20

Kliewer [298], 251, 253, 269, 271, 293, 314, 350, 435
Kliewer [299], 251, 253, 269, 271, 318, 354
Komiya [190], 17, 20, 261
Kopansky [183], 16
Kot [338], 401, 406, 407
Kramer [204], 20, 205, 207, 208, 253, 292, 305, 323,
       325, 334, 335, 340, 344, 355, 357, 362,
       370, 431, 435, 436, 444, 445
Kramer [267], 205, 207, 208, 253, 259, 274, 291,
       340, 354, 361
Kumazawa [168], 16, 232, 242
Kurtenbach [166], 16

Lahouti [175], 16
Lakovic [286], 232
Lamy [206], 21, 293, 294
Lamy-Bergot [279], 232
Lee [190], 17, 20, 261
Lee [341], 401, 405
Lee [348], 402
Li [338], 401, 406, 407
Liang [344], 402, 408
Liew [296], 250, 262, 266, 267, 282, 306, 308, 313,
       322, 325, 343, 347, 355, 439
Liew [354], 406
Linde [170], 16, 93, 231, 232, 242, 428
Lloyd [164], 16, 92, 93, 262–264, 309, 345, 350, 404,
       440
Loan [309], 343
Luby [147], 13, 372, 439
Lynch [282], 232

Ma [191], 17, 20, 261
Ma [282], 232
MacKay [145], 13
MacKay [148], 13, 372, 439
MacQueen [233], 92, 93, 263
Makur [291], 233, 246
Marcellin [178], 16
Masnick [304], 261
Massey [138], 11
Massey [162], 16
Maunder [194], 20, 231, 234, 235, 239, 241, 244,
       247, 254, 255
Maunder [195], 20
Maunder [196], 20, 263, 272, 278, 283
Maunder [197], 20, 402
Maunder [198], 20, 403, 409, 411, 414, 418
Maunder [199], 20
Maunder [200], 20, 21, 291, 297, 300, 305, 306, 314
Maunder [201], 20, 21
Maunder [202], 20, 21, 259, 339, 344, 345, 353, 355,
       356, 358, 363, 364, 367, 368
Maunder [203], 20, 21
Maunder [208], 21
Maunder [273], 231–234, 255, 256, 430

Max [165], 16, 92, 262, 263, 309, 345, 350, 440
Mertins [298], 251, 253, 269, 271, 293, 314, 350, 435
Miller [163], 16
Miller [177], 16
Miller [182], 16
Miller [341], 401, 405
Mitzenmacher [147], 13, 372, 439
Modestino [187], 17
Modestino [189], 17, 20, 261
Montorsi [275], 231, 232, 237, 259, 261, 262, 266,
       291, 294, 305, 307, 339, 408
Montorsi [278], 231, 232, 238, 259, 261, 262, 266,
       291, 294, 305, 307, 339, 430
Murakami [276], 231, 233, 234, 241, 242, 260, 428,
       429

Namekawa [168], 16, 232, 242
Natarajan [153], 15, 232, 234, 261
Neal [145], 13
Netravali [154], 16, 232, 234, 261
Ng [194], 20, 231, 234, 235, 239, 241, 244, 247, 254,
       255
Ng [195], 20
Ng [196], 20, 263, 272, 278, 283
Ng [197], 20, 402
Ng [198], 20, 403, 409, 411, 414, 418
Ng [199], 20
Ng [271], 222
Ng [272], 222
Ng [273], 231–234, 255, 256, 430
Ng [274], 231–234, 255, 256, 430
Ng [293], 234, 237, 238, 252, 269
Ng [300], 253, 269, 305, 347, 355–358, 365, 371,
       444, 446
Ng [350], 402, 410
Ng [353], 406
Ng [355], 410
Nguyen [288], 232, 233

Offer [292], 233, 237, 248, 253, 266, 307, 339, 341,
       343, 347
Olmo [188], 17
Olmo [287], 232
Omura [150], 15

Paccaut [206], 21, 293, 294
Papke [292], 233, 237, 248, 253, 266, 307, 339, 341,
       343, 347
Park [177], 16
Park [182], 16
Park [348], 402
Pei [189], 17, 20, 261
Peterson [339], 401
Phamdo [172], 16
Pollara [308], 305, 340, 345
Pollara [351], 402
Proakis [269], 207
Proakis [352], 405, 407

Qu [189], 17, 20, 261

Rao [153], 15, 232, 234, 261
Raviv [48], 5, 9, 13, 19, 116, 122, 146, 148, 233, 238,
        248, 266, 279, 305, 330, 343, 347, 366,
        368, 371, 408, 430, 437, 439, 447, 453,
        458
Ray-Chaudhuri [302], 260
Reiffen [137], 11
Richardson [313], 372, 439
Ritcey [358], 446
Robbins [154], 16, 232, 234, 261
Robertson [295], 250
Robertson [343], 401
Rosenblatt [305], 262
Ryan [349], 402

Sayood [171], 16
Scanavino [188], 17
Scanavino [287], 232
Schreckenbach [360], 446
Shannon [133], 11, 15, 25, 28, 32–34, 37, 43, 54, 59,
        62, 64, 72, 82, 83, 207, 231, 255, 259,
        269, 319, 376, 428
Shokrollahi [147], 13, 372, 439
Shokrollahi [313], 372, 439
Somerville [152], 15, 259, 262
Spielman [147], 13, 372, 439
Stark [344], 402, 408
Stemann [147], 13, 372, 439
Streit [151], 15, 17, 231, 232, 234–236, 250, 254,
        259, 261, 262
Stuber [359], 446
Subbalakshmi [280], 232
Subbalakshmi [281], 232
Subbalakshmi [283], 232
Subbalakshmi [285], 232
Subbalakshmi [291], 233, 246

Tüchler [149], 13, 20, 21, 259, 274, 276, 287, 291,
        319–321, 335, 339–344, 356–358, 402,
        414, 431, 437, 442, 447
Tüchler [306], 276, 280, 431
Takishima [276], 231, 233, 234, 241, 242, 260, 428,
        429
Tan [359], 446
Tanner [146], 13
Tao [271], 222
Taricco [311], 347, 445
Tee [208], 21
Tee [350], 402, 410
Teh [338], 401, 406, 407
Thitimajshima [142], 13, 16
Thobaben [174], 16, 233, 248, 249
Thobaben [176], 16
Tu [342], 401

Ungerboeck [277], 231, 232, 252, 261, 262, 266, 269,
        282, 430, 444, 446, 447
Urbanke [313], 372, 439

Vaisey [290], 232, 234, 237, 241, 428
Vaishampayan [167], 16, 17
Valenti [345], 402
Vandendorpe [186], 16
Vary [173], 16
Vary [268], 205
Villasenor [286], 232
Villebrun [295], 250
Viterbi [150], 15
Viterbi [193], 19, 246, 430, 439
Vucetic [289], 232, 233

Wada [276], 231, 233, 234, 241, 242, 260, 428, 429
Wang [194], 20, 231, 234, 235, 239, 241, 244, 247,
        254, 255
Wang [195], 20
Wang [196], 20, 263, 272, 278, 283
Wang [197], 20, 402
Wang [207], 21
Wang [208], 21
Wang [271], 222
Wang [273], 231–234, 255, 256, 430
Wang [284], 232
Wang [355], 410
Webb [300], 253, 269, 305, 347, 355–358, 365, 371,
        444, 446
Wintz [166], 16
Wolf [304], 261
Wolfgang [355], 410
Woodard [152], 15, 259, 262
Wozencraft [136], 11
Wozencraft [137], 11

Xiao [289], 232, 233

Yang [194], 20, 231, 234, 235, 239, 241, 244, 247,
        254, 255
Yang [195], 20
Yang [196], 20, 263, 272, 278, 283
Yang [197], 20, 402
Yang [198], 20, 403, 409, 411, 414, 418
Yang [207], 21
Yang [271], 222
Yang [273], 231–234, 255, 256, 430
Yang [353], 406
Yang [355], 410
Yeap [296], 250, 262, 266, 267, 282, 306, 308, 313,
        322, 325, 343, 347, 355, 439
Yeap [354], 406
Yu [284], 232

Zhang [340], 401
Ziemer [339], 401